ADVANCED MATERIALS, TECHNOLOGY AND APPLICATION

Proceedings of the 2016 International Conference on
Advanced Materials, Technology and Application (AMTA2016)

ADVANCED MATERIALS, TECHNOLOGY AND APPLICATION

Proceedings of the 2016 International Conference on Advanced Materials, Technology and Application (AMTA2016)

Changsha, China 18 – 20 March 2016

Editor

Qingzhou Xu

Morehead State University, USA

World Scientific

NEW JERSEY • LONDON • SINGAPORE • BEIJING • SHANGHAI • HONG KONG • TAIPEI • CHENNAI • TOKYO

Published by

World Scientific Publishing Co. Pte. Ltd.
5 Toh Tuck Link, Singapore 596224
USA office: 27 Warren Street, Suite 401-402, Hackensack, NJ 07601
UK office: 57 Shelton Street, Covent Garden, London WC2H 9HE

British Library Cataloguing-in-Publication Data
A catalogue record for this book is available from the British Library.

ADVANCED MATERIALS, TECHNOLOGY AND APPLICATION
Proceedings of the 2016 International Conference on Advanced Materials, Technology and Application (AMTA2016)

ISBN 978-981-3200-46-3

Printed in Singapore

Preface

The 2016 International Conference on Advanced Materials, Technology and Application (AMTA2016) was successfully held in Changsha, China from March 18th to March 20th, 2016. The main objective of this conference is to provide a platform for researchers, academicians and industrial professionals from all over the world to present the results of their research, promote research and development in materials science, and exchange ideas and compare notes on issue of trend and future direction in advanced material research, technology and applications, so that to forge research collaboration.

The AMTA2016 received over 150 submissions, however, after a rigorous process of double peer review, only 59 articles were selected to be included in this proceedings. It covers seven areas of material science - **Chemical Materials**, **Composite and Nano Materials, Polymer and Concrete Materials**, **Structural Materials, Metal and Alloy Materials, Electrical Materials, and Biomaterials**.

Finally, we would like to take this opportunity to express our sincere gratitude to all members of the conference committee and reviewers for their effort and comments on articles under their review. Without them, we would not able to pull out this proceedings so quickly after our conference.

Guest Editor
Dr. Qingzhou Xu (Morehead State University)
On behalf of the AMTA2016 Organizing Committee
Changsha, China

Committees

General Chair

Yun-Kong He,
Ph.D, China University Of Geosciences, China

Co-Chair

Ji-Heng Liu
Ph.D., Harbin University, China

Technical Program Committee

Li-Bao An
Professor, College of Mechanical Engineering, North China University of Science and Technology,China

Hua-Feng Tian
Associate Professor, School of Materials Science and Mechanical Engineering, Beijing Technology and Business University, Beijing China

Qing-Li Zhang
Ph.D, Director of Electromechanical Experimental Center, College of Engineering, Ocean University of China, Shandong, China

Liang-Miao Zhang
Ph.D, Assistant Researcher, Shanghai University, Shanghai, China

Jing-Wei Zhao
Ph.D, University of Wollongong, Australia

Yuan-Xun Li
Professor, Electronic Information Materials and Components, University of Electronic Science and Technology, China

Jing-Ping Fu
Ph.D., Northeast Agricultural University, China

Yu Zhang
Ph.D., Yantai University, China

Wen-Feng Shui
Ph.D., Northeast Dianli University, China

Zi-Hui Shi
Ph.D., Beijing Union University, China

Bao-Hua Dong
Ph.D., Taiyuan University Of Science And Technology, China

Mei Rong
Ph.D., Shanghai Dian Ji University, China

Hao-Yu Zhang
Ph.D., Changchun University of Science and Technology, China

Ruo-Qian Zhou
Ph.D., Anhui University, China

Song-Gui Pi
Ph.D., Shanghai Dian Ji University, China

Shi-Yi Liu
Ph.D., Anhui Jianzhu University, China

Hua Dong
Ph.D., University, China

You-Lv Dou
Ph.D., Inner Mongolia Normal University, China

Dan-Yi Liang
Ph.D., Liaocheng University, China

Pu-Chun Qu
Ph.D., Liaocheng University, China

Yue-Heng Jin
Ph.D., North China University of Technology, China

Fang Yao
Ph.D., Donghua University, China

Cheng-Min Xiong
Ph.D., Beijing Institute of Technology, China

Wein-Qin He
Ph.D., Zhejiang Gongshang University, China

Ning Zhang
Professor, Beijing Union University, China

Sheng-Qi Chen
Ph.D., Department of Polymer Science and Engineering, University of Science and Technology of China, China

KEYNOTE SPEECH

Name: Prof. Deliang Chen
School of Materials Science and Engineering, Zhengzhou University, China

Deliang Chen has been a full professor in Materials Science at Zhengzhou University since 2013. He earned his B.S. and M.S. degrees in Mineral Process Engineering from Central South University in 1999 and 2002, respectively. Then he moved to Shanghai Institute of Ceramics, Chinese Academy of Sciences, where he earned his Ph.D. degree in Materials Science and Engineering in 2005. From April 2005 to March 2007, he moved to Waseda University as a visiting research associate. From April 2007 to December 2012, he worked at Zhengzhou University as an associate professor. From March 2011 to March 2012, he moved to KAIST as a visiting scholar supported by China government. Since March 2012, he returned to Zhengzhou University. He has been the co-author of 120 scientific papers, and has chaired 3 NSFC projects. His current research interests covers chemical sensors, photocatalysts, and synthesis of low-dimensional materials using intercalation chemistry, microwave chemistry and sonochemistry.

KEYNOTE SPEECH

Name: Prof. Yangchuan Ke
College of Science, China University of Petroleum, China

Yang Chuan Ke, Male, Chinese nationality
Earned Ph D in 1996 from Jilin University
Full professor of College of Science, CUPB
Director of CNPC nanochemistry research laboratory
CNPC nanochemistry key laboroatory
Director of nanotechnology of energy resources
State key laboratory of heavy oil
Members of material society, China Petroleum and Chemical Industry Association, and others.
So far, won 3 important technical prizes including 2 second prizes and 1 first prize of China Petroleum and Chemical Industry Association. published 125 papers and 5 books, more than 30 patents, and conference invitation lectures more than 20 times including 5 keynote speakers. Published papers cited for more than 1000 times. A single paper cited for 357 times.

KEYNOTE SPEECH

Name: Prof. Pengsheng Wei
Department of Mechanical and Electro-Mechanical Engineering, National Sun Yat-Sen University, Kaohsiung, Taiwan

Dr. Peng-Sheng Wei received Ph.D. in Mechanical Engineering Department at University of California, Davis, in 1984. He has been a professor in the Department of Mechanical and Electro-Mechanical Engineering of National Sun Yat-Sen University, Kaohsiung, Taiwan (China), since 1989. Dr. Wei has contributed to advancing the understanding of and to the applications of electron and laser beam, plasma, and resistance welding through theoretical analyses coupled with verification experiments. Investigations also include studies of their thermal and fluid flow processes, and formations of the defects such as humping, rippling, spiking and porosity. Dr. Wei has published more than 80 journal papers. He is a Fellow of AWS (2007), and a Fellow of ASME (2000). He also received the Outstanding Research Achievement Awards from both the National Science Council (2004), and NSYSU (1991, 2001, 2004), the Outstanding Scholar Research Project Winner Award from National Science Council (2008), the Adams Memorial Membership Award from AWS (2008), the Warren F. Savage Memorial Award from AWS (2012), and the William Irrgang Memorial Award from AWS (2014). He has been the Xi-Wan Chair Professor of NSYSU since 2009, and Invited Distinguished Professor in the Beijing University of Technology, China, during 2015-2017.

KEYNOTE SPEECH

Name: Prof. Rangsu Liu
Department of Applied Physics, Hunan University, China

Professor Liu Rang-su, he graduated from Department of Physics of Hunan University In 1964. He is working in Department of Applied Physics of Hunan University and as a supervisor of Ph.D. students. Moreover, he is the director of Institute of Physics of Functional Material of Hunan University, the member of Academic Council of the International Conference on Intelligent Processing and Manufacturing of Materials (IPMM) (for six consecutive –12 years). Professor Liu has ever undertaken 14 scientific projects on provincial, ministry and national levels, including 7 projects from National Natural Science Foundation of China. These projects are at the forefront in the field. He has published more than 150 research papers on well-known academic journals.

Table of Contents

Chapter 1

Chemical Materials

The comparative study about chemical adsorption mechanism of CO molecules on Al(101) surface

Qian Liao[1,b], Zheng-Xin Yan*, Wei Liu, Yong-Shen Wang, An Gong,
Qian Chen and Shao-Hua Zhang

College of Science of Xi'an University of Science and Technology,
Xi'an 710054, China
*E-mail: *zhengxinyan163@163.com, [b] lqian2009@163.com*

This research paper uses the Density functional theory of First-principles to investigate the chemadsorption mechanism at top site, bridge site and hollow site of CO on Al(101) surface, for favorable adsorption site. High electrical conductivity and good oxidation of high purity are mainly used in filling interface, materials design and electronics industries. It was important to discover the interaction mechanism between gas molecules and the Al(101) surface by analyzing the adsorption energy, density of states, electronic property and other properties by studying single molecule absorption systems. The results of the absorption energy, density of states and electronic properties show that the adsorption energy at bridge site is the smallest one,-0.2912 eV, the length of C-Al bond was shortened and the total Mulliken population significantly increased. Lastly, the most stable site of the CO adsorption of the surface is the bridge site.

Keywords: Density Functional Theory; Al (101) Surface; CO Molecule; Adsorption.

1. Introduction

Nano-micron aluminum powder is an important energetic material filler, because of its large surface area and high activity. The theoretical research and engineering application about the oxidation mechanism of aluminum surface are considered an important issue. High electrical conductivity and good oxidation of high purity is mainly used in filling interface, materials design and electronics industries.

In recent years, more and more research on Al adsorption surface has been studied extensively. For examples, Lei Xiaoli[1] et al. used the density functional theory to study the properties of structure, energy and electronic when H_2O and O_2 molecules is absorbed on Al(111) surface. They noted that the process about adsorption of H_2O and O_2 molecules on the Al(111) surface are divided into weak chemisorption and strong one Zhang Rong, Huang Kaiyou[2] et al. draw a conclusion that adsorption of Cl^- on a Al(100) surface is mainly the result of orbital interactions between 2*s* and 2*p* of Cl^-orbit and 3p orbit of basal metal by analyzing density of states of adsorption of Cl^- on different sites of

[1] Corresponding author

clean Al(100) surface. Chen Yuhong, Cao Yijie[3] et al. investigated the absorption condition that H_2 molecules has been impacted on the pure Al(110) surface doped by Ti atoms, They have revealed that the H_2 molecules on pure Al(110) surface could not be dissociated, but it really exists a adsorption path to make the H_2 molecules dissociated when they are adsorbed on the Ti atoms above the top site of the Al(110) surface that Ti atoms were doped. However, studies on the adsorption characteristics of Al (110) surface are found rarely, and meanwhile it has significant meaning to understand its structure and properties in the study of catalytic when CO is mainly used as one of probe molecule[4,5,6]. Therefore, there is some value to study CO molecule on adsorption of the Al(101) surface.

In this paper, some molecular simulation and calculation were done on the base of the theory of First-principles method to explore three typical adsorption sites: top, bridge and hollow site when the CO molecules are affected on Al(101) surface. It will have important reference significance to discover the interaction mechanism between gas molecules and the Al(101) surface by analyzing the adsorption energy ,density of states, electronic property and other properties by studying single molecule absorption systems.

2. Calculation

Three kinds of chemisorption sites about adsorption of CO on the Al(101) surface are built and calculated by the Castep module of Material Studio software on the base of first-principles density functional theory (DFT). The interaction between the valence electron and ion core is depicted by the PAW (Projector Augmented Wave),and the relevant parts of electronic exchange are described by the the standard exchange-correlation functional method called PBE[3] in the generalized gradient approximation (GGA).The method of integration in brillouin zone is the Monkhorst-Pack to calculate the special k-point mesh point. During the calculation, k-point grid size is set to 441, the cutoff of energy is set to 400eV and the creation of energy convergence is set to 210-5eV.The setting above well meet the computational accuracy requirements through convergence test. To make the quantity of calculation for every surface affected by other one less, the selected super-cell is the size of 221, the thickness of plane model is five atomic layers, and the thickness of vacuum zones along the Z-direction is 10Å.The results of Zang Lili[7,8] et al. is referred to set the length of bond of C-O and C-Al as 1.15Å and 1.90Å.The top layer of atoms around the balanced position is only allowed to be relax and other atoms is fixed during the adsorption of CO on the Al(101) surface. As a result, the influence of dipole moment from atoms in the lowest layer becomes small as soon as possible and meantime the quantity of calculation is decreased.

3. Result and Discuss

As the figure 1 depicts, the bond length and bond angle of Al-C-O has been changed after the adsorption of CO on the Al(101). The bond lengths of Al-C-O are all stretched for three different adsorption sites and the bond angle is turned into133.322 relative to 130.050 before optimization, which is optimized to achieve a stable state.

Not only considering the type of orientation accumulation and the second time symmetrical structure to the Al (101) crystal, but also its parameters of the model(bond length, lattice constants, et al.) decide three possible sites that the adsorption of CO molecules on the Al(101) surface: top, bridge and hollow sites. Also the inner properties of CO that the electronic negative property of carbon atom is lower than the oxygen atom lead to the latter always provide the charges for the former. The result is the charge property of carbon atoms has shown negative and the oxygen atom is positive, and therefore it often occurs that the carbon atom is bonded to the metal atom when the CO molecules is absorbed on the metal surface[9].The adsorption model case is shown in Figure 1. Adsorption energy for every site in the adsorption process is calculated and the relevant density of states and charge density difference are analyzed based upon the results of above.

3.1. Adsorption energy

It will release some of the energy when gas is adsorbed on one interface, and the energy is called adsorption energy. In the same context, the adsorption energy can reveal the stable characteristics of the gas molecules adsorbed on the surface of the material, the lower its energy the stronger its adsorption capacity and the more stable its system. According to the changes of the total system energy before and after the adsorption, the energy of system is defined as follows [10]:

E_{chem} is the energy of adsorption of CO on the Al(101); $E_{CO/Al(101)}$ represents the total adsorption energy for the whole system; $E_{Al(101)}$ is the meaning of energy of Al (101) surface that a single cell clean plane. The calculated results are shown in Table 1.

The calculations of adsorption energy and bond lengths changes are shown in table 1. The ranks of site is top, bridge and hollow when the CO molecules are absorbed on the Al(101) surface in the terms of length changes of C-Al bonds. Compared to the freedom of CO, the length of C-O bond all become increased, and the most obvious of three sites is the bridge site. The ranking of the total adsorption energy is the top site, the hollow site and bridge site. While from the view of the Mulliken population for three adsorption sites, the orders of C-O bond are top, bridge and hollow site and the ranking of C-Al are top, bridge and hollow site. In other words, the Al-C bond energy has turned stronger and meantime it leads to weaken the bond energy of Al-Al and C-O. The details for the main reason to analyze it will be given from the analysis of density of states behind.

Table 1 Calculation results of CO adsorption at different sites of Al(101) surfaces where $d_{C\text{-}Al}$ is the length of bond between Al atom and carbon atom. d_{CO} is the length of bond between oxygen atom and carbon atom.

Adsorption sites	$d_{\text{C-Al}}(nm)$	$d_{\text{CO}}(nm)$	$E_{chem}(eV)$	Population(e)	
				C-O	C-Al
Top	0.2017	0.1171	-0.0857	1.13	0.54
Bridge	0.2113	0.1207	-0.2912	1.08	0.46
Hollow	0.2238	0.1199	-0.2044	1.07	0.22
Free	-	0.1156	-	0.97	-

$$\Delta E_{chem} = E_{CO/Al(101)} - E_{Al(101)} - E_{CO}$$

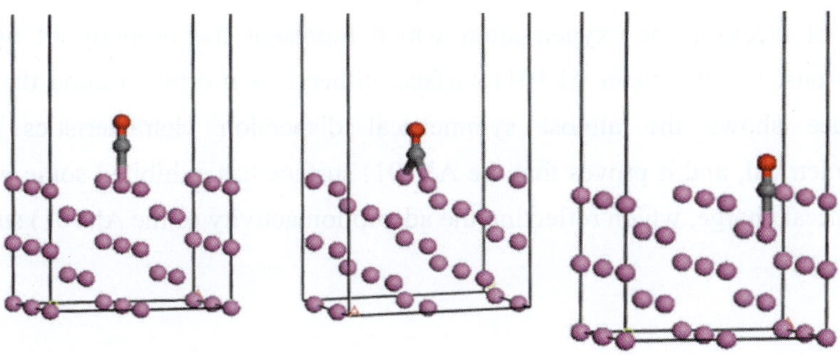

(Top) (Bridge) (Hollow)

Fig.1 Adsorption configuration of Top, Bridge and Hollow sites at CO molecules adsorbed on the Al(101) surface

3.2. Analysis of density of state

The change trend of the PDOS for the CO molecule before and after the adsorption on the bridge of Al(101) surface are described in figure 2. The characteristics of the frontier orbital density of states of free CO molecules before the adsorption is depicted in figure 2(a).The orbit around the Fermi surface is 5σ of CO and next orbits below the Fermi energy are 1π,4σ and 3σ orbit respectively. On the contrary, 2π* orbit and 6σ orbit are ordered above the surface. The orbit of 5σ occupied with lone pair electrons of carbon atom is the highest occupied molecular one, which featured with hybridization of sp_x orbit. The lowest unoccupied orbit of 2p is hybridized by the p_y and p_z orbit of C and O atom, which is characterized with anti-bonding orbit. The orbits of 1π and 2π* are similar to each other, which just present a characteristic of bond making. The 4σ orbit and 6σ orbit are established by the hybridization of sp_x orbit from carbon and oxygen atom. Only the electronic cloud of 4σ orbit is distributed in the middle of two atoms and the electrons are not easy to lose. The characteristic of 6σ orbit to behave is as same as the 2*p* orbit. The orbit of 3σ is constructed by the hybridization of sp_x orbit from oxygen atom, and it is occupied with the lone pair of electrons of oxygen atom which represent the property of negative electrons. For the clean Al(101) surface, either *s* or *p* orbits around the Fermi surface shows the almost symmetrical dispersion characteristics almost symmetrical, and it proves that the Al(101) surface has exhibited some positive electrical charge, which reflecting the adsorption activity of the Al(101) surface.

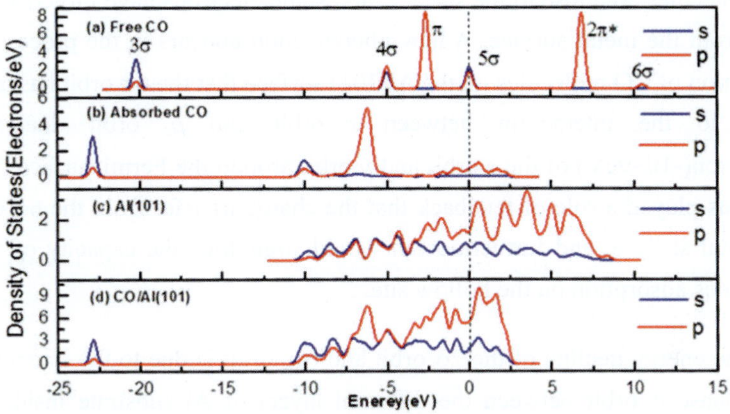

Fig.2 The PDOS and the DOS for the CO molecule before and after the adsorption on the bridge of Al(101) surface

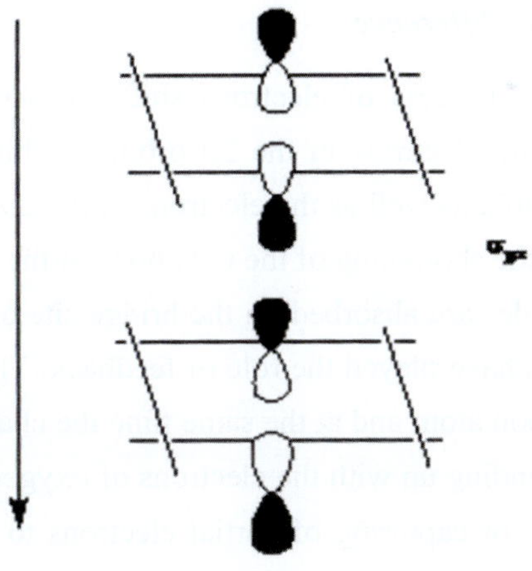

Fig.3 The interaction diagram of basal Al atoms in internal orbit.

The 5σ orbit that the highest occupied molecular one has been matched with the *s* and p_z orbits of the Al(101) surface on the symmetry and meanwhile the energy of them are close. They may work well for the three orbits above when

CO molecules are close to the surface, and some electrons have turned to the 5σ orbit from the metal surface. A new phenomenon appears in the process of the adsorption of CO molecules on the Al(101) surface that the 6σ orbit has made an effect to the interaction between *s* orbit and p_z orbit due to the dispersion(-10~9eV) of the *s* orbit and *p* orbit around the Fermi surface. The 6σ orbit has played a role of feedback that the charge transfer from the 6σ orbit to the metal surface, and furthermore it would strengthen the capacity of the CO molecules adsorption on the hollow site.

The energy decline of the 2σ orbit and σp_z orbit is due to the interaction of same-phase σ orbit between the different layers of Al substrate inside in the process of distortion when the CO molecules are adsorbed on the surface. The result is the reduction of energy to the corresponding orbits and it is shown in figure 3.

3.3. Electronic density difference

The details about the process of electrons shift are shown in the electronic density difference. The electrons in the 2*p* orbit(p_y orbit mainly) of Al atom move to the carbon atom as well as the electrons in the 2*p* orbit of oxygen atom do, which leading to the shortening of the C-O bond at the top site of adsorption. When the CO molecules are absorbed on the bridge site of Al(101) surface, the bond of carbon atoms have played the role of feedback. The electrons of two Al atoms transfer to carbon atom and at the same time the charge of the carbon turn to the oxygen atom, ending up with the electrons of oxygen atom are shift to the carbon atom, because of capturing of partial electrons to the anti-bonding 2π* the length of C-O bond become longer than before. The situation is similar to the bridge site for the adsorption of hollow site, and it is not direct contact. The electrons transferred to the carbon atom near the carbon atom are relatively small, and the same to the transfer process of carbon atom to oxygen atom. As a result, the degree of the length of C-O bond to increase is relatively small than the bridge one.

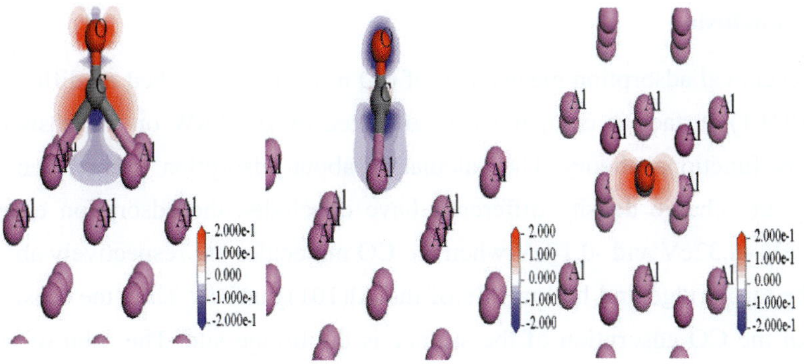

(a)Front (b)Side (c)Overlooking

Fig.4 Electronic density difference of CO molecules adsorbed on the Al(101) surface.

The change of charge density is described for the CO molecules adsorption of bridge site of Al(101) surface stably in figure 4, and a deletion of the charge area represents by the blue region, while the red region indicates that the zone is the accumulation of charge[1].As the figure shows, the electrons have increased significantly along the carbon atom and oxygen atom and the charge near the Al atom are turn to the carbon atom which represents a trend to shift to the $2\pi^*$ orbit because of the migration of charge. A weakened geometry of CO bond that is manifested as the bond length becomes long ,that is, from the size of 0.1150nm to 0.1207nm(figure 1).The decrease of electronic density around the Al atom and the increase of charge around the carbon atom have concluded that the formation of the ionic bonds between CO molecules and Al atom[11,12].The strong force from the p_z orbit of CO and p_z orbit of Al make the increase of energy, then the charge become less for the corresponding track and the charge distribution turn to increase as to the low energy state. This is the reason of reduction of energy distribution in the figure 2(d) compared to the energy distribution in figure 2(c) as well.

4. Conclusion

The chemical adsorption mechanism of CO molecules absorbed on different site of Al(101) surface is comparatively analyzed by the PAW on the basis of the density functional theory. The calculation about adsorption energy, density of states and charge density difference have concluded the adsorption energy is -0.12eV, -0.32eV and -0.17eV when the CO molecules are respectively absorbed on the top, bridge and hollow site of the Al(101) surface. Then the most stable site of the CO adsorption of the surface is the bridge site. The orbit of 6σ has mainly played a feedback role in the interaction between p_z orbit and s orbit by the analysis of the DOS and PDOS for the CO after the adsorption on the bridge of Al(101) surface. Namely the intensity of the CO adsorption on the bridge site of the surface is improved when the electrons are transferred from 6σ orbit to the metal surface. It has formed a more stable chemical bond after the CO molecules adsorbed on the surface of the metal surface and the entire system is indeed strong chemical adsorption. The electrons cloud of Al atoms significantly have moved to the carbon atoms at the process of the CO molecules adsorption on the Al(101) surface and it reveals that it has generated a stable ionic bond between carbon atom and Al atom.

Acknowledgements:

National Natural Science Foundation of China (No.11404257, No.11405127); Science and Technology Department of Shaanxi Province Natural Science Foundation of China (No.2015JM1027).

Reference

1. Xiao-li Lei,Li-xia Yang,Ming-xing Liu,et al.Density functional theory study on the adsorption of H_2O and O_2 on Al(111) surface[J].Computers and Applied Chemistry, 2011,28(6):793-796.

2. Rong Zhang,Kai-you Huang,Pang Zhang,et al.Density functional theory study on the adsorption of Cl^- on Al(100)[J].Materials Science and Engineering of Powder Metallurgy,2010,15(005):433-438.

3. Yu-hong Chen,Yi-jie Cao,Bao-xing Ren. Catalytic effect by the process of the hydrogen absorption on the Ti in the Al(110) surface[M].Ac- ta Phys.Sin.,2010,59(11).

4. Ping Li.Theoretical studies on the mechanism of CO catalytic oxidation on the Au surfaces[D].Henan Normal University,2010.

5. Paul J, Hoffmann F M. CO adsorption on clean and oxidized Al (110)[J]. Chemical Physics Letters, 1986, 130(3): 160-163.

6. Shinn N D, Madey T E. CO Chemisorption on Cr(110): Evidence for a Precursor to Dissociation[J]. Journal of Chemical Physics, 1985, 83(11): 5928-5944.

7. Li-li Zang,Jian-ming Hu,Yu Zhang,et al.Geometry and Electronic Structure of the Adsorption CO Molecules on Different Al_2O_3 Surface[J].Acta Phys. Chim. Sin.,2011,27:2311-2318.

8. Andreoni W, Varma C M. Binding and Dissociation of CO on Transition Metal Surfaces[J]. Physical Review B, 1981, 23(2): 437.

9. Yan-fang Li.First principles study of the interaction between small molecules/atoms and magnesium surfaces[D].Taiyuan University of Technology,2010.

10. Jiang S Y. Adsorption and Oxidation of CO over CeO_2(111) surface[J]. Acta Physico Chimica Sinica, 2009, 25: 1629-1634.

11. Shu-hong Ma.First principal studies for the adsorption of sulfur and CO on the transition mental surfaces[D].University of electronic science and technology,2009.

12. Zhi-xiao Liu.The adsorption and dissociation of oxygen contained gas molecule on uranium surface:a first principal study[D].Hunan University,2012.

Electromagnetic scattering characteristics of ultra-high molecular weight polyethylene (UHMWPE) ropes in different postures

Da-Wei An*,a and Wu-Yi Chen[b]

School of Mechanical Engineering and Automation, Beihang University, Beijing, 100191, China

†*E-mail:* [a]*sarlatin@163.com*

[b]*wychen@buaa.edu.cn*

An improved parallel target suspension structure in the radar cross section (RCS) measurement presenting low RCS in any azimuth requires inclined dielectric strings and moveable string driven motors on the rotator. Larger inclination and moveable range can reduce the RCS but increase the difficulty of manufacture. In the paper, the electromagnetic scattering characteristics of UHMWPE ropes in various inclinations and azimuths are analyzed by both FEKO simulations and measurements in compact range to identify a practicable configuration of the dielectric strings suspension structure, which always keeps a relative low level of the background scattering with acceptable mechanic performance. The results indicate that a slight inclination and a small azimuth away from the normal both no more than 10° can obviously degrade the RCS of UHMWPE rope, and the effect of larger inclination and azimuth shifting is decreased. To build an improved parallel target suspension system using UHMWPE strings for RCS measurements, any normal position to the incidence wave of a single rope should be avoided, to prevent RCS increase to an unacceptable level in this situation.

Keywords: Radar Cross Section; Rope; UHMWPE; Posture.

1. Introduction

The ultra-high molecular weight polyethylene (UHMWPE) rope owns very high specific strength and low dielectric permittivity, so that it may be a kind of suitable sling to support the test target in the radar cross section (RCS) measurement. The target suspension using dielectric strings usually is a parallel structure. The aspect angle of target is changed through the relative motion of multiple strings. If the target needs revolution movement, a rotator can be used to drive the target and strings together. This type of suspension structure has been utilized in some facilities [1-4].

* Corresponding author

The signals of return waves from the strings cannot be subtracted from the total measured signals as it is impossible to measure the string echoes alone in the same status, which would require that the ropes suspend a body with the same characteristic of the test target except it is invisible to the radar wave. So the string echoes can only be reduced in the practice to decrease the disturbance to the target RCS. It is obvious that an inclination of the string can degrade its backscattering because the mainlobe of string echo will be away from the incidence of test wave. However, the inclination also requires a larger radius of the rotator and greater tension on the ropes. Therefore, limited by the mechanical performance, the string inclination cannot increase arbitrarily. It is necessary to find an appropriate inclination with effective reduction of string echo and convenience in the fabrication.

The inclined strings circle with the turntable. In the traditional design [1,4], the motors driving the ropes are fixed on the rotator. It should be noticed that every string in the structure will still be normal to the incidence wave twice as the turntable has a full revolution, when the electromagnetic scattering of the string increases significantly. It may likely be relatively visible in the measurement if the target is stealth. To avoid the peak echo of the perpendicular string, the gear driving the ropes can be movable on the rotator to make the string never be perpendicular to the incidence wave. However, the size of turntable and the configuration of strings are limited by the mechanical performance and other design conditions. The wire-driven gear cannot shift arbitrarily. We should find out the proper activity range with effective reduction of the string scattering for practical design. For the tilt string, it means an appropriate azimuth range away from the vertical direction to the incidence wave.

In summary, it is necessary to investigate the relationship between the RCS of dielectric string and its spatial posture, to study in what gesture its echo will decrease from the peak to an acceptable level. In the paper, the electromagnetic scattering characteristics of UHMWPE ropes in various inclinations and azimuths are analyzed by both simulations and experiments helping to find a practicable configuration of the dielectric strings suspension structure which always keeps a relative low level of the background scattering.

2. Settings of Simulations and Experiments

A rope braided by the UHMWPE fibers was measured in an indoor compact range, whose length was about 2.25 m (the quiet zone height of the compact range was 2 m) and radius was about 1 cm (as the rope had clearance, its strict size was hard to get). The two ends of test rope were tied by another type of UHMWPE ropes whose radius was no more than 2mm, shown in Fig. 1. One thin rope was fastened on a hook in the ceiling of the anechoic chamber, and the other one was fixed on the ground by a weight pressing on it. The combined rope was straightened by tension contributed by the weight. Some points were marked on the ground. When the bottom end of the tightened rope was fixed on the points by the weight, its aspect angle could be calculated according to the height of ceiling, the coordinates of hook and marked points measured in advance. So the control of string gestures was realized by moving the weight position on the ground. The thin rope was measured independently in advance. Its RCS was lower than -25 dB when it was normal to the incident wave, and its RCS signal was concealed by the background signal when the inclination with respect to the vertical is only about 1°. Therefore, the returns of thin rope could be thought as marginal compared with the test rope. Two series of spatial postures were measured, in which one had variable inclinations and the other one had variable azimuths. Limited by the size of quiet zone, the varying inclination range was from 0° to 15° with respect to the vertical. The varying azimuth had a range of 90°, from the parallel position to the incidence of wave to the perpendicular. The inclination was fixed at 15° in the experiments with variable azimuths. Figure 2 demonstrated the posture configuration. The experiment frequencies included the LS band (practically 1.7-2.6 Ghz), X band (practically 8.2-12.4 GHz) and Ku band (12-18 GHz). The measurements were carried out in both the horizontal and vertical polarization. A dielectric rope model with same size was also simulated in FEKO 7.0 to calculate its RCS, although it was uniform but not braided structure limited by the software performance. The relative permittivity of simulated UHMWPE rope model was set as 2.3. The simulation frequency was only 2 GHz restricted by the computing source, which was one of the reasons for what we needed the experiment in the higher frequencies.

3. Analysis of the Results

Figure 3 demonstrates the measurement results. Notice that the RCS is obtained by Fourier transform from the raw sweep-frequency measured data, which can be thought as the integration RCS performance of the total band. A comparison of simulation and measurement results in 2 GHz is recorded in Figure 4.

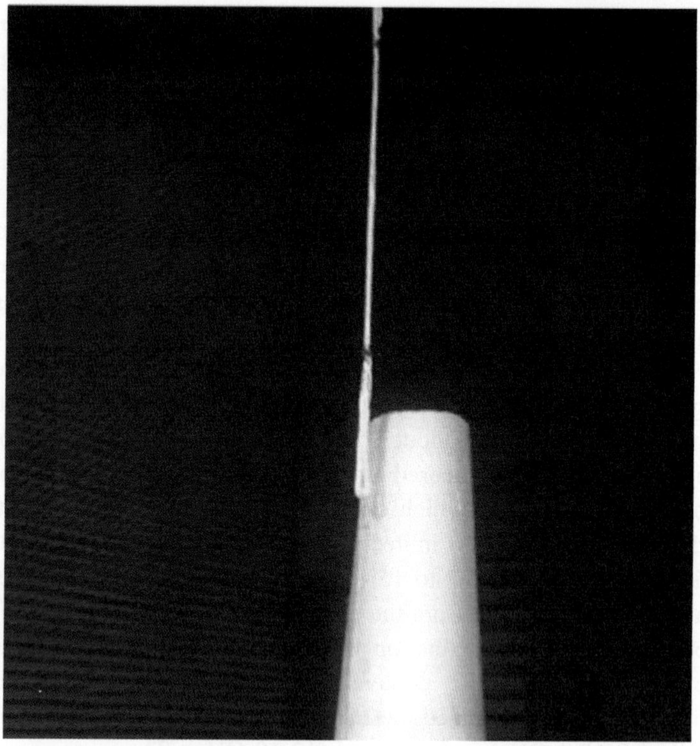

Fig. 1. The UHMWPE rope measured in the compact range.

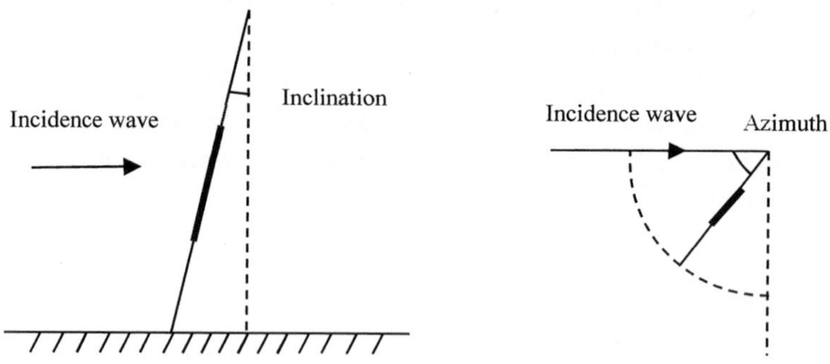

Fig. 2. The spatial posture of measured rope, (a) inclination, (b) azimuth.

According to all the results, it can be found that the rope possesses higher RCS in the vertical polarization than the horizontal polarization. Because the rope exhibits extremely larger length than its radius, and the scattered electric field in the vertical polarization integrates more. Fig. 3 (a), (b) and (c) show that the inclination could significantly degrade the RCS of UHMWPE rope within a small range of about 3°. Larger inclination still can reduce the RCS but the effect is declining. Fig. 3 (d), (e) and (f) indicate a similar trend in the azimuth. The most RCS reduction happens in no more than 10° from the vertical, while the other parts of the traces are nearly linear. The inclination of rope projection on the plane parallel with the incidence of test wave is easy to calculate. When the spatial inclination is 15° and the azimuth is 80°, the inclination of projection is only 2.57°, approaching the most effective range in the results of the variable inclination experiment. It means that the inclination with respect to the incidence wave is the main contribution to the RCS reduction of the UHMWPE rope. The string perpendicular to the test wave must be avoided in the measurement, as its RCS of vertical polarization is higher than -10 dB in all three bands, critically contaminating the target RCS. Figure 4 proves that the FEKO simulation is reliable as its results match the experiment well, so it can be used in the further design of the parallel suspension system.

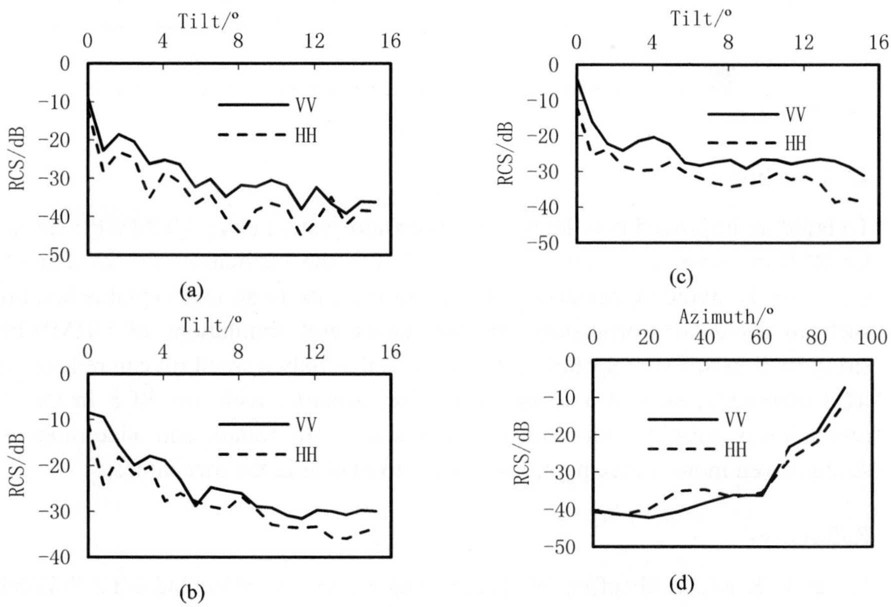

Fig. 3. Measurement results ("VV" and "HH" mean the vertical and horizontal polarization): (a) variable tilts in LS band, (b) variable tilts in X band, (c) variable tilts in Ku band, (d) variable azimuths in LS band, (e) variable azimuths in X band, (f) variable azimuths in Ku band.

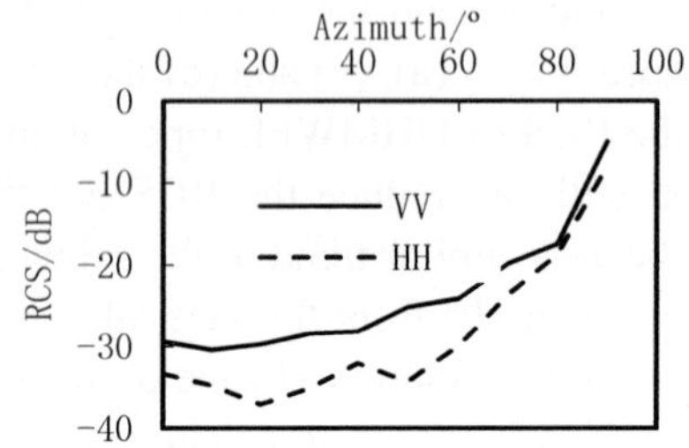

(e)

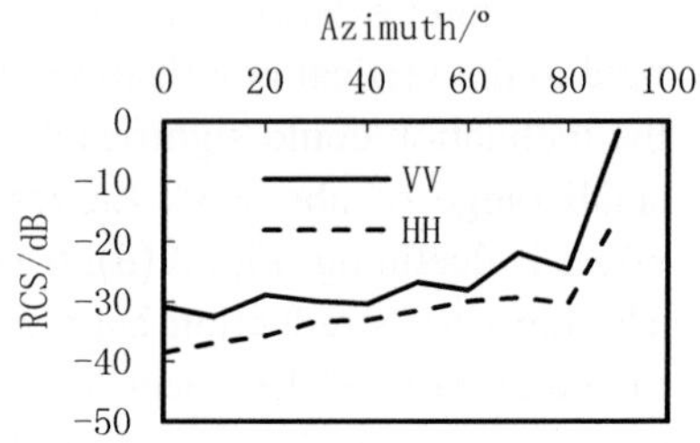

(f)

Fig. 3. (Cont'd)

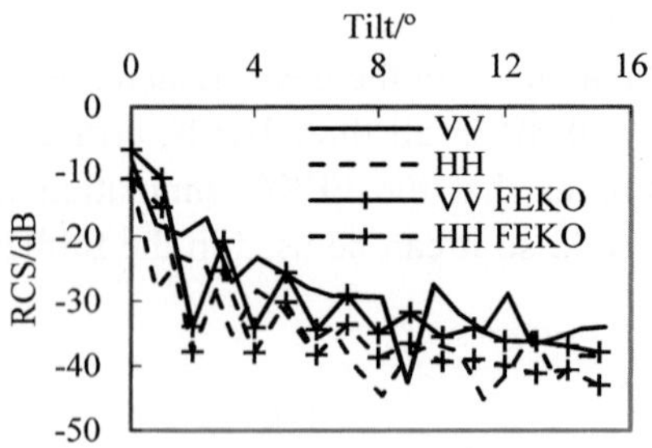

(a)

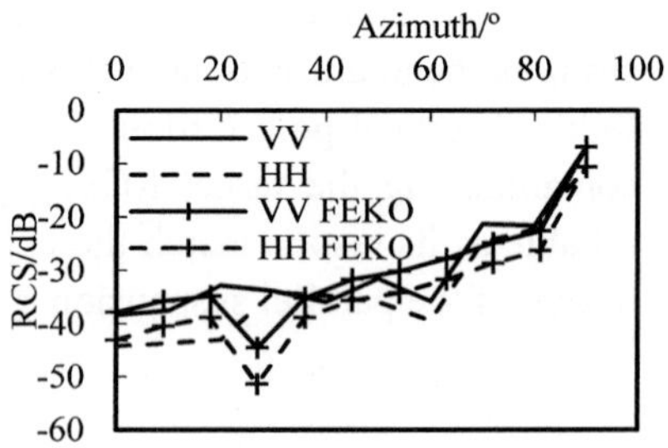

(b)

Fig. 4. Comparison of measurement and FEKO simulation results in 2 GHz: (a) variable tilts, (b) variable azimuths.

4. Conclusion

To build an improved parallel target suspension system using UHMWPE strings for RCS measurements, any normal position to the incidence wave of a single rope must be avoided, because its RCS will increase to an unacceptable level in such an occasion. Fortunately, measurements and simulations of UHMWPE string RCS in different spatial postures prove that only a small tilt can reduce the RCS obviously. So that a string suspension structure with low RCS in the all revolution is possible, not requiring large size of the rotator and wide range of string driven motor movement, and is easy to realize in the mechanics.

References

1. E. F. Knott, J. Shaeffer, M. Tuley, *Radar Cross Section*, 2nd edn.(SciTech Publishing, New York, 2004).
2. E. F. Knott E. F., *Radar Cross Section Measurements*, (SciTech Publishing, New York, 2006).

3. A. Buterbaugh, B. M. Kent, C. Mentzer, et al. Demonstration of an Inverted Stewart Platform Target-Suspension System Using Lightweight, High-Tensile Strings, *IEEE Antennas and Propagation Magazine*, 49, 5 (2007).
4. P. J. Collins, Dielectric strings: A low clutter method for bistatic RCS measurements, in *IEEE International Symposium on Antennas and Propagation (APSURSI'2011)*, (Spokane, USA, 2011).

Synthesis of oil-soluble viscosity reducer for crude oil

Ming-Shan Yang[1,2,*] , Hong-Liang Tong[1,2] and Ming Liang[1]
[1]*Department of Material science and Engineering, Beijing Key Lab of Special Elastomer Composites Materials, Beijing Institutes of Petrochemical Technology,*
Beijing,102617,China;
*E-mail: *yangms001@126.com*

[2]*College of Materials Science and Engineering, Beijing University of Chemical Technology*
Beijing,100029,China.

A major source of our energy consumption is from oil, coal, natural gas and other fossil fuels, out of which, oil is the most popular. Unfortunately, due to human activities such as mining and extraction of oil (especially in the past 100 years), oil resources are running out. Therefore, interest in heavy oil resources has been increasing. However, due to the large viscosity of the heavy oil, it is difficult to drill and transport. In this paper, the preparation of the viscosity reducer has been discussed. Several different types of oil-soluble viscosity reducer for the viscous crude oil were prepared by using solution polymerization method and compared by their physical properties and viscosity reduction effect by Infrared (IR), thermogravimetric (TG) and other testing method. We found that the viscosity reduction effect of the octadecyl acrylate - styrene copolymer is the best and that the viscosity reducer prepared by different polymerization monomers has a certain viscosity reduction effect.

Keywords: Monomer; Copolymer; Viscosity Reducer; Viscosity Reduction Effect.

1. Introduction

The rapid development of today's society is inseparable from the constantly progress of science and technology, but also need to continue to consume more energy. But as so far the vast majority of our energy consumption come from oil, coal, natural gas and other fossil fuels. Among them, the oil resources are the most widely used and popular. While, mankind has been mined for nearly half of the light crude oil resources in the past hundred years. Recoverable conventional oil resources were running out. Therefore, as another resource of

---

* Corresponding author.

the oil resources, namely heavy oil resources, attract more and more people's attention. However, due to the large viscosity of the heavy oil, it becomes very difficult to drill and transport. Therefore, the development and utilization of such resources has been a major problem in the oil industry.

The colloid and asphaltene molecules of the heavy oil contain hydroxyl, amino, carboxyl and carbonyl group which can form hydrogen bonds. And the heteroaryl fused ring plane of the asphaltene molecules overlap each other piled together, which is fixed by the hydrogen bonds between polar groups. So the asphaltene particle forms. Meanwhile, the heteroaryl fused ring plane of the colloid molecules overlaps stack and is fixed by the hydrogen bonds on the surface of the asphaltene particle. Therefore it forms a coating layer of the asphaltene particles. Such particles may also be connected to each other via hydrogen bonds. This has resulted in high viscosity oil. According to the colloid and asphaltene molecules were stacked state level and it will become loose via high temperature and solvent action, the viscosity reducer molecular would penetrate between the colloid or the asphaltene molecule layers. So it destroys the old aggregate original rules, with forming a new loose aggregate. However, due to the content of colloid and asphaltene vary with different regions, the molecular structure and molecular size are different. And therefore the oil soluble viscosity reducer for heavy oil has a certain selectivity.

2. Experiments

2.1. *Materials and reagents*

Acrylamide, with a least 98% purity was purchased from Shandong Po Mo Biological and Chemical Corporation. The viscous crude oil was drilled from the Shandong Shengli oilfield Chengdao 27A wells. Octadecanol (AR) was purchased from Aladdin (China). Acrylic acid, vinyl acetate, styrene, azobisisobutyronitrile(AIBN), benzoyl peroxide, which were analytical reagent were purchased from Fu Chen Tianjin Chemical Reagent Factory. Toluene, acetone, sulfuric acid, which were analytical reagent were purchased from Beijing Chemical Reagent Company.

2.2. *Test and instrumentation*

The IR testing uses Nicolet 6700 Fourier Transform Infrared Spectrometer. Thermal gravimetric analysis uses Mettler-Toledo TGA/DSC1. The viscosity of the test proceeds by DV3T which is produced by the United States Brookfield.

2.3. *Polymerization procedures*

(1) Preparation of octadecyl acrylate

Add a certain amount of stearyl alcohol, toluene, and hydroquinone to a three-necked flask which equipped with a thermometer, a stirrer and a reflux condenser. Then add acrylic acid and catalyst when it heated to 60°C until them full dissolved. When it was heated to reflux for about 3h, attach the trap to dehydrate. When the amount of water is equal to the theoretical value, that is considered the esterification reaction is substantially completed. Distillation under reduced pressure to remove unreacted acrylic acid and solvent. The product was then poured into a beaker, with a mass fraction of 5% NaOH solution was washed away and then washed with deionized water several times until it was neutral. Put it in 40°C vacuum drying 8h, to obtain a waxy solid octadecyl acrylate. The reaction principle shows as following.

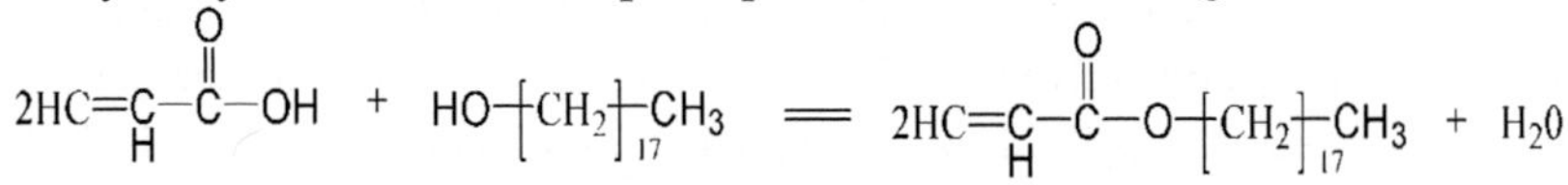

Fig.1. Principle of the reaction of octadecyl acrylate

(2) The synthesis of the viscosity reducer

Add a certain amount of octadecyl acrylate with styrene or another monomer in a clean three-neck flask, and then an appropriate amount of toluene was added to dissolve. After the dissolution of the drug, add benzoyl peroxide (BPO), with stirring and nitrogen for 15 minutes to remove the system of oxygen. Then the reaction system was heated to 85°C and reaction time was 6 hours. The reaction product was then removed and washed with acetone, filtered and then placed in a vacuum oven at 40°C, dried for 12 hours under -0.1MPa conditions to get the target product. Table 1 shows the specific experimental formula.

Table 1. The specific experimental formula.

Composition / Number	Octadecyl acrylate(g)	Styrene(g)	Vinyl acetate(g)	Acrylamide(g)	Maleic Anhydride (g)	BPO(g)
Formula 1	6	0.24	0	0	0	0.050
Formula 2	6	0	0	0	0.23	0.050
Formula 3	8	0	0	0.22	0	0.065
Formula 4	8	0	0.27	0	0	0.065
Formula 5	90	0	2.98	0	3.40	0.770
Formula 6	40	0	4.95	0	2.42	0.240

2.4. *Determination of viscosity reduction rate*

Bring a certain amount of heavy oil which taken from Shandong Shengli Oilfield Well No. 27A Long Island to do the viscosity test. Test temperature is 29°C. Pour the crude oil into the sample adapter, and test the viscosity. Then make the viscosity reducer dubbed the mass fraction of 50% solution. After a ratio of 1/2000 were added and mixed to keep for 10min. Test the viscosity and calculate the viscosity reduction rate. The specific equation shows as following.

$$K=\frac{\eta_0-\eta_1}{\eta_0}\times 100\%$$

K- the viscosity reduction rate,%

η_0-the crude oil viscosity before the reducing agent, cp

η_1- the crude oil viscosity after the reducing agent, cp

3. Results and analysis

3.1. *Analysis of the copolymer*

1.Fourier transform infrared (FTIR) analysis

To determine whether the double bond of the monomer could open and participate in the polymerization reaction. We conducted several experiments, but here only illustrate Formula 1 to explain this phenomenon. Figure 2-4. are the IR spectrum of stearyl acrylate, styrene and stearyl acrylate-styrene copolymer respectively.

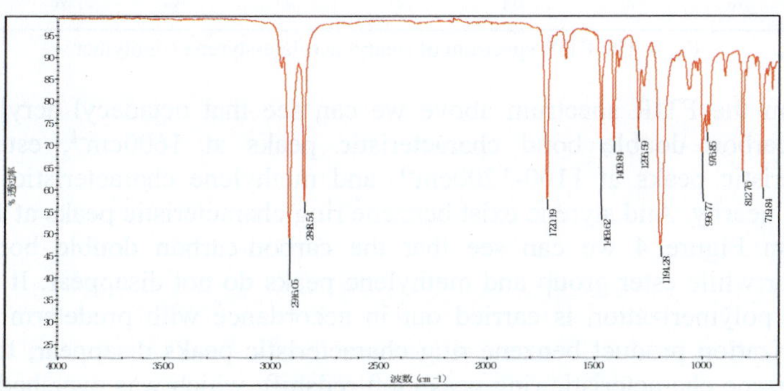

Fig.2. The FTIR spectrum of stearyl acrylate

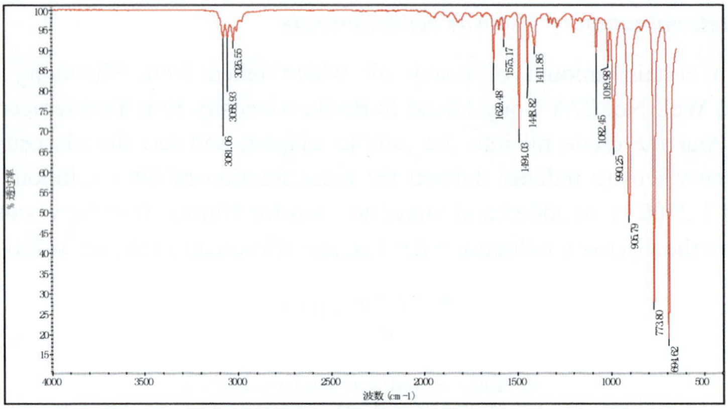

Fig.3.The FTIR spectrum of styrene

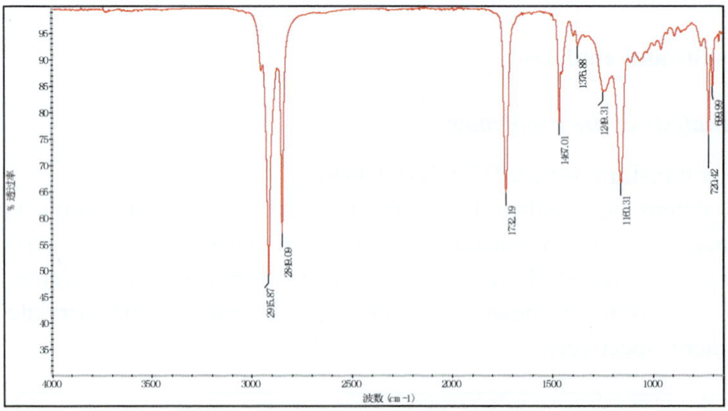

Fig.4. The FTIR spectrum of stearyl acrylate-styrene copolymer

From the FTIR spectrum above we can see that octadecyl acrylate exist carbon-carbon double bond characteristic peaks at 1600cm^{-1}, ester group characteristic peaks at 1100-1200cm^{-1}, and methylene characteristic peaks at 720cm^{-1} nearby. And styrene exist benzene ring characteristic peaks at 1600cm^{-1}. But from Figure 4 we can see that the carbon-carbon double bond peaks disappear while ester group and methylene peaks do not disappear. It indicates that the polymerization is carried out in accordance with predetermined. But polymerization product benzene ring characteristic peaks disappear, there may be a benzene characteristic ring peaks left redshift, which was overshadowed by the peaks at 1700cm^{-1}.

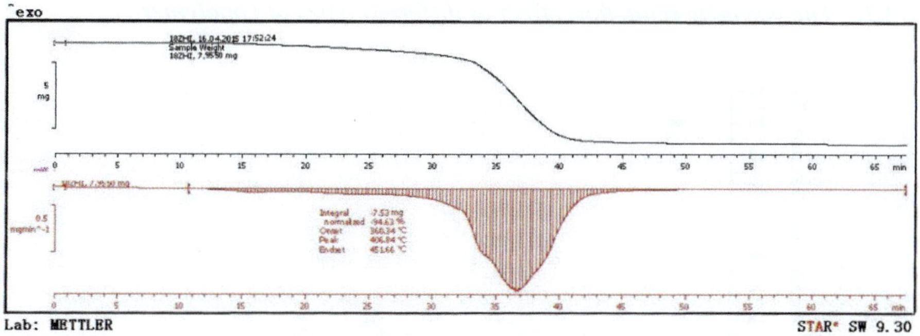

Fig.5. The TG spectrum of stearyl acrylate

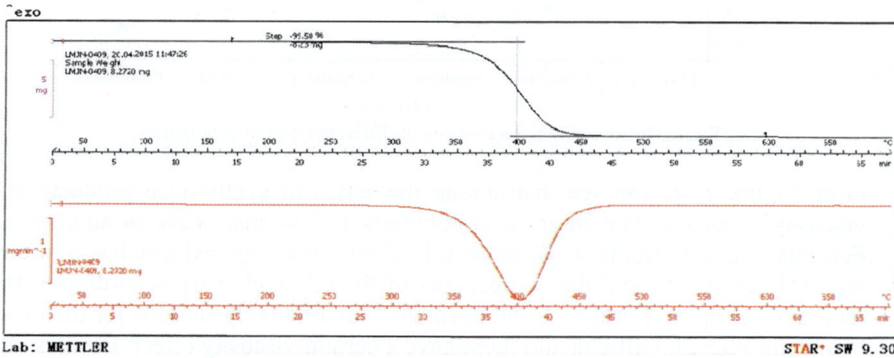

Fig.6.The TG spectrum of stearyl acrylate-styrene copolymer

2. Thermo gravimetric Analysis

Also due to the limited space we only illustrate Formula 1 to explain this.

As can be seen from the TG spectrum, the thermal decomposition temperature of octadecyl acrylate used in the experiments is 406°C nearby, generally higher than the monomers. It can be inferred that the stearyl acrylate is used as the prepolymer in the experiments. And the thermal decomposition temperature of stearyl acrylate-styrene copolymer is close to stearyl acrylate, which is about 400°C. Therefore it can be inferred that the thermal decomposition of stearyl acrylate-styrene copolymer is essentially the break of the ester group among the octadecyl acrylate.

3.2. *The viscosity reduction effect of different types of copolymer*

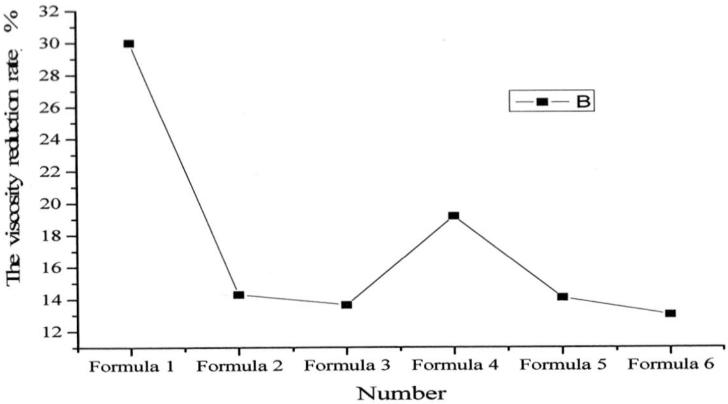

Fig.7. The viscosity reduction rate of different types of copolymers

From Figure 7 we can see that among the series of synthesized products, the viscosity reduction rate of most copolymers is less than 15% in addition to Formula 1 and Formula 4. Because oil-soluble viscosity reducer has a strong regional selectivity, and the components of the oil fields vary according to the region of oil fields. Formula 1 and Formula 4 are relatively suitable for the oil of Shandong Shengli Oilfield, and them have a certain viscosity effect. Besides, the heavy oil we had used in the experiments are also oilfield fluid, obtained after demulsification dehydration. The viscosity additives have been added before drilling, but could not be completely removed after dehydration. Therefore this part of additives also had influence on the viscosity reduction effect.

4. Conclusion

By FTIR and TG characterization, we know that the viscosity reducer can be prepared through the following conditions: octadecyl acrylate used as main prepolymer, toluene used as solvent, benzoyl peroxide used as initiator. Use a solution polymerization process, under 85°C for 6 hours, and then we can get it. The viscosity reducer prepared by different polymerization monomers has a certain viscosity reduction effect.

Acknowledgements:

This work was supported by Graduate Innovation Foundation (15033981002/050), Project of Outstanding Professors and Management Experts(BIPT-POPME-2014) of Beijing Institute of Petrochemical Technology.

References

1. Zhao Lisha, Wu Xiao-chuan, Yi Chen-xi, WANG Wan-qing, Status and Development Trend of Heavy Oil Recovery Techniques. Liaoning Chemical Industry. 2013, 42(4) :363-365.
2. MISHRAMK, RAYMOND G. Pour point depressantsvia anionic polymerization of(meth) acrylicmonomers [P].US5834408,1998.
3. FISCHERJ. Terpolymer of ethylene, vinyl acetate and isobutylene useful as pour point depressants in crude oils[P].US5256166,1993.
4. Jiang Hongwei. Laboratory development of a kind of oil soluable viscosity reducing agent for heavy oil. Petroleum Geology & Oilfield Development.2006 1000-3754.

Preparing, characterization and catalytic application of palladium nanoparticles supported by boron-doped carbon-encapsulated cobalt

Ying Liu, Hai-Jun Ruan, Jian-Wei Yin, Lei Wang and Xun-Gao Zhang[†*]
College of Chemistry and Molecular Sciences, Wuhan University
Wuhan, 430072, China
E-mail: [*]xgzhang66@whu.edu.cn

Boron-doped carbon-encapsulated cobalt (Co@CB) nanoparticles were effectively prepared using an alternating current (AC) arc discharge method and were investigated in detail. A novel magnetic retrievable catalyst of Pd nanoparticles supported on Co@CB (Pd/Co@CB) was developed. The catalytic properties for aerobic oxidation of benzyl alcohol and the recycling performance of the catalyst were investigated. The results of the X-Ray Powder Diffraction (XRD), Raman and X-ray photoelectron spectroscopy (XPS) suggested that about 3.5% of boron doped within graphite network and some Boron-Oxygen bonds were present on the surface of carbon. Pd/Co@CB exhibited high catalytic activity and selectivity for aerobic oxidation of benzyl alcohol under a mild condition. The catalyst can be reused at least 7 cycles, and it can be conveniently recycled under a magnetic field.

Keywords: Heterogeneous Catalyst; Boron-doped Carbon-encapsulated Cobalt; Palladium Nanoparticles; Aerobic Oxidation of Benzyl Alcohol.

1. Introduction

Magnetic metal and metal oxides nanoparticles have many potential applications such as using as a support for nanocatalyst [1,2]. However, owing to their high surface energy, metal or metal oxides nanoparticles are extremely sensitive to environments [2]. Encapsulated with graphitic carbon is an effective route to improve the chemical stability of metal nanoparticles. Since Ruoff et al. discovered carbon-encapsulated LaC_2 nanocrystals [3], many applications of carbon-encapsulated magnetic metal nanoparticles have been reported. For example, its can be designed as magnetic supports for drug targeting [4], high sensitive electrochemical sensors [5], photocatalysts for hydrogen production [6] and the supports for nanocatalyst [7,8], and so on.

Recently, applications of precious metal catalyst in the field of chemical synthesis and fine chemical industry have been intensively investigated [9,10]. It

[†] Corresponding author.

shows advantages in clean and sustainable chemistry because many reactions are easily completed under mild and environmental friendly condition while using these catalysts. Nevertheless, the separation and recycling of precious metal nanocatalysts remain a challenging issue for conventional techniques such as centrifugation or filtration. To overcome this problem, the use of magnetic nano-support has emerged as one of the best solutions [1,7,10]. In this paper, boron-doped carbon-encapsulated cobalt (Co@CB) nanoparticles were effectively prepared by an alternating current (AC) arc discharge method on large scale, and the structure was characterized in detail. A novel magnetic retrievable catalyst of Pd nanoparticles supported on Co@CB (Pd/Co@CB) was developed. The catalytic properties for aerobic oxidation of benzyl alcohol and the recycling performance of the catalyst were also investigated.

2. Experimental

2.1. *Preparation of Co@CB*

Co@CB was prepared by an AC arc discharge method under helium buffer gas. The details of AC arc discharge apparatus and techniques have been described in our earlier work [11]. In this work, two electrodes were prepared as fellows: two spectrum pure graphite rods with diameter of 20 mm and length of 150 mm were drilled holes in axial (15 mm in diameter and 120 mm in length), and filled with the mixture of 105 g Cobalt powder (purity about 99%, 5 ~ 15μm in size) and 6 g amorphous boron powder (purity about 95%, 5 μm in size). According to the remaining weight of graphite rods after arc discharge, the starting weight ratio of C/B/Co was calculated to be 34/6/105 (g). The AC arc discharge experiment was carried out in helium gas of 2×10^{-2} MPa. The discharge current was kept at 700 ~ 900 A, and the voltage ranged from 25 to 35 V. After arc discharge, the powders on the inwall of furnace were collected. Treated with dilute hydrochloric acid for 24 h and washed with deionized water to remove the fused and incomplete encapsulated Co particles, then dried in air at 50°C. The obtained magnetic powders were named as Co@CB. By using this method, dozens of grams of Co@CB can be obtained in one experiment.

As a comparison sample, carbon-encapsulated cobalt (Co@C) was prepared by a similar procedure without loading boron powder.

2.2. *Characterization*

The Co content was measured on an x-ray fluorescence (XRF) spectrometer (S4 Pioneer, Bruker AXS). TEM and HRTEM images were obtained on a JEM-

2010F transmission electron microscope operated at 200 kV. XRD patterns were conformed on a D/max-γA (Rigaku Co. Ltd.) x-ray powder diffractometer with Cu Kα radiation of 0.15406 nm. XPS data were conducted on an x-ray photoelectron spectrometer (Kratos, XSAM 800) with Mg Kα radiation. The Raman spectra were investigated on a RM 1000 (Renishaw Co. Ltd.) laser confocal Raman microspectroscope. The excitation line was 514.5 nm with a power of approximately 0.1 mW on the sample. The magnetic hysteresis loop was measured by means of a vibrating sample magnetometer (VSM-9600, LDJ).

2.3. *Preparation of Pd/Co@CB catalyst*

83.5 mg $PdCl_2$ (AR, Sinopharm Chemical Reagent Co. Ltd), 50 mL of 0.01 mol/L HCl, and 40 mL of deionized water were added to a 100 mL three-neck flask, stirred for 2h at ambient temperature to obtain a yellow-brown H_2PdCl_4 solution (0.02 mol/L). The obtained solution was then transferred to a 500 mL three-neck flask with a reflux condenser, 500 mg PVP (AR, Sino-pharm Chemical Reagent Co. Ltd), 0.5 g Co@CB and 100 mL deionized water were added, the suspension was stirred for 2h at ambient temperature. Then 100 mL ethanol was added and stirred at 80°C for 4 h, cooled to the ambient temperature. After magnetic separation, the product named as Pd/Co@CB was washed sequentially with deionized water and ethanol for four times, respectively, and dried at 50 °C for 12 h.

2.4. *Aerobic oxidation of benzyl alcohol*

1 mmol benzyl alcohol (AR, Sinopharm Chemical Reagent Co. Ltd), 18.6 mg Pd/Co@CB (0.016 mmol Pd) catalyst, 128 mg K_2CO_3 and 12 mL CH_3CN were added to a 25 mL three-neck flask with a reflux condenser. The suspension was stirred magnetically at 80°C under an oxygen flow in a thermostatic water bath for several hours. The product was analyzed by high performance liquid chromatography (HPLC, Agilent 1100) and the conversion of benzyl alcohol and the selectivity for benzaldehyde were calculated by the peak area normalization method.

After the reaction, the solution was transferred into a vessel and a magnet was placed outside of the vessel wall. The Pd/Co@CB catalyst quickly adsorbed onto the wall of the vessel and the catalyst was then collected and washed with ethanol and deionized water for three times, respectively, dried at 50 °C for reuse.

3. Results and discussion

3.1. *Elemental analysis*

The Co content was measured by XRF. The Co content of Co@CB was about 59 wt%, which is obviously lower than that of the loading ratio. The main reason is that Co easily melted and flowed out of the arc area during arc discharge. In addition, some incomplete encapsulated Co particles were also removed by dilute hydrochloric acid treatment. The Pd content of Pd/Co@CB catalyst was detected to be 9.0 wt%, which is very close to the loading ratio. It suggests that $PdCl_2$ was completely reduced by ethanol.

3.2. *XRD pattern analysis*

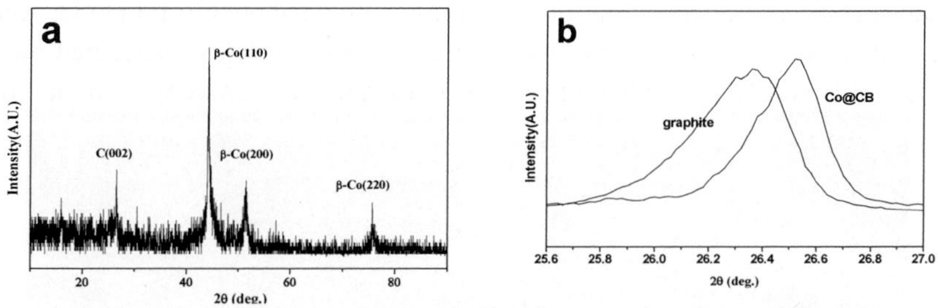

Fig. 1. (a) XRD pattern of Co@CB; (b) The C (002) diffraction peaks of Co@CB and graphite

Fig. 1a shows the XRD pattern of Co@CB. Fig. 1b shows the C(002) diffraction peaks of Co@CB and graphite in the range of 2θ at $25.6^0 \sim 27.0^0$. A diffraction peak at about 2θ: 26.55^0 is associated with the 0.335 nm for (002) interspacing of the carbon shells of Co@CB, which is smaller than the d_{002} value of 0.338 nm (2θ: 26.35^0) of graphite (shown in Fig. 1b). It is well known that the interspacings of the curved carbon shells in MWCNT, onions or metal@C [3] are frequently larger than the d_{002} values of graphite. However, owing to the depleted π electron density between the graphite planes, the d_{002} value of the B-doped graphite will be reduced with the substitution of carbon atoms with boron atoms within the graphene sheets [12]. The XRD results reveal that the carbon in Co@CB is a typical graphite-like structure and the reduced d_{002} value of carbon shells may be attributed to boron atoms incorporated within the graphite network.

In Fig. 1a, the peaks at 2θ values of 44.2^0, 51.4^0 and 75.7^0 can be identified to the (110), (200) and (220) planes of β-Co with a face-centered cubic (fcc) structure (JCPDF 5-681). No peaks for other phases of cobalt such as oxide, carbide or boride alloy are found. It suggests that Co in the as-made Co@CB is in metal state only, which is similar to the case of Co@C without boron doping [11].

3.3. *The morphology observation*

Fig. 2a shows the typical TEM image of the as-made Co@CB. The morphology of Co@CB is similar to that of Co@C without boron doping [11]. The encapsulated Co nanoparticles are spherical, and the diameter is in the range of 5 ~ 50 nm. HRTEM image of Co@CB (shown as the inset of Fig. 2a) shows that (110) interspacing of Co is about 0.22 nm, which is consistent with the XRD result. The thickness of carbon is about several nanometers.

Fig. 2b shows the typical HRTEM image of Pd/Co@CB, the inset shows the enlarge image of Pd particle. It can be seen that the Pd nanoparticle was not encapsulated. The interspacing of 0.22 nm can be attributed to Pd (111). About 100 statistical data showed that the diameter of Pd particles distributed on the surface of Co@CB was in the range of 2-5 nm, and the average was about 3 nm.

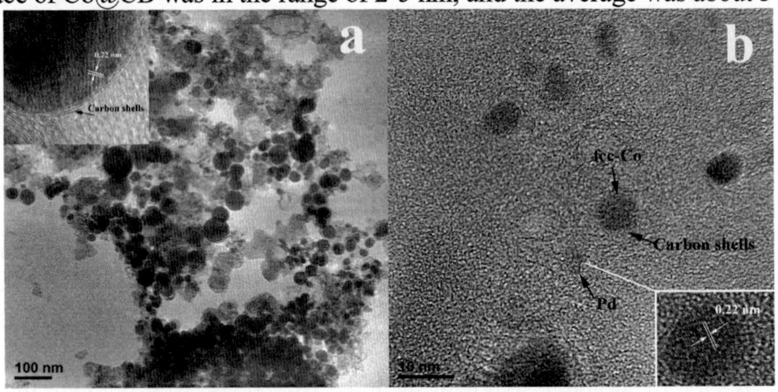

Fig. 2. TEM images of Co@CB (a) and Pd/ Co@CB (b)

3.4. *Raman measurement*

Fig. 3 shows the first- and second-order Raman spectra of the Co@CB and graphite at room temperature. In the first-order Raman spectrum, the peaks located at 1344 cm^{-1} and 1593 cm^{-1} for the as-made Co@CB are associated with the defect-induced feature (D-band) and the vibrations in sp^2 bonded carbon atoms in a two-dimensional hexagonal lattice (G-band) [13,14], respectively. Compared with the graphite, the frequency of the D-band for Co@CB downshifts about 14 cm^{-1}, while the frequency of the G-band upshifts by 11 cm^{-1}. The high intensity of D-band reveals the increasing number of defects in Co@CB. In the second-order Raman spectrum, the G′-band located at 2690 cm^{-1}

with a downshift of 22 cm^{-1} is attributed to the overtone peak of D-band. Similar to boron-doped graphite and boron-doped graphene [13,14], the intensity of G′-band for Co@CB decreases remarkably. The frequency of D+G located at 2938 cm^{-1} has a small downshift of 2 cm^{-1}, which is attributed to the changes in both D- and G-bands. The Raman features of Co@CB are in good agreement with those of boron-doped graphite, boron-doped SWCNTs and boron-doped graphene [13,14]. These effects are attributed to the increasing in-plane lattice constant and a weakening of the B-C in-plane bond strength relative to the corresponding C-C bond [14].

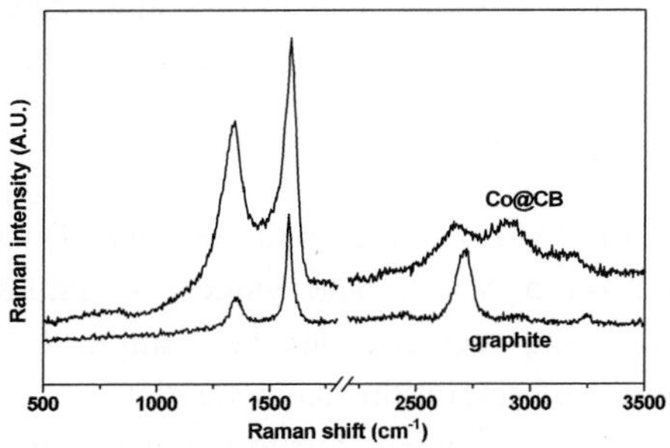

Fig. 3. Raman spectra of Co@CB and graphite

3.5. *XPS analysis*

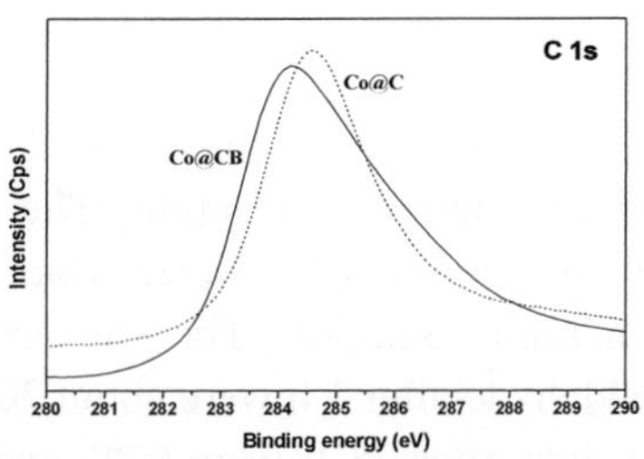

Fig. 4. XPS spectra of the C 1s of Co@CB and Co@C.

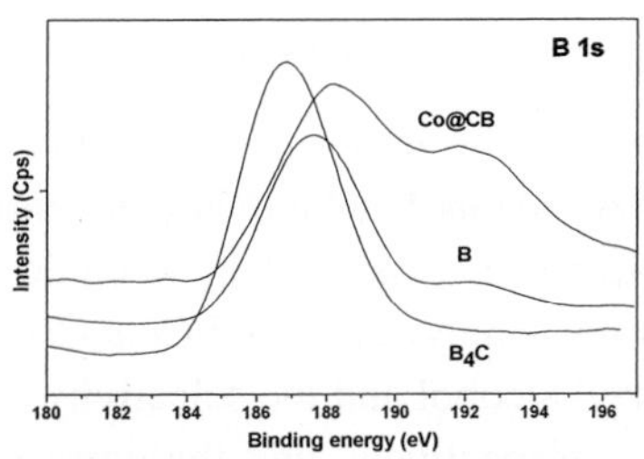

Fig. 5. XPS spectra of B 1s of Co@CB, B and B_4C.

Further information about the structure of the as-made Co@CB is obtained from XPS analysis. As seen in Fig. 4, the C 1s peak in Co@C without boron is located at binding energy of 284.6 eV, which is in good agreement with that of pure graphite. For the as-made Co@CB, the C 1s peak shifts to lower binding energy at 284.3 eV. The lowering of C 1s binding energy for Co@CB may be due to the lowering of Fermi level because of the redistributed π-electrons in the graphite layer [15].

Fig. 5 shows the XPS spectra of B 1s energy level of Co@CB, B powder and B_4C powder (purity about 97%, 250 μm in size, commercial product). B 1s peak in B powder is located at binding energy of 187.6 eV with a shoulder of 192 eV ascribed to the oxide compounds on the surface. B 1s peak in B_4C powder appears at binding energy of 186.8 eV. B 1s binding energy of the as-made Co@CB is at 188.2 eV which indicates that the binding state of boron in Co@CB is quite different from that of B_4C or elemental boron. We believe that the B 1s peak at 188.2 eV for Co@CB is attributed to the sp^2-hybridized boron atoms substituted for carbon atoms in graphite plane. This is in good agreement with the literatures [16] and also supported by our XRD and Raman results. A strong broad shoulder peak at around 192 eV for Co@CB is assigned to boron oxides formed on the carbon surface. The boron content of Co@CB is calculated to be about 3.5 at% by the relative sensitivity factor method.

3.6. *Magnetization measurement*

The magnetization curve (*M-H*) of the as-made Co@CB sample was measured at room temperature on a VSM. The values of saturation magnetization, remanent magnetization and coercivity are 38.3 emu/g, 6.7 emu/g and 195 Oe, respectively. The values of saturation magnetization is lower than that of bulk cobalt, but is closed to that of cobalt ferrite nanoparticles [2]. The data reveal that the as-made Co@CB is a typical ferromagnetic material, which can be easily separated from a liquid medium by an external magnet.

3.7. *Catalytic performance*

The catalytic performance of Pd/Co@CB was investigated for the oxidation of benzyl alcohol, which is frequently used as a template reaction. The results indicated that benzaldehyde could be high selectively synthesized from benzyl alcohol catalyzed by Pd/Co@CB under a mild condition. The conversion of benzyl alcohol and the selectivity of benzaldehyde after 1 h were about 70% and 99%, respectively, and the corresponding data after 23 h were 97% and 98%, respectively. Pd/Co@CB exhibited not only a high activity for the oxidation of benzyl alcohol, but also an remarkable advantage of easy magnetic-separation.

The recycling performance is also a key issue for industrial consideration. Table 1 shows the recycling performance of the catalyst in eight reaction cycles. The catalyst remained a high catalytic activity after 7 cycles, the conversion kept about 90%. Although the conversion was slightly down to 73% after recycling eight times, the selectivity of benzaldehyde remained unchanged. The recycling performance of Pd/Co@CB is much better than that of Pd/Fe@C without boron doping [8]. We propose that the excellent recycling performance may be attributed to the increasing defects, B-O and B-C bonds formed on the surface of carbon shells, where boron atoms may act as the active sites for Pd nanoparticles anchoring.

Table 1. The recycling performance of Pd/Co@CB catalyst for benzene alcohol oxidation reaction.

Cycle number	1st	2nd	3rd	4th	5th	6th	7th	8th
Conversion (%)	90.0	92.7	92.0	90.8	91.0	91.7	89.3	73.0
Selectivity (%)	99.2	97.6	98.9	99.3	99.6	99.0	99.1	98.4

Reaction conditions: 1 mmol benzyl alcohol, Pd/Co@CB (Pd 1.6 mol% relative to benzyl alcohol), 128 mg K_2CO_3, 12 mL CH_3CN, 1 atm O_2, 80°C, 7 h.

4. Conclusions

We have developed a novel magnetically retrievable nanocatalyst support Co@CB, which could be effectively prepared by an AC arc discharge method on a large scale. Boron atoms were conformed doping within the outer carbon shells of carbon-encapsulated cobalt. Pd nanoparticles supported on Co@CB exhibited high catalytic activity and selectivity for aerobic oxidation of benzyl alcohol and an excellent recycling performance. The catalyst could be easily separated from the liquid reaction system by an external magnet. The application of other precious metal nanoparticles supported on Co@CB is now under way.

Acknowledgments

The authors would like to thank Hubei province natural science foundation of China (No. 2013CFB291)

References

1. R.B.N. Baig, R.S. Varma, Magnetically retrievable catalysts for organic synthesis, *Chem. Commun.* **49**, 752(2013).
2. M. Demirelli, E. Karaoğ lu, A. Baykal, H. Sözeri, E. Uysal, Synthesis, characterization and catalytic activity of $CoFe_2O_4$-APTES-Pd magnetic recyclable catalyst , *J. Alloys Comp.* **582**, 201(2014).
3. R.S. Ruoff, D.C. Lorents, B. Chan, R. Malhotra, S. Subramoney, Single crystal metals encapsulated in carbon nanoparticles, *Science* **259**, 346(1993).
4. M. Bystrzejewski, A. Huczko and H. Lange, Arc plasma route to carbon-encapsulated magnetic nanoparticles for biomedical applications, *Sensors and Actuators B* **109**, 81 (2005).

5. L.J. Yang, H.Y. Xiong, X.H. Zhang, S.F. Wang, X.G. Zhang, Direct electrochemistry of glucose oxidase and biosensing for glucose based on boron-doped carbon-coated nickel modified electrode, *Biosens. Bioelectron.* **26**, 3801(2011).
6. T.Y. Peng, X.H. Zhang, P. Zeng, K. Li, X.G. Zhang, X.G. Li, Carbon encapsulation strategy of Ni co-catalyst: Highly efficient and stable Ni@C/CdS nanocomposite photocatalyst for hydrogen production under visible light, *J. Catal.* **303**, 156(2013).
7. S.C. Tsang, V. Caps, I. Paraskevas, D. Chadwick and D. Thompsett, Magnetically Separable, Carbon-Supported Nanocatalysts for the Manufacture of Fine Chemicals, *Angew. Chem. Int. Ed.* **43,** 5645 (2004).
8. H. Zhang, Y. Liu, X.G. Zhang, Selective oxidation of benzyl alcohol catalyzed by palladium nanoparticles supported on carbon-coated iron nanocrystals, *Chin. J. Catal.* **32**, 1693(2011).
9. G.M. Scheuermann, L. Rumi, P. Steurer, W. Bannwarth, R. Mülhaupt, Palladium nanoparticles on graphite oxide and its functionalized graphene derivatives as highly active catalysts for the Suzuki–Miyaura coupling reaction, *J. Am. Chem. Soc.* **131**, 8262(2009).
10. T. Tian, Y. Liu, X.G. Zhang, Bimetallic synergistic Au/CuO-hydroxyapatite catalyst for aerobic oxidation of alcohols, *Chin. J. Catal.* **36**, 1358(2015).
11. J. Ling, Y. Liu, G.M. Hao, X.G. Zhang, Preparation of carbon-coated Co and Ni nanocrystallites by a modified AC arc discharge method, *Mater. Sci. Eng. B* **100**, 186(2003).
12. C. T. Hach, L. E. Jones, C. Crossland, P.A Thrower, An investigation of vapor deposited boron rich carbon – a novel graphite-like material – part I: the structure of BC_x (C_6B) thin films, *Carbon* **37**, 221(1999).
13. Y. Hishiyama, H. Irumano, Y. Kaburagi, Y. Soneda, Structure, Raman scattering, and transport properties of boron-doped graphite, *Phys. Rev. B* **63**, 245406-1(2001).
14. K. McGuire, N. Gothard, P.L. Gai, M.S. Dresselhaus, G. Sumanasekera, A.M. Rao, Synthesis and Raman characterization of boron-doped single-walled carbon nanotubes, *Carbon* **43**, 219 (2005).
15. G. Hennig, Diffusion of Boron in Graphite, *J. Chem. Phys.* **42**, 1167-(1965).
16. T.Shirasaki, A. Derré, M. Ménétrier, A. Tressaud, S. Flandrois, Synthesis and characterization of boron-substituted carbons,*Carbon* **38**,1461 (2000) .

Polyaniline fiber bundles synthesized by interface oxidative polymerization with aqueous solutions containing dumbbell-structure ionic liquid

Hai-Yan Wang*, Xiao-Ting Zhang, Ni Wang, Ding-Jun Zhang and Ran Fen
School of Materials Science and Engineering, Lanzhou University of Technology
Lanzhou 730050, China
*E-mail: *wanghy03@lut.cn*

Nanostructured polyaniline(PANI), particular one-dimensional PANI, has gain relatively more attention as opposed to other conducting polymers because they possess more prominent in the large surface area and porous structure. In our work, a dumbbell-structured ionic liquid composed of pyridines was used for PANI synthesis. The structure and performance of samples were studied and the PANI bundles were synthesized by interface oxidative polymerization with aqueous solutions containing dumbbell-structure ionic liquid. The structure and performance was explored by scanning electron microscopy (SEM), Fourier-transform infrared (FTIR), wide-angle X-ray, thermogravimetric analysis and conductivity test. The results showed that dumbbell-structure ionic liquid was the structure-directing agent to construct PANI fiber bundles which were synthesized at different mole ration of ionic liquid to aniline. The mole ratio of ionic liquid to aniline exhibits different effects on doped degree, crystallinity, thermal stability and conductivity of the samples. Their conductivity maintained at similar level for 10^{-4}~10^{-5} S cm^{-1}, but the highest decomposition temperature was 442 °C. This work provides a simple approach to fabricate nanostructure polyaniline with controllable morphology.

Keywords: Polyaniline; Ionic Liquid; Morphology; Interfacial Polymerization; Fourier-transform infrared (FTIR) ; Scanning electron microscopy (SEM).

1. Introduction

Nanostructured polyaniline(PANI), particular one-dimensional PANI, has received more interests in contrast with other conducting polymers because they possess more specific surface area, more controllable morphology, and lower cost. These characteristics result in a diverse range of applications such as chemical sensors, hydrogen storage, supercapacitors, anticorrosion coating, electromagnetic shielding, and transparent flexible electrodes [1-5]. The synthesis methods of one-dimensional nano-PANI have magic progress in recent nearly twenties years [6-9]. The routes involve hard (zeolites) or soft templates

* Corresponding author

(organic surfactants, polyelectrolytes) as well as polymer additives [10-11]. They act as structure-directing agents, but facile and effective transition of the morphology remains an important challenge.

In recent years, ionic liquids (ILs) had played an important role in polymer synthesis because they participated in various intermolecular interactions such as Coulombic, Van der Waals, dispersive, n-p and p-p interactions and hydrogen bonds [12,13]. They usually form large nanostructures through hydrogen bonding in the liquid state, which are of very interest in the preparation of various nanoparticles [14]. Majda Žigon added ILs into reaction mixture and controlled PANI morphology toward nanowires or various complex two- and three-dimensional structures. And the UV–Vis, IR and Raman spectra of PANI nano-tube showed that the used ILs do not affect the chemical structure of PANI, which indicates that they do not affect the formation of PANI molecules but only their assembling during polymerization[15]. In addition, Ionic liquids are electrically conductive, thermally stable, low vapor pressure and a green alternative to volatile organic solvents. In our work, dumbbell-structure ionic liquid composed of pyridines was used for PANI synthesis. The structure and performance of samples were studied.

2. Experimental

2.1. Chemicals

Aniline (Beijing Chem. Co.) was distilled under reduced pressure prior to use. Ammonium peroxodisulfate(APS), 1, 6-dibromohexane, pyridine (A.R., Beijing 3rd Chemical Reagents Factory) and other reagents (from Tianjin Chem. Co.) were used as received.

2.2. Ionic liquid synthesis

1, 6-dibromohexane(2.44 g) and pyridine(1.4 g) were added into a certain volume of ethanol solvent. The reaction was then allowed to proceed under agitation for 24 h at 50 °C. The product distillated used ethyl acetate to wash for many times. Then the product was added into the solution of sodium bisulfate and ethanol to complete the ion exchange reaction under the stirring condition at 80 °C for 6 h. Dumbbell-structure ionic liquid power was obtained after drying. The structure of the ionic liquid was shown in Figure 1. There was almost no impurity peak on H^1-NMR spectrum. It indicated that the purity of ionic liquids power was more than 95%.

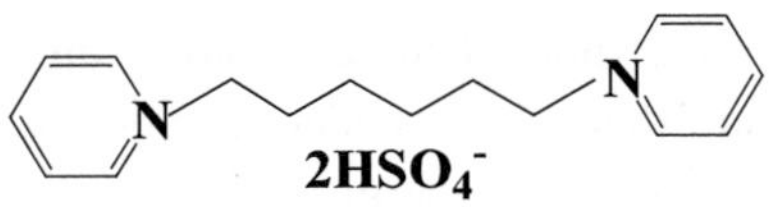

Fig. 1 the dumbbell-structure of ionic liquids

2.3. Polyaniline synthesis

In the typical synthesis procedure, the volume ratio of oil phase to water was maintained as 1:1. The 0.5 mL of aniline was dissolved into 50 mL of tetrachloromethane in a vessel. The mole ratio of aniline to APS was 1:1. APS and ILs were dissolved into distilled water, and then the solutions were carefully poured onto. The reaction was carried out at the phase boundary of the two phases at 0~5 °C. After 10 h, the resulting dark-green precipitates were filtered and rinsed with deionized water and ethanol for several times to remove the unreacted chemicals and aniline oligomers. The obtained products were dried in vacuum oven at 40 °C for 24 h. The molar ratios of ionic liquid to aniline were 0.015, 0.025 and 0.05, so the products were named PANI-0.015, PANI-0.025 and PANI-0.05, respectively.

2.4. Characterization

Morphologies of the as-prepared PANI were investigated with a 6700F field emission scanning electron microscope (SEM). The Fourier-transform infrared (FTIR) spectra were recorded in the range of 4000~400 cm^{-1} with a spectral resolution of 1 cm^{-1}, using pressed KBr pellets. The crystalline phase of PANI were analyzed by A wide-angle X-ray powder scattering (RigakuD/max-2500 made in Japan) using Cu K line ($\lambda = 1.54056 A$). The thermogravimetric analysis of PANI were carried out using STA 449C (Japan) thermogravimetric instrument analyzer in the temperature range of 25~800 °C with a heating rate of 10 °C min^{-1} under a nitrogen. The conductivity of PANI was measured using TH2818 Automatic Component Analyzer at ambient temperature employing the two-probe method on a pressed pellet according to the formula: $\sigma = L/RS$. Here σ, L, R and S referred to electrical conductivity, thickness, resistance and touching area of the sample, respectively. The pellets were obtained by subjecting the powder sample to a pressure of 10 MPa. The reproducibility of the result was checked by measuring the resistance three times for each pellet.

3. Results and discussion

Figure 2. depicts representative SEM images of PANI-0.015, PANI-0.025 and PANI-0.05. All PANI exhibited cluster structure looking like fibrous flower bundles in Figure 2a, b, and c. But there was somewhat difference in the morphology as the molar ratio of ionic liquid to aniline increased. It was seen form Figure 2d, e, and f that PANI cluster became big and the diameter of PANI fibers became thin with the increase in the mole ratio of ionic liquid to aniline. The diameters of PANI fibers were from ~300 nm (Figure 2d and e) to 150~300 nm (Figure 2f). Pyridinium type ILs possess large head groups with well delocalized positive charge and capability of **π-π** interactions, which is the main reason for the self-assembly of fibrous flower bundles.

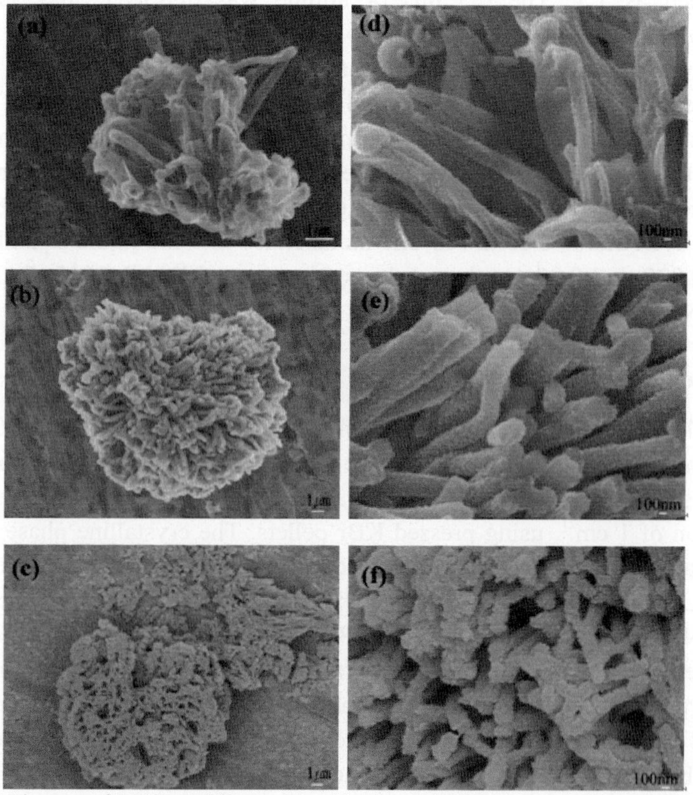

Fig. 2. SEM images of PANI at different magnification (PANI-0.015: a, d; PANI-0.025: b, e; PANI-0.05: c, d)

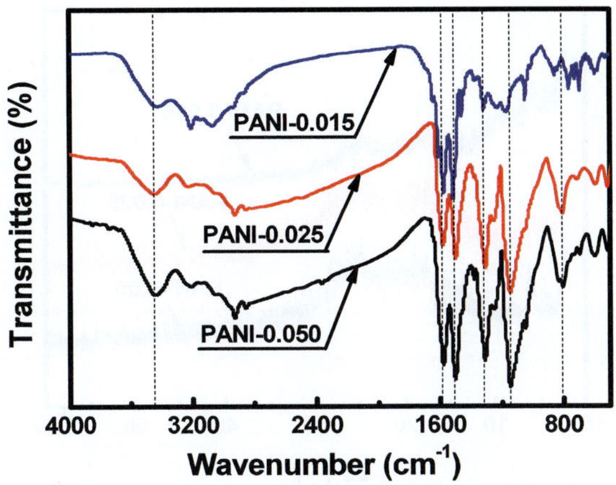

Fig. 3 Fourier transform infrared spectra of polyaniline powder

The FTIR spectra in Figure 3 of PANI samples were almost identical with the FTIR spectrum of PANI obtained in the absence of any ILs by other works [16-17]. Almost identical IR spectra suggested the similarity of the chemical structure of all samples, which indicated that the used ILs did not influence significantly the molecule structure and oxidation state of PANI. The band assignment is as follows: The characteristic band at about 1580 cm^{-1} was assigned to the vibration of quinonoid rings, while the band near 1502 cm^{-1} was attributed to benzenoid ring stretching vibrations. The peak at 1314 cm^{-1} corresponds to C-N stretching vibration of the benzenoid ring. The relatively high intensity observed in this region of the spectrum, with the maximum at 1158 cm^{-1}, assigned to aromatic C-H in-plane bending. Out-of-plane deformations of C-H bonds on 1,4-disubstituted rings are located in the region of 800~880 cm^{-1} with themaximum at 828 cm^{-1}.The band around 3447 cm^{-1} in the spectrum of the PANI base was assigned to the -NH- stretching vibration of secondary amine groups. The bands at 2911 cm^{-1} and at 2583 cm^{-1} were assigned to C-H stretching vibration of the benzene rings. It was noteworthy that the weak absorption peak at 1158 cm^{-1} and band at 812 cm^{-1} for PANI-0.015 indicated less electron delocalization than for PANI-0.025 and PANI-0.05. The band assigned to the -NH- stretching vibration magnified with the increase in mole ratio of ionic liquid to aniline, which illustrated that hydrogen bonding interaction enhanced.

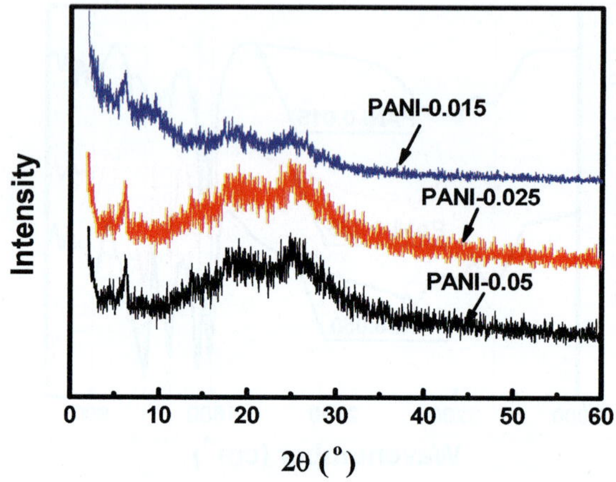

Fig. 4 XRD scattering patterns of PANI powder.

X-ray diffraction (XRD) was used to further probe the structure of PANI nanostructures as shown in Figure 4. Two broads at 2θ=19° and 25.5° were ascribed to the periodicity parallel and perpendicular to the polymer chains, which indicated partially crystalline as previously reported [18-19]. The peaks at 2θ=6° was assigned to the repeat unite of polyemeraldine chains, which was originated from the formation of high crystalline materials [20]. It was obvious that the peaks intensity magnified with the increase of the mole ratio of ionic liquid to aniline. These results implied that the crystallinity of PANI grew with the increase mole ratio of ionic liquid to aniline.

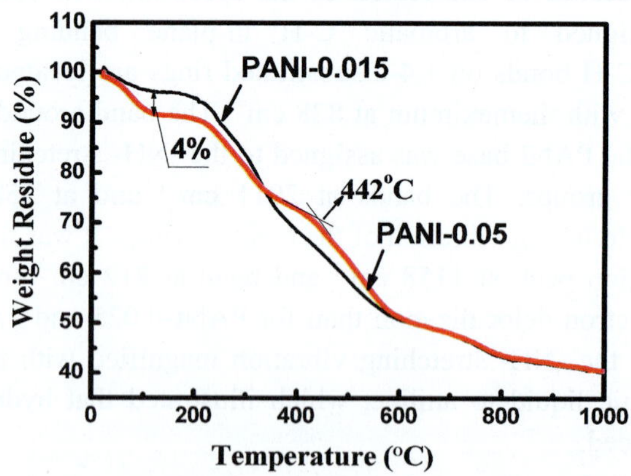

Fig. 5 TGA thermograms for PANI-0.015 and for PANI-0.05.

The thermal behaviors of both PANI-0.015 and PANI-0.05 were studied by TGA analysis under nitrogen in Figure 5. Both two curves were similar. The thermal degradation illustrated firstly to the loss of water in range of 30~120 °C. The weigh loss of PANI-0.05 was about 4% more than that of PANI-0.015, which is consistent with high ILs content in PANI-0.05 reactive solution. As rapid weight loss was found in temperature range of 200~351°C and 200~340 °C, this was consequence of structural decomposition of the polymer and elimination of dopant molecules. After those, the DTA curve of PANI-0.05 was more flat than that of PANI-0.015 before 442 °C, where thermal decomposition of PANI-0.05 took place. It implied that PANI-0.05 possessed better thermal stability than PANI-0.015, attributing to the fact that more ionic liquid involved increased the interactions from the molecular chain.

Table 1 Conductivity of PANI samples

Samples	PANI-0.015	PANI-0.025	PANI-0.05
Conductivity(S cm-1)	1.008×10^{-5}	1.09×10^{-4}	8.696×10^{-5}

Conductivity of All samples synthesized at the different mole ratio of ionic liquid to aniline kept at similar level for 10^{-4}~10^{-5} S cm^{-1}. The conductivity of PANI-0.05 was highest, which presented a good agreement with above results of structure characterization.

4. Conclusion

PANI fibre bundles were synthesized through an interface oxidative polymerization. In this reactive system, the dumbbell-structure ionic liquid played two important roles in structure-directing agents and doped acid. All of PANI samples prepared at different mole ration of ionic liquid to aniline showed the morphology of fiber bundles. The mole ratio of ionic liquid to aniline had also remarkable effect on doped degree, crystallinity, thermal stability and conductivity of the samples. PANI synthesized at 0.05 molar ratio of ionic liquid to aniline showed better thermal stability (442 °C). Their Conductivity kept at similar level for 10^{-4}~10^{-5} S/cm. This work provided a simple approach to fabricate nanostructure polyaniline with controllable morphology.

Acknowledgements

The project was supported by the Natural Science Foundation of China (No. 51363014), and Natural Science Funds of the Gansu Province (1308RJYA038).

References

1. D.S. Dhawale, A. Vinu, C.D. Lokhande., Stable nanostructured polyaniline electrode for supercapacitor application, *J. Electrochim. Acta.*, 56: 9482-9487(2011).
2. Y. Shi, L.J. Pan, B.R. Liu, Y.Q. Wang, Y. Cui, Z.A. Bao, G.H. Yu, Nanostructured conductive polypyrrole hydrogels as high-performance, flexible supercapacitor electrodes, *J Mater. Chem. A*, 2(17): 6086-6091 (2014).
3. K.S. Ryu, K.M. Kim, S.G. Kang, G.L. Lee, J. Joo, S.H. Chang, Electrochemical and physical characterization of lithium ionic salt doped polyaniline as a polymer electrode of lithium secondary battery, *J Synth. Met.*,110: 213-217 (2000).
4. M.M. Ayad, G. El-Hefnawey, N.L. Torad, J. Hazard, A sensor of alcohol vapours based on thin polyaniline film and quartz crystal microbalance, *J Mater.*, 168: 85-88(2009).
5. M.M. Ayad, N.L., Quartz, crystal microbalance sensor for detection of aliphatic amines vapours, *J Torad. Sens. Actuat. B*, 147: 481-487(2010).
6. Mahdi Aleahmada, Hamidreza Ghafouri Taleghania, Hossein Eisazadehb, Preparation and characterization of PAn/NiO nanocomposite using various Surfactants, *J Synth. Met.*,161: 990–995(2011).
7. Marija Radoi˘cic′ a, Zoran S˘aponjic′, Jovan Nedeljkovic′, Gordana C′ iric′-Marjanovic′, Jaroslav Stejskal, Self-assembled polyaniline nanotubes and nanoribbons/titanium dioxide nanocomposites, *J Synth. Met.*,160: 1325-1334 (2010).
8. J. Bhadra, D. Sarkar. Self-assembled polyaniline nanorods synthesized by facile route of dispersion polymerization. *J Mater. Lett.* 2009, 63: 69–71.
9. A. Rahy, D. J. Yang, Synthesis of highly conductive polyaniline nanofibers, *J Mater. Lett.*, 62: 4311–4314(2008).
10. H.W. Park, T. Kim, J. Huh, M. Kang, J.E. Lee, H. Yoon, Anisotropic growth control of polyaniline nanostructures and their morphology dependent electrochemical characteristics, *ACS Nano.*, 6: 7624-33 (2012).

11. Z.W. Xuan, L.D. Lu, Facile synthesis of 1-(N-butyl-1,8-naphthalimide-4′-yl)-3-(4-methoxyl-phenyl)-5-phenyl-pyraoline/polyaniline core–shell nanofibers and polyaniline nanotubes, *J Mater. Lett.*, 65:754–756 (2011).
12. C. Chiappe, D. Pieraccini, Ionic liquids: solvent properties and organic reactivity. *J Phys. Org. Chem.*, 18:275-297 (2005).
13. P. Kubisa, Application of ionic liquids as solvents for polymerization processes. *Prog. Polym. Sci.*, 29:3-12 (2004).
14. Gao H, Jiang T, Han B, Wang Y, Du J, Liu Z. Aqueous/ionic liquid interfacial polymerization for preparing polyaniline nanoparticles. *Polym.*, 45: 3017-9 (2004).
15. David Pahovnik, Ema Žagar, Ksenija Kogej, Jiři Vohlídal, Majda Žigon, Polyaniline nanostructures prepared in acidic aqueous solutions of ionic liquids acting as soft templates, *J Europ. Polym.*, 49, 1381-1390(2013).
16. H.M. Zhang, Y.P. Li, X.H. Wang, J. Li, F.S. Wang, A facile route to hollow microspherical polyaniline, *Polym.*, 52: 4246-4252 (2011).
17. A.K. Nath, A. Kumar, Ionic liquid based polymer electrolyte dispersed with dedoped polyaniline nanorods, *Solid State Ionics*, 253:8-7 (2013).
18. Z.D. Zujovic, C. Laslau, G.A. Bowmaker, P.A Kilmartin, A.L. Webber, S.P. Brown, J.T. Sejdic, Role of Aniline Oligomeric Nanosheets in the Formation of Polyaniline Nanotubes *Macromolecules*, 43: 662-670 (2010).
19. Huang Zhengzheng, Liu Enhui, Shen Haijie, et al. Preparation of polyaniline nanotubes by a template-free self-assembly method, *J Mater. Lett.*, 65(15): 2015-2018 (2011).
20. X.Z. Wang, X.F. Wang, Y.T. Wu, L. Bao, H. Wang. Interfacial synthesis of polyaniline nanostructures induced by 5-Sulfosalicylic acid, *J Mater. Lett.*, 64: 1865-1867(2010).

Sulfadiazine removal from aqueous solutions by hydrous manganese oxides

Wei-Xiao Zhu[1,2], Shuai Liang[1,2], Yi-Li Wang[1,2] and Hong-Jie Wang[1, 2,*]

[1] *College of Environmental Science and Engineering, Beijing Forestry University Beijing 100083, China*

[2] *Beijing Key Lab for Source Control Technology of Webber Pollution Beijing 100083, P.R.China*

E-mail: [*]whj_99@hotmail.com

The adsorption behavior of sulfadiazine (SDZ) onto hydrous manganese dioxide (HMO) is systematically investigated in this study. SDZ removal is highly pH dependent and ionic strength has slightly effect on the uptake of SDZ. The Webber-Morris equations model shows that SDZ adsorption by HMO is controlled jointly by the film-diffusion and intraparticle-diffusion. The adsorption isotherms at varied temperatures are well described by Langmuiir model, and thermodynamic parameters ($\Delta H>0$, $\Delta S>0$ and $\Delta G<0$) reveal that the adsorption of SDZ on HMO is spontaneous and endothermic process. The uptake process of SDZ by HMO is mainly ascribed to the hydrogen bonding between the nitrogen (N) of aniline and the surface hydroxyl groups of HMO.

Keywords: Adsorption; Sulfadiazine; Hydrous manganese dioxide; hydrogen bonding; Langmuiir model

1. Introduction

Sulfadiazine (SDZ), as one of the most widely used antibacterial veterinary drugs and human medicine in sulfonamides [1], shows high resistance to biological degradation, and has frequently been detected in waste Webber, surface Webber, and ground Webber samples [2]. It is important to develop efficient treatment technologies to control the concentration level of sulfonamides in aquatic environments. Hydrous manganese dioxide (HMO), as one of the important composite oxides, widely exists in soil and sediment and plays an important role in the process of soil and Webber self-purification with the poor crystallinity, the size of the nanometer level, and the large surface area and high activity [3]. According to previous studies, there are few studies conducted for the removal of sulfonamides and the corresponding mechanism needs to be further detected.

Therefore, the objectives of the work are to evaluate the adsorption behavior of the sulfadiazine (SDZ) by HMO in aqueous solutions and to explore the corresponding mechanism.

[*] Corresponding author. *Tel & Fax: +86 10 6233 6900.*

2. Material and Methods

Sulfadiazine (SDZ, $C_{10}H_{10}N_4O_2S$, 98%) was purchased from Aladdin Industrial Corporation (Shanghai, China). HMO was synthesized by redox of KMnO4 and $MnCl_2{\cdot}4H_2O$ with the molar ratio of 2:3 according to the previous works [4]. The concentrations of SDZ were analyzed by High Performance Liquid Chromatography (HPLC, Agilent Eclipse 1260, Agilent Technologies, USA) using a XDB-C18 column (4.6×50 mm, 2.7 μm). The detection was performed by UV absorption at a wavelength of 270 nm. The mobile phase used as eluent was methanol and acidified Webber (0.1% acetic acid) (30:70). The injection volume was 20.0 μL and the flow rate was 0.1 mL·min-1, respectively. The column temperature was maintained at 30°C [5].

3. Results and Discussion

3.1. *pH and ionic strength study*

The effect of pH and ionic strength on SDZ removal is evaluated and the results are illustrated in Figure 1. The SDZ removal efficiency decreases steadily with the increase of solution pH from 4.0 to 10.0 and the maximum adsorption removal efficiency (98 %) is acquired at pH 4.0 with maximum capability of 1778 μg/g. Meanwhile, the level of ionic strength has negligible effect on adsorption capacity, which indicates the adsorption of SDZ on HMO is ionic strength independent and the adsorption mechanism of SDZ is inner-sphere complexation rather than ion exchange.

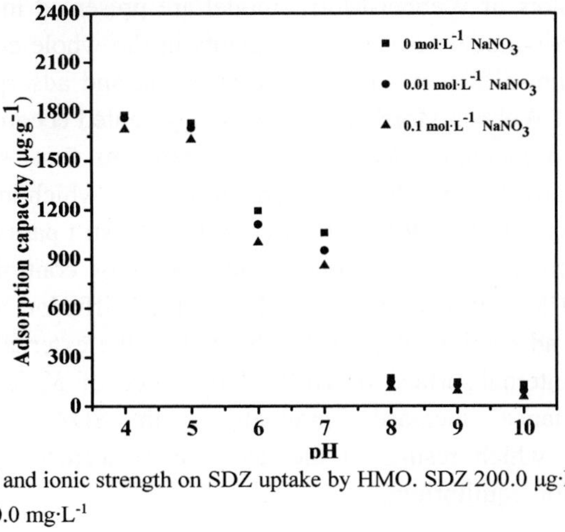

Fig. 1. Effects of pH and ionic strength on SDZ uptake by HMO. SDZ 200.0 $\mu g{\cdot}L^{-1}$; $NaNO_3$ 0.01-0.1 $mol{\cdot}L^{-1}$; Dosage, 100.0 $mg{\cdot}L^{-1}$

SDZ, has two ionization constants ($pK_{a,1} = 1.6$, $pK_{a,2} = 6.48$) [6] , is an amphoteric molecule with multiple ionizable functional groups and may exist as anionic form (SDZ), neutral form(SDZ^0), or cationic (SDZ^+) form at different pH values. Meanwhile, the manganese dioxide and iron hydroxide are found to have an isoelectric point is about 4.3 [7]. Namely, the surface of manganese dioxide carried almost negatively charged at $4.3 < pH < 10.0$, therefore, the adsorption capacity decreases with the increasing of the fraction of the anionic form (SDZ). At pH=4.0, the adsorption capacity reached maximum while the predominating sulfadiazine species is neutral form (SDZ^0), which demonstrates that the corresponding adsorption mechanism is ascribed to not only electrostatic interactions between sulfadiazine and the manganese dioxide but also may involve chemical adsorption.

3.2. *Adsorption kinetics*

The variation of the adsorbed SDZ on HMO with prolonged contact time is shown in Figure 2. The adsorption capacity of SDZ on HMO increases sharply within the initial 2 h at different temperatures then continues with a slower rate and finally tends to reach equilibrium. Meanwhile, the temperature has obvious positive effect on the uptake of SDZ. Figure 2a shows that the adsorption kinetic data are described by pseudo first-order, pseudo second-order, Elovich. and the pseudo second-order equation fits well for SDZ adsorption process by HMO (R^2=0.992). The values of the calculated activation energy (E_a) for SDZ adsorption onto HMO is 89.0 $kJ \cdot mol^{-1}$, which demonstrates the dominating chemisorption mechanism [8]

Diffusion plots of Webber-Morris-model are presented in Figure 2b. Plots of Webber-Morris-model have three segments in the whole adsorption process that represent film-diffusion, intraparticle-diffusion and adsorption equilibrium [9]. The plots of Webber-Morris-model were segmented according to piecewise linear regression method. K_1, K_2, and K_3 refer to first, second and third segments, respectively. And the maximum value is K_1, which indicates that film-diffusion process is fastest. While the second line doesn't pass the origin, which suggests that the internal diffusion is not the only control step and SDZ adsorption by HMO is controlled jointly by film-diffusion and intraparticle-diffusion. After adsorption sites reach saturated on the external surface, SDZ is adsorbed into internal surface of HMO. The value of K_3 is minimum when diffusion resistance increases gradually with SDZ diffusion to the internal surface, which results in the decrease of diffusion rate and finally reaches adsorption equilibrium.

3.3. *Adsorption isotherms*

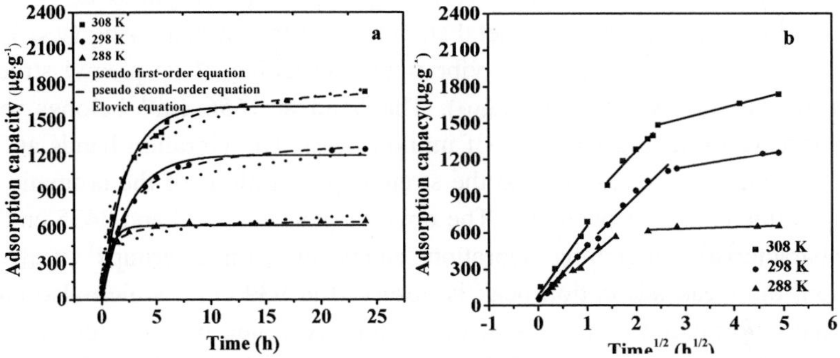

Fig. 2. Adsorption kinetics of SDZ on HMO Dosage,100.0 mg·L^{-1}; SDZ, 200 μg·L^{-1}; $NaNO_3$, 0.01 mol·L^{-1}; 24 h; pH 4.0±0.2 (HMO)

Adsorption isotherms of SDZ on HMO at different temperatures are presented in Figure 3. The nonlinear Langmuir and Freundlich equations are used to depict the isotherms. The corresponding parameters show that the experimental equilibrium adsorption data are well described by Langmuir model (R^2=0.972). The maximum adsorption capacities of SDZ onto HMO at the optimum pH are calculated to be 3100, 4069, 5680 μg·g^{-1} at 288, 298 and 308 K, respectively. Meanwhile, the parameters of $\Delta G<0$ and $\Delta H>0$ and $\Delta S>0$ reveal the spontaneous and endothermic process.

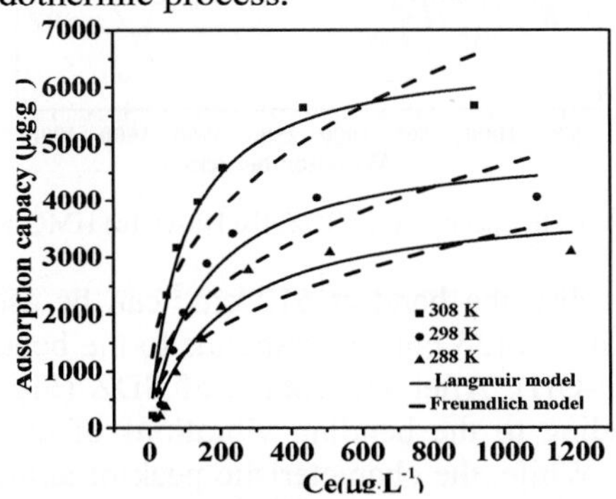

Fig. 3. Adsorption isotherms of SDZ on HMO Dosage, 100.0 mg·L^{-1}; 24 h; $NaNO_3$, 0.01 mol·L^{-1}; pH 4.0±0.2 (HMO)

3.4. *FTIR analysis*

The FTIR spectra of virgin and SDZ loaded adsorbents are shown in Figure 4. To SDZ (Figure 4a), the peak at 1149 and 1328 cm^{-1} are ascribed to the symmetric and the asymmetrical stretch of the sulphonyl group, coupled to the C_{-SO2-} stretching vibration, respectively [10]. The vibration bands at 1411 and 1487 cm^{-1} originate from the skeletal stretching of the aromatic ring ($\nu_{C=C}$). The peak at 1590 cm^{-1} is assigned to vibrational stretching of its phenyl structure conjugated to the NH_2 group (ν_{-NH2}). The band at 1656 cm^{-1} belongs to the characteristic stretching vibration of imino (ν_{-NH-}). The vibration bands at 3016 and 2925 cm^{-1} can be assigned to the stretching vibrations of the aromatic ring (ν_{Ar-H}), amine (ν_{N-H}), respectively. The obvious peaks at 3350 and 3435 cm^{-1} are the characteristic stretching vibration bands of aniline groups (ν_{-NH2}). In addition, the weak adsorption peak is observed at 939 cm^{-1}, which also comes from the deformation vibrations of the carbon hydrogen bonds in the aromatic and heterocyclic rings ($\delta_{=C-H}$). Other weak peaks including 846, 680 and 617 cm^{-1} can be attributed to the bending vibrations of carbon hydrogen bonds ($\gamma_{=C-H}$) .

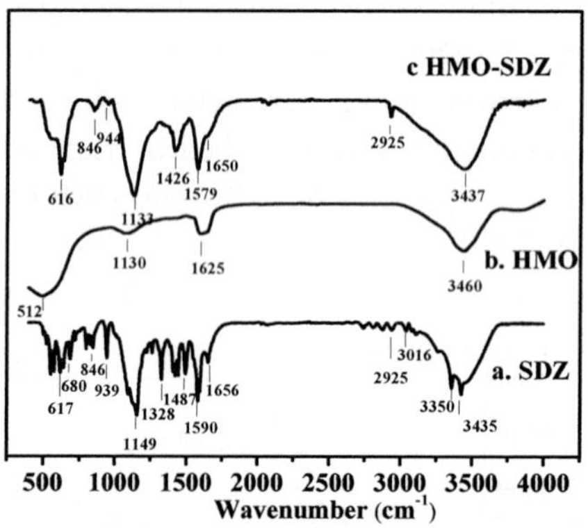

Fig. 4. FTIR spectra of (a) SDZ, (b) HMO, (c) HMO-SDZ

To HMO (Figure 4b), the band at 512 cm^{-1} can be ascribed to the Mn-O vibration [11]. The peak at 1133 cm^{-1} corresponds to the bending vibration of the hydroxyl groups (Mn-OH). After the uptake of SDZ (Figure 4c), the peak at 1625 cm^{-1} corresponding to the bending vibrations of the hydroxyl group on HMO is disappeared. While, the characteristic peak of amino (-NH-) of SDZ at 1656 cm^{-1} has no obvious change. But the characteristic peak of aniline ($-NH_2$) of SDZ at 1590 cm^{-1} has obvious red-shift to 1579 cm^{-1}. It indicates the interaction of the $-NH_2$ and the surface hydroxyl groups of HMO may accounts for the uptake of SDZ [12].

4. Conclusion

Hydrous manganese dioxide (HMO) presents preferable ability in the uptake of sulfadiazine (SDZ). The analysis of SEM and XRD shows that HMO is amorphous form and has high surface area. SDZ removal is highly pH dependent and ionic strength has slightly effect on the uptake of SDZ. The uptake ability of SDZ obviously decreases with the increasing of pH during pH from 4.0 to 10.0. The Webber-Morris equations model shows that SDZ adsorption by HMO is controlled jointly by film-diffusion and intraparticle-diffusion. The adsorption isotherms at varied temperatures are well described by Langmuiir model, and thermodynamic analysis reveals spontaneous endothermic process of SDZ adsorption onto HMO. The uptake process of SDZ by HMO is mainly ascribed to the hydrogen bonding between the nitrogen (N) of aniline and the surface hydroxyl groups. Therefore, hydrous manganese dioxide is a promising and effective candidate for SDZ removal from aqueous solutions.

Acknowledgments

This work was supported by Major Science and Technology Program for Webber Pollution Control and Treatment of China (2015ZX07204-002) and the National Natural Science Research Fund (No. 51578067).

References

1. S. Thiele-Bruhn, Pharmaceutical antibiotics compounds in soil-a review, *Plant Nutr. Soil. Sci.* **166**, 145-167 (2003) .
2. H. Y. Niu, D. Zhang, S. G. Zhang, X. L. Zhang, Z. F. Meng, Y. Q. Cai, Humic acid coated Fe3O4 magnetic nanoparticles as highly efficient Fenton-likecatalyst for complete mineralization of sulfathiazole, *J. Hazard. Mater.* **190**, 559-565 (2011) .
3. H .W.Wang, D. Y. Zhang, S. Y. Mou, W. J. Song, F. A. Al-Misned, M. G. Mortuza, X. L. Pan, Simultaneous removal of tetracycline hydrochloride and As(III) using poorly-crystalline manganese dioxide, *Chemosphere*. **136**, 102-110 (2015) .
4. G. S. Zhang, H. J. Liu, J. H. Qu, W. Jefferson, Arsenate uptake and arsenite simultaneous sorption and oxidation by Fe-Mn binary oxides: Influence of Mn/Fe ratio, pH, Ca^{2+}, and humic acid, *J. Colloid. Interf. Sci.* **366**, 141-146 (2012) .
5. F. Mor, F. S. Kocasari, G. Ozdemir, B. Oz, Determination of sulphonamide residues in cattle meats by the Charm-II system and

validation with high performance liquid chromatography with fluorescence detection, *Food. Chem.* **134**, 1645-1649 (2012) .

6. W. B.Yang, F. F. Zheng, X. X. Xue, Y. P. Lu, Investigation into adsorption mechanisms of sulfonamides onto porous adsorbents, *J. Colloid. Interf. Sci.* **362**, 503-509 (2011) .
7. Y. X. Xia, L. Y. Meng, Y. J. Jiang, Y. S.Zhang, , Dai, X. X., Zhao, M. J. Facile preparation of MnO2 functionalized baker's yeast composites and their adsorption mechanism for Cadmium, *Chem. Eng. J.* **259**, 927-935 (2015)
8. X. Li, K. He, B. C. Pan, S. J. Zhang, L. Lu, W. M.bZhang, Efficient As(III) removal by macroporous anion exchanger-supported Fe-Mn binary oxide: Behavior and mechanism, *Chem. Eng. J.* **193-194**, 131-138 (2012).
9. W.J. Weber, J.C. Morris, Proceedings of the International Conference on Water Pollution Symposium, (Pergamon Press, Oxford, 1962), pp. 231–266.
10. A. Ungurean, M. Oltean, L. David, N. Leopold, J. P. P. Ramalho, V . Chisile, Adsorption of sulfamethoxazole molecule on silver colloids: A joint SERS and DFT study, *J. Mol. Struct.* **1073**, 71-76 (2014).
11. J. B. Lu, H. J. Liu, X. Zhao, W. Jefferson, F. Cheng, Q. H. Qu, Phosphate removal from Webber using freshly formed Fe – Mn binaryoxide: Adsorption behaviors and mechanisms. *Colloids. Surf. A. Physicochem, Eng. Asp.* **455**, 11-18 (2014) .
12. Z. M. Qiang, X. L. Bao, W. W. Ben, MCM-48 modified magnetic mesoporous nanocomposite as an attractive adsorbent for the removal of sulfamethazine from Webber, *Webber. Res.* **47**, 4107- 4114 (2013) .

Phosphate Removal from wastewater with Ferric and Manganese Oxides coated Zeolite

Xue-Pu Li[1,2], Shuai Liang[1,2], Tao Chen[3] and Hong-Jie Wang[*,1,2]

[1]*Beijing Key Lab for Source Control Technology of Water Pollution*
Beijing 100083, China
*E-mail address: *whj_99@hotmail.com*

[2]*College of Environmental Science and Engineering, Beijing Forestry University*
Beijing 100083, China

[3]*Research Center for Environmental Policy Planning & Assessment of Zhengzhou University*
Zhengzhou, 450002, China

To control the eutrophication of water bodies, phosphate discharge is nowadays being strictly limited. A novel material, ferric and manganese binary oxides coated zeolite (FMBO-Z), was synthesized by in-situ redox and co-precipitation method and used for phosphate removal from waste water. The phosphate adsorption is highly pH dependent and acid condition is advantageous for phosphate uptake by FMBO-Z. The adsorption data are well fit by the Freundlich isotherm model. The maximal adsorption capacity of FMBO-Z is 3.70mg/g at pH 6.0, which is two times higher than that of the unloaded zeolite. Kinetic data correlates well with the pseudo-second-order kinetic model, suggesting the chemisorption mechanism. The coexisting anions of carbonate and sulfate have no obvious effect on phosphate removal by FMBO-Z, while silicate and arsenate reveal a significantly negative influence on phosphate adsorption.

Keywords: Phosphate Removal, Adsorption, Ferric Oxides, Manganese Oxides.

Introduction

As one of the main elements causing eutrophication, an excessive intake of phosphate in water bodies may spur an abnormal growth of algae and aquatic plants[1], which poses a risk of eutrophication to the receiving water bodies. To control the excess discharge of phosphate, a number of treatment techniques, are conducted. Among all of these available approaches, adsorption is regarded as one of the most promising approaches to remove phosphate from water and wastewater, due to its simplicity, high efficiency and cost-effectiveness [2]. A number of sorbents have been tested for their phosphate removal abilities, among those sorbents, a novel Fe–Mn binary oxide (FMBO) sorbent has been successfully developed, exhibiting high effectiveness in removing phosphate [3].

* Corresponding author

The FMBO power offers very fast kinetics and enhanced sorption capacity [4]. However, it cannot be well used in fixed-bed or other flow-through systems because of its weak mechanical strength and propensity to aggregate, which is expectable to lead to high pressure drop and poor hydraulic characteristics[4,5]. To overcome these limitations, such synthesized materials are usually loaded onto surface of normal filtrate materials.

Zeolite, as a low-cost and easily obtainable material, has characteristics of large surface area, strong capabilities for ion exchange and adsorption due to their particular tetrahedral pore framework, which is one of the commonly used substrates in water treatment. Especially, metal oxides loaded zeolites, such as hydrous ferric oxides [6], aluminum oxides [7], manganese oxides [8], etc., are wildly evaluated.

Therefore, the purposes of the paper are to develop a ferric and manganese binary oxides loaded zeolite (FMBO-zeolite) and to investigate the adsorption behavior of phosphate by FMBO-zeolite through batch adsorption experiments.

1. Material and methods

1.1. *Adsorbent preparation*

The preparation of ferric and manganese binary oxides loaded zeolite (FMBO-Z) was according to the following procedure: after cleaning and drying, Zeolite in 20-30 mesh was immersed in 75 mL solution of $FeSO_4$ (0.075 mol/L). A magnetic stirrer was used to ensure sufficient metal ions diffusion into the zeolite structure. Then 75 mL solution of $KMnO_4$ (0.025 mol/L) was added into the reactor, meanwhile 0.1 mol/L NaOH or HCl was applied to adjust the pH at 4.5-5.0. After 1h, the system was filtrated by a 20 mesh sieve. The solid was washed several times with distilled water and then dried at 60 °C for 24h for future use. Nature zeolites without FMBO were also prepared under similar experimental conditions.

2. Batch adsorption experiments

To investigate the influence of initial pH on PO_4^{3-} uptake by FMBO-zeolite, the experiments were performed under initial pH from 2.0 to 10.0, with 2.5 g/L FMBO-zeolite and 1.0 mg/L PO_4^{3-} solutions in 250ml glass vessels for 24 h of equilibrium time at 298K. In kinetics tests, adsorption experiments were carried out in 250 mL glass vessels with 1.0 mg/L PO_4^{3-} and 2.5mg/L FMBO-zeolite under 298K and the speed of 130 rpm. The isotherm experiments were conducted at pH 6 to evaluate adsorption capacities of PO_4^{3-} at constant temperatures of 288, 298 and 308 K, respectively. The dosage of FMBO-zeolite was 2.0 g/L and the initial PO_4^{3-}concentrations were 1.0-500 mg/L. The effects of common coexisting anions in water, i.e., carbonate, sulfate, silicate and arsenate, were investigated at pH 6.0 by adding a defined amount (2.5 g/L) of FMBO-zeolite into solutions which contained 1.0 mmol/L of the coexisting anions and 1.0 mg/L of PO_4^{3-}. The vessels were shaken for 6.0 h at 298 K.

3. Results and discussion

3.1. *Effect of pH*

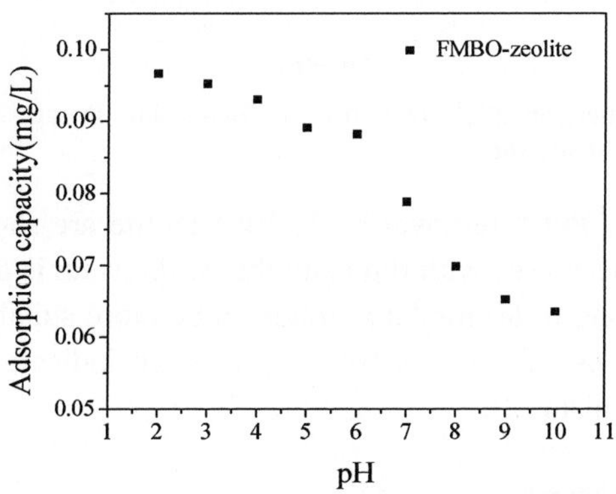

Fig. 1. Effect of initial solution pH on PO_4^{3-} adsorption by FMBO-zeolite. Dosage 2.5 g/L, initial PO_4^{3-} concentration 1.0 mg/L, 24 h, 298K.

As can be seen in Figure 1, with the solution pH increased from 2 to 6, a slight decrease in PO_4^{3-} removal is observed, accompanied by a sharp decline of removal efficiency at pH≥7. Acidic conditions are advantages for PO_4^{3-} uptake by FMBO-zeolite, which is ascribed to the static attraction of the positively charged FMBO-zeolite surface and the negatively charged phosphor anions. Oppositely, with the increase of solution pH, the surface of FMBO-zeolite is dominated by negative charge, which in turn causes the drop of PO_4^{3-} adsorption [9].

3.2. *Adsorption kinetics*

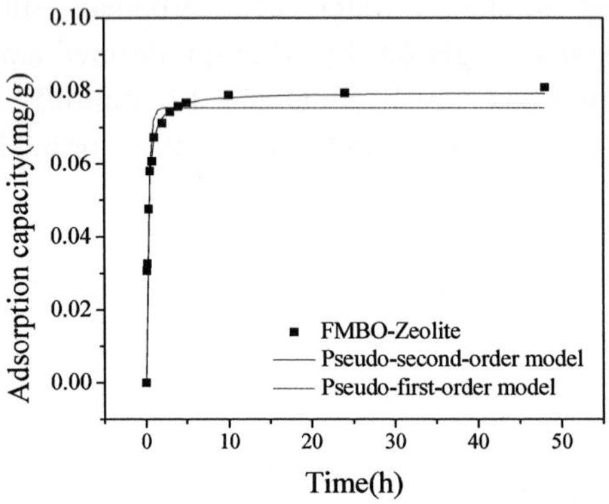

Fig. 2. Effect of contact time on PO_4^{3-} sorption on FMBO-zeolite. Dosage 2.5 g/L, initial PO_4^{3-} concentration 1.0 mg/L, pH 6.0, 298K.

The kinetic results of PO_4^{3-} removal by FMBO-zeolite are shown in Figure 2. The adsorption rate decreases with time and the equilibrium is almost reached in 5 h. The pseudo-second order model is found to be more suitable for fitting the kinetic data than the pseudo-first order model, which indicates the dominating chemisorption mechanism [2].

3.3. *Adsorption isotherms*

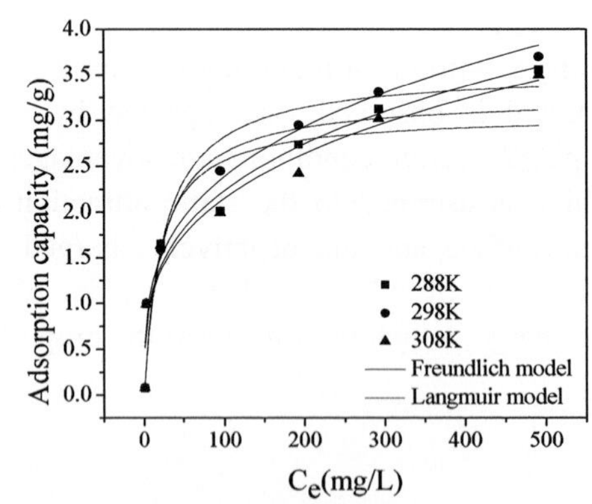

Fig. 3. Adsorption isotherms of PO_4^{3-} at different temperatures. Dosage 2.5 g/L, initial PO_4^{3-} concentration 1.0-500 mg/L, 24 h, pH 6.0

Adsorption isotherms of PO_4^{3-} on FMBO-zeolite at different temperatures are presented in Figure 3. The nonlinear Langmuir and Freundlich equations are used to depict the isotherms. The experimental equilibrium adsorption data are well described by Freundlich model, which assumes a heterogeneous surface and a multilayer adsorption occur on the surface of FMBO-zeolite with an energetic non-uniform distribution[10].

The maximum adsorption capacities of PO_4^{3-} on FMBO-zeolite are calculated to be 3.5, 3.56, 3.7 mg/g at 288, 298 and 308 K, respectively. Meanwhile, the negative value of ΔG and the positive values of the ΔH and ΔS reveal the spontaneous and endothermic uptake process.

3.4. *Influence of the co-existing anions*

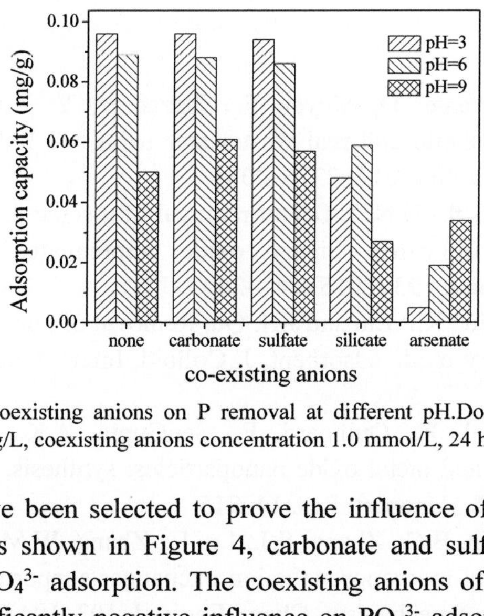

Fig. 4. Effects of coexisting anions on P removal at different pH.Dosage 2.5 g/L, initial PO_4^{3-} concentration 1.0 mg/L, coexisting anions concentration 1.0 mmol/L, 24 h, 298K.

Four anions have been selected to prove the influence of coexisting anions at different pH. As shown in Figure 4, carbonate and sulfate have no obvious impacts upon PO_4^{3-} adsorption. The coexisting anions of silicate and arsenate revealed a significantly negative influence on PO_4^{3-} adsorption because of the competition for the similar adsorption sites [11]. But the negative efficiency of PO_4^{3-} caused by silicate and arsenate onto FMBO-zeolite decreased when the pH was raised.

4. Conclusions

A ferric and manganese binary oxides impregnated zeolite (FMBO-zeolite) was successfully fabricated and FMBO-zeolite exhibited better PO_4^{3-} removal efficiency than that of the unloaded natural zeolite. Acid solution was advantageous to PO_4^{3-}adsorption and the adsorption equilibrium could be reached quickly. The maximal sorption capacities of FMBO-zeolite were 3.50, 3.56 and 3.70 mg/g at 288, 298 and 308K, respectively. The presence of silicate and arsenate had negative effects on PO_4^{3-} adsorption, while, carbonate and sulfate had no obvious impact on PO_4^{3-} uptake. In view of the relatively high PO_4^{3-} adsorption capacity, quick equilibrium and the resistance to the effect of the coexisting anions, FMBO-zeolite showed the possibility and feasibility acted as a potential phosphate scavenger.

Acknowledgments

This work was supported by Major Science and Technology Program for Water Pollution Control and Treatment of China (2015ZX07204-002) and the National Natural Science Research Fund (No. 51578067).

References

1. C. Barca, C. Gérente, D. Meyer, F. Chazarenc, Y. Andrès, Phosphate removal from synthetic and real wastewater using steel slags produced in Europe, Water Res. 46 (2012) 2376–2384.
2. Qi J.Y, Zhang G.S, Li H.N, Efficient removal of arsenic from water using a granular adsorbent:Fe–Mn binary oxide impregnated chitosan bead. Bioresource Technol. 193 (2015) 243–249.
3. G.S Zhang, H.J.Liu , R.P.Liu and J.H. Qu, Removal of phosphate from water by a Fe–Mn binary oxide adsorbent, J. Colloid. Interf. Sci.335 (2009) 168–174
4. Sarkar, S., Guibal, E., Quignard, F., SenGupta, A.K., 2012. Polymer-supported metals and metal oxide nanoparticles: synthesis, characterization, and applications. J. Nanopart. Res. 14, 715
5. Li, X., He, K., Pan, B.C., Zhang, S.J., Lu, L., Zhang, W.M., 2012. Efficient As(III) removal by macroporous anion exchanger-supported Fe-Mn binary oxide: behavior and mechanism. Chem. Engin. J. 193–194, 131–138
6. Xie J, Wang Z, Wu D.Y and Kong H.N, Synthesis and properties of zeolite/hydrated iron oxide composite from coal fly ash as efficient adsorbent to simultaneously retain cationic and anionic pollutants from water.
7. Diana Guaya,César Valderrama, Adriana Farran , Chabaco Armijos, José Luis Cortina, Simultaneous phosphate and ammonium removal from aqueous solution by a hydrated aluminum oxide modified natural zeolite, Chem. Eng.J.

8. Weihua ZOU, Lei ZHAO and Runping HAN,Removal of Uranium (VI) by Fixed Bed Ion-exchange Column Using Natural Zeolite Coated with Manganese Oxide,Chinses J Chem. Eng.
9. Zhang G.S., Qua J.H,, Liu H.J , Liu R.P, Wua R.C. Preparation and evaluation of a novel Fe–Mn binary oxide adsorbent for effective arsenite removal. Water Res. (2007) 1921 – 1928.
10. Sanna Hokkanen, Eveliina Repo a, Song Lou, Mika Sillanpää, Removal of arsenic(V) by magnetic nanoparticle activated microfibrillated cellulose. Can J Chem. Eng. 260 (2015) 886–894
11. Chang F.F, Qu J.H, Liu H.J , Liu R.P , Zhao X. Fe–Mn binary oxide incorporated into diatomite as an adsorbent for arsenite removal: Preparation and evaluation. J. Colloid. Interf. Sci. 338 (2009) 353–35

Progress in producing technology of propylene oxide

Shu-Yong Shang[†*] and Li Mei
School of Chemistry and Life Science, Chengdu Normal University
Chengdu Sichuan, 611130, P. R. China
[†]E-mail: ssyandmltcdj@163.com

Peng Fu
Institute of Natural Medicine & Green Chemistry, School of Chemical Engineering and Light Industry, Guangdong University of Technology, Guangzhou 510006, China
E-mail:yesfupeng@126.com

Propylene oxide is an important chemical material, which is mostly used to produce polyether polyol, propylene glycol, etc. This paper compares the producing technologies of propylene oxide employed, which include chlorohydrin method, co-oxidation method, cumene hydroperoxide method, direct oxidation of hydrogen peroxide (HPPO) and direct oxygen method etc. HPPO method, due to its advantages in the aspects of energy saving and environmental protection, is the most promising technology in the industrial production of propylene oxide at present.

Keywords: Propylene Oxide, Chlorohydrin Method, Co-oxidation Method, Cumene Hydroperoxide Method, Direct Oxidation of hydrogen peroxide (HPPO)

1. Introduction

Propylene oxide (PO) is a colorless, tasteless and flammable volatile poisonous liquid at ordinary temperatures and pressures. It's the main raw materials of non-ionic surface active agent and pesticide emulsifier, which has the widely application in chemical industry, light industry, medicine and textile industries and so on. Nowadays the technologies, which use propylene as the raw material to manufacture PO, can be divided into the traditional crafts and modern cleaner production crafts. The former includes the chlorohydrin method and the co-oxidation method, and the co-oxidation method includes co-oxidation of isobutane method (PO/TBA) and co-oxidation of ethylbenzene (PO/SM

---

* Corresponding author

method). The latter includes cumene hydroperoxide method (CHP), direct oxidation of hydrogen peroxide (HPPO) and direct oxygen method.

1. Traditional Producing Technology

1.1. *Chlorohydrin Method*

Chlorohydrin method was invented by the American Union Carbide company in the 1930s, and it is the main method to produce PO. Nowadays about 40% yield of the world's PO has used this method[1]. Chlorohydrin method includes three steps, *i.e.*, the chlorohydrination of propylene, the saponification of lime milk and the distillation of the products. The main chemical reaction equations are showed as follows:

1) Chlorine alcoholization reaction

$$C_3H_6 + Cl_2 + H_2O \rightarrow C_3H_6ClOH + HCl \qquad (\textit{main reation})$$

$$C_3H_6 + Cl_2 \rightarrow C_3H_6Cl_2; C_3H_6 + Cl_2 + C_3H_6ClOH \rightarrow ClC_3H_6OC_3H_6Cl + HCl \ (\textit{side reation})$$

2) Saponification reaction

$$2C_3H_6ClOH + Ca(OH)_2 \rightarrow 2C_3H_6O + CaCl_2 + 2H_2O \qquad (\textit{main reaction})$$

$$2C_3H_6ClOH + Ca(OH)_2 \rightarrow 2C_3H_6(OH)_2 + CaCl_2 \qquad (\textit{side reaction})$$

$$2C_3H_6ClOH + Ca(OH)_2 \rightarrow 2C_3H_6O + CaCl_2 + 2H_2O; 2HCl + Ca(OH)_2 \rightarrow CaCl_2 + 2H_2O \ (\textit{side reaction})$$

The process of chlorohydrin method is shown in Figure 1.

At 40~90°C and normal pressure, a propylene, chlorine and water react in a hypochlorous acid reactor. the reaction product, the mixtures of chlorine propanol, is separated in a propylene splitter to wipe off some incompletely reacted propylene. Then mix the liquid phase with some excess calcium hydroxide solution, and send them into the saponification reactor for reaction to generate the crude products of PO. Removing the water, dichloropropane and other organic impurities by two steps of distillation, at last, the products of PO can be got[2]. Chlorohydrin method has some advantages such as the mature technology, the simple process, the large operation in accordance with elasticity and good product selectivity, while it also has some disadvantages such as a large quantity of emissions of the "three wastes", the serious pollution and the strong corrosion to the equipment.

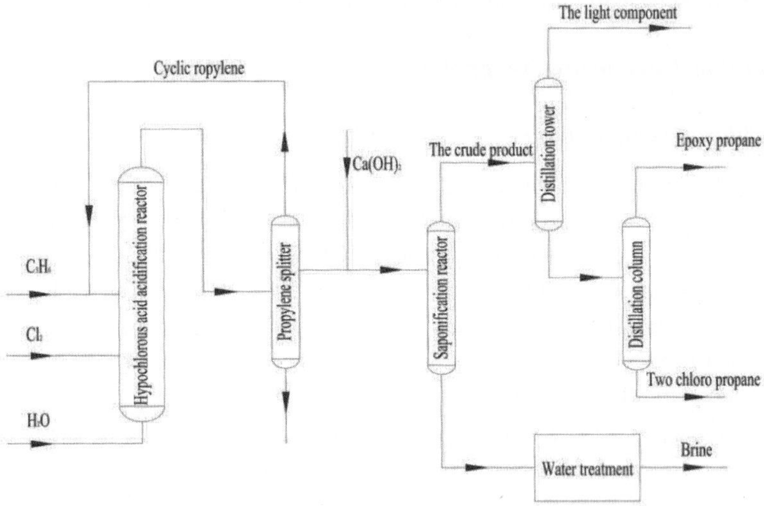

Fig. 1. The process of chlorohydrin method

Due to above deficiencies, some foreign enterprises have developed and applied a series of improved processes of the alkali saponification, however it still can't fundamentally solve the problems of resource utilization and environmental protection problem, and at the same time the costs of the production has been greatly increased. Therefore many foreign enterprises have eliminated this method. Some chlorohydrin enterprises, including the American Dow chemical, have begun to use new environment-friendly direct oxidation technologies to construct new PO devices[4]. Besides, the National Development and Reform Commission have listed the chlorohydrin as a restricted way on the development process in the catalogue for the guidance of industrial structure adjustment in 2011. Thus we can see the chlorohydrin method will gradually withdraw from the stage of history over a period of time in the future.

1.2. *Co-oxidation Method*

Co-oxidation method was developed by the Auckland company in the United States. It has accounted for about 57% total outputs of PO in Europe since 1969[1]. This method overcomes the disadvantages of the chlorohydrin method, such as the strong corrosion to equipment and a large quantity of emission of wastewater. Besides, it has another advantage, the low product costs., It still exists some inevitable problems such as the complex process, the high equipment cost and the large construction investment. Only when the market demand of PO and household products are matched well, its advantages can be shown. According to the differences of the raw materials and the household products, it can be divided into co-oxidation of isobutane method(PO/TBA method) and ethylbenzene co-oxidation (PO/SM method).

1.2.1. *Co-oxidation of Isobutane Method (PO/TBA Method)*

PO/TBA method mainly employs propylene and isobutane as the raw materials and the whole reaction process includes the oxidation of isobutane and the epoxidation of propylene. Firstly, the tertiary butyl hydroperoxides are prepared by the oxidation reaction between propylene and air at 120~130°C and 3.0~5.0 MPa. Secondly, at 130°C and 3.0-5.0 MPa, just mentioned product and propylene can occur epoxidation reaction with the help of the molybdenum catalyst and the PO and tertiary butyl alcohol (TBA) are produced. Because the TBA can be easily converted into the methyl tertiary-butyl ether (MTBE), which exist the risk to human health and water pollution, this process is gradually replaced by other safe processes.

1.2.2. *Ethylbenzene Cooxidation Method (PO/SM Method)*

The PO/SM method was invented by the Alcoa Company in U.S.. This method employs propylene and superoxide of ethylbenzene as the raw materials to produce PO. The process mainly includes the epoxidation of ethylbenzene and the oxidation of propylene. In this reaction PO and alpha phenyl ethanol are generated at the same time and the latter can produce the styrene under the condition of the dehydration in the gas phase or liquid phase. The main chemical reaction equations are shown as follows:

1) Ethylbenzene oxidation reaction

$$C_8H_{10} + O_2 \rightarrow C_8H_{10}O_2 \qquad \text{(main } \textit{reaction}\text{)}$$

$$2C_8H_{10}O_2 \rightarrow 2C_8H_9O + H_2O;\; 2C_8H_{10}O_2 \rightarrow 2C_8H_8O + O_2 \quad \text{(side } \textit{reaction}\text{)}$$

$$2C_8H_9O + O_2 \rightarrow 2C_8H_8O + 2H_2O; 2C_8H_8O + 3O_2 \rightarrow 2C_7H_6O_2 + 2HCOOH \text{ (side } \textit{reaction}\text{)}$$

2) Propylene epoxidation reaction

$$C_8H_{10}O_2 + C_3H_6 \rightarrow C_3H_6O + C_8H_9O$$

The process of the PO/SM method is shown in figure 2.

Firstly, under the conditions of oxidation reactor, 140°C, 0.25MPa and catalysts, ethylbenzene and air can produce the ethylbenzene hydrogen peroxide. Secondly, in ethylbenzene separator the ethylbenzene and other components can be removed. Thirdly, in epoxidation reactor, under the effects of catalyst, previous reactant can react with propylene and produce crude products of PO. Lastly, through separation tower of PO, propylene will be moved out and refined PO can be gained by rectifying the top components of PO's tower. And the recycled ethylbenzene and styrene can also be gained by rectifying the bottom components of thylbenzene styrene separation tower and styrene distillation tower.

The PO/SM method is a relatively advanced PO production process, as the first selection of the new large-scale PO device technologies, it has occupied about 50% outputs of the global PO at present [6]. The Chinese national offshore oil corporation combined with Shell built a 250 kt/a PO device with this technology in Guangdong, which broke the previous pattern that the production of PO simply rely on the cooxidation method in China, and also opened up the reform of the domestic production of PO. In 2010 the Sinopec Zhenhai Refining & Chemical Limited by Share Ltd and the United States Lyondell company together invested in the construction of 285 kt/a PO project, and this project is the world's largest PO/SM production device at present [3].

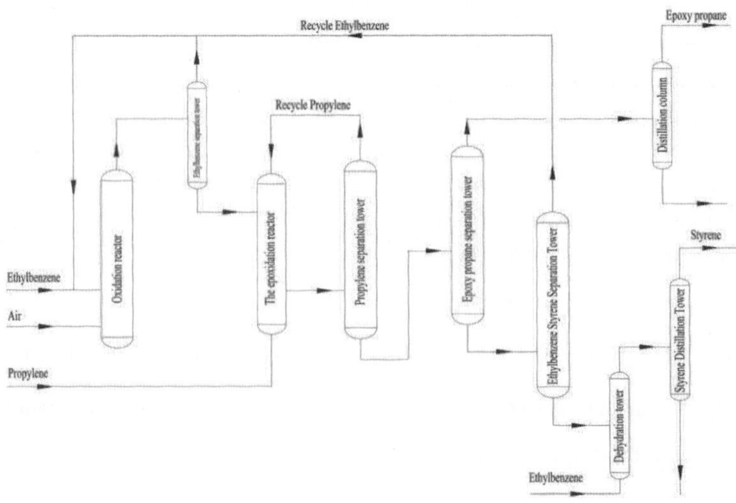

Fig. 2 The PO/SM process flow diagram

2. Modern technology of cleaner production

In view of the various defects existing in the traditional production technologies as well as the increasing awareness to the use of resources and energy and environmental protection and so on, in recent years many world manufacturers were committed to research some new technologies which are simple process, less by-products, more efficiently using of energy and protection of environmental. These advanced production technologies developed recently include CHP, HPPO and oxygen directly method.

2.1. *Cumene hydroperoxide method*

This method was developed by Japanese Sumitomo chemical company, and it used oxygen to oxidate cumene to product CHP with a titanium based catalyst in a fixed bed reactor. Then under the effect of the titanium silicalite molecular sieve catalyst, CHP and propylene begin to react in liquid phase in a fixed bed reactor, and PO and second methyl benzyl alcohol (CMA) can be generated some time later. After rectificating the PO in the crude products, dehydrate the CMA to the α-methyl styrene hydrogenation, the CMA can be obtained after the latter was hydrogenated, and the CMA will be oxidized into CHP.

In fact the CHP is a kind of improved cooxidation methods, its major characteristic is that the ethylbenzene is replaced by the cumene isopropylbenzene, which can be recyclable and have no side products. This method has the advantages of less pollution, less construction investment. The Sumitomo Chemical company and Chiba in Japan built a set of 200 Kt/a PO device based on this method in 2003. Besides the Petro Rabigh company, the joint venture with the Sumitomo and Saudi Aramco, built a set of 200 Kt/a PO device with this method, and it was put into use in 2009 March. This method is expected to be widely used in the future.

2.2. *Direct oxidation of hydrogen peroxide method*

HPPO is a kind of clean production technology developed in recent years, and it has the advantages of simple processes, high product yield, energy saving and high efficiency. This method firstly employs titanium silicon molecular sieve as the catalyst and use methanol as the solvents at normal temperature and low pressure conditions. Secondly, propylene and hydrogen peroxide begin to react directly in a tubular fixed bed reactor. Lastly, the crude products of PO in the products of PO can be separated by refining process, at the same time the propylene and methanol solvent can be recycled in the process[6]. HPPO process is shown in Figure 3. The main chemical reactions process equation includes:

$$C_3H_6 + H_2O_2 \rightarrow C_3H_6O \text{ (propylene oxide)} + H_2O \text{ (main } \textit{reaction}\text{)}$$

$$C_3H_6O + H_2O \rightarrow C_3H_8O_2 \text{(Propylene glycol)}; 2H_2O_2 \rightarrow 2H_2O + O_2 \text{ (side } \textit{reaction}\text{)}$$

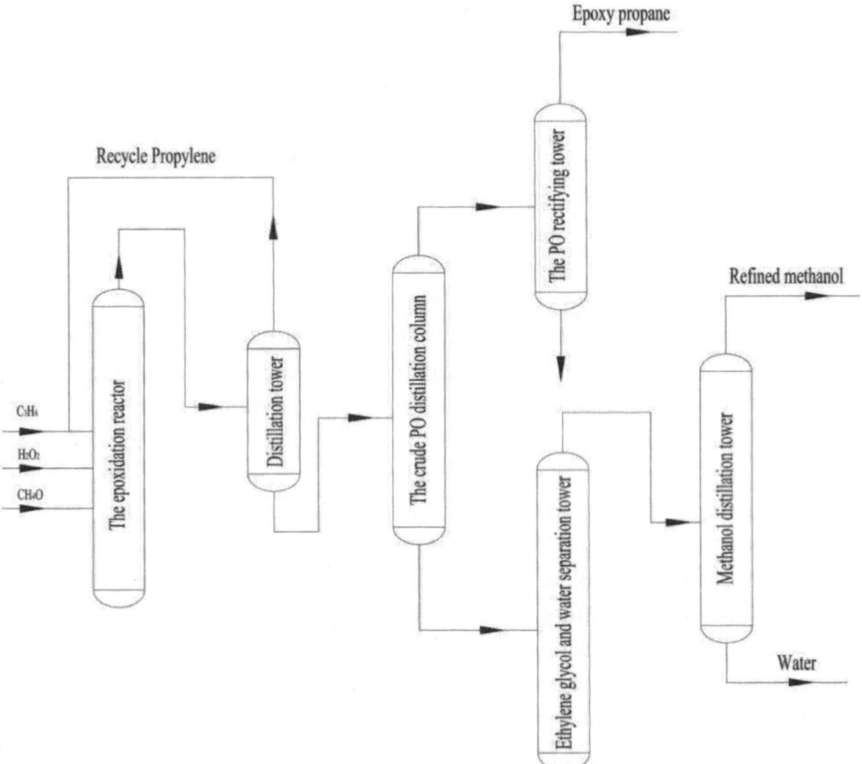

Fig. 3 HPPO process

At present the most mature industrial technology is in the Dow/BASF and Evonik/Uhde. The Dow chemistry and BASF developed a set of 300 Kt/a PO device, which have been put into production in the Belgian Antwerp in 2009. In addition, the Dow chemistry and Solvay planted to build a new set of 390 Kt/a PO device in Thailand Ma TA Fu, which have been put into production in 2011.

From the view of the domestic situation, there are many significant progresses in PO production. Shanghai Research Institute of petroleum and chemical industry, combined with Dalian Institute of Chemical Physics, etc, have developed a domestic technology of the PO. Tianjin Dagu chemical company, in cooperation with the East China University of Science and Technology, successfully run a test device of 150 kt/a PO based on HPPO. The development of the domestic technology solves some serious pollution problems of domestic existing chlorohydrin and realizes clean production, and provides a strong support to PO production technology upgrade, which break the monopoly of the technology that developed countries in China for a long time.

2.3. *Direct oxygen method*

Theoretically speaking, propylene reacting with oxygen to produce PO directly is the most ideal process, and it has the highest atomic economy. Because of the advantages of this process, no byproducts and no expensive reagent, this process has attracted great attention. America Lyondell is developing a technology which can directly oxidize propylene, hydrogen and oxygen into PO. Its basic principle is that hydrogen reacting with oxygen to make hydrogen peroxide with a silicate supported palladium titanium double function catalyst, and then the hydrogen peroxide can react with propylene to produce PO. However, there are still some technical bottlenecks, for example, the yield of the products need to improve further. In order to further improve this process to industrialization the company has built a unit construction in the states of the Faria Pennsylvania America. At present, this technology is still in experimental stage.

3. Technical comparison

The several technologies used widely in industry at present are compared briefly, as shown in Table 1 and Table 2.

From table 1 can be seen that, HPPO has a higher yield and lower product unit consumption compared with other two kinds of technologies, and there is no by-products, which makes the products refining process to be more energy saving and utilities consumption can be saved about 40%, so it has obvious advantages, as well as a clean production process for high efficiency and low cost.

4. Conclusion

At present, PO producing technologies almost are chlorohydrin and co-oxidation method, which account for more than 90% of the world's total capacity.

Table 1. Advantages and disadvantages of all kinds of process

Method	Advantages	Disadvantages
Chlorohydrin	Mature technology, simple process, operating with elasticity, good product selectivity; safety and stability	"Three wastes" emissions, pollution is serious; equipment corrosion and serious
Cooxidation	Environmental pollution is small, less equipment corrosion, good economic benefit	Variety of raw materials, complex process, high equipment requirement and infrastructure cost
CHP	Less environmental pollution, less equipment corrosion, no household products, equipment, less investment	Produce more alpha methyl styrene production process, need to increase recycling device
HPPO	Simple process, energy saving and high efficiency, high yield and conversion	Solvent methanol has certain risk, hydrogen peroxide transport

Table 2. Comparison of Chlorohydrin method, PO/SM method and HPPO

Name		Chlorohydrin	PO/SM	HPPO
PO/Yield(%)		88	91~93	94~95
Consumption of raw materials	Propylene(t/tPO)	0.8~0.9	0.8	0.8
	Chlorine(t/tPO)	1.3-1.5		
	Quick lime(t/tPO)	1.0		
	Hydrogen(t/tPO)		2.73	
	Ethylbenzene(t/tPO)		3.2	
	Caustic soda(t/tPO)		0.0315	
	Octane(t/tPO)		0.001	
	Hydrogen peroxide(t/tPO)			0.72
By-product	Two chloro propane(t/tPO)	0.10~0.27		
	Two chloro diisopropyl ether(t/tPO)	0.022		
	Styrene(t/tPO)		2.60	
	Propylene glycol(m^3/tPO)			0.026
Utilities consumption	Electric(kW/h)	600	550	330
	Fresh water(m^3/tPO)	40		
	Circulating water(m^3/tPO)	400	945	330
	Desalted water(m^3/tPO)	4	2	0.2
	Process gas(Nm^3/tPO)	125	135	184
	Steam(t/tPO)	0.1(1.5MPa) 4.5(0.5MPa)	12(1.0MPa)	3.5(2.0MPa)
	Total	800	1680	650

Table 2. (Cont'd)

Environmental protection	Exhaust gas(Nm^3/tPO)	45	160	205
	Wastewater(t/tPO)	40; COD: 1200mg/L	0.3; COD: 3000mg/L	0.8; COD: 140mg/L
	Waste residue(kg/tPO)	2000	50	0.4
Equipment investment		1.0	2.5	1.5
Product cost		1.0	1.05	0.9

In recent years, with increasingly concern to environment protection, resource utilization, moreover with transformation and improvement of traditional process, producing technologies development of epoxy propane have been developed. HPPO, which is a clean production process for high efficiency and low consumption, is a new force suddenly rises and gradually enters the industry, and it have been put into production in many plants and achieved many good results. It is the most advanced technology in the industrial applications, and it will obtain the widespread application in the future.

Acknowledgements:

This work is supported by Innovative program of Chinese National College Students (201210641109, 201410641003), Scientific Technology Foundation of Sichuan, P. R. China (2012GZ0114), Scientific Technology Foundation of Chengdu Normal University, P. R. China .

References

1. Cai jie, Wu lina. *China Chlor-Alkali*, 5: 2-5 (2009)
2. Xue Zuyuan. Abatracts of chemical enginering design, 19(5): 3-10 (2009)
3. Yang lin,Cai ting. *Guangzhou Chemical Industry*, 38(6): 28-29 (2010).
4. Luo Chengxian. *Sino-Global Energy*, 15(5): 71-76 (2010)
5. Zhu Liuqin. *Advances in fine petrochemicals*, 13(10): 39-41 (2010)
6. Zhang Zhifeng, Zhao Chunmei. *Chemical industry*, 27(12): 4-47 (2009)
7. Yu Jiankun, Li Zhong. Chemical Propellants and Polymeric Materials, 9(5): 8-22 (2011)
8. Sun Ronghua, Lu Xuguo. *Techno-Economics in Petrochemicals*, 23(5): 39-40 (2007)
9. Zhang Jian, Xie Yu, Niu Zhimeng. *Science and technolog in chemical industry*, 18(3): 75-79 (2010)
10. Jiao Yilin. *Guangzhou Chemical Industry*, 41(3): 7-8 (2013)

Experimental research on relative permeability of fractured-vuggy carbonate reservoirs

Hong-Mei Ren [1,2]

[1]*East China University of Science and Technology*
Shanghai, 2000237, P.R.China

[2]*Exploration and Development Research Institute, PetroChina Huabei Oilfield Company*
Renqiu, 062552,P.R.China,

Ping Guo[3*], Hui-Min Zhang[3], Jian-Fen Du[3] and Zhou-Hua Wang[3]

[3]*State Key Laboratory Of Oil And Gas Reservoir Geology And Exploitation (Southwest Petroleum University), d. 8, indu District,*
Chengdu City, Sichuan Province, 610500[†], .R.China

[*]*Email: guopingswpi@vip.sina.com*

For the complex pore structure and various connection modes in fractured-vuggy carbonate rock, there exist no generally accepted flow theories and only a few relative permeability curves for this kind of reservoir. In the displacement experiment of fracture-vug system, connection modes of vugs were divided into two extreme ways: vertical and horizontal. Fractures and vugs were artificially made on outcrop carbonate core which contained no natural fractures and oil-water/gas-water/oil-gas relative permeability experiments were carried on the core respectively with simulated oil, water and gas. For fractured-vuggy reservoirs with vertically connected vugs, the recovery effect is good because of gravity segregation. Besides, the larger the vertical viscosity difference is, the better displacement effect is. For fractured-vuggy reservoirs with horizontally connected vugs, the displacement efficiency depends on the connection position of fracture and vug, and it is generally worse than core with vertically connected vugs. Thus, the viewpoints on relative permeability with effects of gravity drive have been applied and confirmed in single fracture-vug unit in field pilots of displacing oil with water and displacing water with oil, which results in good recovery effect and has

[*] Corresponding author.

important significance on the promotion and application of oil recovery enhancement for fractured-vuggy reservoirs.

Keywords: Fractured-vuggy Carbonate Reservoirs; Unstable Method; Relative Permeability Curves.

1. Introduction

With complex distribution of fracture-vug system and connection modes, the fractured-vuggy carbonate reservoirs have strong heterogeneity as well as characteristics of dual porosity network and complicated flow. Generally, this kind of reservoir contains storage system as matrix, fracture and vug in which complex flow can be formed, resulting in relative low oil recovery[1]. Therefore, a correct knowledge of flow mechanism in fractured-vuggy carbonate reservoir and solving problems with relative permeability has important significance for the development of this kind of reservoir[2].

As important experiment information, relative permeability curve, lays foundation for development program. Relative permeability experiment researches have been carried out in succession by many oilfields in recent years and there exist a lot of views[3-8] and deep understanding on the shape and cause of the relative permeability curves, but still few analysis on curves of fractured-vuggy reservoir[9-13].

This paper mainly focuses on complex fracture-vug structure of fractured-vuggy carbonate reservoir. Vug is the main storage space, while fracture is the main flow path. Here we divide the connection modes of vugs into two extreme ways as vertical and horizontal: vertically connected ones represent the case that vugs are connected by high dip fractures, while horizontally connected represent the low dip case, other cases in between.

This research emphasizes fracture-vug system of fractured-vuggy reservoirs, and the matrix nearly bears no poroperm capacity. It is impossible to obtain applicable full diameter cores embodying representative porosity and connection modes of vug from real fractured-vuggy carbonate reservoirs and small cores can be unrepresentative. Therefore, full diameter outcrop cores with diameter near 10 cm were used to prepare fractured-vuggy cores with man-made fractures and caves. Water-gas, oil-gas and oil-water relative permeability curves are measured on those cores. In the experiment, cores were horizontally and vertically placed to represent the above-mentioned connection modes respectively. Based on the comparison analysis of the testing results, we have drawn conclusions on the influence of different connection modes on behavior of relative permeability curves.

2. Experiment

2.1. *Experimental methods*

According to the Oil and Gas Industry Standards SY/T5345-2007, relative permeability curves of cores with man-made fractures were measured on 103 Gas-Water/Oil-Gas/Oil-water Relative Permeability Testing Device which mainly composes of pressure control system, core holding system, measurement system, etc.

Before testing, the core was dried and vacuated firstly in order to be 100% saturated with formation water or distilled water and then weighed to calculate the actual amount of water saturated in core. Next, core was put into the full diameter core holder. Gas-water relative permeability tests were carried out under appropriate displacing pressure difference based permeability of the core. Data like atmospheric pressure, temperature, displacing pressure difference, gas/water production rate at each moment were recorded during the experimental process.

2.2. *Core preparation*

After repeated argument of collocation mode of fracture-vug system, it is hard to say which kind of mode is more representative. But from aspect of the overall distribution and collocation law, the following principle of fractured-vuggy reservoir core has been established: vugs are not filled up with medium; outcrop cores are used; matrix flow is not considered; scale of the hole can be 15% of porosity; connection mode of fracture-vug system is vertical or horizontal and permeability of fracture is about 50 mD. The core was split in the middle and three holes were made on the surface of fracture with one fracture as a channel. The fractured-vuggy reservoir core model is shown in Figure 1.

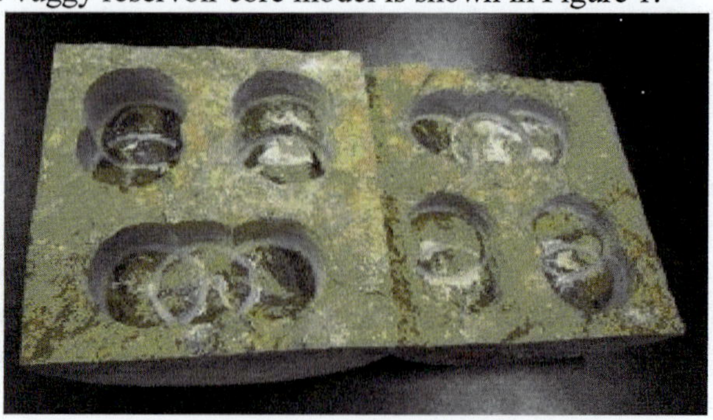

Fig.1. Core with vertically connected fracture-vug system

Fractured-vuggy relative permeability tests were conducted on just one core. Basic data of the core is shown in Table 1.

Table 1 Basic data of fractured-vuggy core

Length, cm	Diameter, cm	Porosity, %	Permeability, mD
11.117	9.965	16.83	48.26

2.3. *Experimental fluid*

Commercial nitrogen was applied as the displacement fluid. The water sample was prepared based on the analysis data of formation water. Simulated oil was prepared according to the viscosity ratio of oil and water in reservoir condition.

3. Results of relative permeability curve tests on fractured-vuggy core

In order to simulate the gravity effect, for situation of vertically connected vugs, gas-water relative permeability was tested by injecting water at the top of the core and producing gas from bottom, oil-gas relative permeability by injecting gas at the top and producing oil from bottom and oil-water relative permeability by injecting water at the bottom and producing oil from top (Figure 2A). Gas-water/oil-water/oil/gas relative permeability tests were carried out on horizontally connected-vug core (Figure 2B). All tests were at a room temperature without considering influence of irreducible water. Testing results are shown as Figure 3 to Figure 5.

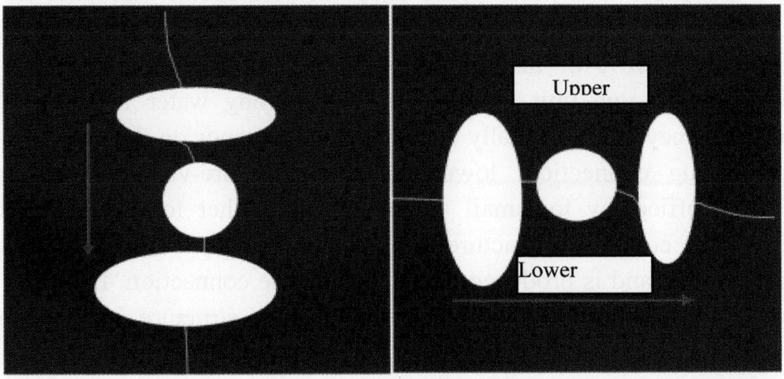

A. vertically placed B. horizontally placed

Fig.2. Connection mode of fractured-vuggy model

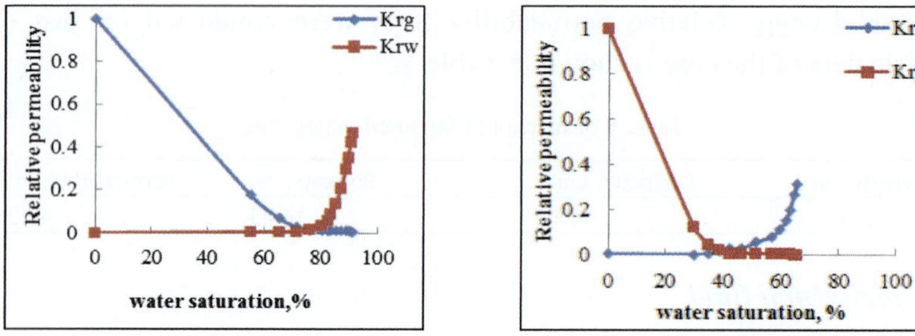

A. vertically placed B. horizontally placed

Fig.3. Gas-water relative permeability characteristics in fractured-vuggy reservoir

Figure 3 respectively shows displacing results on vertically and horizontally connected cores by injecting water after being saturated with gas to study flooding effect with strong water flooding. For vertically connected core, irreducible gas saturation is 8.87% and displacement efficiency is 91.13%, which indicates good displacement efficiency of bottom water flooding along as gravity flooding. Gas production efficiency for large unfilled unit with strong bottom water presents good gas production efficiency. For horizontally connected core, water flows into the lower part of fracture-vug connection and gas above the outlet gas-water interface cannot be produced, resulting in higher residual gas saturation of 34.22% and displacement efficiency of 65.78%. Relative permeability curves of vertically and horizontally connected cores share the same gas phase and water phase endpoint permeability, because fracture system is the main flow path of the cores. But endpoint relative permeability of water is only 40-50% of that of gas. Gas recovery of core with vertically connected fracture-vug unit is still high with strong water flooding. Water flooding efficiency of horizontally connected one depends on the volume below the fracture-vug connection: lower connected facture-vug unit has lower displacement efficiency for small volume, while higher for upper connected situation as the connecting fracture is located at the top of vug, where water flows downwards and is produced until reaching the connection. But there may exist several gas-water interfaces for the complex structure of fracture-vug system, which can cause a big difference in the testing result.

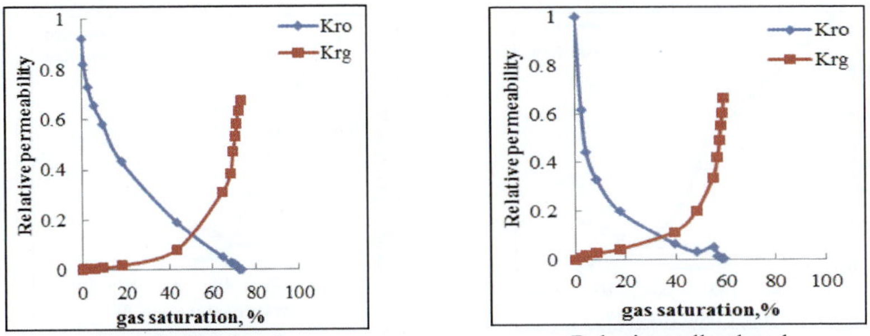

A. vertically placed B. horizontally placed

Fig.4. Oil-gas relative permeability characteristics in fractured-vuggy reservoir

Figure 4A is the displacing relative permeability curves when gas is injected at the top and oil is produced oil from bottom for vertically placed core to simulate the situation where energy of fracture-vug system is supplied by expansion of gas at the top or injecting gas flooding efficiency and analyze the displacement efficiency of gas flooding. Residual oil saturation is 26.69% and displacement efficiency is 73.31%. The testing results show good effect and ability to enhance oil recovery by gas flooding in fracture-vug unit along with some technology. The case where water is first injected, and then gas to displace oil in Tahe oilfield confirms this is an effective EOR method.

While core is placed horizontally as Figure 4B, gas cannot flow into the upper part of fracture-vug connecting system and displace oil blow the gas-oil interface, which obtains residual saturation of 41.07% and displacement of 58.93%. Because the endpoint values depend on permeability of fracture, endpoint permeability of gas and oil for vertically placed core are consistent with those of the horizontally connected one. While endpoint relative permeability of gas is only 70% of oil, which shows better mobility of gas and possibility of gas channeling. For gravity effect, oil recovery is high for gas flooding on vertically placed core. While water flooding efficiency of horizontally connected one depends on the volume below the fracture-vug connection: lower connected facture-vug unit has higher displacement efficiency for small volume, while lower for upper connected situation as the connecting fracture is located at the top of vug, where gas flows upwards. But there may present a big difference in the testing result for several gas-oil interfaces of complex structure of fracture-vug system.

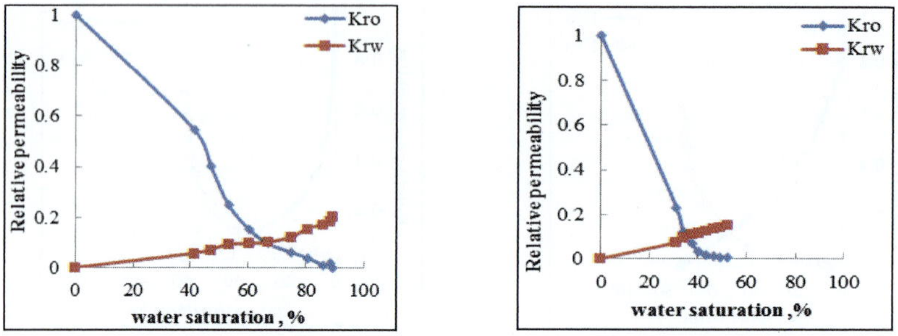

A. vertically placed B. horizontally placed

Fig.5. Oil-water relative permeability characteristics in fractured-vuggy reservoir

Figure 5 shows displacing results on vertically and horizontally connected cores by injecting water at the bottom of the core and producing oil from the top to study flooding effect with strong water flooding. For vertically connected core, irreducible oil saturation is 10.85% and displacement efficiency is 89.15%, which indicates that for big unfilled vug unit with strong bottom water flooding, oil recovery can be high as oil presenting in the upper part because of density difference. For horizontally connected core, residual oil saturation is 48.02% and displacement efficiency is 51.98%. Because water flooding efficiency of core with horizontally connected vugs depends on the volume below the fracture-vug connection: lower connected facture-vug unit has higher displacement efficiency of water flooding for small volume, while lower for upper connected situation as the connecting fracture is located at the top of vug, where water flows upwards. But there may exist several oil-water interfaces for the complex structure of fracture-vug system and there may be a big difference in the testing result. Similarly, the endpoint values depend on permeability of fracture, endpoint permeability of water and oil for vertically placed core are consistent with that of the horizontally one. On the basis of above analysis, oil recovery of water flooding for core with vertically connected vugs is higher than that of horizontally connected ones.

From Figure 3 to Figure 5, displacement efficiency is higher for cores with vertically connected vugs in the above 3 displacement process if making full use of gravity effect. Displacement efficiency is between 52%-91%, which signifies that viscosity ratio has a big influence on gravity effect. For the viscosity difference between oil and water is the smallest, the gravity effect is the weakest. While viscosity difference between gas and water is the largest, recovery efficiency is the highest for gas displaced by water. For core with horizontally connected vugs in which displacement efficiency depends on the fracture-vug connection point, displacement efficiency values of the core in this research are low as fractures are located in the middle of vugs. There is a big difference in residual saturation between cores with vertically connected and horizontally connected vugs, but a little difference in relative permeability of displacing phase because of the big contribution of fractures as the main flow path. While there is a big difference in relative permeability of different displacing phase: the highest is gas relative permeability in flooding process of oil by gas, then that of water in flooding of gas by water, and finally that of water in flooding of oil by water.

Theoretically, the effect of gravity drive is very good. The displacement efficiency can even reach 100%, only if the interface is stable, but this cannot happen in actual reservoir. This experiment has also presented a situation closer to the reality than ideal one. Although this research is relatively ideal, it still has provided reference for development of fractured-vuggy reservoirs. Pilot test in the condensate gas well Tazhong#1 has shown great effect by water flooding. Besides, field pilot tests of oil recovery by injecting water and gas in Tahe Oilfield have already succeed and been promoted as the main EOR method in a field-scale.

4. Conclusions

If taking full advantage of gravity segregation, development effect of fractured-vuggy reservoir with vertically connected vugs can be good and displacement efficiency of this kind of reservoir depends on the vertical viscosity difference, larger the difference, better the efficiency. Technology such as displacing oil with water, displacing gas with water in fracture-vug unit can be applied to realize gravity effect.

Displacement efficiency of fractured-vuggy reservoir with horizontally connected vugs depends on fracture-vug connection and is generally lower than that of vertically connected one.

There is a small difference in endpoint permeability of the same phase whether the vugs are vertically connected or horizontally connected, which indicates that the main flow path is fracture system.

Although the newly developed experiment method in this paper do not establish the fracture-vug system same as real reservoir pore structure, researches on relative permeability of the core with vugs vertically and horizontally connected by fracture still be of great importance and viewpoints proposed on this research has been applied and confirmed in production.

Acknowledgment

Foundation item: Supported by PetroChina Major Science and Technology Project(2014E-35-05).

References

1. Nailing Xiu, Research on flow mechanism in fractured-vuggy carbonate reservoir, Graduate School of the Chinese Academy of Science,2008
2. Shilun Li, Development of gas field and condensate gas field (Beijing,2004).
3. Min Yi, Ping Guo and Liangtian Sun ,The experiment of water displacing gas permeability in unstable method, Natural Gas Industry. 2007,27(10): 92-94.
4. Zhihong Kang, Principium experiment of oil seepage driven by water in fracture and vug reservoir of carbonate rocks (Chengdu,2006) pp.87-90.
5. Yang Xiao, Zhenzhong Cai, Tongwen Jiang, Jilei Feng, Jianfei Xie, Research on relative permeability character of water displacing gas in fractured-vuggy carbonate reservoir. (Sichuan Chengdu ,2012) pp.87-93.
6. Song Peng and Ping Guo, Physical simulation of exploiting fractured-vuggy carbonate gas condensate reservoirs by water injection.2014, pp.645-649.
7. Ping ping Shen, Bin Zhu, Xianbin Li, Yushu Wu, An experimental study of the influence of interfacial tension on water-oil two phase relative permeability ,2010,pp. 505-520.
8. Saeed Shad, Brij B Maini, Ian D Gates, Effect of fracture and flow orientation on two-phase flow in an oil-wet fracture relative permeability curves and flow structures ,SPE 132229-MS.
9. P Q Lian, L S Cheng, C Y Ma, The characteristics of relative permeability curves in naturally fractured carbonate reservoirs, SPE 154814-PA, 2012.
10. Akhlaghinia Manoochehr, Torabi Farshid Chan Christine W, Effect of Temperature on Two-Phase Relative Permeabilities of Heavy Oil, 2013 1185-1193.vol.27, No.2.
11. Zhonghong Chen ,Zunjing Ma , Xiaoyu Sha, Shouchun Zhang, Youshu Bao, The effects of high pressure on oil-to-gas cracking during laboratory simulation experiments, Sichuan ,2014. 143-162.

12. S.Y. Mo S.L. He G. Lei S.H. Gai Z.K. Liu. Effect of the drawdown pressure on the relative permeability in tight gas, 2015, pp.264-271.
13. T. Babadagli S. Raza,X. Ren K. Develi, Effect of surface roughness and lithology on the water–gas and water–oil relative permeability ratios of oil-wet single fractures, 2015.

The solution-processed film of Tetra[Trifluoroethoxy]-Phthalocyaninato Cobalt(II) in nitrogen dioxide sensing

Zhen Dong[a], Yu-Ying Jiang[a], Xia Kong[b] and Yan-Li Chen[*a,b]

[a]*School of Chemistry and Chemical Engineering, University of Jinan*
Jinan 250022, China

[b]*School of Science, China University of Petroleum (East China)*
Qingdao, 266580, China
E-mail:[*] yanlichen@upc.edu.cn, [*]chm_chenyl@ujn.edu.cn

Nitrogen dioxide (NO_2) detection with a high sensitivity in room-temperature has become a research focus due to the increasing attention on environmental issues. Herein, a solution processed tetra[trifluoroethoxy]-phthalocyaninato cobalt(II) complex $CoPc(OCH_2CF_3)_4$ (**1**) was designed and synthesized, successfully, and fabricated into solution-based thin film by using a simple, low-cost quasi-Langmuir Shäfer (QLS) method. The gas sensing behavior of the QLS film of **1** toward electron acceptor gas, NO_2, was investigated in the concentrations of 2-10 ppm at ambient temperature. Unexpectedly good sensitive responses to NO_2 gas are obtained for the ultra-thin solution processed $CoPc(OCH_2CF_3)_4$ film in a fast response/recovery cycle of only 1/5 min at room-temperature. This work should provide an efficient way to fabricate low-cost and reliable gas sensors through the combination of molecular design and film fabrication technique.

Keywords: Solution-processed Film; Phthalocyanine; NO_2 Gas Sensor; Nitrogen Dioxide detection;

1. Introduction

The industrial development in the past decade together with the improvement of quality of life and transit mobility has increased enormously the amount of toxic gases released into the atmosphere, with dramatic consequences for both the environment and the public health. Toxic gas detection has led to the development of selective and sensitive gas sensors to protect the environment and biological systems. The electrical conductivity of metal phthalocyanines (MPcs), which are mostly *p*-type organic semiconductors, is known to vary

---

[*] Corresponding author.

substantially in the presence of oxidizing or reducing gases due to the change in charge carrier concentration upon adsorption of gas molecules. Comparing to inorganic semiconductor that is only sensitive at high temperature (200-400 °C) with the risk of gas explosion in the existing environment, MPcs are of sensitive at low temperature, low-cost and easy fabrication properties, enabling it a promising candidate for NO_2 sensing in the practical environment [1-2]. In this manuscript, a tetra[trifluoroethoxy]-phthalocyaninato cobalt(II) complex $CoPc(OCH_2CF_3)_4$ (**1**) was design and synthesized, **Scheme 1**. Introduced numbers of trifluoroethoxy on the periphery of phthalocyaninato cobalt(II), not only provide the interaction sites of central-metal but more importantly increases and/or adjusts its solubility in organic solvents, ensuring the fabrication of gas sensor devices through a simple solution process. A high sensitive NO_2 gas sensor with rapid response and recovery at room-temperature was obtained based on $CoPc(OCH_2CF_3)_4$ (**1**) thin film fabricated by a simple quasi-Langmuir Shäfer (QLS) method. The present work will provide an efficient way to fabricate low-cost and high-sensitive gas sensors through the combination of molecular design and film fabrication technique[3-4].

2. Experiment

2.1. *Preparation of complex $CoPc(OCH_2CF_3)_4$ (1)*

4-(Trifluoroethoxy)phthalonitrile:A mixture of trifluoroethanol (1.50 g, 15 mmol),4-nitrophthalonitrile (0.85 g, 5 mmol), and K_2CO_3 (2.05 g, 15 mmol) were heated with magnetic stirring in 100 ml of a round-bottomed flask at room temperature in DMF (30 ml) for 48 h. The cooled reaction mixture was poured into cold water (80 ml) with stirring. The white precipitate was filtered off, washed with water until the filtrate was neutral, and dried in vacuum at 40°C. Yield: 1.15 g (90%).

Tetra[trifluoroethoxy]-phthalocyaninato cobalt(II) (**1**): A mixture of compund 4-(trifluoroethoxy)phthalonitrile (0.29 g, 0.50 mmol), $Co(CH_3COO)_2\cdot 4H_2O$(0.032 g, 0 .125 mmol) and dried pentanol (2 ml) was heated and stirred at 115 ℃ for 3.5 h in a vacuum sealed, pressure-resistant glass tube of 10 ml. The cooled reaction mixture was poured into ethanol with stirring. After stirring for 1h, the resulting black-green mixture was evaporated under reduced pressure, and the residue was chromatographed on a silica gel column using CH_2Cl_2 as eluent. Repeated chromatography followed by recrystallization from $CHCl_3$ and methanol gave compound tetra[trifiuoroethoxy]-phthalocyaninato cobalt(II) $CoPc(OCH_2CF_3)_4$ (**1**).

Repeated chromatography followed by recrystallization from $CHCl_3$ and CH_3OH gave pure compound as dark blue powder. Yield: 42 mg (21%). ^1H NMR (400 MHz, $CDCl_3$): δ[ppm]: 7.91 (d, 4H, Pc-α_1), 7.45 (d, 4H, Pc-α_2), 7.17 (d, 4H, Pc-β_1), 4.48 (s, 8H, -CH_2CF_3). UV–vis [CH_2Cl_2; λ max (band)]: 315 nm (Soret-band), 598 nm (Q_1-band), 659 nm (Q_2-band); MALDI-TOF MS: an isotopic cluster peaking at *m/z* 960.4, Calcd for $C_{40}H_{20}Eu_2CoF_{12}N_8O_4$,$[M]^+$,963.55 .

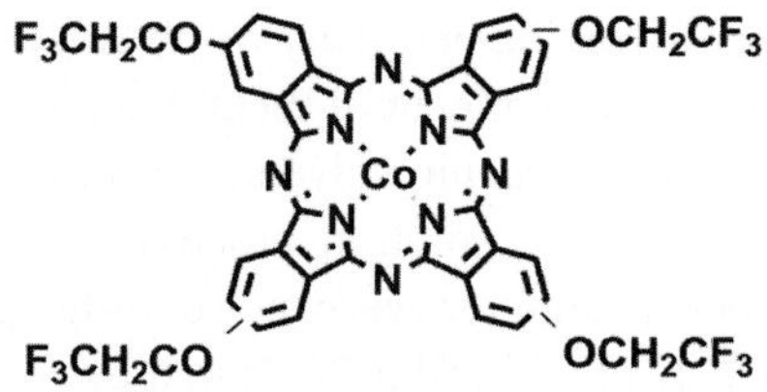

Scheme 1. Schematic molecular structures of $CoPc(OCH_2CF_3)_4$ (**1**).

2.2. *Fabrication of the thin film of (1)*

The ITO substrate is prepared in advance, with the interdigitated microelectrode arrays was etched, resulting ITO interdigitated electrodes (IDEs) (Figure 1A). Each array was composed of 50 microelectrodes measuring 0.1 mm wide and 6 mm long, the insulating gap between the neighboring electrodes around 0.075 mm. This microsystem geometry was chosen because it allowed for the measurement of conductivity changes between the two arrays and because it optimized the signal delivered.

Quasi-Langmuir-Shäfer (QLS) film[10]: 5.0 mL of chloroform solution of tetra[trifluoroethoxy]-phthalocyaninato cobalt(II) (**1**) (1×10^{-6} mol L^{-1}) was put into a cylindrical glass container (diameter, 9.5 cm; height, 1.5 cm; volume, 106.3 cm^3), then 65 mL of water was slowly added onto the container (caution: the amount of water added cannot completely cover the surface of the chloroform solution so as to maintain a path for the evaporation of the solvent, $CHCl_3$). During the solvent evaporation, the molecules gradually assembled to form some fine nanostructures at the $CHCl_3$/water interface and in turn the densely packed film on the water surface until complete evaporation of $CHCl_3$. The film can be easily transferred from the water surface onto a ITO substrate by horizontal lifting, i.e., the slide was horizontally and carefully lowered onto the film surface and then raised.

3. Results and discussion

3.1. *Current–voltage (I–V) characteristics*

The phthalocyanine derivatives with uniform membrane structure would be promising candidates for applications in electronic devices. To demonstrate the potentials of QLS film of compound **1**, the I–V curve was carefully measured. From the current-voltage (*I-V*) characteristic shown in Figure 1B, the high electronic conductivity of the QLS film of **1** is calculated to be around 2.71×10^{-3} S/m, which is partly attributed to strong intermolecular interactions between the adjacent molecules and higher ordered crystalline molecular arrangement in the QLS film of $CoPc(OCH_2CF_3)_4$ on the basis of the previous observations and literatures[5-7]. In addition, intrinsic organic semiconducting nature of **1** also play an important role for the excellent conductivity of the QLS film of **1**, according to the HOMO/LUMO (-5.56/-4.12 eV) energy levels of **1** estimated from its first oxidation and the first reduction potentials: Eox1 = 1.12 V; Ered1 = -0.32 V vs. SCE obtained by differential pulse poltammograms (DPV) measurement.

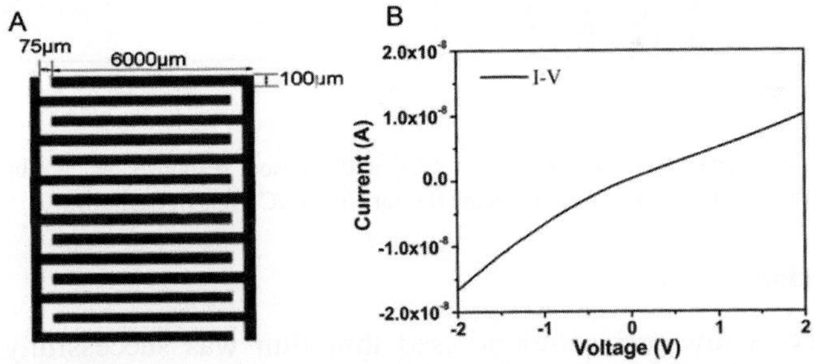

Fig. 1. Schematic configuration of the ITO IDEs (A) and I-V curve measured for the QLS film of **1** deposited on ITO IDEs (B).

3.2. *Gas sensing study*

Response of the QLS film-based sensor from **1** deposited onto the ITO IDEs was determined from the time-dependent current plots when exposed to NO_2 gas at room temperature, recorded by Agilent B2901 precision source/measure unit. As shown in Figure 2A, when the QLS film of **1** was exposed to NO_2 gas atmosphere with the concentration of 2-10 ppm (with a duty cycle where the dynamic exposure period is fixed at 1 min and the recovery period at 5 min), the conductivity decrease during exposure and increase to its initial value during recovery at room temperature, indicating a *n*-type responded behavior toward

NO_2. Upon exposure to the oxidizing NO_2 gas (electron-acceptor), the electron concentration near the surface will be weakened since NO_2 traps majority charge carriers, resulting in the decrease in conductivity.

In order to quantitatively analyze the sensor responses, the percent current change was calculated for each concentration. We define relative response intensity $S = (\Delta I/I_0) \times 100\%$, where I_0 is the baseline current value, and $\Delta I = I_g - I_0$; I_o is the current value at the beginning of an exposure/recovery cycle and I_g is the current value at the end of the 1 min exposure period.[8-10] In the present case, the sensor response of the QLS film of $CoPc(OCH_2CF_3)_4$ is linear with respect to the concentrations of NO_2 in the range of 2-10 ppm with a slope (the sensitivity) of 1.48 $\%ppm^{-1}$ (Adj. R^2 = 0.99). Thus, such film can be used to quantitatively analyze the concentration of NO_2. The present $CoPc(OCH_2CF_3)_4$ film-based sensor implied its potential in the particle application.

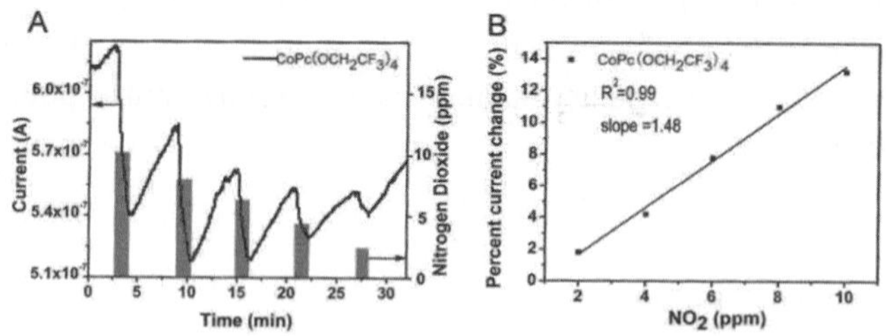

Fig. 2. The time-dependent current plots (A) and sensor response as a function of the NO_2 concentration in the range of 2-10 ppm (B) for $CoPc(OCH_2CF_3)_4$ QLS film

4. Conclusion

In summary, a low-cost solution-based thin film was successfully fabricated based on a novel $CoPc(OCH_2CF_3)_4$ compound by a simple QLS method. The QLS film exhibited good electrical conductivity. Exposed to NO_2 gas atmosphere, the $CoPc(OCH_2CF_3)_4$ film displayed excellent sensitivity to NO_2 in the range of 2-10 ppm at room temperature. Since the sensor monitored the NO_2 without high temperature and complicated producing process, it will be anticipated that have a significant application potential in low-cost gas sensors.

Acknowledgments

This work was financially supported by the National Natural Science Foundation of China (No. 21371073), Research Foundation from China University of Petroleum (East China) (Y1510051).

References

1. R. Ehrlich, M.C. Henry, Chronic toxicity of nitrogen dioxide. I. Effect on resistance to bacterial pneumonia, *Journal of Occupational Medicine* **34** 12 (1970) .
2. K.Mukala,J.Pekkanen,P.Tiittanen,S.Alm,R.Salonen.J.Tuomisto,Personally measured weekly exposure to NO_2 and respiratory health among preschoolchildren, *European Respiratory Journal* **13** 1411–1417 (1999).
3. S. Pochekailov et al. / *Sensors and Actuators B* **169** 1–9 (2012)
4. Y, Chen.; M, Bouvet.; T ,Sizun; Y ,Gao.; C, Plassard; E,Lesniewska; J,.Jiang, Facile Approaches to Build Ordered Amphiphilic Tris (Phthalocyaninato) Europium Triple-Decker Complex Thin Films and Their Comparative Performances in Ozone Sensing. *Phys. Chem. Chem. Phys.*, **12**, 12851−12861.(2010)
5. Y Chen, D Li,a N Yuan,J Gao, R Gu, G Luc and M Bouvet Tuning the semiconducting nature of bis(phthalocyaninato) holmium complexes via peripheral substituents.*J. Mater. Chem.* **22**, 22142–22149(2012)
6. D Gao,X Zhang, X Kong, Y Chen, and J Jiang (TFPP)Eu[Pc(OPh)8]-Eu[Pc(OPh)8]/CuPc Two-Component Bilayer Heterojunction-Based Organic Transistors with High Ambipolar Performance *Applied Materials & Interfaces* 7, 2486−2493(2015)
7. G . Guillaud, J.Simon, J.Germain, Metallophthalocyanines:gas sensors resistors and field effect transistors, *Coordination Chemistry Reviews* 178 1433–1484. (1998)
8. P. Moseley, Solid state gas sensors, *Measurement Science and Technology* **8** 223–237(1997).
9. M. Bouvet, Phthalocyanine-based field-effect transistors as gas sensors, *Analytical and Bioanalytical Chemistry* **384** 366–373(2006).
10. J Gao, G Lu,J Kan,Y Chen, M Bouvetb, Solution-processed thin films based on sandwich-type mixed (phthalocyaninato)(porphyrinato) europium triple-deckers: Structures and comparative performances in ammonia sensing *Sensors and Actuators B* **166– 167** 500– 507 (2012).

High strain rate tensile response of epoxy resin system

†Hai-Xia Hu, Xiang -Yang Chen, Zhe-Bo Zhou, Jin-Zhang and Lun -Ye Sun
College of Mechanical Engineering, Anhui University of Science and Technology, Huainan, 232001, P.R.C
E-mail: †huhx2006@163.com

Lin-Mao Ye
State Key Laboratory of Turbulence and Complex System, College of Engineering, Peking University, Beijing, 100871, P.R.C
E-mail: lmleaf001@gmail.com

The high strain rate tensile behaviors of epoxy resin filled with Polyaryletherketone (PAEK) powders have been studied on TS-1000 model high speed material apparatus under dynamic loading conditions. Stress-strain curves of epoxy resin system under high strain rate loading were obtained. The results of the study indicated that the fracture elongation and the tensile strength of epoxy resin casting system which toughened by the PAEK powders are higher than those of pure epoxy resin matrix at the strain rate of 3500/s.

Keywords: Epoxy Resin; High Strain Rate; Dynamic Mechanical Properties; Fracture Surface; PAEK powders

1. Introduction

Polymers are becoming increasingly used in aerospace structural applications, where they experience complex, non-static loads. Correspondingly, the mechanical properties at high strain rates are of increasing importance in these applications [1]. Epoxy resins is high molecular polymer with epoxy group, it is characterized by high strength, high elastic modulus, strong bond ability, and excellent chemical stability [2], which are widely used in the fields of aerospace, automotive, electronic, and chemical engineering [3, 4].

Polymer composite structures undergo different loading conditions during their service life. In recent years, some articles mentioned the dynamic behavior of epoxy resin and the composites [5-8]. M.S. Sohn et al. [5] studied the

† Corresponding author.

mechanisms and energy dissipation of carbon fiber reinforced epoxy composites under impact and high strain rate conditions. K.Y. Zeng et al. [6] studied the morphology, mechanical properties and fracture behavior of epoxy-alumina nanocomposites. They found that the tensile modulus, tensile strength and fracture toughness are affected by the geometry of the particles. Hui Zhang et al. [7] studied the fracture properties of epoxy resins filled with different amounts of soft rubber particles and/or rigid silica nanoparticles. Amos Golat et al. [8] studied the mechanical response of E-862 and PR-520 resins in tensile and shear loadings. They found that the toughened PR-520 resin has a higher failure stress in tension and a higher maximum stress in shear compared to the untoughened E-862 resin.

In this study, we investigated the high strain rate tensile response of an epoxy resin casting system which toughened by the PAEK powders. The morphologies of the impact fracture surfaces were studied. The effects of PAEK powders on the dynamic mechanical properties of the epoxy resin material were discussed. It was expected that this study would provide guidance for the application and broaden the use of bicomponent high performance epoxy resin.

2. Experimental

2.1. *Materials*

In this experiment, the epoxy resin samples were prepared in advance. And the diglycidyl ether of bisphenol A- (DGEBA) based epoxy resins was supplied by Wuxi Synthetic Resin (Jiangsu, P.R.C). Tetraglycidy 1-4, 4′-diaminodiphenylmethane (TGDDM) with EEW of 125 was provided by the Shanghai Institute of Synthetic Resins (Shanghai, P.R.C). Curing agent DDS with amino hydrogen equivalent weight of 62 was provided by Shanghai Reagent (Shanghai, P.R.C). And Beijing Institute of Aeronautics Materials (Beijing, P.R.C) provided the PAEK powders.

To make the epoxy resin, the obtained transparent homogeneous mixture, these materials were mixed at the ratio of DGEBA: TGDDM: DDS=2: 3: 2. The loading level of PAEK powders to epoxy precursors (mixture of epoxy monomer and hardener) is 10 phr. The above resins were poured into the rectangular molds. Cast bars were fabricated with the neat resin or the blends for morphological and high strain rate tensile tests. Cured epoxy samples were recorded as BE and BE/PAEK-10 samples.

2.2. *Test*

High strain rate tensile responses of the epoxy resin casting systems were studied on the TS-1000 model high speed material apparatus at the strain rate of 3500/s. The incident I, reflected R and transmitted T signals obtained at a strain rate of 3500/s using strain gauges. Prior to each testing, the surfaces of the samples were cleaned with alcohol dipped cotton. Then, the samples were dried at 50°C for 1 h to remove the moisture on the surface layer and the cooled in desiccators to ambient temperature. Three repeated tensile tests were carried out for each specimen so as to minimize the data scattering, and the average values were taken. The ambient temperature was roughly 25°C and the relative humidity about 52%.

After impact, the fracture surfaces of the samples were coated with a thin layer of gold powders, and then observed with an S-4800 scanning electron microscope (SEM) at an accelerating voltage of 5 kV to understand the high strain rate tensile characteristics and damage mechanisms.

3. Results and discussion

3.1. *Tensile performance*

The stress strain curves of the pure bicomponent epoxy resin and the BE/PAEK-10 samples at room temperature are obtained using one-dimensionalwave theory. The plots are presented in Figure 1. for a strain rate of 3500/s. It can be seen that the stress strain curve of the pristine bicomponent epoxy resin was short than that of the BE/PAEK-10 samples. This could be due to the higher crosslink degree of pure epoxy resin.

From Figure 1, it also can be seen that the fracture elongation of the pure bicomponent epoxy resin was lower than that of the BE/PAEK-10 specimen. That is to say, the fracture elongation of the BE/PAEK-10 sample increased with the addition of PAEK powders. The reason is that PAEK powders dispersed in the polymer matrix uniformly and can improve the mechanical properties. This indicated that plastic PAEK powders dispersed in epoxy resin matrix were very effective to improve the toughness of materials.

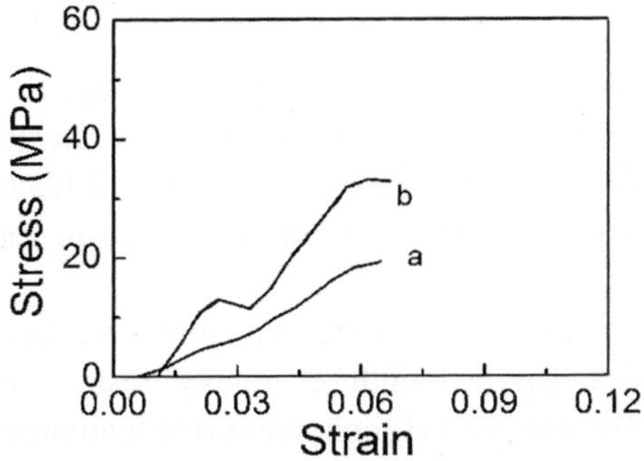

Fig. 1 Stress strain curves of (a) Pristine BE and (b) BE/PAEK-10 samples

3.2. *Morphologies of the fracture surfaces*

In order to further study the tensile behavior of epoxy resin casting samples, SEM analysis was performed on the specimens after impact tested under room temperature. The morphologies of the impact fracture surfaces of epoxy resin casting samples when the strain rate is about 3500/s were shown in Figure 2. It can be seen that shell texture are observed on the fracture surface of pure epoxy resin matrix, as is shown in Figure 2 (a). And the fracture surface is relatively smooth, which indicated that the typical brittle characteristic of the pure epoxy resin matrix. By contrast, the morphology of the BE/PAEK-10 sample obviously show that there is an increase in the fracture surface roughness due to the addition of the PAEK powders, as is shown in Figure 2(b). This is due to the PAEK phase separated and dispersed uniformly throughout the epoxy resin matrix in the process of solidification. Excellent interface connection between the dispersed phase PAEK particles and matrix was formed.

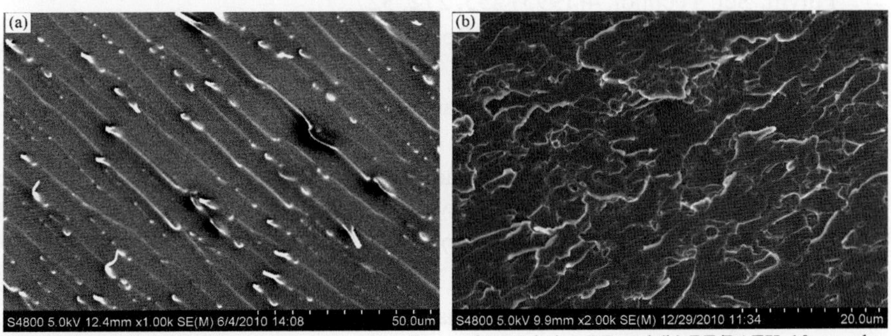

Fig. 2 Morphologies of the impact fracture surfaces of (a) Pristine BE and (b) BE/PAEK-10 samples

4. Conclusions

This study presents the dynamic tensile properties of epoxy resin casting system on the high speed material tester under high strain rate loading. Stress strain curves of epoxy resin casting systems were obtained. Effects of PAEK powders on tensile mechanical properties of the epoxy resin casting system were discussed.

The present study shows that there is a significant increase in the fracture elongation and the tensile strength of epoxy resin casting system which toughened with PAEK powders at high strain rate loading compared with those of pure epoxy resin samples at the strain rate of 3500/s.

Acknowledgement

This work was supported by the Anhui provincial Natural Science Foundation of P.R.C (No.1508085QE94), the young and middle aged academic backbone and the Doctoral Scientific Research Fund of Anhui University of Science and Technology. Authors thank Prof. X.S. Yi and Dr. M. Zhang of the National Key Laboratory of Advanced Composites in Beijing Institute of Aeronautics Materials for their helps.

References

1. A.M. Waas, N. Takeda, J. Yuan and S.H. Lee, in: Proceedings 12th Annual Technical American Society for Composites Conference, edited by R.F. Gibson, G.M. Newaz, Dearborn (1997).
2. C. Jarlskog, C.P. Violation (World Scientific, Singapore, 1988). J. R. Fried: Polym. Sci. Technol. 4(1995) 255, 272, 327.
3. P. Chen and S. P. Liu, Epoxy Resins, Chemical Industry, PressBeijing, P.R.C 1997.
4. ASM, ASM International (American Mathematical Society, Materials Park, USA, 18 (1992).
5. M.S. Sohn and X.Z. Hu, Impact and high strain rate delamination characteristics of carbon fibre epoxy composites. Theoretical and Applied Fracture Mechanics 25(1996) 17-29.
6. S.H. Lim, K.Y. Zeng and C.B. He, Morphology, tensile and fracture characteristics of epoxy-alumina nanocomposites. Mater. Sci. and Eng. A, 527(2010) 5670-5676.

7. L. C. Tang, H. Zhang, S. Sprenger, L. Ye, and Z. Zhang, Fracture mechanisms of epoxy-based ternary composites filled with rigid-soft particles, Compos. Sci. and Technol. 72(2012) 558-565.
8. G. Amos, K. G. Robert, D.R. Gary, Strain Rate Sensitivity of Epoxy Resin in Tensile and Shear Loading, NASA Technical Reports Server (NTRS), 2005-213595.

Preparation and characterization of organic segments modified graphene oxide obtained by thermal expansion

Hong-Bo Liu *†, Wu-Ying Zhang and Feng Lin

School of Applied Chemistry and Biological Technology, Shenzhen Polytechnic, Shenzhen 518055, People's Republic of China
E-mail: † lhb206@163.com

The graphene oxides were prepared form graphite by thermal expansion, and the polyurethane acrylates containing the reactive NCO (PACN) were synthesized. Then the polyurethane acrylates modified graphene oxide were prepared by NCO of PACN reacting with the hydroxyl groups of the graphene oxides. The polyurethane acrylates modified graphene oxide was characterized by Fourier transform infrared spectrometer (FTIR), scanning electron microscope (SEM) and Raman spectra. The FTIR spectra showed that the NCO of PACN reacted with the hydroxyl groups of graphene oxide. The measurement of SEM and Raman spectra showed the structure and surface morphology of polyurethane acrylates modified graphene oxide did not change. The polyurethane acrylates modified graphene oxide will be benefit to prepare the polymer/graphene nanocomposites

Keywords: Organic Segments, Graphene Oxide; Thermal Expansion; NCO groups; FTIR Spectra; Graphene Nanocomposites

1. Introduction

Graphene sheets have attracted tremendous attention from both the theoretical and experimental scientific communities in recent years, due to its unique physical properties like superior mechanical properties, flexibility, high specific surface area, extraordinary thermal conductivity and excellent electronic transport properties [1].

When the graphene was dispersed in polymer as nanofiller, the polymer/graphene nanocomposites show superior mechanical, thermal, gas barrier, electrical and flame retardant properties compared to the neat polymer. However, the improvement in the physicochemical properties of the polymer/graphene nanocomposites depends on the distribution of graphene layers in the polymer matrix as well as interfacial bonding between the graphene layers and polymer matrix. Interfacial bonding between graphene and the host polymer dictates the final properties of the graphene reinforced polymer nanocomposite [2,3].

* Corresponding author.

It is well known that the graphene is not compatible with organic polymers and it is difficult to form polymer/graphene nanocomposites. The graphene are heavily oxygenated to prepare the graphene oxide sheets. In contrast, the graphene oxide can alter the van der Waals interactions significantly and be more compatible with organic polymers because of its bearing hydroxyl, epoxide, diols, ketones and carboxyls functional groups [4-6].

Some organic segments can modified the graphene oxide through its hydroxyl, epoxide, diols, ketones and carboxyls functional groups. In this paper, the graphene oxide was prepared by thermal expansion. Then, the polyurethane acrylates containing the reactive NCO groups (PACN) were synthesized. The graphene oxide was modified by the polyurethane acrylates by the NCO reacting with hydroxyl, its structure and properties were measured.

2. Experimental

2.1. *Materials*

Graphite, concentrated sulfuric acid, hydrogen peroxide, hydrochloric acid, potassium permanganate, sodium nitrate, barium chloride, hydroquinone, polyethylene glycol (PEG-400), 2,4-toluene diisocyanate(TDI), 2-hydroxyethyl acrylate (HEA), di-n-butyltin dilaurate (DBTDL) were purchased from Guangzhou Chemical Regents Co. All the materials in this study were used as received without further purification.

2.2. *Preparation of graphite oxide*

The concentrated sulfuric acid (115ml, 98%)and sodium nitrate($NaNO_3$, 2.5g) were added in the beaker and were stirred under ice bath. Then the graphite was added when the temperature was below 5^0C in system and still stirred 30min. The potassium permanganate (15g) was slowly added and still stirred 60min under ice bath. The beaker was placed in constant temperature water bath at 40 ^0C and still stirred 30min, and the 220mL deionized water was added in the beaker. The temperature was controlled under 90 ^0C. The hydrogen peroxide (5%,35ml) was added, and the color of solution was changed from brown black to light brown. The product was filtered and washed by 5% hydrochloric acid until the SO_4^{2-} didn't detect by barium chloride solution. Then the product was washed by deionized water and dried in the oven at 50 ^0C.

2.3. *Preparation of graphene oxide materials*

The graphene oxide was prepared by thermal expansion. The graphite oxide was placed in pipe furnace under nitrogen protection, and the temperature rise to 1050 °C for 1-2min.

2.4. *Preparation of the polyurethane acrylates containing the reactive NCO (PACN)*

The 2,4-toluene diisocyanate (TDI, 0.6mol) were added in the 1000mL round-bottomed flask. While stirring, the 2, 4-toluene diisocyanate was heated to 50°C, followed by adding PEG-400 (0.3mol) containing 0.1 wt% DBTDL by dropping. Then the mixture was heated to 65°C and maintained for 2.5h. The PACN was obtained when the HEA (0.3mol) containing 0.06 wt% hydroquinone were added by dropping in the above reaction system and the reaction was continued 2h.

2.5. *Preparation of the polyurethane acrylates modified graphene oxide*

The polyurethane acrylates modified graphene oxide were prepared by mixing graphene oxide and PACN. The standard chemical titration method was used to determine the content of NCO groups of the prepolymer during the reaction process. The reaction was conducted at 65 °C until the NCO groups could not be detected by titration method.

2.6. *Measurement*

The polyurethane acrylates modified graphene oxide were measured by Fourier transform infrared spectrometer (Analect RFX-65A, FTIR), scanning electron microscope (Hitachi S-3000H, SEM) and Raman spectra (Advantage 532).

3. Results and Discussion

The graphene oxide was prepared by thermal expansion, and its measurements were showed the other paper. The SEM spectra of the graphene oxide are shown in Figure 1. It shows the surface morphology and structure of the graphene oxide. The graphene oxide prepared by thermal expansion would lose large number of active functional groups, such as hydroxyl, epoxide, diols and ketones, et al. The decrease of active functional groups in graphene oxide was disadvantage to chemical modification. The polyurethane acrylates containing the reactive NCO (PACN) were prepared and modified the graphene oxide in this paper.

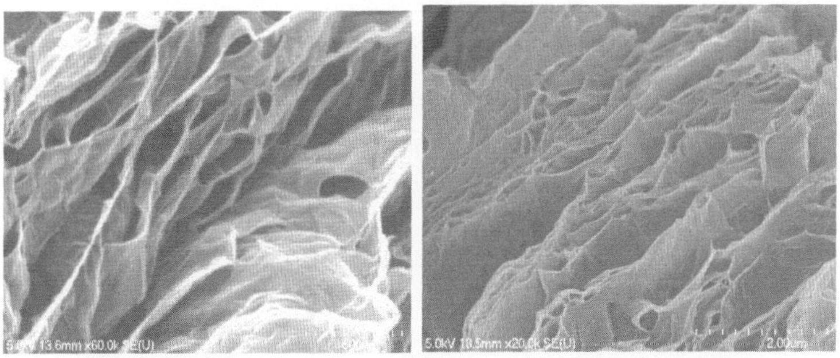

Fig. 1. The SEM of graphene oxide prepared by thermal expansion

The FTIR spectra of the polyurethane acrylates containing the reactive NCO (PACN) is shown in Figure 2. It could also prove that the synthetic oligomer was the target PACN. There existed evident absorption peaks of carbonyl of carboxylic ester was at 1730cm^{-1}. The characteristic C=C absorption of the unsaturated acrylic groups was at 1635 cm^{-1}, and the absorption peaks of NCO groups was at 2270 cm^{-1}.

In order to investigate the change of NCO groups, the polyurethane acrylates modified graphene oxide was analyzed by FTIR. Its spectra was shown in Figure 3. The graphene oxide obtained by thermal expansion existed some defects. For example, the detachment of graphene oxide layers was not sufficient, and the active functional groups at graphene oxide surface would decomposed. Those defects would effected the quality of FTIR and the chemical modification. The quality of FTIR of the polyurethane acrylates modified graphene oxide (the polyurethane acrylates modified graphene oxide) was not very good, but the change of absorption peaks was evident. The absorption peaks of NCO groups (2270 cm^{-1}) were not detected. The results in Figure 3 showed that the reaction of NCO groups was essentially complete in the preparation of the polyurethane acrylates modified graphene oxide.

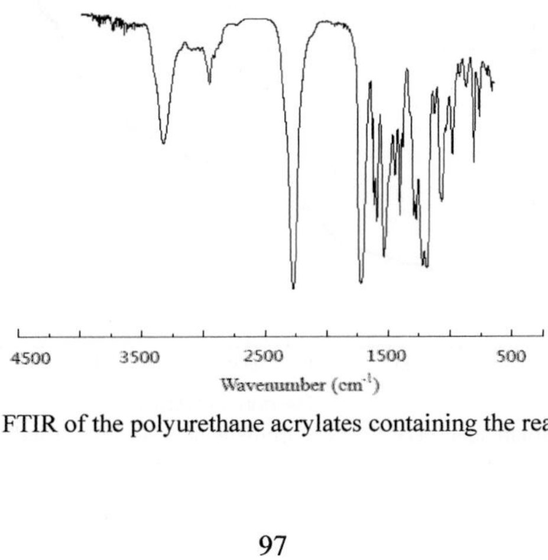

Fig. 2. The FTIR of the polyurethane acrylates containing the reactive NCO

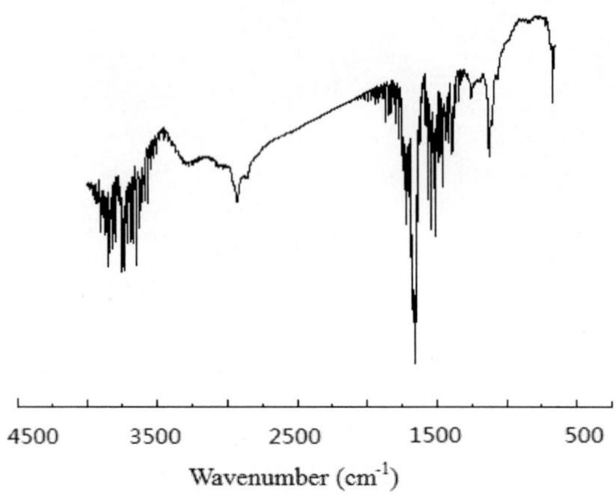

Fig. 3. The FTIR of the polyurethane acrylates modified graphene oxide

The polyurethane acrylates modified graphene oxide were characterized by Raman spectra and SEM, and the measurement were shown in Figure 4, Figure 5. There are two evident absorption peaks at 1420 cm^{-1} and 1590 cm^{-1}. The Raman spectra of the polyurethane acrylates modified graphene oxide showed that the organic segments did not change the structure of the graphene oxide. The SEM of the polyurethane acrylates modified graphene oxide also indicated that that the organic segments did not change the surface morphology of the graphene oxide. The other properties of the polyurethane acrylates modified graphene oxide need more experiment and would reported the other papers.

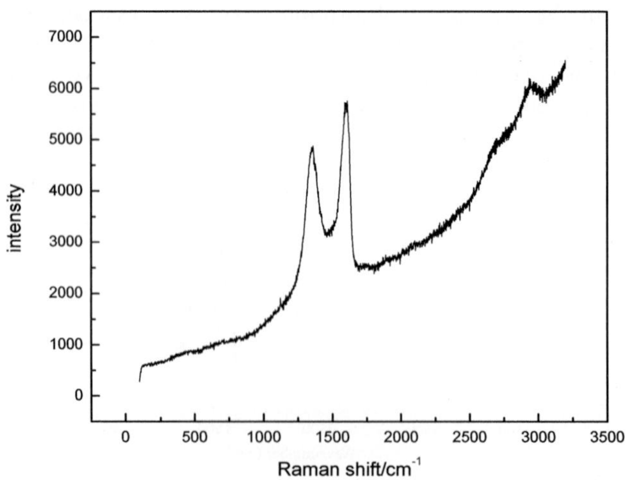

Fig. 4. The Raman spectra of the polyurethane acrylates modified graphene oxide

Fig. 5. The SEM of the polyurethane acrylates modified graphene oxide

4. Conclusion

The graphene oxides were prepared by thermal expansion, and the polyurethane acrylates containing the reactive NCO (PACN) were synthesized. Then, the polyurethane acrylates modified graphene oxide were prepared. The FTIR of PACN and the polyurethane acrylates modified graphene oxide were measured, the results showed the polyurethane acrylates modified the graphene oxide. The Raman spectra and SEM of graphene oxides were tested, the results indicated that the structure and surface morphology of polyurethane acrylates modified graphene oxide did not change.

Acknowledgements:

This work is supported by the Shenzhen Science and Technology Foundation (JCYJ 2014 0702 1720 29937)

References

1. R. Daniel, S.P. Dreyer, W.B. Christopher, The chemistry of graphene oxide. Chem. Soc. Rev. 39, 228(2010).
2. T. Ramanathan, A.A. Abdala, S. Stankovich, et al. Functionalized graphene sheets for polymer nanocomposites. Nat. Nanotechnol. 3, 327(2008).
3. S. Stankovich, D.A. Dikin, G.H.B. Dommett, et al. Graphene based composite materials. Nature, 442, 282(2006).
4. H. Kaczmarek, A. Podgorski. Photochemical and thermal behaviours of poly (vinyl alcohol)/graphite oxide composites. Polym. Degrad. Stabil. 92,939(2007).

5. R. Hao, W. Qian, L. Zhang, Aqueous dispersions of TCNQ-anion-stabilized graphene sheets. Chem. Commun. 2008, 48, 6576(2008).
6. Y. Xu, H. Bai, G. Lu, et a1. Flexible graphene films via the filtration of water-soluble noncovalent functionalized graphene sheets. J. Am. Chem. Soc. 130,5856(2008).

Effect of texture morphology on lubrication and sealing property of textured valve core

Chuan-Li Wang [a], Tao He[*,b], Hai-Shun Deng [c] and Ding Ma [d]
School of Mechanical Engineering, Anhui University of Science and Technology Huainan, 232001, China
[a]chlwang@aust.edu.cn, [b]taoheaust@163.com, [c]dhs1998@163.com, [d]mading127@126.com

In order to improve the lubrication and friction performance of the valve core, surface texture was introduced in the surface design of coal mine water hydraulic valve core. However, research of the valve core surface texture design has rarely been conducted. In this paper, the pressure distribution and leakage characteristics of 4 kinds of texture morphologies –cylinder, sphere, square and triangle, have been researched and the shear friction and bearing force of the textures were analyzed. Geometry models of valve core texture were established, and the flow field of the core textures was simulated by computational fluid dynamics. The results show that the suitable core texture morphology can produce liquid film pressure support and effectively improve the lubrication characteristics of valve core. The shear friction and bearing force are different under the different texture morphologies. All the topographies have good load-carrying properties except for the triangular topography. Sphere morphology has the minimum shear friction and leakage which falls by nearly 7% as compared to other morphologies. Its shear friction also reduces by about 35%, when the area ratio S is 36%.

Keywords: Surface Texture; Valve Core; Hydrodynamic Lubrication; Sealing Property.

1. Introduction

Surface texture has the characteristics of secondary lubrication, storing wear debris and lubricating fluid, end face sealing and micro-hydrodynamic carrying, etc. [1-3]. Coal mine water hydraulic valve has the characteristics of bad wear and lubrication, large leakage amount, and easy to be damaged by wear particle. So in order to improve the lubrication and friction performance of the valve core, surface texture was introduced in the surface design of coal mine water hydraulic valve core. At present which has been applied in the fields of mechanical seal [4], bearing [5], internal combustion engine cylinder [6], piston pump port plate [7], and noise suppression [8], et al.

[*] Corresponding author

But the research to the valve core surface texture design is rarely reported. The lubrication and friction performances of the texture are greatly affected by the texture morphology [9]. The pressure distribution and leakage characteristics of 4 kinds of texture morphologies, such as cylinder, sphere, square and triangle, have been researched and the shear friction and bearing force of the textures analyzed.

2. Simplified Geometric Model

The structures of the surface texture core and single texture morphology are as shown in Figure 1. Pressure on two sides of the clearance is p_n and p_0 respectively. U refers to valve core speed. The walls of left and right surface texture are in the periodic boundary condition with pressure dropping. No slip boundary used in the top wall, so does the bottom. Pressure gradient can be computed from the pressure difference (p_n-p_0) of the two ends and the sealing length of the clearance L_0. Figure 2. shows that the sketch maps of the texture morphology structure, as cylinder, sphere, square and triangle.

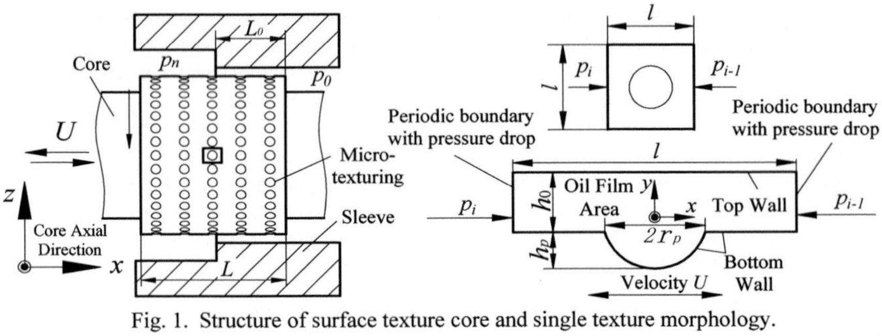

Fig. 1. Structure of surface texture core and single texture morphology.

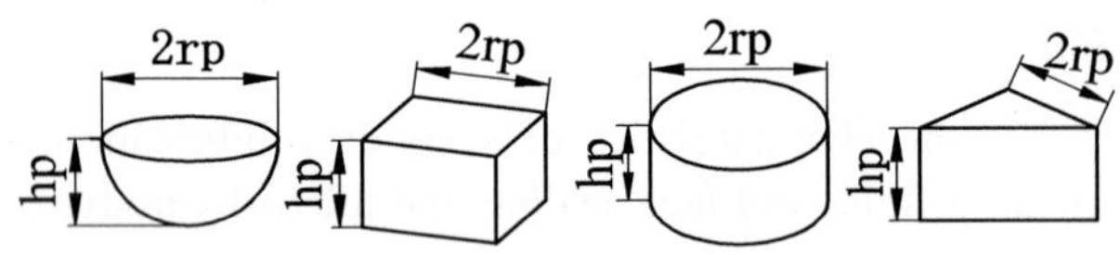

Fig. 2. Sketch maps of the texture morphology.

3. Simulation Parameters

According to the technical demands of water hydraulic, the friction pair clearance of smooth area h_0=5μm, and the document [7] h_p=5μm is selected. In the light of the Eq. (1) and Eq. (2) the formula of concentric ring gap flow respectively, Δp=0.8kPa/μm and Reynolds number Re=12.5 get calculated. Besides l=100μm, L_0=10mm, U=2.5m/s, p_0=0 and p_n=8MPa are all chosen with the Eq. (3), the value of equivalent radius r_p gotten under the different area ratio S, which as shown in Table 1.

$$\frac{\partial p}{\partial x}=\frac{p_n-p_0}{L_0}. \tag{1}$$

$$Re=\frac{U}{\gamma}\delta. \tag{2}$$

$$S=\lambda\frac{r_p^{\;2}}{l^2}. \tag{3}$$

Where, δ for friction pair clearance, $\delta=h_p$, λ for characteristic parameter of texture morphology, γ for kinematic viscosity of water.

Table 1. Value of equivalent radius r_p.

Area ratio	Equivalent radius r_p/ μm			
S/ %	sphere ($\lambda=\pi$)	cylinder ($\lambda=\pi$)	square ($\lambda=4$)	triangle ($\lambda=3^{1/2}$)
9	16.9	16.9	15	22.8
16	22.6	22.6	20	30.4
25	28.2	28.2	25	38
36	33.9	33.9	30	45.6

4. Results and Analysis

4.1. *Bearing force distribution of the texture morphologies*

As shown in Figure 3, Y direction (normal to the motion direction) velocity contour map (sphere, S=36%) is obtained by the Fluent simulation. The more intensive the contour is, the bigger the acceleration. The change of Y velocity is concentrated in texture entrance (divergent wedge) and exported (convergence wedge). Y velocity near entrance is along $+Y$ direction, near export along $-Y$ direction, so when the fluid flowing into convergence wedge produces positive pressure and into divergent wedge produces negative pressure.

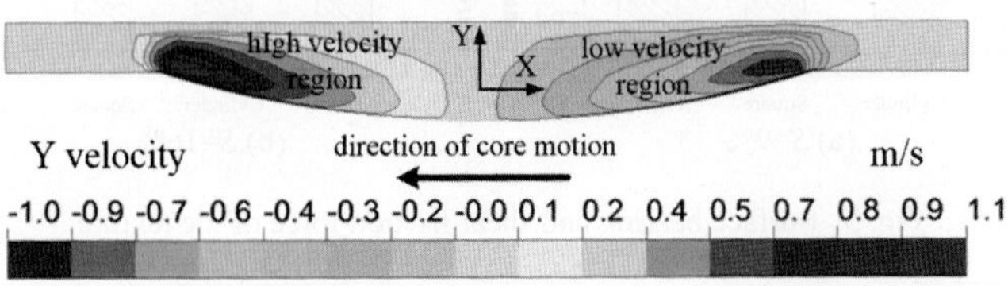

Fig. 3. Y velocity contour map (sphere S=36%).

Pressure distribution curves on each texture morphology center section are shown in Figure 4. The positive and negative pressure peak appears nearby the divergent and convergence wedges respectively. The intersection point coordinates of curves with the x axis are greater than zero for all the morphologies. Their positive pressure areas are significantly greater than the negative and would produce hydrodynamic lubrication effect except for the triangle morphology. That's because its' negative influence area significantly greater than positive area and net pressure value is negative, although the maximum pressure of the triangle morphology is big in the convergence wedge.

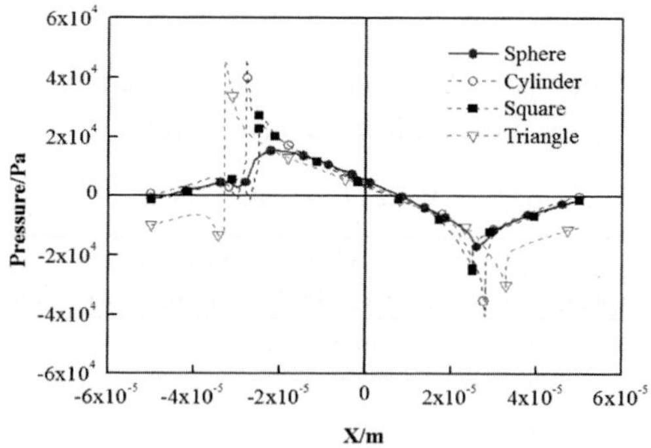

Fig. 4. Pressure distribution curve.

4.2. *The shear friction and bearing force of the texture morphologies*

As shown in Figure 5, integrating the pressure along the x axis to get the bearing force and calculate the shear friction according to the law of Newton internal friction. Figure 5 shows that the bearing forces of all the texture morphologies are greater than zero except triangle morphology.

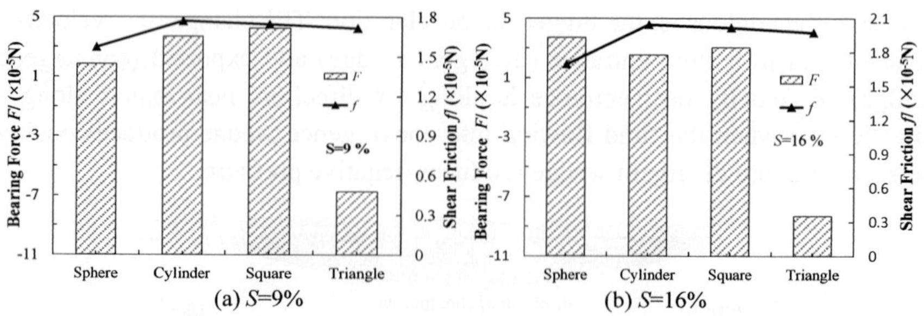

Fig. 5. Surface bearing and shear friction force of the texture.

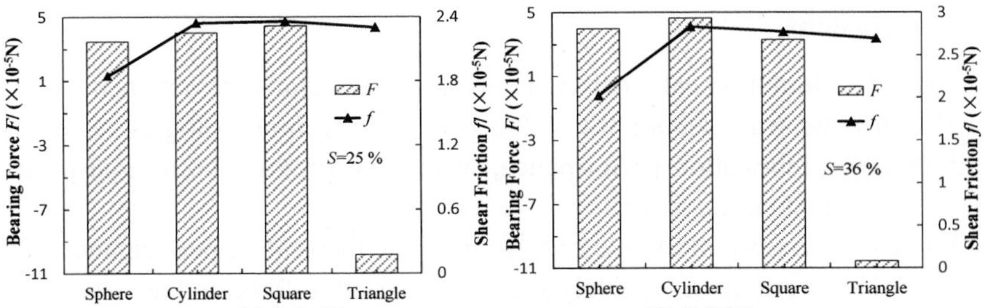

(c) S=25%

(d) S=36%

Fig. 5. (Cont'd).

For the shear friction, sphere morphology is relatively small, the others are approximately equal each other and the advantages of the sphere are becoming ever more obvious with the increase of area ratio S. For example, when S=36 % the friction force of the sphere is 2×10^{-5}N, which is reduced by 35% relative to others. Besides, the shear friction force of the core surface texture has the trend of increasing along with the increase of area ratio S.

4.3. *The leakage of the texture morphologies*

For As shown in Figure 6, the textured valve core does not have the function of hydrodynamic sealing, when the area ratio S = 9%, the leakage of textured core and smooth core are similar. But with the increase of area ratio S, the leakage of texture morphology increases gradually. In addition, sphere morphology has the minimum leakage in all the morphologies, and the advantage is becoming ever more obvious with S increasing. When S =36%, the leakage of sphere morphology gets fell by nearly 7% relative to other morphologies.

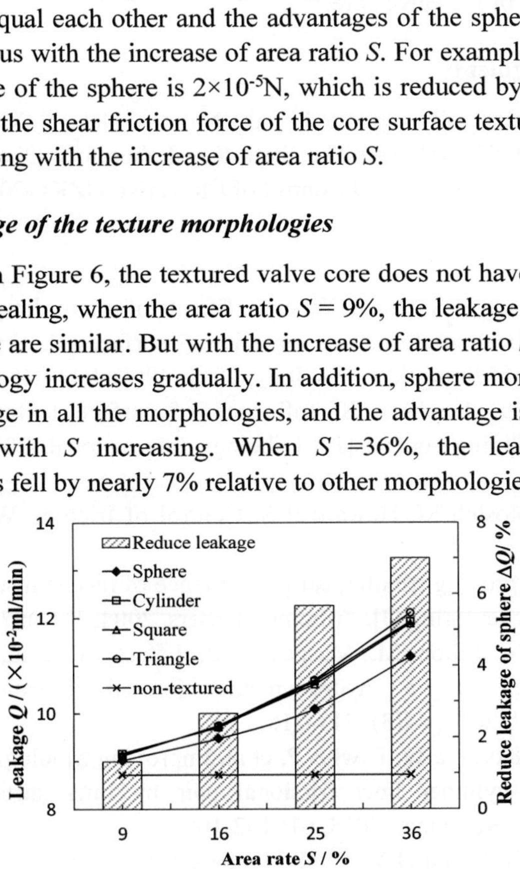

Fig. 6. Leakage rate of the texture.

5. Conclusions

Through the simulation calculation, the follow conclusions are obtained:
(1) Besides the triangle, all the valve core texture morphologies have good hydrodynamic lubrication performance, which can produce both large positive pressure and the bearing force.
(2) Valve core texture morphology can't improve the sealing performance of the core, but the sphere has a relatively small shear friction force and leakage.
(3) Sphere morphology has the advantages of processing convenient and its area ratio can reach 78.5%. So sphere can be selected to improve the lubrication and sealing property of the valve core.

6. Acknowledgments

The work was financially supported by the National Natural Science Foundation of China (No. 51575002) and The State Key Laboratory Open Foundation of Fluent Power Transmission and Control of China (No.GZKF-201205).

References

1. Yu H W, Deng H S, Huang W, et al. Effects of micro-dimple arrangements on tribological performance of sliding surfaces[J]. Journal of China University of Mining & Technology, 2011, 40(6): 943-948. (in Chinese)
2. Pettersson U, Jacobson S. Influence of surface texture on boundary lubricated sliding contacts[J]. Tribology International, 2003, 36(11): 857-864.
3. Suh N P, Mosleh M, Howard P S, Control of friction, Wear, 175 (1994) 151-158.
4. Etsion I. Improving tribological performance of mechanical components by laser surface texturing [J]. Tribology Letters, 2004, 17(4): 733-737.
5. Tala-Ighil N, Fillon M, Maspeyrot P. Effect of textured area on the performances of a hydrodynamic journal bearing[J]. Tribology International, 2011, 44(3): 211-219.
6. Grabon W, Koszela W, Pawlus P, et al. Improving tribological behaviour of piston ring–cylinder liner frictional pair by liner surface texturing[J]. Tribology International, 2013, 61: 102-108.
7. Deng H S, Qi S, Yu H W, et al. Numerical analysis for textured port plate of axial piston pump[J]. Transactions of the Chinese Society for Agricultural Machinery, 2011, 42(6):203-207. (in Chinese)

8. Ge X H, Mo J L, Liu M Q, et al. Suprressing friction-induced squeal noise at interface by surface texturing[J]. China Surface Engineering, 2015, 28(3): 102-107. (in Chinese)
9. Ji J H, Fu Y H, Hua X J, et al. Tribological properties of 45 steel surface with V-grooves[J]. China Surface Engineering, 2014, 27(4): 107-111. (in Chinese)

Microwave absorbing properties of MnAlY magnetic powder

Zhen-Zhong Wang, Pei-Hao Lin [*†], Shun-Kang Pan , Zi-Qiang Qiao and Li-Chun Cheng
School of Materials Science and Engineering,
Guilin University of Electronic Science and Technology,
Guilin 541004,China[†]
E-mail: [†] gllph2002@163.com

Microwave absorbing materials have been widely utilized in areas such as controlling radioactive pollution, civic field, military field, etc. MnAl alloy can be used as microwave absorbent. It has properties such as good thermal stability, light in weight and corrosion resistance. But as permanent magnets, MnAl has shortcomings such as their difficulties of fabrication and high costs that prevent them from being used. In this paper, we will discuss the effects of changing relative content of Y on absorption properties of Al8Mn5 alloy. MnAlY powder was prepared by the process of vacuum levitation melting and high-energy ball milling. The morphology and phase structure of the powder were analyzed by Scanning Electron Microscope (SEM), X-ray diffraction (XRD) and the effect of the Y content on microwave absorbing properties of MnAl powder was investigated by a vector network analyzer. The results show that the powders mainly consist of Al8Mn5 phase with different Y content, the resonant frequency and absorption peak of ε'' and μ'' moves towards lower frequency region simultaneously. The (Al8Mn5)0.98Y0.02 exhibits preferably comprehensive microwave absorbing properties in the range of frequency between 8.5 GHz to 18 GHz. The value of the minimum reflectivity and absorption peak frequency with the coating thickness of 1.4 mm, are -43.4 dB, 12.7 GHz respectively.

Keyword: MnAlY Powder; High-Energy Ball Milling; Absorbing Material; Electromagnetic Parameters; Microwave-Absorbing properties.

1. Introduction

With the rapid development of science and technology, the application of electromagnetic technology has brought material civilization to the

---

[*] Corresponding author

communities. Hence, the microwave absorbing materials becomes the hotspot in recent years [1]. Microwave absorbing materials for controlling the radioactive electromagnetic pollution is a pressing matter at the moment. In the civil field, absorbing material has been widely used in the fields of radio, television, radar, microwave anechoic chamber and electronic devices. In the military field, stealth weapon plays an important strategic role in wars in future [2-7]. In view of this utilization, excellent microwave absorbing materials should exhibit strong microwave absorption properties over a wide frequency range and need to be thin and light in weight. It has an important meaning to the application of magnetic microwave absorbing materials [8-11].

MnAl alloy can be used as microwave absorbent. It has properties such as good thermal stability, light in weight and corrosion resistance [12-14]. But as permanent magnets, MnAl has shortcomings such as their difficulties of fabrication and high costs that prevent them from being used as permanent magnets. A great value would be brought to either theory or practice through proper treatments as further improvement in properties. In this paper, we will discuss the effects of changing relative content of Y on absorption properties of Al_8Mn_5 alloy.

2. Experiments

The samples were prepared by Mn, Al and Y with the purity of no less than 99.50%, according to the formula $(Al_8Mn_5)_{1-x}Y_x$(x=0; 0.005; 0.02; 0.04). The powder was smelted by vacuum are furnace under the protection of high purity argon. The ingots were sealed in a quartz tube, heat treated at 1000 °C for 48 hrs, then quenched into an ice-water mixture. The alloys were mechanically milled for 12 hrs under the protection of ethanol by using a planet ball mill (QM-lSP) with a rotational speed of 300 r/min. The ball to powders weight ratio was fixed at 15:1. The alloys were ground into powders and characterized by X-ray diffraction (XRD, D8-ADVANCE). The morphology was examined by scanning electron microscopy (SEM, JSM-5610-LV). Then prepared AlMnY powder was mixed with paraffin in the volume ratio of 30:70. The mixture was made into a coaxial ring with a thickness of 3.5 mm, of which the outside and inside diameters were 7 mm and 3mm, respectively. Then the complex permeability and the complex permittivity of the mixtures were determined using an Agilent 8722ES vector network analyzer (VNA) in the frequency band of 2–18 GHz.

3. Result and Discussion

3.1. *Analysis of X-ray diffraction (XRD)*

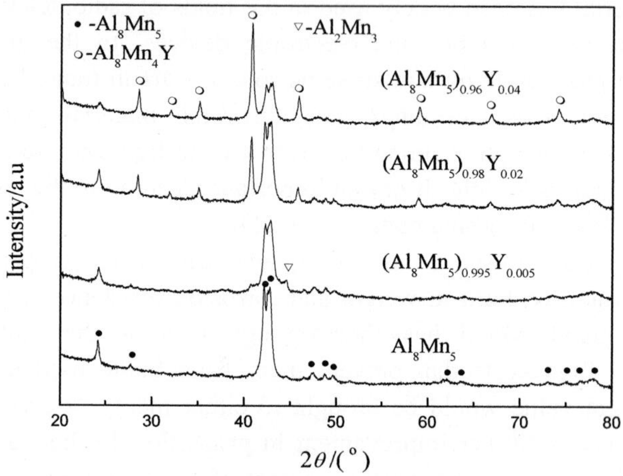

Fig. 1 XRD patterns of powder

The XRD patterns of powders are demonstrated in Fig. 1. It shows that Al_8Mn_5 powder is mainly composed of Al_8Mn_5 phase, The phase compositions of $(Al_8Mn_5)_{0.995}Y_{0.005}$ powders includes Al_8Mn_5 phase besides a spot of Al8Mn4 phase and Al_2Mn_3 phase. $(Al_8Mn_5)_{0.98}Y_{0.02}$ and $(Al_8Mn_5)_{0.96}Y_{0.04}$ powder is mainly composed of Al8Mn5 phase and Al_8Mn_4Y phase. According to Scherrer formula[15], the grain sizes of powder Al_8Mn_5 phase are 26 nm, 23 nm, 25 nm and 26 nm, respectively. The cell parameters and cell volume of Al_8Mn_5 phase are shown in Table 1.

Table. 1 Lattice parameters of powder Al_8Mn_5 main phase

Alloy powder	main phase	cell parameters (nm)			cell volume (nm^3)
		a	b	c	
Al_8Mn_5	Al_8Mn_5	1.266	1.266	1.588	2.2037
$(Al_8Mn_5)_{0.995}Y_{0.005}$	Al_8Mn_5	1.2647	1.265	1.587	2.1986
$(Al_8Mn_5)_{0.98}Y_{0.02}$	Al_8Mn_5	1.2657	1.266	1.585	2.1993
$(Al_8Mn_5)_{0.96}Y_{0.04}$	Al_8Mn_5	1.2638	1.264	1.587	2.1946

It can be observed that with addition of Y, the lattice parameters and cell volume of Al_8Mn_5 phase of powder decrease. The atomic radius of Y (2.27 Å) is greater than the atomic radius of Al (1.82 Å) and Mn (1.79 Å) which indicate that Y is not dissolve into Al_8Mn_5, form a separate Al_8Mn_4Y phase only.

3.2. *Analysis of scanning electron microscopy (SEM)*

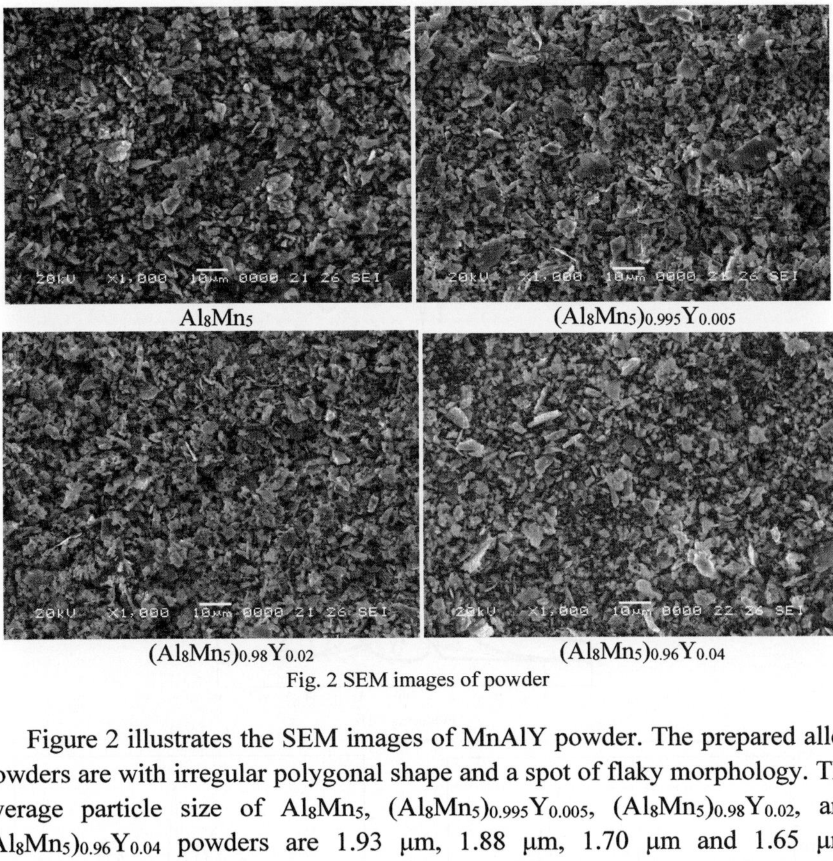

Al_8Mn_5 $(Al_8Mn_5)_{0.995}Y_{0.005}$

$(Al_8Mn_5)_{0.98}Y_{0.02}$ $(Al_8Mn_5)_{0.96}Y_{0.04}$

Fig. 2 SEM images of powder

Figure 2 illustrates the SEM images of MnAlY powder. The prepared alloy powders are with irregular polygonal shape and a spot of flaky morphology. The average particle size of Al_8Mn_5, $(Al_8Mn_5)_{0.995}Y_{0.005}$, $(Al_8Mn_5)_{0.98}Y_{0.02}$, and $(Al_8Mn_5)_{0.96}Y_{0.04}$ powders are 1.93 μm, 1.88 μm, 1.70 μm and 1.65 μm, respectively, measured by microscope. which reveals that with the addition of Y, the average particle size of alloy powders is decrease.

3.3. *Analysis of electromagnetic parameter*

The relative complex permittivity and permeability of the MnAlY were investigated, and the results are shown in Fig. 3, which illustrate the permittivity and permeability as a function of the frequency in the range of 2-18 GHz. It shows that the resonant frequency of Al8Mn5 powder moves towards lower frequency from 12.8GHz to 8.2GHz. It is mainly because of the decrease of grain sizes and average particle size and the increase of crystal defects, which makes the conductivity (σ) of powders decrease. The value of resonant frequency of ε'' is proportional to σ/ε' [16], thus the resonant frequency moves towards lower frequency, with the increase of Y. As Al_8Mn_4Y phase of $(Al_8Mn_5)_{0.96}Y_{0.04}$ increases more dramatically than $(Al_8Mn_5)_{0.98}Y_{0.02}$, which makes the resonant frequency moves towards lower frequency at 8.2 GHz.

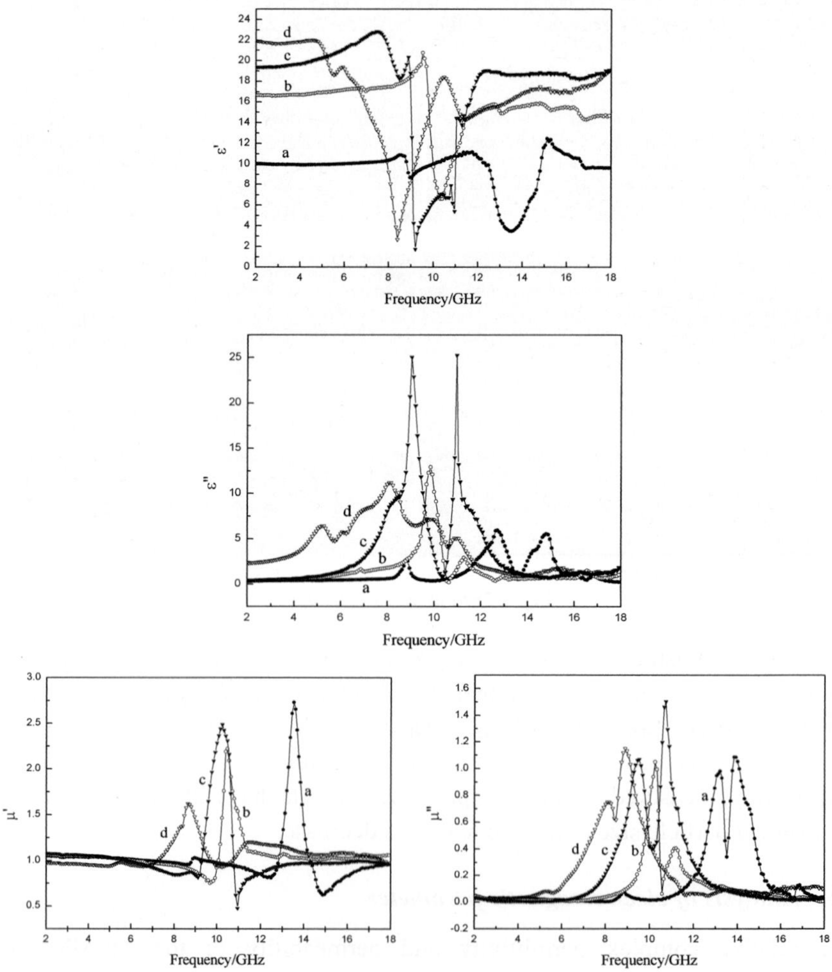

Fig. 3 Electromagnetic parameters of sample
a- Al8Mn5; b- $(Al8Mn5)_{0.995}Y_{0.005}$; c-$(Al8Mn5)_{0.98}Y_{0.02}$; d-$(Al8Mn5)_{0.96}Y_{0.04}$

As seen from the complex permeability curve and the complex permittivity curve, the resonant frequency of ε'' increases with the Y content is the same as the resonant frequency of μ'' increases with the Y content. The μ'' of Al_8Mn_5, $(Al_8Mn_5)_{0.97}Ni_{0.03}$, $(Al_8Mn_5)_{0.95}Ni_{0.05}$ and $(Al_8Mn_5)_{0.91}Ni_{0.09}$ powders at 13.2GHz, 10.2GHz, 9.5GHz, and 8.2GHz shows the resonant frequency, respectively. According to Snoek formula [17]

$$(\mu_s - 1)f_r = \frac{2}{3}\gamma M_s \tag{1}$$

In this formula, f_r; μ_s; γ and M_s are respectively represented as the natural resonance frequency, initial permeability, gyromagnetic ratio and saturation magnetization. With the addition of Y, the saturation magnetization (*Ms*) of the powder was also decreased the same as f_r of powder. The value of μ_s can be approximate to the value of μ' at 2 GHz, as shown in Fig. 3, the value of μ_s changed little with different contents of Y, which make the resonant frequency of imaginary part of complex permeability (μ'') moves towards lower frequency. In this view, with the addition of Y, the resonant frequency moves towards lower frequency.

3.4. *Analysis of microwave absorbing properties*

The reflectivity (R) of the powders were computed from Formula[18]:

$$R = 20\lg\left|\frac{\sqrt{\frac{\mu_r}{\varepsilon_r}}.\tanh(j\frac{2\pi fd}{c}\sqrt{\mu_r\varepsilon_r})-1}{\sqrt{\frac{\mu_r}{\varepsilon_r}}.\tanh(j\frac{2\pi fd}{c}\sqrt{\mu_r\varepsilon_r})+1}\right| \tag{2}$$

where μ_r is the different permeability, ε_r is the relative dielectric constant, d is the thickness of the absorber, f is the frequency of electromagnetic waves, c as the velocity of electromagnetic wave in free space (the speed of light), j is the imaginary unit.

The reflectivity of the $(Al_8Mn_5)_{1-x}Y_x$(x=0, 0.005, 0.02, 0.04) powders are computed from Formula (2) . With different Y content when coating thickness d =1.4 mm, in the frequency ranged from 2 GHz to 18 GHz is demonstrated in Figure 4. The $(Al_8Mn_5)_{1-x}Y_x$(x=0, 0.005, 0.02, 0.04) powders have several absorption peaks in the frequency range of 8.5-18 GHz, and the absorption peaks frequency moves towards lower frequency region. These characteristics are same with the resonant frequency of ε'' and μ'' decrease with the addition of

Y and these behaviors mean that the dielectric loss and magnetic loss affect the reflectivity of the powder. The minimum reflectivity values of Al_8Mn_5, $(Al_8Mn_5)_{0.995}Y_{0.005}$, $(Al_8Mn_5)_{0.98}Y_{0.02}$, $(Al_8Mn_5)_{0.96}Y_{0.04}$ are -13.1 dB, -9.8 dB, -43.4 dB, -8.9dB respectively. The reflection loss of microwave absorption materials achieves 90%.

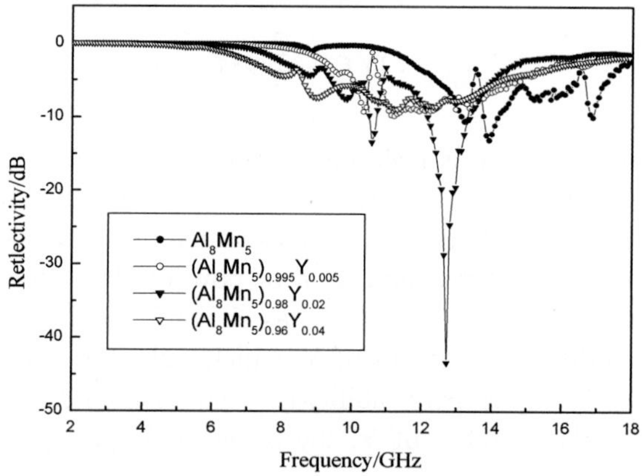

Fig. 4. Reflectivity of MnAlY powder coating

The $(Al_8Mn_5)_{0.98}Y_{0.02}$ powder, have 4 absorption peaks in the range of 9 GHz to 18 GHz regions and the corresponding reflectivity is -43.4 dB, which shows that the absorbing properties can affect through the introduction of Y. This shown the $(Al_8Mn_5)_{0.98}Y_{0.02}$ exhibits preferably comprehensive microwave absorbing properties in the range of 8.5-18 GHz.

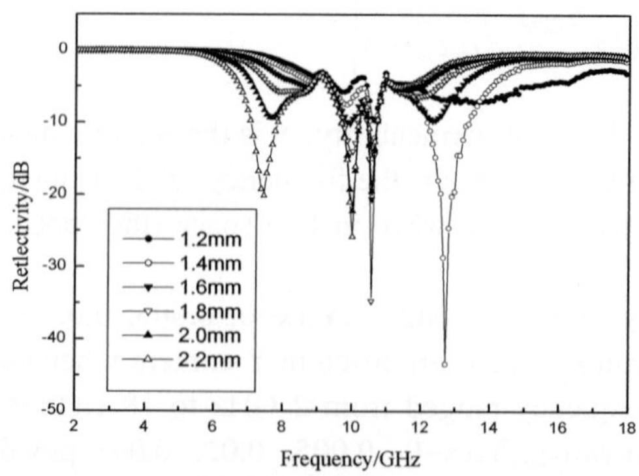

Fig. 5. Reflectivity of (Al8Mn5)0.98Y0.02 powder coating with different thickness

Fig. 5 shows the reflectivity of $(Al_8Mn_5)_{0.98}Y_{0.02}$ powder coating with different thickness in the frequency band of 2 GHz to 18 GHz. As Fig. 5 shows the lowest reflection loss peak of the different coating thickness is with different frequency band. When the thickness of coating increased from 1.2 mm to 2.2 mm, the reflectivity curve of coatings was multiple peak curve in the whole measurement range.

The $(Al_8Mn_5)_{0.98}Y_{0.02}$ power with 1.8 mm thickness has the lowest reflection loss peak, with the value is -34.7 dB (microwave absorption rate 99.966%) in frequency band of 10.3GHz to 11.0 GHz. The $(Al_8Mn_5)_{0.98}Y_{0.02}$ thickness which power is 1.4 mm has the least reflection loss peak is -43.4dB (microwave absorption rate 99.995%) in frequency band of 11.0 GHz to 14.8 GHz. When the coating of $(Al_8Mn_5)_{0.98}Y_{0.02}$ thickness is 1.4 mm at 12.7 GHz, the value of relative input impedance $Z=|\ Z_{in}/Z_0\ |=1$ (Z_0 as air resistance, Z_{in} as input impedance of powder), powder achieved complete loss of electromagnetic waves, while the value of $|\varepsilon_r/\mu_r|$ is 20.17.

From table 2 shows the value of ε'', μ'' and loss factor ($tg\delta$) of Al8Mn5 powders are largest at 12.7 GHz. But as a result of the relationship among the powder of electromagnetic energy electromagnetic loss and ε'', μ'' and $tg\delta$ are proportional [19], so $(Al_8Mn_5)_{0.98}Y_{0.02}$ powder in electromagnetic loss of electromagnetic wave is the largest at 12.7 GHz. Seen from Fig. 4 the $(Al_8Mn_5)_{0.98}Y_{0.02}$ powder with 1.4mm thickness has the best absorbing ability at the 12.7GHz, but the reflection loss of Al_8Mn_5 powder is just -6.1dB (microwave absorption rate 74.453%). In the view of this, the absorbing mechanism of AlMnY powders on the electromagnetic wave related to electromagnetic loss within the absorbing coatings and effect of coating thickness on the interference of electromagnetic wave [20].

Table. 2 Electromagnetic loss parameter in sample at 12.7 GHz

sample	ε''	μ''	tgδ
Al_8Mn_5	5.956	0.596	1.437
$(Al_8Mn_5)_{0.995}Y_{0.005}$	0.53	0.114	0.144
$(Al_8Mn_5)_{0.98}Y_{0.02}$	2.532	0.146	0.29
$(Al_8Mn_5)_{0.96}Y_{0.04}$	1.523	0.051	0.142

4. Conclusions

This work demonstrated that the $(Al_8Mn_5)_{1-x}Y_x$(x=0; 0.005; 0.02; 0.04) alloys with tunable Y content were prepared by arc smelting and high energy ball milling method. Al_8Mn_5 powder is mainly composed of Al_8Mn_5 single phase, The phase compositions of$(Al_8Mn_5)_{0.995}Y_{0.005}$ powders includes Al_8Mn_5 phase besides a spot of Al_8Mn_4 phase and Al_2Mn_3 phase.$(Al_8Mn_5)_{0.98}Y_{0.02}$ and$(Al_8Mn_5)_{0.96}Y_{0.04}$ powder is mainly composed of Al_8Mn_5 phase and Al_8Mn_4Y phase. The lattice parameters, morphology, magnetic properties of the samples were carefully research and discussed. the lattice parameters and cell volume of Al_8Mn_5 phase of powder decreased with the increasing amount of Y substitution. The saturation magnetization decreased as the amount of Y substitution increased. In addition, The $(Al_8Mn_5)_{0.98}Y_{0.02}$ power with 1.8 mm thickness has the lowest reflection loss peak, with the value is -34.7 dB (microwave absorption rate 99.966%) in frequency band of 10.3 GHz to 11.0 GHz. The $(Al_8Mn_5)_{0.98}Y_{0.02}$ thickness which power is 1.4 mm has the least reflection loss peak is -43.4 dB (microwave absorption rate 99.995%) in frequency band of 11.0 GHz to 14.8 GHz. Overall, the absorption properties of MnAlY powder can be affected by electromagnetic loss and the coating thickness.

Acknowledgements

This paper has been made possible by a grant from the National Natural Science Foundation of China (51361007) and Natural Science Foundation of Guangxi (2013GXNSFAA019295).

References

1. Tan Eeling, Wang Zhong. Nano stealth radar absorbing materials[J]. Winged Missiles Journal,2009,3:57-59.
2. Liu Danli, Liu pingan, Yang Qingsong, Tang guowu, Zhao Liying, Zeng fancong. Research Status and prospect of Wave Absorbing Materials[J]. Materials Review, 2013, 17: 74-78.
3. Li Binpeng, Wang Chengguo, Wang Wen. Progress of Electromagnetic Wave Absorbing Materials Based on Carbon[J]. Materials Review, 2012, 26(4): 9.
4. Wang Chen, Gu Jialin, Kang Feiyu. Research Progress in Theoretical Designs of Microwave Absorbing Materials[J]. Materials Review, 2009, 23(5): 5.
5. Tan Hongbin, Ma Xiaoling. Research Progress in Nano Ferrite of Microwave Absorbing Materials[J]. Journal of Ceramics, 2008, 29(1): 77.
6. Li Baodong, Li Qiaoling, Zhang Cunrui. New Research progress of Ferrite Wave Absorbing Composites[J]. Materials Review, 2008, 12: 226.
7. Fazakas E, Varga L K, Mazaleyrat F. Journal of Alloys and Compounds[J], 2007, 434–435: 611.

8. Qiu Qin, Zhang Yanqing, Zhang Xiong. Research progress of microwave absorption materials[J]. Electronic Components and Materials, 2009,28(8):78-81.
9. Kubacki R, Ferenc J, Przesmycki R, et al.The Nanocrystalline FeSiBCuNb Finemet Absorption Properties at Microwaves[J]. IEEE Transactions on Electromagnetic Compatibility, 2012,45(1):93-100.
10. Feng Y B, Qiu T. Preparation, characterization and microwave absorbing properties of FeNi alloy prepared by gas atomization method[J]. Journal of Alloys and Compounds, 513 (2012) :455–459.
11. Jia Haipeng, Su Xunjia, Hou Genliang, et al. Progress in Fabrication and Microwave Absorption Capacity of Graphene-based Magnetic Nanocomposites[J]. Journal of Materials Engineering, 2013,(5):89-93.
12. DENG Longjiang, ZHOU Peiheng, WENG Xiaolong. Functional Materials[J], 2005, 2(4): 8.
13. Obi O, Burns L, Chen Y J.IEEE Transactions on Magnetics [J], 2013, 49(7): 3372.
14. Yang Yingshan, Han Junde. The Magnetic Body Manual[M].Beijing:Metallurgical Industry Press, 1984: 162.
15. Liao Libing,Li Guowu. X-ray diffraction method and application[M].Beijing: geological Press,2008: 113-115.
16. Liao Shaobin. Ferromagnetic science (Next volume) [M].Beijing: Science Press, 1988: 3−88.
17. TANG Liyun;CHI Xiao;WEI Jianqiang;WANG Bochong;WANG Tao;LI Fashen;Electromagnetic parameters and microwave absorption of Fe50Ni5 0/methylmethacrylate composites[J]Journal of Magnetic Materials and De vic-es,2013,6,P:10-12
18. Liao Shaobin, Yin Guangjun. The absorption and reflection of absorbing material on electromagnetic wave[J]. Aerospace Materials and Technology, 1992, 2: 16-20.
19. Liao Shaobin. Ferromagnetic science (Next volume) [M].Beijing: Science Press, 1988: 3−88.
20. Yi Haibo. Microwave absorbing properties of planar rare earth intermetallic compounds powders/paraffin composites [D].Lanzhou: Lanzhou University, 2012.

Chapter 2

Composite and Nano Materials

Composite photocatalyst $CdIn_2S_4$-$In(OH)_3$ reducing carbon dioxide to methyl formate

Xiao Shao[1], Wei-Yue Xin[2], Xiao-Hong Yin[1,*,†] and Shuai-Hua Ma[1]
[1]*Tianjin Key Laboratory of Organic Solar Cells and Photochemical Conversion, School of Chemistry and Chemical Engineering, Tianjin University of Technology, Tianjin, 300384, China*
[†]*E-mail:yinxiaohong@tjut.edu.cn*

[2]*School of Chemical Engineering, East China University of Science and Technology, Shanghai, 200237, China*

The progress of science and technology makes the carbon dioxide emissions increase and result in global warming. Therefore, developing a novel utilization of carbon dioxide has become an urgent issue. In this article, the $CdIn_2S_4$-$In(OH)_3$ catalysts are synthesized by hydrothermal method and used to photocatalytically reduce carbon dioxide. The hydrothermal method was used to synthesize a series of micro spherical $CdIn_2S_4$-$In(OH)_3$ photocatalysts, which were characterized by X-ray diffraction, UV–Vis absorption spectra, transmission electron microscope and N_2 sorption analysis at 77 K. The results showed that the catalyst was hydrothermally synthesized for 32 hours at 140°C, to be endowed with the best activity for photocatalytically reducing carbon dioxide. It also showed that formation rate of methyl formate was higher.

Keywords: Hydrothermal Synthesis; $CdIn_2S_4$-$In(OH)_3$; Photocatalysis; Reduction of CO_2.

1. Introduction

The progress of science and technology makes the carbon dioxide emissions increase, and results in global warming. Therefore, developing novel utilization of carbon dioxide has become an urgent issue[1]. Photocatalytic reduction of CO_2 is a promising long-term solution to CO_2 utilization that employs sunlight as a renewable energy input[2,3]. In 1979, Honda et al has reported that photoreduction CO_2 to organic compound like HCOOH etc.[4]. In recent years, a series of compounds of Indium have been reported, among them $In(OH)_3$ is a semiconductor with a band gap of 5.15eV, which has been applied in photocatalysis[5,6] and solar energy[7]. $CdIn_2S_4$ was also reported due to its

[*] Corresponding author.

narrow band gaps[8]. However, single $CdIn_2S_4$ or $In(OH)_3$ can't satisfy the needs of reduction of CO_2. In this article, the $CdIn_2S_4$-$In(OH)_3$ catalysts are synthesized by hydrothermal method and used to photocatalytically reduce carbon dioxide.

2. Experimental Section

2.1. *Preparation of photocatalysts*

The synthesis process of the $CdIn_2S_4$-$In(OH)_3$ is described as follows: a set of $InCl_3 \cdot 4.5H_2O$, $CdCl_3 \cdot 2.5H_2O$, thiourea, which were analytical reagents and purchased from Tianjin Benchmark chemical Reagent Company, were dissolved in 55mL deionized water and magnetically stirred 15 minutes to obtain a yellow liquid. Then we poured the liquid to the 75ml autoclave with a Teflon liner, The autoclave was maintained at 140°C for 16h, 32h, 48h, 75h, and then cooled down to room temperature naturally. The obtained precipitates were collected by centrifugal, washed with distilled water and ethanol for several times, and dried under vacuum at 80°C over night

2.2. *Characterization of catalysts*

The as-prepared samples were characterized by X-ray powder diffractometer with Cu Kα radiation at a scan rate of $2min^{-1}$. The morphologies and sizes of the products were identified by field emission scanning electron microscope (FE-SEM, JEOL-JSM6700F) and transmission electron microscope (TEM, G2F20). The BET of APP V-Sorb to measured the specific surface areas, and spectra of UV-vis were received by Shimadzu UV-2550 and using $BaSO_4$ as a reflectance standard.

2.3. *Evaluations of photocatalytic activities*

The photocatalytic reductions of carbon dioxide were evaluated in a batch slurry reactor made of quartz. 0.02g photocatalyst and 20mL methanol were put into the reactor, the cooling water was passed through the jacket outside to keep the reaction temperature constant. Carbon dioxide was bubbled into the reactor at 200mL/min for 30 minutes to replace the air. Subsequently the reactor was sealed and a 250W high-pressure mercury lamp with an intensity of $2500\mu W/cm^2$ on the liquid surface was used to irradiate, after 10h, we analyzed the products with gas chromatography of SCION 456-GC equipped with a hydrogen flame detector (FID). Meanwhile, for comparison, the experiment was also carried out by irradiating the methanol without photocatalyst, or with photocatalyst in dark, the products were undetectable. The blank test implied that the reaction depended on catalyst and light irradiation.

3. Results and Discussion

Figure 1 shows the XRD patterns of the as-obtained samples of $CdIn_2S_4$-$In(OH)_3$ for 16h, 32h, 48h, 75h at 140°C. At short reaction time, Peak of cubic $In(OH)_3$ is obvious, which is good agreement with card JPCDS-16-0161, when 2θ is 22.262 and 31.680, which response to the (200) and (220) facets, but the cubic $CdIn_2S_4$ is weak. With the time of hydrothermal synthesis prolonged, the peak of $CdIn_2S_4$ became more and more stranger, and at 14.136°(111)and 23.180°(220) had a sharp peak, which accordance with the card JPCDS-27-0060, and indicated that L - cysteine as sulfur source could strengthen the combination ability of In^+ and OH^- and $In(OH)_3$ could be rapidly generated in appropriate temperature, while the S could combine with In^+ to increase the percent of $CdIn_2S_4$.

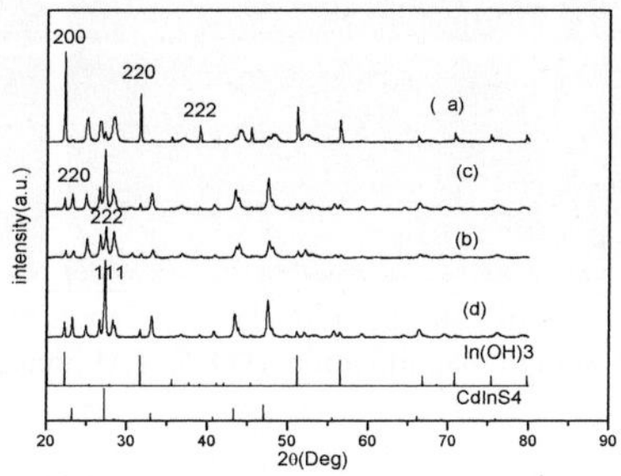

Fig. 1. The XRD pattern of $CdIn_2S_4$-$In(OH)_3$ of 16h(a), 32h(b), 48h(c), 75h(d) at 140°C

In Figure 2 the morphologies and microstructure of the samples were characterized by SEM and TEM. At different hydrothermal time, the samples showed different crystallinities. With the increase of time, particles in the sample agglomerated much seriously. Two photos in Figure 2a showed the bad crystallinity of particles for hydrothermal synthesis of 16 h, where the spherical particles was unfinished product with mean size 7.858 μm. At 32 h(Fig 2b) the spherical aggregate better, the surface was more solid, the average particle size is about 5.611 μm and specific surface area is larger, as revealed at Table 2. It can provide more active site, so the activity also higher than other groups. At 48h the radius of samples about 4μm, and the time longer, the specific surface area of the sample smaller, surface structure also has a bigger difference.

The BET specific surface areas of the $CdIn_2S_4$-$In(OH)_3$ were listed in Table 1. With the increase of hydrothermal synthesis time, the specific surface area became more and more small. At 16h (Figure 2a.) there has some flakes in the samples, and the ball is loose. At 32h (Figure 2b.) the ball tights together and becomes perfect, As time goes on, it tight together heavily.

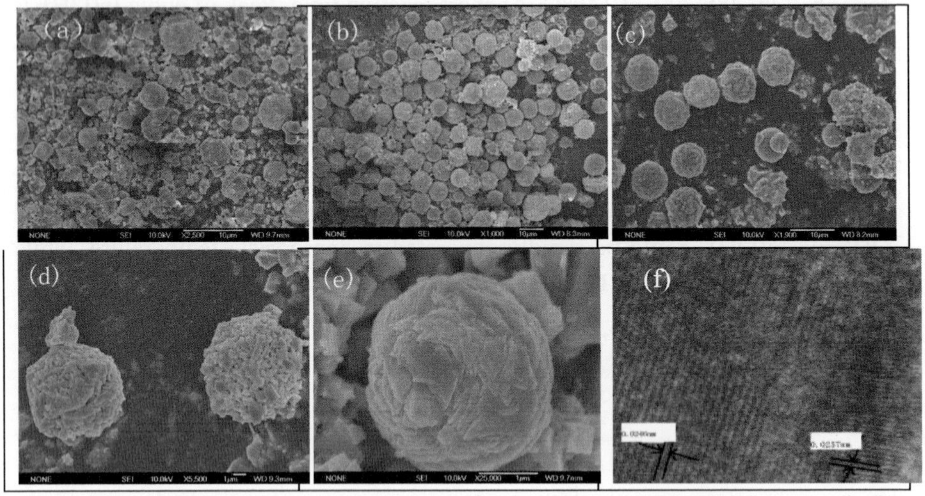

Fig. 2. SEM and HRTEM images of $CdIn_2S_4$- $In(OH)_3$ at different hydrothermal time: (a) 16h, (b) 32h, (c) 48h, (d) 75h, (e) 32h at high magnification, (f) HRTEM images of $CdIn_2S_4$- $In(OH)_3$ at 32h hydrothermal time

Table 1. The specific surface areas of the $CdIn_2S_4$-$In(OH)_3$ synthesized at 140°C

Photocatalyst $CdIn_2S_4$-$In(OH)_3$	16h	32h	48h	75h
Specific surface area (m^2/g)	32. 37	23. 79	16. 25	14. 37

Figure 3 is the UV-vis diffuse reflection spectra of the catalysts. It can be seen that the $CdIn_2S_4$-$In(OH)_3$ has a very strong absorption ability to visible and ultraviolet light. At 140°C, 16h, 32h, 48h, 75h, the absorption edges is 603nm, 760nm, 631nm, 611nm, respectively. Using the formula $Eg=1240/\lambda$calculate the band gap value is 2.06ev, 1.63ev, 1.97ev, 2.03ev. From this we can see the composite material is a kind of photocatalyst with wide spectral response. At 140°C, 32h it has the highest catalytic activity, then with the extension of the crystallization time, the catalyst activity sharply reduce, this may because at 140°C, 32h it has the complete crystal shape, longer reaction time can make the particles together, which reduces the catalytic activity.

Figure 4 is the photocatalytic activity, we can see that with the reaction time increase, the activity is increased after decreased, that is because different reaction time, compounds in the sample proportion is different, at140°C, 32h to achieve the best proportion. Activity is best, and good crystal shape, specific surface area is larger and more active site. After 32 h, as time increases, the sample composition changes, crystal shape changes, it affects the activity of the sample, resulting in a decrease of active site, the activity reduce.

In this experiment, we use methanol as sacrificial reagent and solvent to separate the hole-election pairs.

In order to know the mechanism of the catalytic, the blank experiments were carried out. we use the Ar instead of the CO_2, and then we do it without CO_2, without catalysts and without light. We can see that the product (MF) is even nothing.

All the results showed that only two materials of the products, the difference is the percentage. We guess that due to the narrow band gap of 1.63eV, it is more active to reducing CO_2 to formic acid. Meanwhile, some of formic acid come from methanol oxidation. All of them react with methanol to pruduce Methyl formate(MF). In other conditions, the band gap were much wider, it is not enough to reduce CO_2, most of the formic acid may come from oxidation, so the producion of MF is less. At the same time, large surface area and the absorption of the light are attribute to the photocatalytic activity.

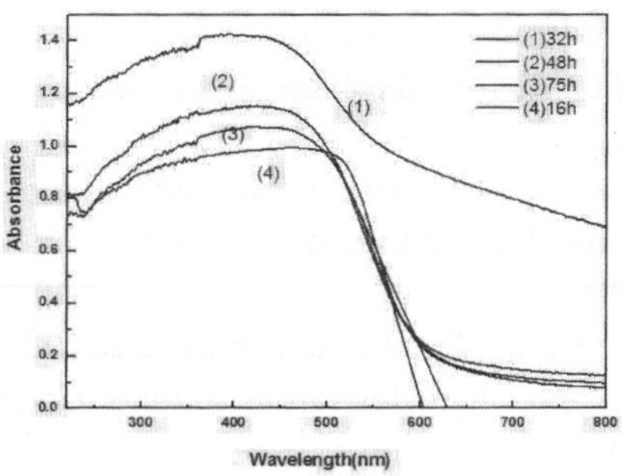

Fig. 3. the UV-vis diffuse reflection spectra of the catalysts at140°C, 16h,32h,48h,75h.

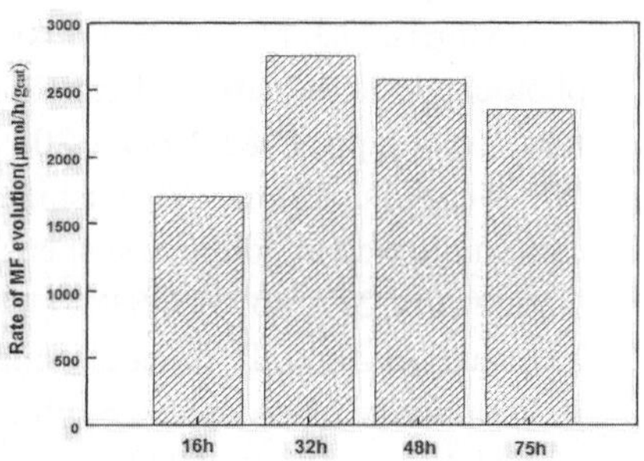

Fig. 4. Photocatalytic activities of $CdIn_2S_4$ $In(OH)_3$ at different hydrothermal time for CO_2 reduction

4. Conclusion

Microspheric $CdIn_2S_4$-$In(OH)_3$ catalyst were prepared by hydrothermal synthesis, whose hydrothermal time and temperature could result in different band gaps of semiconductors. At 140°C, 32h, the perfect spherical morphology, large surface area and more activity site to absorb the light contributed to the highest production of the MF. However, the formic acid comes from CO_2 reduction and methanol oxidation, we need to do further research to know the mechanism of the formation of MF.

Acknowledgements

We gratefully acknowledge financial support by the National Natural Science Foundation of China(NSFC)(NO.21176192) and the Tianjin natural science foundation(NO.12JCZDJC294000).

References

1. C.Liu, K. DDubois, M. ELouis, et al, Photocatalytic CO_2 Reduction and Surface Immobilization of a Tricarbonyl Re(I) Compound Modified with Amide Groups, ACS Catal Vol.3, pp. 655–66, (2013).
2. K. B, Llorente M, Froehlich J, et al, Photochemical and Photoelectrochemical Reduction of CO_2, Annual Review of Physical Chemistry ,Vol.63, pp. 541–569, (2012).
3. A.J. Morris, G.J. Meyer, E. Fujita, Molecular approaches to the photocatalytic reduction of carbon dioxide for solar fuels, Acc. Chem. Res ,Vol.42, pp. 1983-1994, (2009).
4. T.Inoue, A.Fujishima, S.Konishi, et al, Photoelectrocatalytic reduction of carbon dioxide in aqueous suspensions of semiconductor powders, Nature ,Vol.277, pp. 637-638 , (1979).
5. XH Liu, LB Zhou, R Yi, et al, Single-Crystalline Indium Hydroxide and Indium Oxide Microcubes: Synthesis and Characterization, *J. Phys. Chem. C* ,Vol.47, pp. 18426–18430, (2008).
6. ZB Lei, GJ Ma, MY Liu, et al. Sulfur-substituted and zinc-doped $In(OH)_3$:a new class of catalyst for photocatalytic H_2 production from water under visible light illumination. Journal of Catalysis, 237 (2006).

7. Bayón R, Maffiotte C, Herrero J, Chemical bath deposition of indium hydroxy sulphide thin films: process and XPS characterization, Thin Solid Films ,Vol.353, pp. 100–107, (1999).
8. WL Jiang, XH Yin, F Xin, Preparation of $CdIn_2S_4$ microspheres and application for photocatalytic reduction of carbon dioxide, Applied Surface Science ,Vol.288, pp. 138-142, (2014).

Comparison of the reinforcing efficiency of bis-(3-triethoxysilylpropyl)tetrasulfane, bis-(3-triethoxysilylpropyl)disulfane and 3-propionylthio-1-propyl trimethoxylsilane for NR/silica composites

Hua-Long Peng,Ying-Ying Wang, Xiao-Ran Xie, Jian-Guo Liu and Yuan Zhu
Jiangsu qixiang high-tech materials co., LTD, Jiangsu huaian salinization new material industrial park Lee Bay Road 19
Huaian, Jiangsu, China
E-mail: penghualong@cheeshine.com

Bifunctional Sulfur-containing silane such as bis-(3-triethoxysilylpropyl)tetrasulfane (TESPT), bis-(3-triethoxysilylpropyl)disulfane (TESPD) can chemically react with both silica and rubber to obtain significant reinforcement for vulcanizate. A novel silane-3-propionylthio-1-propyl trimethoxylsilane (PXT) synthesized by us was used to compare with TESPT and TESPD in modifying the surface speciality of silica and reinforcing natural rubber(NR)/silica composites. The effect of Sulfur-containing silane on the surface characteristic and morphology of silica were studied. Results show PXT gets the best comprehensive dynamic and abrasion properties of all.

Keywords: Sulfur-containing Silane; Surface Modification; Abrasion; 3-propionylthio-1-propyltrimethoxylsilane; Dynamic Mechanical Properties.

1. Introduction

Silica is one of the most important inorganic filler applied to reinforce tire tread to improve their mechanical properties[1-3] and dynamic mechanical properties such as rolling resistance, wear resistance and wet traction. However silanol (Si-OH) groups on silica's surface leads to the poor rubber-filler interaction and strong particle-particle interaction giving rise to high tendency for filler agglomeration in the rubber matrix[4]. Sulfur-containing silane coupling agents can greatly improve reinforcing efficiency of silica by enhancing the rubber-filler interaction via chemical linkage[5]. They possess two functionally active end groups such as alkoxy group and sulfur-containing group. The alkoxy group (such as $-OCH_3$ or $-OC_2H_5$) will react with silanols on the surface of silica to form stable siloxane linkage. The sulfur-containing group can participate in vulcanization leading to chemical linkage with the rubber. The sulfur-containing silanes have strong coupling between rubber and silica and high reinforcing efficiency for NR/inorganic filler composites.

In this study, a novel blocked mercaptosilane-PXT synthesied by us was utilized to compare with TESPT and TESPD for investigating the surface modification of silica and properties of NR/silica composites. The chemical structure of sulfur-containing silanes are given in Figure 1. It would be expected that PXT can present superior modifying efficiency because of its small molecular structure and more reactive methoxyl group.

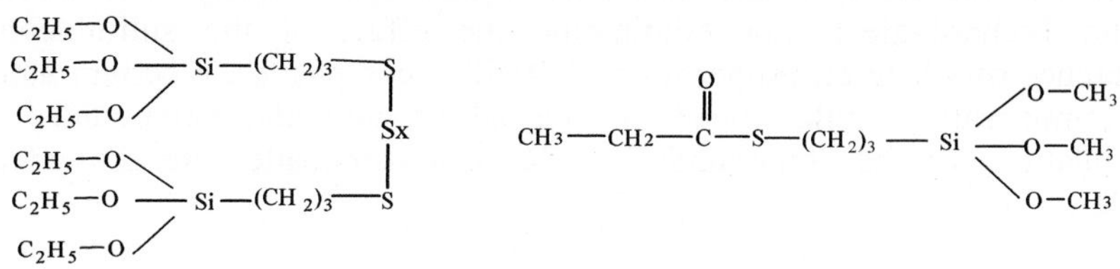

(a)(It'sTESPT as x=2 and it'sTESPD as x=0) (b)

Fig. 1. The chemical structures of TESPT,TESPD (a) and PXT (b)

2. Experimental part

2.1. *Materials*

Natural rubber ISNR-3 was used; Precipitated silica with specific surface area of 140 m^2/g was purchased from Jiangxi huiming Co.Ltd. China. TESPT and TESPD was purchased from Nanjing Shuguang chemical Group. PXT was supplied by our laboratory; Other materials were of industrial grade.

2.2. *Modification of silica with sulfur-containing silane*

The modification of silica was carried out in a three-neck flask equipped an overhead stirrer a thermometer. Precipitated silica, silane and drops of acid were added into flask to react in alcohol solvent at a designated temperature for a certain time. After the reaction, the modified silica was filtered and extracted by alcohol for 48h, and dried to constant weight.

2.3. *Characterization of neat and modified silica*

The particle size and distribution of unmodified and modified silica solution were characterized by a laser particle analyzer LS13 320 of Beckman Coulter. The refraction index of silica was 1.456. Alcohol was used as the flowing liquid. The carbon and hydrogen element content of silica were analyzed by ELEMENTAR vario elemental analysis instrument (German). The silica morphology was investigated by a JEOL JEM-100SX Transmission electron microscope using an acceleration voltage of 80 kV.

2.4. *Preparation of NR/silica composites*

NR, precipitated silica and others ingredients were mixed for 15min at room temperature in an φ160*330 mm open mill. The compounds were vulcanized in an electrically heated press at 143 °C for the optimum cure time (t_{90}), which was previously determined with rubber process analyzer (RPA 2000, Alpha Technologies). For eliminating the effect of the sulfur content difference of silane on properties of NR/silica composites, a certain amount of compensatory sulfur needs to be added into the compounds. The formulations of the compounds of NR/silica composites are presented in Table 1.

Table 1. The formulations of compounds of NR/silica composites.

Compounds (phr)	1#	2#	3#	4#	5#	6#
NR	100	100	100	100	100	100
Silica	30	30	30	30	30	30
Zinc oxide	5	5	5	5	5	5
Stearic acid	2	2	2	2	2	2
Antioxidant 4010NA	1.5	1.5	1.5	1.5	1.5	1.5
Accelerator CBS	1.8	1.8	1.8	1.8	1.8	1.8
Accelerator DPG	0.5	0.5	0.5	0.5	0.5	0.5
PXT	0	2	0	0	2	0
TESPT	0	0	2	0	0	0
TESPD	0	0	0	2	0	2
S	1.6	1.6	1.6	1.6	1.6+0.22	1.6+0.21

phr: parts of per hundred rubber

2.5. *The measurement of abrasion property and dynamic mechanical properties*

Akron abrasion loss was evaluated by a Gao-Tech abraser according to BS 903/A9-1988. The temperature dependence of Tanδ for NR/silica composites was determined by rubber process analyzer (RPA 2000, Alpha Technologies). The condition of measurement is at a frequency of 10Hz, a strain of 14% and a temperature range from 50 to 120°C. The dynamic mechanical spectra of the samples were obtained by using TA 2890 (TA Instrument USA) dynamic mechanical analyzer (DMA). The specimens were analyzed in tensile mode at a constant frequency of 3.3 Hz, a strain of 0.1%, and a temperature range from -90 to 50°C at a heating rate of 5 °C /min. Mechanical loss factor (tanδ) was measured as a function of temperature for the samples under identical conditions.

3. Results and discussions

3.1. *Particle size and distribution of silica*

As shown in Figure 2, all the four curves of particle size and distribution have two peaks. The one below 1 μm belongs to the individual silica particles, the other between 1.5 and 11μm is ascribed to the silica aggregates. By the surface modification with sulfur-containing silane, the volume ratio of single particles increases while the content and size of aggregates reduce. The mean size of neat and TESPT, TESPD and PXT modified silica in alcohol solvent are 3.686 μm, 3.366 μm, 3.316 μm and 3.263 μm respectively, suggesting that PXT possess superior modifying efficiency.

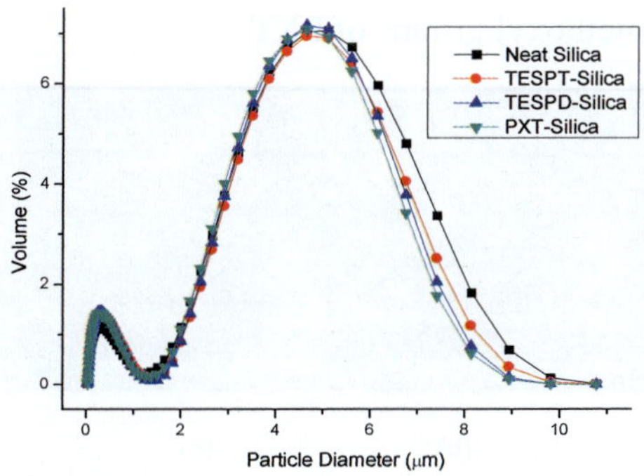

Fig. 2. Particle size and distribution of neat and three kinds of modified silica

3.2. *Element content of silica*

Table 2. Element content of carbon and hydrogen of silica

Symstem	Neat-Silica	TESPT-Silica	TESPD-Silica	PXT-Silica
Element content of Carbon(%)	0.48	0.63	1.05	1.26
Element content of Hydrogen(%)	1.10	1.11	1.15	1.19

As presented in Table 2, After the modification, the content of carbon and hydrogen on silica improve due to the organic coupling agent coating. The surface of PXT modified silica contains the most content of carbon and hydrogen element among three kinds of silanes indicating the most pronounced modifying effect of PXT. It is attributed to that it's easier for the methoxyl groups on PXT to alcoholysis and react with hydroxyl group on silica than the ethoxyl groups on TESPT and TESPD do.The other reason is TESPT and TESPD have too much steric hindrance to react efficiently with silica owing to their bigger structures.

3.3. *Morphology of silica*

TEM photomicrograph of neat silica is shown in Figure 3(a) and Figure 4(a), neat silica particles aggregate together easily. The modification by silane coupling agent reduces the agglomeration and improve the separation and distribution. From the high magnification photo of silica, it can be seen that the size of silica aggregates gets smaller. Compared with TESPT and TESPD, PXT exhibits more effective modification and gets more even separation and distribution of particles. This can be also attributable to the small structure and more reactive methoxyl groups of PXT.

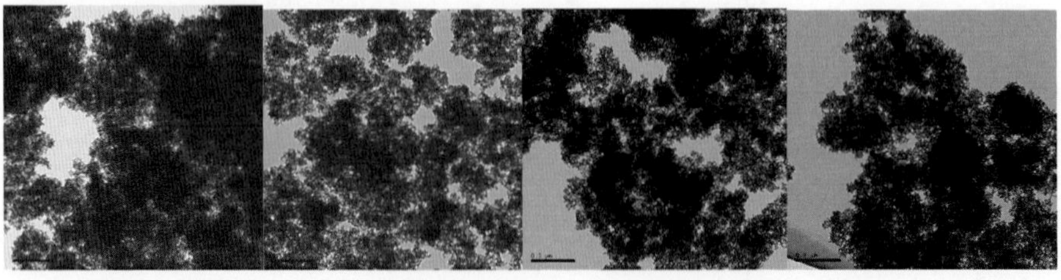

(a) (b) (c) (d)

Fig. 3. TEM photomicrograph of four kinds of silica at low magnification: (a) neat silica; (b) PXT modified silica; (c) TESPT modified silica; (d) TESPD modified silica

(a) (b) (c) (d)

Fig. 4. TEM photomicrograph of four kinds of silica at high magnification: (a) neat silica; (b) PXT modified silica; (c) TESPT modified silica; (d) TESPD modified silica

3.4. *RPA temperature sweep of NR/silica composites*

Shown in Figure 5 is plots of loss factor versus Temperature for NR/silica composites. With the increase of temperature, Tanδ drops gradually. This can be explained by rubber shell theory that silica can absorb layers of rubber to form rubber shell aroud the particles because of the large specific surface area of nano-scale silica particles. The molecular chains in rubber shell are strongly absorbed by fillers leading into its poor mobility and being still in glass transition zone, which enhances the whole composite's Tanδ. The temperature rise makes moieties of the molecular chains in rubber shell more feasible resulting in the drop of loss factor of material. It can be also seen from Figure 5 that silane-modified and sulfur-added composites possess lower Tanδ than unmodified material does. The loss factor of composite depends on filler networks, the mobility of molecular chains and the friction between filler and chains[6].

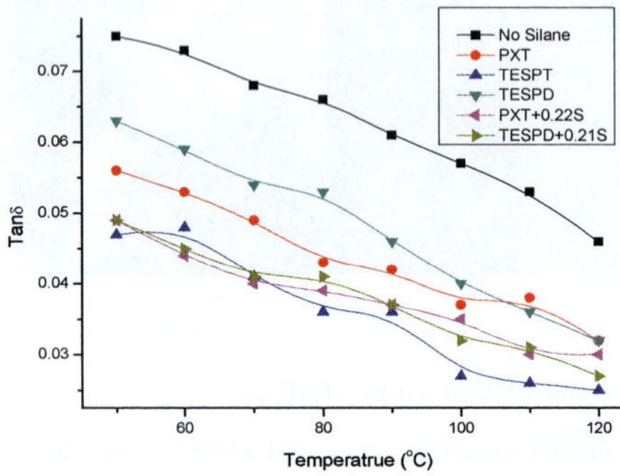

Fig. 5. Plots Tanδ of versus Temperature for NR/silica composites

3.5. *Dynamic mechanical analysis of NR/silica composites*

An ideal material that meet the requirements of a high performance tire should get a lower Tanδ at 60°C and a higher Tanδ at 0 °C in order to reduce rolling resistance and to obtain high skid resistance. Figure 6 and Table 3 shows that the glass transition temperature (T_g) of NR/silica composites rise with addtion of sulfur-containing silane. This is attributed to that the bifunctional groups of three kinds of sulfur-containing silanes can interact with silica and rubber to increase the content of bound rubber on the silica surface whose tg is higher than polymer matrix. Figure 6 and Table 3 also show that the Tanδ peak of composite with PXT and 0.22phr compensatory sulfur is the higest. This can be explained by that the bound rubber chains absorbed by silica is still in glass transition region while the others rubber stay in elastic state above Tg. PXT enhances the filler-matrix interaction and compensatory sulfur improves the crosslinking density. These astrict the mobility of rubber chains and make more rubber stay in glass transition area resulting in the increase of Tanδ. As presented in Table 3, all the sulfur-containing silane can enhance Tanδ at 0 °C and reduce Tanδ at 60 °C to obtain high skid resistance and low rolling resistance. And NR/silica composite with PXT gets better comprehensive dynamic properties than others.

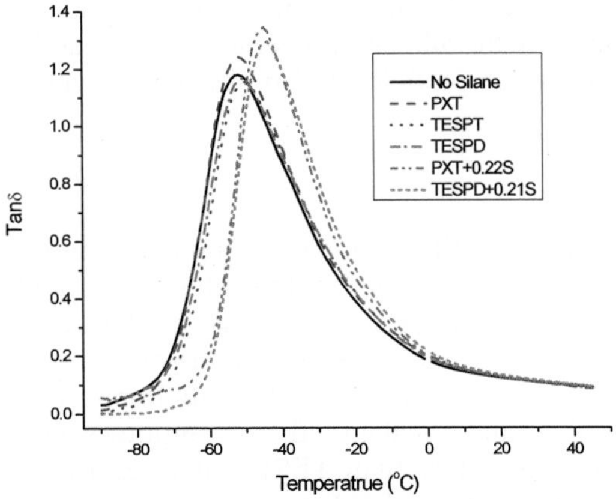

Fig. 6. DMA curve of NR/silica composites

Table 3 Tg and Tan δ at Tg, 0°C and 60°C of NR/silica composites

System	No Silane	PXT	TESPT	TESPD	PXT+0.22S	TESPD+0.21S
Tg(°C)	-52.3	-52.0	-50.4	-50.8	-45.1	-44.2
Tanδ at Tg	1.178	1.242	1.156	1.166	1.345	1.295
Tanδ at 0°C	0.183	0.198	0.196	0.194	0.202	0.216
Tanδ at 60°C*	0.073	0.053	0.048	0.059	0.044	0.045

* Tanδ at 60°C was gained from Figure 5.

3.6. *Abrasion property of NR/silica composites*

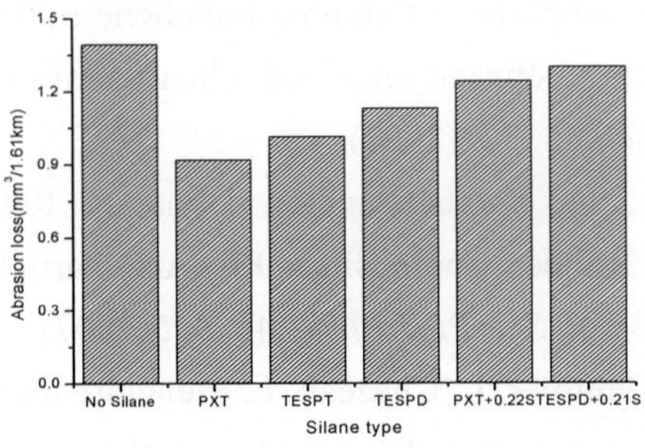

Fig. 7. Effect of Silane type on abrasion loss of NR/silica composites.

The abrasion property is one of the most important properties for the lifetime of tire tread. As can be seen from Figure 7, three kinds of sulfur-containing silane coupling agent can improve the abrasion resistance much. NR/silica composite with PXT is most of all for abrasion resistance. It is attributed to that PXT have the more efficiency to modify the silica surface than the other two. Thus, it react with rubber and silica more readily to reinforce the interaction between silica particles and rubber chains and reduce the volume of abrasion loss.

4. Conclusion

The effect of sulfur-containing silane coupling agent on surface speciality of silica and properties of NR/silica composites were studied. Particle size determination, element analysis and TEM photomicrograph shows that sulfur-containing silane can easily react with the hydroxyl group on surface of silica particles. Three kinds of silane can all improve the separation and distribution of silica and weaken silica's agglomeration. PXT possesses superior modifying efficiency than TESPT and TESPD do. It is concluded from the determination of dynamic mechanical properties that sulfur-containing silane can strengthen the reinforcement of silica for NR composites to improve the skid resistance and to reduce the rolling resistance. PXT gets the optimum comprehensive dynamic mechanical properties of all. For the abrasion propeties, three kinds of silane all improve the abrasion resistance and PXT is better than others.

References

1. P. Sae-oui, C. Sirisinha, U. Thepsuwan and K. Hatthapanit, Influence of silica loading on the mechanical properties and resistance to oil and thermal aging of CR/NR blends, *J. Appl. Polym. Sci*104, 3478(2007).
2. Ansarifar, L.Wang, R.J. Ellis, S.P. Kirtley and N. Riyazuddin, Enhancing the mechanical properties of styrene-butadiene rubber by optimizing the chemical bonding between silanized silica nanofiller and the rubber, *J. Appl. Polym. Sci*105, 322(2007).
3. S. Thongsang, N. Sombatsompop, Dynamic Rebound Behavior of Silica/Natural Rubber Composites: Fly Ash Particles and Precipitated Silica,*J. Macromol. Sci. Pt B:Phys.* ,46, 825(2007)
4. J. Ramier, C. Gauthier, L. Chazeau, L. Stelandre and L.Guy,Payne effect in silica-filled styrene-butadiene rubber: influence of surface treatment, *J. Polym. Sci.: Pt B: Polym. Phys.* , 45, 286(2007).

5. P. Sae-oui, C. Sirisinha, U. Thepsuwan and K. Hatthapanit, Comparison of reinforcing efficiency between Si-69 and Si-264 in a conventional vulcanization system., *Polym. Test.* 23,871(2004)
6. B. Maridass and B.R. Gupta,Effect of carbon black on devulcanized ground rubber tire-natural rubber vulcanizates: cure characteristics and mechanical properties, *J. Elastom. Plast.* ,38,211(2006).

Experimental study on enhancing foam stability with nanoparticles

Ai-RongWang[*,†]
Faculty of Energy Resource of Chengdu University of Technology
Chengdu, Sichuan, China
[†]*E-mail: 4263834@qq.com*

Fei Liu, Xin-Feng Li
PetroChinaLiaohe Oilfield, Panjin,
Liaoning, China
E-mail: liufei412@163.com

Guo-Qiang Ouyang, HaoWang
PetroChinaTarim Oilfield,Kuerle,
Xinjiang,China

Nanomaterial is developed through the liquid phase reaction, with an average grain size of 29.3nm and near-neutral surface wettability. As a foam stabilizer, this nanomaterial, whose optimal concentration is 0.08%, works synergistically with PAM and improves foam stability. It is also called "ultrafine particle material" and is formed by nanoparticles. From the results observed, the synergy between this nanomaterial as a foam stabilizer and PAM in stabilizing foams is great, and can assist to improve foam stability. Moreover, in the dynamic water drainage test using foams, the liquid-carrying capability of foams significantly improves with nanoparticles.

Keywords: Nanoparticle; Foam; Stability; Liquid-carrying Capability.

1. Introduction

The nanomaterial, also called ultrafine particle material, is formed by nanoparticles. The nanoparticle generally refers to the particles with a grain size of 1~100nm. It can be hydrophilic or hydrophobic, and is able to form the protective layer at oil-water or oil-gas interface, which is similar to surfactant. The nanoparticle with appropriate surface wettability can adhere to the oil-water and air-water interface. That is to say, it has surface activities [1-2]. The surface wettability of particles, namely the contact angle at the fluid interface, θ (usually refers to the contact angle measured with water), is an important parameter representing the surface activity of particles, like the hydrophile-lipophile balance value of surfactant.

[*] Corresponding author

Similar to adhering of surface-active nanoparticles to the oil-water interface of stabilized emulsions, the surface-active nanoparticle can also stay at the gas-water interface and therefore stabilize the foam. Research shows foam stabilization with solid particles has two mechanisms [3-8]. One is that particles are fully dispersed in water due to no affinity between them and liquid-gas interfaces, and gather at the balance boundary after foams are formed, which hence decrease the foam coalescence. However this kind of foam stabilization is limited. The other one is that the particle itself is surface-active and gathers at the liquid-gas interface. This foam system is well stabilized. When the water contact angle is 90°, namely half of the particle is in the liquid phase and the other half in gas, foam reaches the best stabilization.

2. Development and performance evaluation of nanoparticles

2.1. *Development of nanoparticles*

The raw material of the nanomaterial is shown in Table 1.

Table 1. Experimental material and chemical agent

Item Name	Purity	Manufacturer
Raw nanomaterial(TEOS)	Chemical reagent	Chengdu CR & G
Surfactant(CTAB)	Analytical reagent	Chengdu CR & G
Glycerol(C3H8O3)	Analytical reagent	Chengdu CR & G
Deionized water(H2O)	--	Laboratory made
Ammonia water(NH4OH)	25%	Chengdu CR & G

The three-necked, 250ml flask was kept in thermostat water bath, after installed with the reflux condensing tube and thermometer. Appropriate amount of mixed solution of glycerol, deionized water and ammonia water were added into the flask and stirred for 10min at the defined temperature. Then appropriate amount of TEOS should be fully and quickly added into the reaction solution mentioned at one time. The stirring rate should be kept at 400rpm at the defined temperature. The surfactant, CTAB, was added after the reaction lasts for specific time in the water bath, and then the reaction went on. The nanomaterial was obtained after the high-speed centrifugal separation, three-time washing with deionized water and drying (Figure 1.).

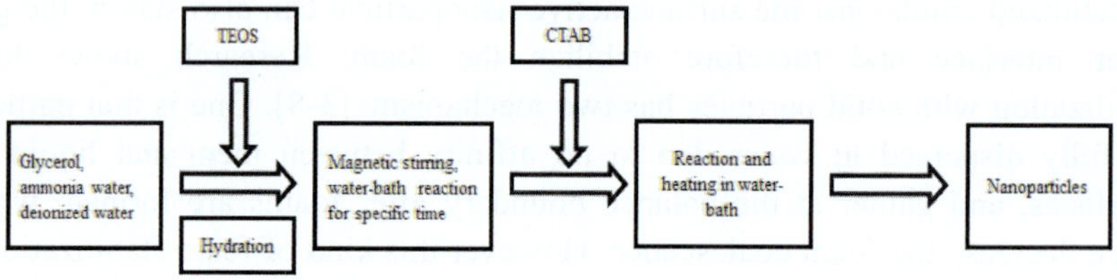

Fig.1. Synthetic route of nanoparticles

2.2. *Nanoparticle characterization*

0.2g nanoparticles were put in 10ml distilled water and ultrasonic dispersion was conducted for 5min. The grain sizes of nanoparticles were measured with the scanning electron microscope (SEM) and the nanoparticle analyzer. The corresponding scanning images are shown in Figure 2.

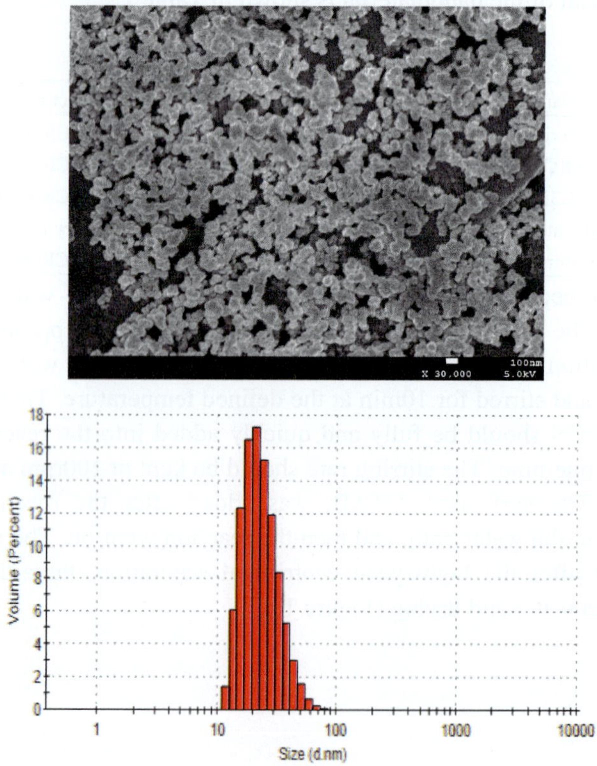

Fig.2. SEM image and grain size distribution of nanoparticles

It can be seen from Figure 4 that the grain sizes of nanoparticles are all below 100nm, with an average size of 29.3nm. However, the nanoparticle, to some degree, has coalescence capability in water solution, due to Van der Waal force and surface wettability.

2.3. *Surface property of nanoparticles*

Nanoparticles were made into thin slices under 7MPa with the tablet machine. In the static drop contact angle tester, XED-SJP, the whole process of the liquid drop approaching the slice was filmed. The contact angle of water at the nanoparticle surface then was determined using computers from the image of the moment when the contact between the liquid and slice was stable (Figure 3).

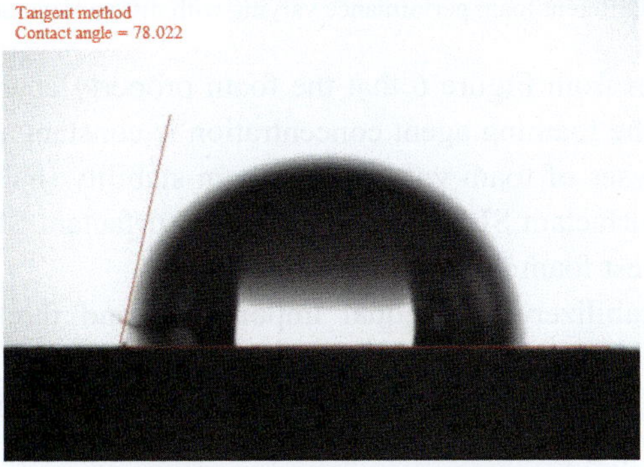

Fig.3. Surface wettability of nanoparticles

As is shown in Figure 3, the surface contact angle of nanoparticles is 78° and the surface wettability of the nanomaterial is near-neutral.

3. Research on foam property enhancing with nanoparticles

Since the synergistic effects of mixed-surfactant are much better than those of a single surfactant in terms of foam performance, it is preferred to mix the SLSN and SRBH, which are anionic surfactant and zwitter-ionic surfactant, respectively, with fixed total concentration of 3%. By varying the mixing ratio (mass ratio) we can get different foam performance, as shown in Figure 4.

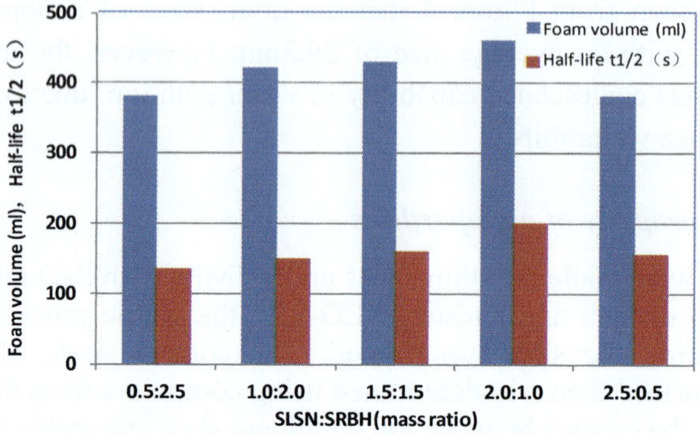

Fig.4. Different foam performance varying with different nanomaterial

It can be seen from Figure 6 that the foam property changes with different mix ratios when the foaming agent concentration is constant at 3%. Through the comparative analyses of foam volume and foam stability (half-life), the optimal ratio of anionic surfactant SLSN to zwitter-ionic surfactant SRBH is 2:1, which will achieve the best foam volume and stability.

The foam stabilizer is of equal importance, and thickening agents are usually used to increase the viscosity of the liquid phase and decrease foam drainage speed. CMC and PMC were used as foam stabilizers and experimental results are shown in Table 2.

Table 2. Effects of polymeric thickening agents on the system

Concentration(%)	CMC		PAM	
	Foam Volume(ml)	$T_{1/2}$(S)	Foam Volume(ml)	$T_{1/2}$(S)
0.1	510	643.3	510	863.7
0.2	525	683.5	540	888.3
0.3	530	715.1	595	913.5
0.4	555	776.5	585	974.5
0.5	565	784.2	575	981.1

The foam stabilizing capability of nanoparticles were tested by changing its addition amount, with a mix ratio of anionic surfactant SLSN and zwitter-ionic surfactant SRBH, 2:1, a total concentration of 3% and 0.3% foam stabilizer, PAM. The results are shown in Figure 5.

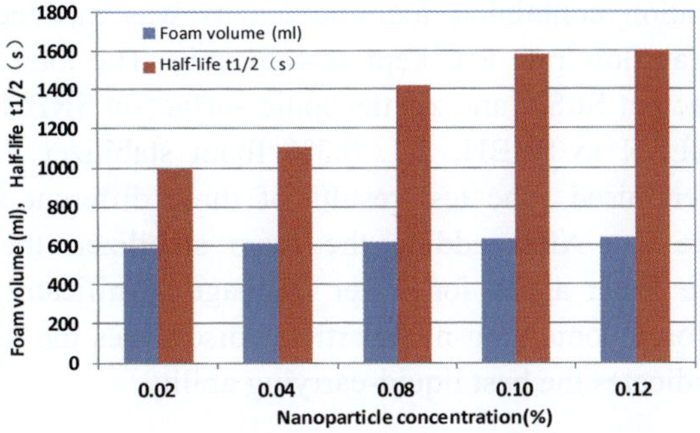

Fig. 5. different foam performance varying with the mixing ratio with the total concentration 3%

The nanoparticle can significantly improve the foam stability, because nanoparticles can adhere to the foam interface, improve the film strength of foam interfaces and avoid untimely ruptures of interface films. Regarding the addition amount, the addition amount of nanomaterial should below 0.1%, because the increase range of foam stability declines.

4. *Experimental study on water drainage performance of foam containing nanoparticles*

The foam drainage is an assistant technique for water drainage, where surfactants are added into the well. The treatment agent combines with water after being put into the well, and is easy to disperse and foam with the disturbance of gas stream. Water changes into foam, which has a relatively lower density. In this way, the gas amount that requires lifting decreases and hence gas can be recovered from gas wells where water is not able to be continuously carried out. The simulation apparatus is shown as Figure 6.

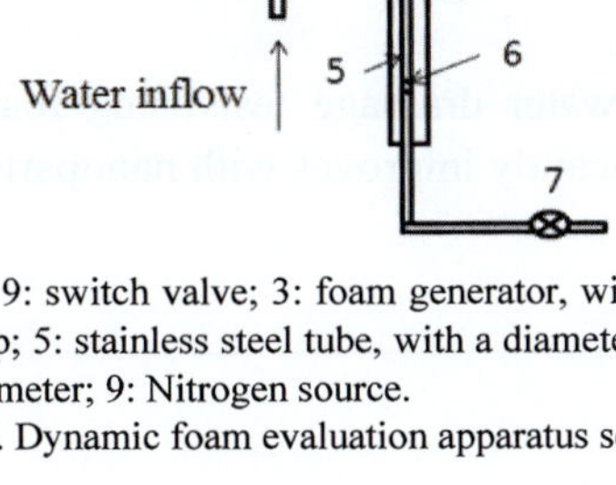

1: fluid flowmeter; 2, 7, and 9: switch valve; 3: foam generator, with the length of 1.5m and inside diameter of 5cm; 4: foam trap; 5: stainless steel tube, with a diameter of 2cm; 6: small cap (stainless steel, four holes); 8: gas flowmeter; 9: Nitrogen source.

Fig.6. Dynamic foam evaluation apparatus schematic

200 ml solution containing foaming agents was injected into the foam generator. The gas flow rate was kept at 4~5L/min. The total concentration of the anionic surfactant SLSN and zwitter-ionic surfactant SRBH was 3%, with a mass ratio of SLSN to SRBH, 2:1. 0.3% foam stabilizer PAM and 0.08% nanoparticles were used. The test results of three different formulations are shown in Figure 7. After adding the foam stabilizer, the liquid-carrying capability of the foam agent for water drainage significantly is remarkably improved. The foam containing nanoparticles discharges the largest volume of liquids, which indicates the best liquid-carrying ability.

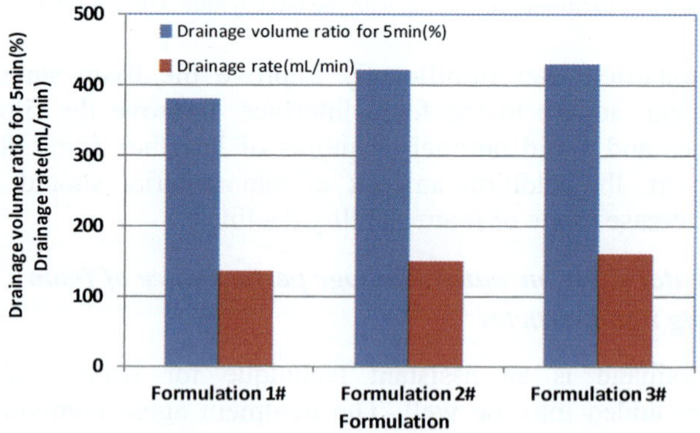

Formulation 1#: total concentration of SLSN and SRBH was 3%, with a ratio of SLSN/SRBH, 2/1; Formulation 2#: based on Formulation 1#, 0.3% foam stabilizer PAM was added; Formulation 3#: based on Formulation 2#, 0.08% nanoparticles were added.

Fig.7. Dynamic liquid-carrying capability of foam

5. Conclusion

(1) The nanomaterial has been developed through the liquid phase reaction, and characterized: the grain size is below 100nm with an average of 29.3nm and the surface contact angle is 78°which indicates neutral wettability.

(2) The synergy between this nanomaterial as a foam stabilizer and PAM in stabilizing foams is great, and can assist to improve foam stability. The optimal addition amount is 0.08%.

(3) In the dynamic water drainage test using foams, the liquid-carrying capability of foams significantly improves with nanoparticles.

References

1. LUOTaotao, OUYangwei, SUZhigang. Study on the Emulsion Stability of Oil-base Drilling Fluid Treated with In-situ Activation Nano Material[J].Drilling Fluid & Completion Fluid, 2014, 31(4): 5-7.
2. Hunter T N, Pugh R J, Franks G V, et al. The role of particles in stabilising foams and emulsions[J]. Advances in Colloid & Interface Science, 2008, 137(2):57–81.
3. DINGLiguo. Research on preparation and application of nano-Sio2[D]. Harbin Engineering University, 2004: 20~21.
4. Binks B P. Particles as surfactants-similarities and differences[J]. Current Opinion in Colloids & Interface Science, 2002, 7(02):21–41.
5. FEIHaihong. Laboratory Experimental Screening of Foaming Agent for Gas Well Production with Water Foaming and Withdrawing in Yancheng[J]. Oilfield Chemistry, 2006, 23(4): 329-333.
6. JIANGZeyin, TANGYongfan, LINDejian, et al. The Research and Application of Foam-dewatering Technology in Pingluoba Gas Reservoir of northwest of Sichuan Gas Field[J]. Chemical Engineering of Oil & Gas, 2007, 36(4): 324-327.
7. CHENG Liang, LUO Taotao, YANG Lin, et al. The Study of Compound Foam Flooding System of NaCMC and Nonionic Surfactant[J]. Chemical Engineering of Oil & Gas, 2006, 35(5): 403-406.

Enhanced stability of benvitimod loaded nanoemulsion based on the PIT method

Zhao-Huan Niu, Ya-Jing Fu, Xiao-Yuan Li, Yue Zhang*†, Wen-Jing Li and Xiao-He Liu
School of Chemical and Pharmaceutical Engineering, Hebei University of Science and Technology,
Shijiazhuang, 050018, Yuhua East Road, China
†E-mail: yuezhang@hebust.edu.cn

Benvitimod is a drug, whose chemical name is 3,5-Dihydroxy-4-isopropylstilbene. Its structure is similar to that of hydroxystilbene and it has a lot of pharmacological activities that have gained a great amount of attention in recent years. However, it has disadvantages such as poor stability and low water-solubility. Hydrophile-Lipophile Balance (HLB) of surfactant is an important factor on the preparation of nanoemulsion. In this study, Benvitimod nanoemulsion was prepared by mixed surfactants with the HLB adjusted to 13.98. Benvitimod could be encapsulated in the nanoemulsion to improve the stability and water-solubility and enhance its skin permeability. The prepared nanoemulsion was characterized by transmission electron microscopy and laser scattering analyzer. The formulation of 3,5-dihydroxy-4-isopropylstilbene nanoemulsion was optimized by using the pseudoternary phase diagram. Tween 80 and Span 60 (9:1) were used as the mixed surfactant and 1,2-propylene glycol and liquid paraffin were used as cosurfactant and oil phase respectively. The prepared Benvitimod nanoemulsion was clear and transparent by appearance with a small viscosity. The optimized Benvitimod nanoemulsion has stable physical and chemical properties.

Keywords: 3,5-Dihydroxy-4-isopropylstilbene; Nanoemulsion; Benvitimod.

1. Introduction

3,5-Dihydroxy-4-isopropylstilbene, drug name Benvitimod, with a typical structure of hydroxystilbene and has a lot of pharmacological activities attracted much more attention[1]. In the prior clinical trials, Benvitimod was proved having satisfactory effective and safety. While Benvitimod has shortcomings such as poor stability and low water-solubility. Being used as a parenteral drug, low water-solubility of Benvitimod always results in low skin penetration and bioavailability. Recently, the emergence of new technologies has a great interest in developing a nanocarrier system to improve the therapeutic properties of a drug. As a transparent or translucent liquid delivery system, nanoemulsion, with the droplet size below 100 nm, is physical stable has been developed to improve the stability and water-solubility for many years[2]. HLB of surfactant is an important factor on nanoemulsion preparation. In this study, Benvitimod

* Corresponding author.

nanoemulsion was prepared by mixed surfactants with adjusting HLB to 13.98. Benvitimod could be encapsulated in the nanoemulsion to improve the stability and water-solubility and enhance its skin permeability.

2. Experimental Section

2.1. *Chemicals and excipients*

Benvitimod (shown in Figure 1.) was synthesized in our laboratory according the literature[3]. Tween 20, Tween 60, Tween 80, Span 20, Span 60 and Span 80 were purchased from Tianjin Yongda Chemical Reagent Development Center (Tianjin, China). Liquid paraffin was food grade obtained from Alddin. Ethanol, 1,2-propanediol and isopropyl alcohol were all of analytical grade purchased from Shijiazhuang Modern Reagent Co (Shijiazhuang, China). The water was twice distilled.

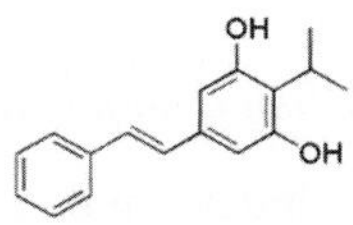

Fig.1. Structure of Benvitimod

2.2. *Development and evaluation of the Benvitimod nanoemulsion*

2.2.1. *Preparation of nanoemulsion*

For the preparation of nanoemulsion, Phase Inversion Temperature (PIT) was employed[4]. Benvitimod was solubilized in ethanol in a vial. Oil medium and surfactant were weighed accurately and added into the vial with stirring and heating. The distilled water was added drop-wise until a stable, colorless, and transparent or translucent nanoemulsion was formed.

2.2.2. *Selection of mixed surfactant and cosurfactant*

Surfactants generally consist of hydrophilic head and hydrophobic tail. It is important to define the most suitable surfactants for nanoemulsion systems. To reach this goal, we have introduced the mixed surfactant to obtain the approprite HLB (8-16) to improve the solubility parameters and surface performance[5].

Cosurfactant and oil were fixed as ethanol and liquid paraffin respectively. Surfactants of Tween 20, Tween 60, Tween 80, Span 20, Span 60 and Span 80 were designated as the mass ratio of 1:9, 2:8, 3:7, 4:6, 5:5, 6:4, 7:3, 8:2 and 9:1 and were compounded with each other. The mass ratio of the mixed surfactant and cosurfactant (Km) was fixed as 3:1. The mass ratio of liquid paraffin to the mixture of mixed surfactant and cosurfactant (Smix) were designed ranging from 1:9 to 9:1 to explore the mixed surfactant. The mixed surfactant could be determined from the above study. When the mixed surfactant were certain, we performed the oil and the mass ratio of oil to Smix were the same to the above experiment. 1,2-propanediol and isopropyl alcohol were choosed to match with the liquid paraffin and mixed surfactant with the Km of 3:1 to select the cosurfactant.

2.2.3. *Optimization of the formulation*

When the mixed surfactant and the cosurfactant were determined. We changed the Km with 4:1 and 5:1 to plot the pseudoternary phase diagram to determine the optimun formulation of Benvitimod nanoemulsion. The formulation will be determined according to the size of nanoemulsion region in the phase diagrams.

2.2.4. *Characterization of Benivitimod nanoemulsion*

The pH, viscosity, refractive index and structure type of Benvitimod nanoemulsion was measured. The average droplet size of Benvitimod nanoemulsion was detected by Laser Scattering Analyzer (NanoS90; Malvern Instruments, Malvern, Worcestershire, UK). Determination of the morphology of Benvitimod nanoemulsion was performed by using TEM (H-7650; Hitachi, Tokyo, Japan).

2.2.5. *Stability study of Benivitimod nanoemulsion*

The optimized Benvitimod nanoemulsion was subjected to stability studies of high humidity (the humidity of 92.5%), high temperature (60°C) and high light experiment (4500±500 LX) as the Chinese Pharmacopoeia guidelines. The content of Benvitimod in the stability test was analyzed by HPLC (LC-10ATVP; Shimadzu, Kyoto, Japan) at a flow rate of 1 $mL \cdot min^{-1}$ with the injected volume of the sample was 20 μL at 318 nm wavelengths at 30 °C using a Phenomenexluna-C18 column (4.6×250 mm, 5 μm). The mobile phase was a mixture of acetonitrile, methyl tertiary butyl ether and water (V/V/V=38/5/57). The linear concentration of Benvitimod ranged from 2.28 $mg \cdot L^{-1}$ to 22.8 $mg \cdot L^{-1}$ (r=0.99937).

3. Results and Discussion

3.1. *Formulation determination*

From pesudo-ternary phase diagrams Figure 2. it is revealed that the largest areas of nanoemulsion appear when T-80/S-60 are used as mixed surfactant, cosurfactant is 1,2-propylene glycol, Liquid paraffin is served as the oil phase and Km is taken as 3:1. This nanoemulsion formulation with the oil solubilized up to 15.00% is with the minimal concenration of Smix (37.40%).

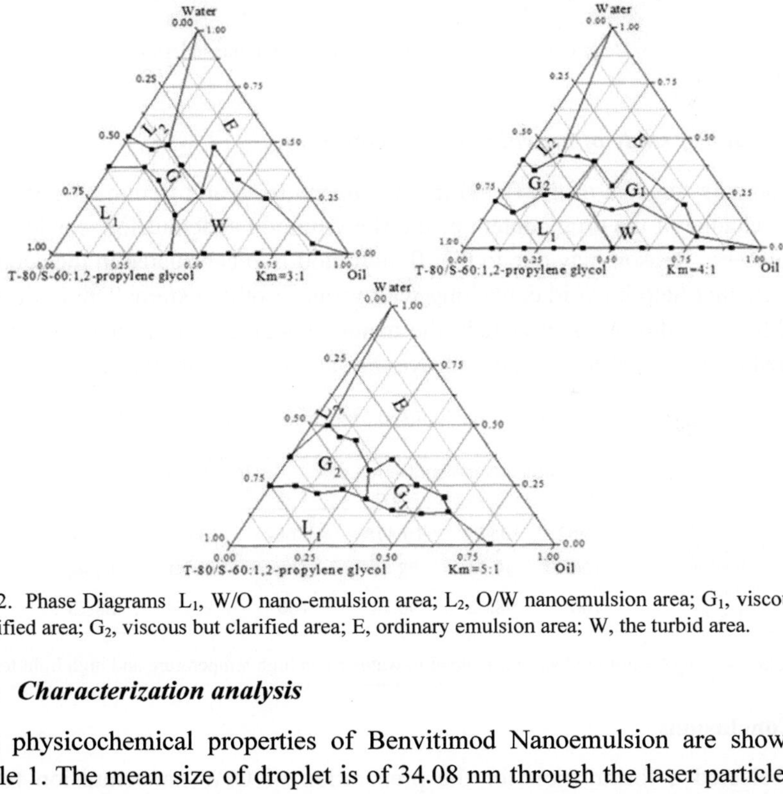

Fig. 2. Phase Diagrams L_1, W/O nano-emulsion area; L_2, O/W nanoemulsion area; G_1, viscous but opacified area; G_2, viscous but clarified area; E, ordinary emulsion area; W, the turbid area.

3.2. *Characterization analysis*

The physicochemical properties of Benvitimod Nanoemulsion are shown in Table 1. The mean size of droplet is of 34.08 nm through the laser particle size analyzer (Figure 3A). The morphology and surface structure of the Benvitimod nanoemulsion has been shown in Figure 3B where the droplet is round and has a smooth margin.

Table 1. Physicochemical parameters of Benvitimod nanoemulsion (n=3)

Parameter	Result
pH	6.96±0.02
Viscosity/cP	3.25±0.02
Refractive index	1.3875±0.002
Structure type	O/W

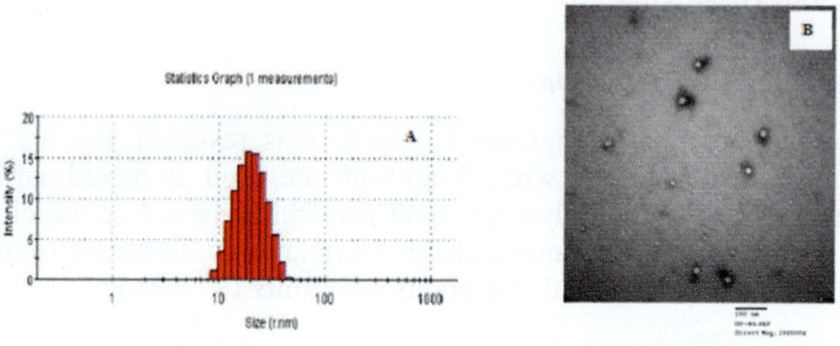

Fig. 3. Distribution of Benvitimod nanoemulsion droplets (A) and transmission electron microscope photograph (B)

3.3. *Stability study of benvitimod nanoemulsion*

To compare the data of the Benvitimod nanoemulsion and Benvitimod in Table 2, the change of the drug content of the Benvitimod nanoemulsion was less than the placebo. Reason may due to that Benvitimod is encapsulated in the centre of droplets that help it avoid contacting with water or other external environmental conditions. It illustrates obviously the nanoemulsion as a drug delivery system can reduce the degradation and improve the stability of Benvitimod.

Table 2. Stabilities of Benivitimod nanoemsulsion and the placebo

Items	Benvitimod Nanoemulsion			Benvitimod*		
	0 d	5 d	10 d	0 d	5 d	10 d
92.5% Humidity/g	16.462	16.471	16.478	13.682	13.685	13.687
60°C/%	100	99.54	98.47	100	97.11	94.96
4500±500 LX/%	100	98.62	96.74	100	95.54	88.73

* Placebo drug Benvitimod was suspended in water when high temperature and high light test.

4. Conclusions

We selected a formulation for the preparation of O/W nanoemulsion for an atopic dermatitis and psoriasis drug Benvitimod. The prescription was determined as follows: Tween 80 and Span 60 (9:1) as the mixed surfactant, 1,2-propylene glycol and Liquid paraffin as the cosurfactant and oil phase separately. The mass ratio of T-80/S-60 to 1,2-propylene glycol (Km) was 3:1 and the Smix to oil was 7:3. The prepared Benvitimod nanoemulsion was clear and transparent by appearance with a small viscosity. The optimized Benvitimod nanoemulsion has stable physical and chemical properties.

Acknowledgments

We express our gratitude for Hebei Research Center of Pharmaceutical and Chemical Engineering, Pharmaceutical Molecular Chemistry Key Labouratory of Ministry Technology and Hebei Province Key Laboratory of Molecular Chemistry for Drug. Support funds from Hebei University of Science and Technology are also appreciated.

References

1. K.J. Hu and M.W. John, Antibiotic production in relation to bacterial growth and nematode development in Photorhabdus-Heterorhabditis infected Galleria mellonella larvae, *Fems Microbiol Lett* **189**, 219-223 (2000).
2. Y. Sun, Z.Y. Xia, and J.K. Zheng, Nanoemulsion-based delivery systems for nutraceuticals: Influence of carrier oil type on bioavailability of pterostilbene, *J Funct Foods* **13**, 61-70 (2015).
3. Y.Q. Li, Q.L. Song, and P.P. Cheng, Synthesis of Hydroxystilbene Derivatives and their Antioxidant Activity by DPPH Method, *Chem Res Chinese U* **22**, 742-746 (2006).
4. R.M. Walker, E.A. Decker and D.j. McClements, Physical and oxidative stability of fish oil nanoemulsions produced by spontaneous emulsification: Effect of surfactant concentration and particle size. *J Food Eng* **164**, 10-20 (2015).
5. B.V. Hassas, F. Karakaication: Effect of surfactant concentration Relationship between hydrophilic lipophilic balance (HLB) of surfactants and coal rank, *Int J MinerProcess* **133**, 97-10(2014).

Immobilizing surfactant molecules for oil displacement by both intercalation and polyacrylamide nanocomposite process

Wei Zhang, Qian Zhou, Yang-Chuan Ke*†, Jing-Qi Ji, Qing-Chun Deng, Chen-Chen Yang, Yang Pei and Jia-Dong Zhang

Nanotechnology Center of Energy Resources, State Key Lab of Heavy Oil Processing, College of Science, China University of Petroleum
Beijing, China
†E-mail: nclpcnc@vip.sina.com

It is difficult for conventional polymer flooding to achieve a good profile control because most oilfields have been flooded with water for a long period of time and the long-term water injection erosion will lead to increase in heterogeneity of the reservoirs. Effective encapsulation of 3-(N,N-dimethylpalmitylammonio) propanesulfonate-cetyltrimethyl ammonium bromide-montmorillonite particles (SB-CTAB-MMT) with polyacrylamide (PAM) polymers was achieved by reactive potassium persulfate (KPS) initiator and methylene-bis -acrylamide (MBA) cross-linker system in an inverse phase suspension polymerization. The characterizations of electron microscopies, X-ray diffraction (XRD), and BZY-2 automatic surface tension apparatus showed that, in the encapsulation process, MMT layer exfoliations presented PAM/SB-CTAB-MMT nanocomposite microspheres, and tuned their particle size distribution. The nanocomposite microsphere samples had molecular mass over 5×10^6, an average particle size from 10 μm to 15 μm with narrow size distribution, and smooth surface morphology. XRD patterns clearly proved MMT layer exfoliations in the polymer matrix, which was consistent with TEM analysis. These nanocomposite microspheres showed low interfacial tension, which may be the candidate materials for oil recovery in oil and gas engineering.

Keywords: Inverse Phase Suspension Polymerization; Polymer Nanocomposite Microspheres; Particle Size; Interfacial Tension.

1. Introduction

The exploration, drilling and recovering engineering for oil and gas became more and more difficult in deep formation. It is necessary to enhance oil recovery in developed techniques [1].

Most oilfields have been water-flooded for a long period of time, the long-term water injection erosion leads to the increased heterogeneity of the reservoir. In this case, injected water will channel into production wells along

* Corresponding author.

the high permeability layers easily. As a result, it is difficult for conventional polymer flooding to achieve a good profile control, and about 65-77% of remaining oil is left in un-swept areas [2]. Thus, the best method to enhanced oil recovery is sweeping remaining oil from un-swept areas.

Many technologies have been applied to sweep remaining oil from un-swept areas, such as: surfactant flooding [3]; polymer flooding [4]; alkaline-surfactant-polymer (ASP) [5]; surfactant-polymer (SP) [6]; and foam flooding [7]. Alkaline-surfactant-polymer (ASP) is one of the most cost-effective methods, and ASP has got high recovery rates of above 20% in some oilfields like Daqing oilfield [8]. One of key advantages of ASP is the possession of a combined chemical phase behavior of the injected surfactant. Therefore, significant developments have made ASP flooding a viable option for field enhanced oil recovery and more attractive than other methods [9]. Meanwhile, disadvantages identified during the implementation process of the ASP flooding technology, such as, severe scaling in the injection lines and strong emulsification of the produced fluid [10] also limited its further application in the field.

To face the problems, surfactant-polymer (SP) flooding technology was developed, and SP flooding was regarded as a very potential enhanced oil recovery technology which has been applied in the western part of Gudong 7th reservoir, and has achieved a good development effect [11].

In this paper, PAM/SB-CTAB-MMT nanocomposite microspheres have been synthesized successfully. In the process, PAM has coated SB-CTAB-MMT, as a result, the SB was well protected from being absorbed by rock and loss during migration. The morphology, surface, and interfacial tension of these nanocomposite microspheres were characterized by scanning electron microscopy (SEM), X-ray diffraction (XRD) and BZY-2 automatic surface tension apparatus.

2. Experimental

2.1. Raw materials

The details of the raw materials were provided in Table 1.

Table 1. Experimental materials

Materials	purity	Supplier
Cetyltrimethyl ammonium bromide (CTAB)	0.98	Tianjing Guangfu
3-(N,N-Dimethylpalmitylammonio) propanesulfonate (SB)	0.99	Aladdin
Acrylamide (AM)	0.98	Tianjing Guangfu
Potassium persulfate (KPS)	0.98	Tianjing Guangfu
Span-80	0.98	Tianjing Guangfu
N,N-Methylene bisacrylamide (MBA)	0.99	Aladdin
Unmodified sodium montmorillonite (Na-MMT)	-	Huai An Saibei
Deionized water	-	Lab-made

2.2. Immobilize surfactant molecules (SB-CTAB-MMT)

Na-MMT (10 g) was immersed in 200 mL deionized water for 30 min in a 500 mL four-necked flask. SB (1.88 g) and CTAB (1.88 g) were dissolved in 100 mL deionized water in a 250 mL beaker, and then continuously added dropwise to the Na-MMT dispersion which was stirred at 75°C. The cationic exchange was carried out at 75°C for 8 h under ventilation of nitrogen gas. Afterwards, the sample was filtrated and washed several times with deionized water, until the Br^- ion was not detected. The sample was dried in vacuum oven at 75°C for 24 h, and then were ground to powder and stored for further use.

2.3. Synthesis of PAM/SB-CTAB-MMT nanocomposite microspheres

First of all, the AM monomers (14.2 g) and SB-CTAB-MMT(0.14 g) solute with deionized water (30mL) were taken into a 100 mL beaker, and a solution of MBA was also added slowly to the beaker to form a mixture of AM/MBA monomers solution at constant stirring rate by 450 $r\cdot min^{-1}$ at room temperature for 30 min. Secondly, the mixture solution was added slowly to the 500 mL four-neck flask with the cyclohexane (80mL) and was continuously stirred for 30 min at 50°C in order to adsorb between the SB-CTAB-MMT and monomers better without phase separation. Finally, the KPS initiator was taken into the mixture by the rate 500 $r\cdot min^{-1}$. The reaction was carried out in nitrogen gas and maintained at 65 °C for 4 h. The produced sample was filtered and rinsed with ethanol for several times, and dried in vacuum at 80 °C for 24 h.

2.4. Characterizations

Fourier transform infrared spectroscopy (FTIR) spectra of samples were obtained using a FTS-3000 spectrophotometer (American Digilab). X-ray diffraction (XRD) analysis was performed on Bruker D8 Advance X-ray diffractometer (STADI P, Germany). To calculate the distance between the silicate layers, Bragg's law was applied to [12].

$$2d \sin \theta = n\lambda \tag{1}$$

where λ is the wavelength of the X-ray, d is the diffraction inter-planar distance (nm), 2θ is diffraction angle (°). Scanning electron microscope (SEM, HITACHI, SU8010) was used to study the surface morphology of samples. The layer exfoliation and dispersion morphology of MMT layers in the polymer matrix was determined by transmission electron microscopy (TEM, JEM-2100, Japan). Surface and interfacial tensions were determined from surface tension measurements by BZY-2 automatic surface tension apparatus.

3. Results and Discussion

3.1. FTIR analysis

FT-IR spectra of the pristine Na-MMT, SB-CTAB-MMT, pure PAM polymer and its nanocomposite samples were investigated and shown in Figure 1.

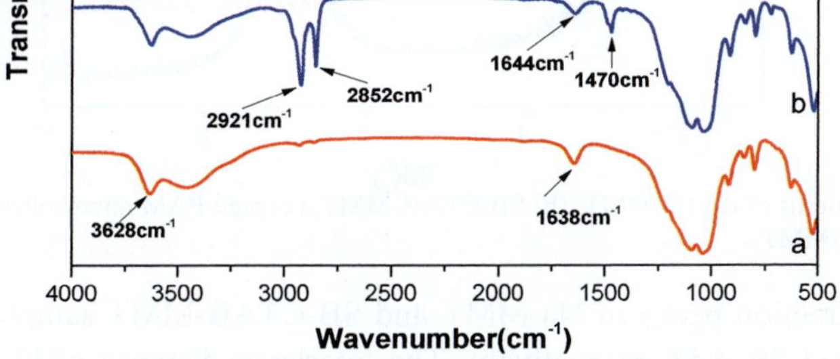

Fig. 1. FT-IR spectra of (a)Na-MMT, (b)SB-CTAB-MMT, (c)pure PAM microspheres, and (d)PAM/1 wt% SB-CTAB-MMT nanocomposite samples

In Fig. 1a for Na-MMT, transmittance band at 3,628 cm^{-1} represented O-H stretching vibration for Al-OH and Si-OH in silicate layers. The band at 1,638 cm^{-1} was representative of a bending vibration in water. In Fig. 1b for SB-CTAB-MMT, strong bands at 2852 cm^{-1} and 2921 cm^{-1} represented the symmetrical $-CH_2$ and asymmetric $-CH_2$ telescopic vibrations, respectively, which characterized the featured organic molecular chains of quaternary ammonium. For the pure PAM microspheres shown in Fig. 1c, transmittance bands at 3,390 cm^{-1}, 3,203 cm^{-1} represented N-H stretching and a strong band at 1,676 cm^{-1} represented C-O stretching. In Fig. 1d for the nanocomposite microspheres, transmittance bands at 3,630 cm^{-1} represented -OH stretching vibration of silicate layers. These indicated the presence of MMT layers in the final nanocomposite microspheres. Therefore, FT-IR investigations elucidated that SB-CTAB-MMT had been dispersed in the PAM microspheres.

3.2. XRD analysis

XRD patterns of the pristine Na-MMT, SB-CTAB-MMT, pure PAM microspheres, and its nanocomposites with (1 wt%) SB-CTAB-MMT loading was shown in Figure 2.

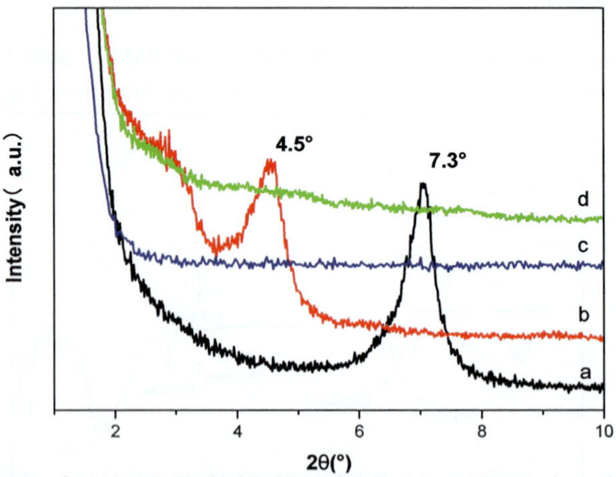

Fig. 2. XRD patterns of (a) Na-MMT, (b) SB-CTAB-MMT, (c) pure PAM microspheres, (d) PAM/1 wt% SB-CTAB-MMT

The diffraction peaks of Na-MMT and SB-CTAB-MMT samples appeared at $2\theta=7.3°$ and $2\theta=4.5°$, respectively. The interlayer distance of Na-MMT and SB-CTAB-MMT samples from the Bragg's equation is 1.21 nm and 1.96 nm, respectively (Figure 2a and b). Figure 2c showed that the pure PAM microspheres have no peaks. In Fig. 2d, no distinct peaks were observed in XRD patterns of the nanocomposite samples. The strong peak in 2θ=4.5° for the typical interlayer spacing of SB-CTAB-MMT had disappeared. This indicated that PAM-MMT nanocomposites were obtained, and MMT tactoids were considered to be exfoliated and dispersed inside the polymer matrix at the molecular level. However, XRD pattern cannot reliably distinguish the existence of intercalated MMT platelets from the exfoliated structures inside polymer matrices. Thus, TEM direct observation of these structures will be adopted to provide the evidence of exfoliation or intercalation levels inside polymer samples.

3.3. *SEM and TEM characterization*

SEM and TEM of nanocomposite microspheres were shown in Fig. 3.

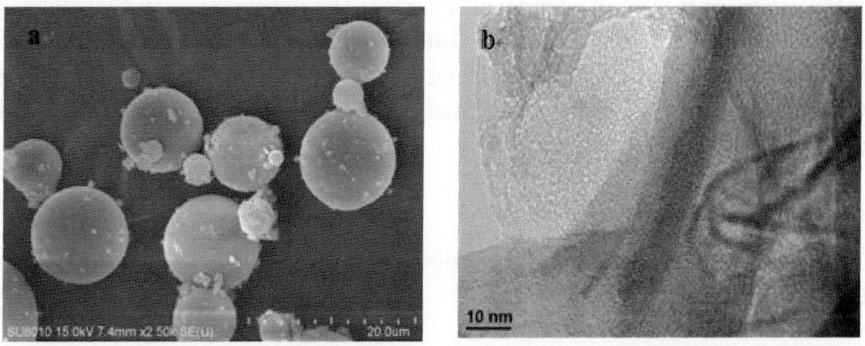

Fig. 3. (a) SEM images of PAM/1 wt% SB-CTAB-MMT; (b) TEM images of PAM/1 wt% SB-CTAB-MMT

SEM morphology and surface behavior of the PAM nanocomposite microspheres with SB-CTAB-MMT loading (1 wt%) was shown in Figure 3a. It is obvious that these microspheres were spheres, irrespective of the components, and their sizes ranged from 10 μm to 15 μm. Also, the surface morphologies of these nanocomposite microspheres were rough and the dispersity of microspheres was good. In the TEM picture Figure 3b, the dark lines represented the intersections of the MMT sheets which were the cross-sections of the MMT layers marked with arrows, and it can be seen the layers of SB-CTAB-MMT were random.

3.4. *Interfacial tension reducation*

The solution interfacial tension of SB-CTAB-MMT and PAM nanocomposite microspheres with SB-CTAB-MMT loading (1 wt%) was investigated. The interfacial tension between SB-CTAB-MMT and crude oil reached level of 1×10^{-1} $mN\cdot m^{-1}$ at 0.35 $g\cdot L^{-1}$ SB-CTAB-MMT concentration. The interfacial between PAM nanocomposite microspheres and crude oil reached 8.3 $mN\cdot m^{-1}$.

4. Conclusion

PAM/SB-CTAB-MMT nanocomposite microspheres were successfully synthesized by inverse phase suspension polymerization. XRD displayed disappearance of d_{001} peak in the nanocomposite patterns. This indicated that regularity of MMT layers was demolished and therefore exfoliated in polymer matrix. TEM morphology also presented the exfoliation and intercalation of MMT layers of these nanocomposite samples. The nanocomposite microspheres had an average particle size from 10 μm to 15 μm, and smooth surface behavior by SEM images. The solution interfacial tension of nanocomposite microspheres was greatly reduced, which made them potential material for enhancing oil recovery in oil and gas engineering.

Acknowledgments

This work is supported by Innovation Fund of China National Petroleum Corporation (2014D-5006-0214).

References

1. Chen Y. M. Modern Reservoir Management; China University of Petroleum Press: Dongying, Shandong, PRC, 2007.
2. Sang Q, Li Y J, Yu L. Enhanced oil recovery by branched-preformed particle gel injection in parallel sand pack models. Fuel 2014, 136, 295-306.
3. Kwan Min Ko, Bo Hyun Chon, Sung Bum Jang. Surfactant flooding characteristics of dodecyl alkyl sulfate for enhanced oil recovery.Journal of Industrial and Engineering Chemistry. 2014, 20(1), 288-233.
4. Hua-Xing Chen, Hong-Ming Tang, Xiao-ping Gong. Effect of partially hydrolyzed polyacrylamide on emulsification stability of wastewater produced from polymer flooding. Journal of Petroleum Science and Engineering. 2015, 133, 431-439.

5. Abass A, Olajire. Review of ASP EOR (alkaline surfactant polymer enhanced oil recovery) technology in the petroleum industry: Prospects and challenges. Energy. 2014, 77, 963-982.
6. Jian Hou, GuangmingPan, XuejiaoLu. The distribution characteristics of additional extracted oil displaced bysurfactant-polymer flooding and its genetic mechanisms. Journal of Petroleum Science and Engineering. 2013, 112, 322-334.
7. Zhang GY, Seright RS. Conformance and mobility control: foam versus polymers. In: SPE 105907, paper presented at international symposium on oilfield chemistry, Houston, Texas, USA, 28 February-2 March, 2007.
8. Shutang G, Qiang G. Recent progress and evaluation of ASP flooding for EOR in Daqing oil field. In: Proceeding of the SPE EOR Conference at Oil and Gas West Asia, Muscat, Oman, 11-13 April, 2010; SPE 127714.
9. Lu S, Li RF, Miller CA. Alkaline/surfactant/polymer processes: wide range of conditions for good recovery. Soc Pet Eng J 2010; 15(2): 282-293.
10. Gao P, Towler B, Li Y. Integrated evaluation of surfactant-polymer floods. In: SPE EOR Conference at Oil & Gas West Asia, Muscat; 2010.
11. Gao P, Towler B. Investigation of polymer and surfactant-polymer injections in south Slattery Minnelusa Reservoir, Wyoming. J Pet Explor Prod Technol 2011; 1(1): 23-31.
12. Y Jin, X S Sun, J Xue. X-Ray Diffraction Analysis Technology. Beijing: Nation defence industry press (2008) (in Chinese).

Preparations, conversions and functions of AM-SAS-SSS copolymers by an inverse micro-emulsion process

Qing-Chun Deng, Yang-Chuan Ke*†, Qian Zhou, Wei Zhang, Jia-Dong Zhang and Jingqi Ji

Nanotechnology Center of Energy Resources, State Key Lab of Heavy Oil Processing, College of Science, China University of Petroleum
Beijing, China
†E-mail: nclpcnc@vip.sina.com

Microemulsions are isotropic, transparent or translucent, and thermodynamically stable multicomponent fluids, normally composed of an aqueous component, an amphiphile as surfactant, and frequently a co-surfactant. The microemulsions have unique properties such as low viscosity, high stability, and transparency. Their uniformly dispersed smaller droplets have made them attractive media for polymerization. In this paper, the inverse microemulsion copolymerization kinetics of acrylamide (AM), sodium allylsufonate (SAS), and sodium 4-styrenesulfonate (SSS) initiated by ammonium peroxodisulfate (APS) and sodium bisulfite redox initiators in Span-80/Tween-80 mixture emulsifier system were reported. This study aims to elucidate the copolymerization conversion mechanism in the microemulsion media. The inverse microemulsion copolymerization factors of emulsifiers, reaction time and temperature, and SAS /SSS mass ratio on the copolymer conversion were investigated using gravimetric method. High copolymer materials are achieved, and these additives will apply to modifying working fluids for oil and gas engineering.

Keywords: Inverse Microemulsion; Copolymerization; Redox Initiators.

1. Introduction

Water-soluble polymers as functional polymer have applied as important stabilizers, flocculants, absorbents, and aqueous viscosity modifiers in latex paint, coating, and even tertiary oil drilling and recovery engineering.

Microemulsions are isotropic, transparent or translucent, and thermodynamically stable multicomponent fluids, normally composed of an aqueous component, an amphiphile as surfactant, and frequently a co-surfactant (usually an alcohol of intermediate chain length). The microemulsions have such unique properties as low viscosity, high stability, and transparency. Their uniformly dispersed smaller droplets have made them attractive media for polymerization [1–5].

Up to now, many water-soluble polymers have been prepared by inverse microemulsion polymerizations, most of them are AM-based ones. Renteria [6] investigated the inverse microemulsion polymerization of acrylamide in a paraffinic solvent stabilized by a mixture of nonionic surfactants, and studied the

* Corresponding author

factors of the initiator composition, HLB, aqueous phase and monomer percentage on the final polyacrylamide conversion and molecule mass. Barton [7, 8] studied the kinetics of free-radical homopolymerization and copolymerization of acrylamide (AM) and styrene (S) initiated by water-soluble ammonium peroxodisulfate (APS), or toluene soluble dibenzoyl peroxide (DBP) in inverse microemulsion toluene/S/AOT (sodium bis(2-ethylhexyl) sulfosuccinate) /water/AM system. The inverse microemulsion polymerization leads to forming partially hydrophobic cross-linked polymer particles of tailored chemical composition, crosslinking degree and size. Kaneda [9] investigated a series of poly(dimethylacrylamide-co-2-acrylamido-2-methyl-1propanesulfonic acid) microgels polymerized in single-phase reversed micelles of about 50 nm in diameter.

However, the inverse microemulsion copolymerization process and kinetics of acrylamide, sodium allylsufonate and sodium 4-styrenesulfonate using sorbitan monooleate (Span-80) and polyoxyethylene sorbitan monooleate (Tween-80) as mixed emulsifiers has not yet been studied.

In this paper, reported was the inverse microemulsion copolymerization kinetics of acrylamide (AM), sodium allylsufonate (SAS), and sodium 4-styrenesulfonate (SSS) initiated by ammonium peroxodisulfate (APS) and sodium bisulfite redox initiators in Span-80/Tween-80 mixture emulsifier system. This study has aim at elucidating the copolymerization conversion mechanism in the microemulsion media. The inverse microemulsion copolymerization factors of emulsifiers, reaction time and temperature, and SAS /SSS mass ratio on the copolymer conversion were investigated using gravimetric method. High copolymer materials are achieved, and these additives will apply to modifying working fluids for oil and gas engineering.

2. Experimental

2.1. *Materials*

Acrylamide(Am), analytical grade; sodium allylsufonate (SAS), analytical grade; Sodium 4-styrenesulfonate (SSS), analytical grade; Sorbitan monooleate (Span-80) and polyoxyethylene sorbitan monooleate (Tween 80), analytical grade; ammonium persulfate (APS) and sodium bisulfite, analytical grade; were used without further purification.

2.2. *AM-SAS-SSS copolymers by inverse micro-emulsion polymerization*

A series of AM–SAS–SSS copolymers with different mass ratio of SAS/SSS, emulsifiers, different reaction time and reaction temperature were prepared by the following procedure. Typically, acrylamide (11 g) was dissolved in 40 mL distilled water and then sodium allylsufonate (SAS) (11 g) and sodium 4-styrenesulfonate (2 g) were added to the above monomer solution and the mixed solution was stirred at room temperature for 30 min. Then, the solution was added dropwise to the inverse microemulsions formed by 40 g cyclohexane, 4 g Span80, and 2 g Tween 80. The water bath was heated slowly to 40°C with mild stirring after redox initiator APS (55 mg) and sodium bisulfite (15 mg) was introduced to the above inverse microemulsions. After about 3 h of the reaction, the resulting product was washed several times with absolute ethanol, dried at 80°C to a constant weight ,then milled and screened. However, this procedure is conventional and the reaction conversion isn't maximum. In this article, we should expound the optimal ingredient.

2.3. *Copolymer conversion measurement*

Copolymer conversion was characterized by gravimetry method. That is, the copolymer samples were withdrawn from the reacting system at given times and interval, and were cleaned by washing with ethanol several times to remove the surfactant. Samples were then weighed and vacuum oven-dried.

The conversion percentage (C%) of the inverse microemulsion polymerization was defined and determined by weight and calculated as below.

$$C\%=(W1/W)\times 100\% \quad (1)$$

Where, W1 is weight of dry polymer, W is weight of monomers.

2.4. *Characterizations*

Infrared spectra were taken with a 510 Nicolet FTIR spectrometer.

3. Results and discussions

3.1. *Reaction time and temperature effects on the copolymer conversion*

Figure 1 depicted the conversion as a function of reaction time and temperature for the copolymerization of (AM-SAS-SSS) copolymers in a micro-emulsion stabilized with Span 80/Tween 80 mixture emulsifiers.

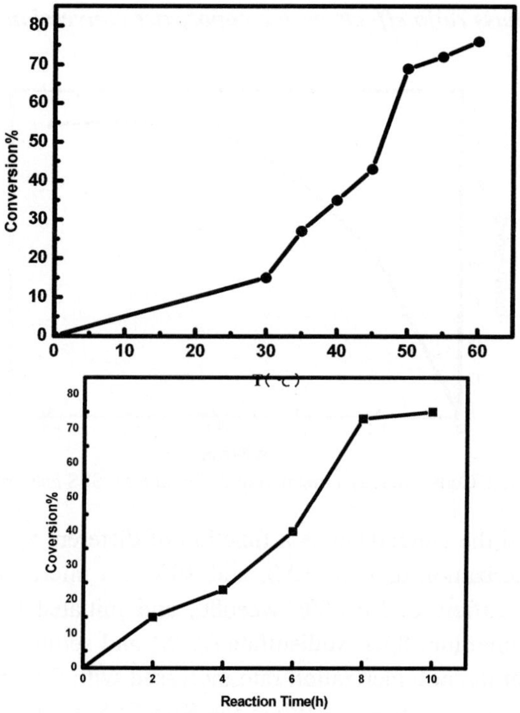

Fig. 1. Conversion as a function of different reaction time(left) and temperature(right).

It could be seen that the process was initially very rapid, reaching a maximum value of at about 75% polymerization conversion at about 55°C and 8 hours. The reaction time and temperature were closely related to the copolymerization. After these reaction time and temperature due to the monomer in the feed nearly expended, the polymerization rate was subdued until the process stopped. Besides, when the reaction temperature surpass the value of about 50 °C, even approach 60 °C, the process was excessively drastic and the reaction was uncontrollable.

3.2. *SAS/SSS mass ratio effects on the copolymer conversion*

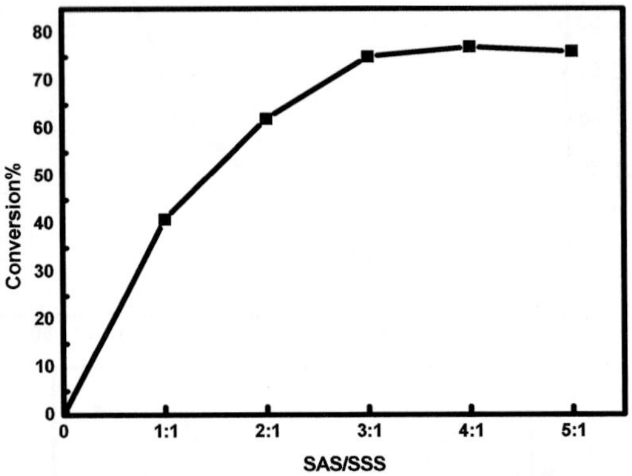

Fig. 2. Conversion as a function of different SAS/SSS mass ratio

Figure 2 depicted the conversion as a function of different SAS/SSS mass ratio for the copolymerization of AM, SAS, and SSS in a microemulsion stabilized with mixed emulsifiers of Span80/Tween80, and initiated by redox initiators composed of ammonium 8peroxodisulfate (APS) and sodium bisulfite. It could be illustrated that the polymerization rate increased with SAS/SSS mass ratio to speed up, at preliminary process. When the SAS/SSS mass ratio approximated 3:1, the polymerization rate was gentle.

3.3. *Span80/Tween80 ratio effect on the copolymer conversion*

The surfactant HLB (hydrophilic liophilic balance) values versus the conversion and polymerization rate were indicated in Figure 3.

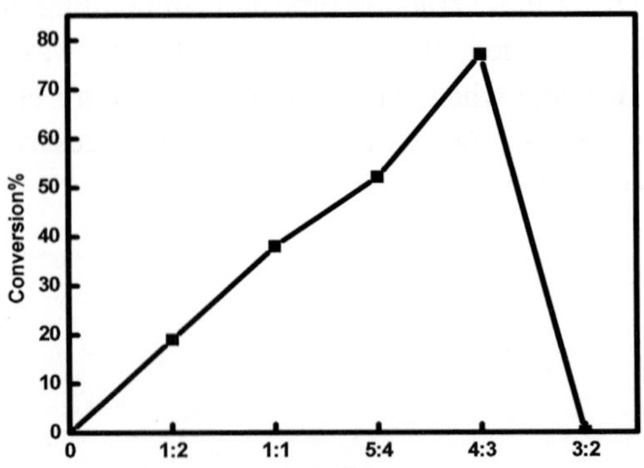

Fig. 3. Conversion curve as a function of Span80/Tween80 ratio

HLB value is calculated as follows:

$$HLB = U_{span80}HLB_{span80} + U_{tween80}HLB_{tween80} \quad (2)$$

where U_{span80} is the mass ratio of Span 80, $U_{tween80}$ is the mass ratio of Tween 80, HLBSpan is the HLB value of Span 80, HLBTween is the HLB value of Tween 80.

According to our experimental results, working at a lower HLB value (8.89) leads to higher conversion. The presence of mixed emulsifiers with different alkyl chain lengths offered some possibilities to regulate the polymerization process with respect to a certain structure of interface. A lower HLB value implied a change in the composition of the surfactant in the microemulsion where more hydrophobic Span 80 and less hydrophilic Tween 80 were used. This favored the formation of inverse microemulsions and formed more monomer solubilized reverse micelles. Therefore, the system formed relatively larger inverse microemulsion area which was suitable for microemulsion polymerization, and conversion increased in lower HLB value. However, further discrease in Span/Tween mass ratio to 3/2 with HLB value of 8.5, the system became unstable, turbid and no inverse microemulsions were formed.

3.4. *FTIR spectrum study*

The typical FTIR spectra of (AM–SAS–SSS) copolymer were shown in Fig. 4.

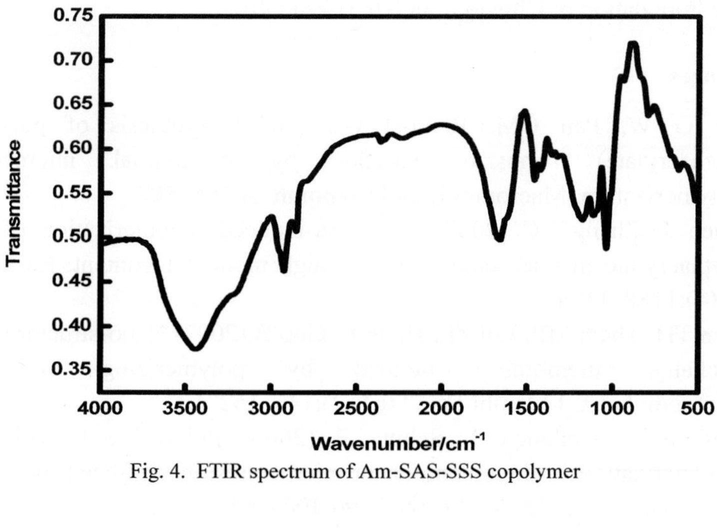

Fig. 4. FTIR spectrum of Am-SAS-SSS copolymer

C–H asymmetric and symmetric stretching peaks, appeared at about 2,925 and 2,854 cm^{-1}, respectively. Stronger absorption peaks at 3,400 cm^{-1} was attributed to NH2 groups. Absorptions at 1,664 cm^{-1} resulted from the overlap of C=O stretching (CONH2) and phenyl stretching. 1,448 cm^{-1} was attributed to asymmetric stretching of -SO2 group. This indicated that sulfonate and amide, groups were incorporated in the copolymer molecules.

4. Conclusions

An inverse microemulsion polymerization technique applies to preparing copolymers of acrylamide (AM), sodium allylsufonate (SAS), and sodium 4-styrenesulfonate (SSS).The factors of emulsifiers, reaction temperature, AM/AA mass ratio, SSS concentration, and initiator concentration on the final copolymer conversion were presented. Such conversion curves exhibited two intervals and confirmed the character of microemulsion polymerization. AM, SAS, and SSS chain segments were characterized in the copolymer molecules. Such copolymers have enhanced water-soluble, viscosifying, and temperature tolerance effect which will find application to varied fields, especially petroleum engineering.

Acknowledgments

This work is supported by Innovation Fund of China National Petroleum Corporation(Grant No.2014D-5006-0214), National Natural Science Foundation of China(Grant No.51274223), and Major Program of the National Natural Science Foundation of China(Grant No.51490650)

References

1. He G W, Pan QM, Rempel GL (2003) Synthesis of poly(methyl methacrylate) nanosize particles by differential microemulsion polymerization. Macromol Rapid Commun 24:585–588
2. Chen J, Zhang ZC (2007) Radiation-induced polymerization of methyl methacrylate in microemulsion with high monomer content. Eur Polym J 43(4):1188–1194
3. Lim TH, Tham MP, Liu ZL, Hong L, Guo B (2007) Nano-structured proton exchange membranes molded by polymerizing bi-continuous microemulsion. J Membr Sci 290(1–2):146–152
4. Katime I, Arellanoa J, Schulz P (2006) Poly(n-hexyl methacrylate) polymerization in three-component microemulsion stabilized by a cationic surfactant. J Colloid Interface Sci 296:490–495

5. Barton J, Capek I (2000) Acrylamide and butyl acrylate polymerization in Winsor IV (w/o) and Winsor I (o/w) microemulsions. Macromolecules 33:5353–5357
6. Renteria M, Munoz M, Ochoa JRL, Cesteros C, Katime I (2005) Acrylamide inverse microemulsion polymerization in a paraffinic solvent: Rolling-M-245. J Polym Sci Part A Polym Chem 43:2495–2503
7. Barton J, Juranicˇova V (2000) Polymerization of acrylamide in styrene containing inverse microemulsions: polymerization kinetics and polymer product composition studies. Polym Int 49:1483–1491
8. Barton J, Kawamoto S, Fujimoto K, Kawaguchi H, Capek I (2000) Preparation of partly hydrophobized, crosslinked polyacrylamide particles by terpolymerization of acrylamide/N, N-methylenebisacrylamide/styrene in inverse microemulsion. Polym Int 49:358–366
9. Kaneda I, Sogabe A, Nakajima H (2004) Water-swellable polyelectrolyte microgels polymerized in an inverse microemulsion using a nonionic surfactant. J Colloid Interface Sci 275:450–457

Enhancing the sensitivity and linearity of piezoresistivity utilization of artificial electronic skin by combinative surface treatments of multi-walled carbon nanotubes

Biao Zhang[*†,a], Bu-Yin Li[b] and Sheng-Lin Jiang[c]

School of Optical and Electronic Information, Huazhong University of Science and Technology, No. 1037, Luoyu Road,
Wuhan 430074, China
[a]E-mail:95347380@qq.com, [b]libuyin@mail.hust.edu.cn, [c]jsl@mail.hust.edu.cn

The effects of multi-walled carbon nanotubes (MWNTs) on the properties of poly(dimethyl siloxane) (PDMS) nanocomposites by using combinative surface treatments to enhance the sensitivity and linearity of piezoresistivity are investigated in the present study. Fourier transform infrared (FT-IR) analysis, electrical, piezoresistive properties are studied. Based on MWNTs/PDMS nanocomposites, a simple yet highly sensitive pressure sensor is developed in pressure regimes (0-100 kPa), which show high sensitivity and wide linearity of S=22.16±1.67×10^{-3} kPa^{-1} (R^2=0.94591) at 0.4 wt.%. According to our knowledge, the pressure sensitivity is only next to a structured pressure sensor (approximately 25×10^{-3} kPa^{-1}), and higher than the typical sensitivity of 2×10^{-6} to 20×10^{-3} kPa^{-1} reported for other sensors. These nanocomposites have promising potential applications in artificial electronic skin.

Keywords: MWNTs/PDMS Nanocomposites; Piezoresistive Properties; Combinative Surface Tratments; Artificial Electronic Skin.

1. Introduction

Flexible pressure sensors have emerged as a highly active field due to their promising applications in artificial electronic skin [1]. Moreover, nanocomposite materials needing for sufficient sensitivity in pressure regimes (< 1-100 kPa) has become a hot topic recently [2]. However, a challenge is poor sensitivity and narrow linearity of piezoresistivity, and so on [3].

The majority of flexible pressure sensors are based on capacitance or piezoelectricity, while piezoresistivity lies at simplicity in device structure and fabrication [4]. Despite this, only limited successful examples were reported on

* Corresponding author.

piezoresistivity of flexible pressure sensors [5]. Pressure-sensitive silicone rubbers (especially the PDMS) are the typically piezoresistive sensors that are most widely applied in flexible pressure sensors [6].

Due to the excellent properties of multi-walled carbon nanotubes, it has been applied as filler in nanocomposites [7]. In order to improve the piezoresistivity of high sensitivity and wide linearity, most studies have sought to lower the percolation threshold [8]. Therefore, a new method using combinative surface treatments has been proposed as a key strategy.

In the present work, the commercially Hydroxy-functionalized Multi-walled carbon nanotubes (MWNTs-OH) were π-π stacked by poly(phenylmethyl siloxane) (PPMS) and wrapping processes using vinyl poly(dimethyl siloxane) (Vi-PDMS). The importance of this study stems from the fact that this new method using combinative surface treatments and its nanocomposites is commercially available and hence is of significant interest for successful applications of artificial electronic skin.

2. Experimental

A solution containing 1 gram of MWNTs-OH and 2 gram of PPMS in 50 gram of chloroform was ultrasonicated for 1 hour. The resulting PPMS-treatmented MWNTs-OH (P-MWNTs-OH) were then added 0.8 gram of Vi-PDMS in it via ultrasonication for 2 minutes (VP-MWNTs-OH).

A certain amount of MWNTs-OH, P-MWNTs-OH and VP-MWNTs-OH mixture solutions were added into the PDMS (Part-A: Base) for mixing and degassing. And then, the PDMS (Part-B: Curing agent) was added and mixed. The mixtures were then casted into molds and were cured at 120°C for at least 30 minutes, and samples are 20×10×4 mm.

FT-IR were recorded using a VERTEX 70 FTIR spectrometer (Bruker, Germany). The electrical measurements were conducted using a four-point test and a multimeter agilent 34401a was used as measuring instrument. Mechanical measurements were using a testing machine (MTS Exceed 40.SANS.China).

3. Results and discussion

3.1. *FT-IR analysis of combinative surface treatments of MWNTs-OH*

Figure 1(a) shows a peak at 3610 cm^{-1} that is ascribed to O-H stretching. Both the P-MWNTs-OH (Figure 1(b)) and VP-MWNTs-OH (Figure 1(c)) shows weak C-H stretching vibration peaks in the range of 3000-3100 cm^{-1}, which are characteristic of the phenyl group. Several peak intensities in the range of

1000-1300 cm^{-1} are ascribed to the Si-O stretching and the peak at 789 cm^{-1} is assigned to phenyl-ring substitution bands. Figure 1(c) shows a weak peak appearing at 1644 cm^{-1} is attributed to $CH=CH_2$ stretching vibration, which confirms the reaction of Vi-PDMS with the P-MWNTs-OH.

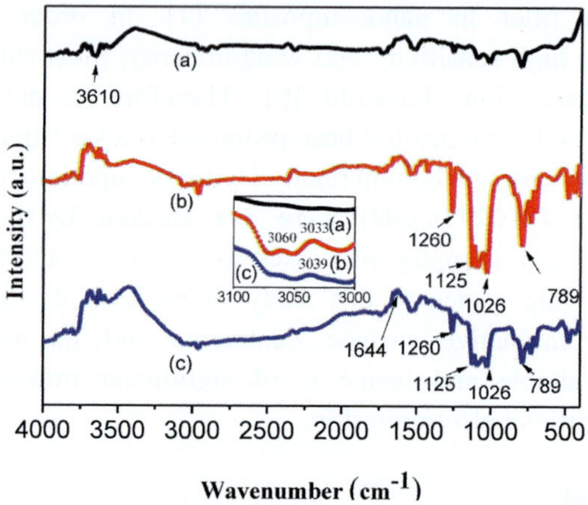

Fig. 1. FT-IR spectra of combinative surface treatments: (a) MWNTs-OH; (b) P-MWNTs-OH; (c) VP-MWNTs-OH.

3.2. *Electrical conductivity of combinative surface treatments of MWNTs-OH/PDMS nanocomposites*

As shown in Figure 2, the incorporation of the VP-MWNTs-OH from 0.2 to 0.4 wt.% increased the electrical conductivity almost nine orders of magnitude, from $\sigma=5.0\times10^{-12}$ to 3.78×10^{-3} S/m. Poor electrical conductivity of the P-MWNTs-OH and MWNTs-OH is observed with the addition of 0.5-1.5 wt.% and 10-20 wt.%, respectely. Therefore, the VP-MWNTs-OH/PDMS nanocomposites showed a lower percolation threshold (0.386 wt.%) than that of the P-MWNTs-OH/PDMS nanocomposites (1.498 wt.%) and the MWNTs-OH/PDMS nanocomposites (18.2 wt.%).

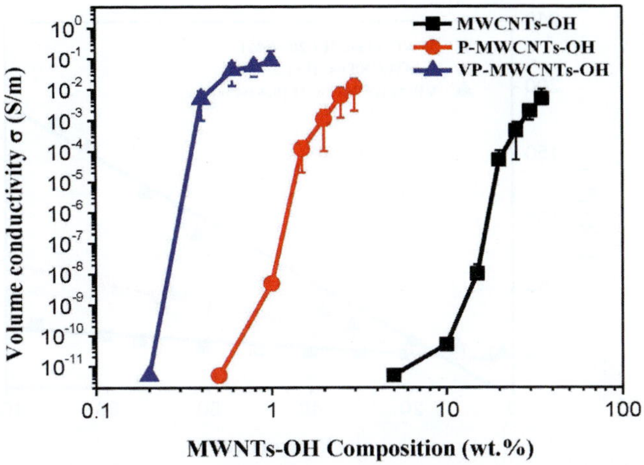

Fig. 2. The electrical conductivity of combinative surface treatments of MWNTs-OH/PDMS nanocomposites: (a) MWNTs-OH/PDMS; (b) P-MWNTs-OH/PDMS; (c) VP-MWNTs-OH/PDMS.

3.3 Piezoresistive properties of combinative surface treatments of MWNTs-OH/PDMS nanocomposites

Figure 3 shows the piezoresistive properties of MWNTs-OH/PDMS nanocomposites at 20 wt.% is low with a sensitivity of S=2.65±0.38×10^{-3} kPa^{-1} (R^2=0.82739). Besides, the addition of MWNTs-OH to PDMS at 1.5 wt.% shows an increase in the sensitivity (S=8.09±0.60×10^{-3} kPa^{-1} (R^2=0.94749)). Particularly, the VP- MWNTs-OH show a significant improvement in the sensitivity $((S = 22.16 \pm 1.67 \times 10^{-3} kPa^{-1} (R^2 = 0.94591))$ at 0.4 wt.%. Based on our knowledge, the resistive pressure sensitivity of this unstructured nanocomposites is only next to the record value in a structured pressure sensor (approximately 25×10^{-3} kPa^{-1}) [9], and it is higher than the typical sensitivity of 4×10^{-5} to 20×10^{-3} kPa^{-1} reported for other sensors [10-12].

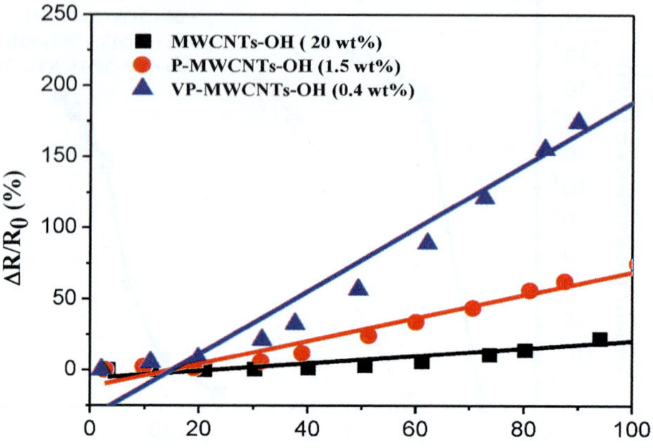

Fig. 3. The percent resistance change on the applied pressure for combinative surface treatments of MWNTs-OH/PDMS nanocomposites at slightly above percolation threshold content.

4. Conclusions

In the present work, the major finding from this study are highlighted as follows:
1. The new method using combinative surface treatments can significantly lower the percolation threshold of nanocomposites.
2. The percolation threshold of the nanocomposites containing VP-MWNTs-OH (percolation threshold: 0.386 wt.%) decreased due to the new method using combinative surface treatments.
3. The sensitivity of the nanocomposites containing VP-MWNTs-OH ($S = 22.16 \pm 1.67 \times 10^{-3} kPa^{-1} (R^2 = 0.94591)$) increased significantly, is only next to a structured pressure sensor (approximately 25×10^{-3} kPa^{-1}), and higher than the typical sensitivity of 2×10^{-6} to 20×10^{-3} kPa^{-1} reported for other sensors.

Acknowledgements

Biao Zhang would like to acknowledge the financial support from Innovation Fund for the International Science and Technology Cooperation Project Independent of HUST (project number: 2013ZZGH014). We also thank for the Analytical and Testing Center of HUST for the technical support: FT-IR test.

Appendix

The volume conductivity (σ) was then computed use the equation:

$$\sigma = \frac{1}{\rho} = \frac{L}{R \times A} \tag{A.1}$$

where ρ is volume resistivity, A is cross-sectional area and L is length. Here, the relationship between the ΔR/R0 and P (kPa) can be described:

$$\frac{\Delta R}{R_0} = K + S\mathrm{P} \quad (S = \frac{\delta(\Delta R/R_0)}{\delta \mathrm{P}} \quad (R^2)) \tag{A.2}$$

Where K are the constants, S also reflects the sensitive degree of resistive change-pressure (kPa^{-1}), R^2 is goodness of fit with reflects the linear degree.

References

1. G. Schwartz, B.C. Tee and Z. Bao, Flexible polymer transistors with high pressure sensitivity for application in electronic skin and health monitoring, Nat. Commun. **4**, 1 (2013).
2. S.C.B. Mannsfeld, B.C-K. Tee and Z. Bao, Highly sensitive flexible pressure sensors with microstructured rubber dielectric layers. *Nat. Mater.* **9**, 859 (2010).
3. Y. Zang, F. Zhang, C. Di and D. Zhu, Advances of flexible pressure sensors toward artificial intelligence and health care applications, *Mater. Horiz.* **2**, 140 (2015).
4. T. Yamada, Y. Hayamizu and K. Hata, A stretchable carbon nanotube strain sensor for humanmotion detection, *Nat. Nanotechnol.* **6**, 296 (2011).
5. C. Pang, G-Y. Lee and K-Y. Suh, A flexible and highly sensitive strain-gauge sensor using reversible interlocking of nanofibers, *Nat. Mater.* **11**, 795 (2012).
6. C.L. Choong, M.B. Shim and U.I. Chung, Highly stretchable resistive pressure sensors using a conductive elastomeric composite on a micropyramid array, *Adv. Mater.* **26**, 3451 (2014).
7. S. Park, M. Vosguerichian and Z. Bao, A review of fabrication and applications of carbon nanotube film-based flexible electronics, *Nanoscale* **5**, 1727 (2013).
8. Y. Shu, H. Tian and T.L. Ren, Surface-modified piezoresistive nanocomposite flexible pressure sensors with high sensitivity and wide linearity, *Nanoscale* **7**, 8636 (2015).

9. S. Jung, J.H. Kim and D-H. Kim, Reverse-Micelle-Induced Porous Pressure-Sensitive Rubber for Wearable Human–Machine Interfaces, *Adv. Mater.* **26**, 1 (2014).
10. H-K. Lee, S. Chang and E. Yoon, A flexible polymer tactile sensor: fabrication and modular expandability for large area deployment, *J. Microelectromech. S.* **15**, 1681 (2006).
11. Y-M. Huang, N-C. Tsai and J-Y. Lai, Development of tactile sensors for simultaneous, detection of normal and shear stresses, Sensor. Actuat. A-phys. 159, 189 (2010).
12. F.M. Yaul, V. Bulovic and J.H. Lang, A flexible underwater pressure sensor array using a conductive elastomer strain gauge, J. Microelectromech. S. 21, 897 (2012).

Adsorption differences of acidic, neutral and alkaline amino acids onto nano/micron COM and COD crystals

Kai Yu, Xiao-Ling Wen, Yi-Ming Ding and Jian-Ming Ouyang*†

Department of Chemistry, Jinan University,
Guangzhou, 510632, China
†E-mail: toyjm@jnu.edu.cn

In this study, the adsorption differences of different kinds of amino acids (acidic L-glutamic acid (L-Glu), neutral L-threonine (L-Thr) and alkaline L-histidine (L-His)) onto calcium oxalate monohydrate (COM) and calcium oxalate dihydrate (COD) crystals with sizes of 50 nm, 100 nm, 1 μm, 3 μm and 10 μm were investigated using spectrophotometry. The study of adsorption differences of amino acids on calcium oxalate has rarely been conducted; hence, we studied it with different pH values onto COM and COD crystals. The amount of adsorbed amino acids was determined using the depletion method and the centrifugation method was applied. At pH=7.3, COM crystals mainly adsorbed acidic L-Glu and have less adsorption to L-Thr and L-His, while COD mainly adsorbed alkaline L-His. Results of this study showed that when the size of crystal increases, the adsorption amount of crystals decreased. This was attributed to the decreases of specific surface area of larger crystals. The adsorption amount of COM was greater than that of COD with the same size because the specific surface area of COM is greater than COD. Our results help to further illustrate the formation mechanism of calcium oxalate stones and provide inspiration for inhibiting its recurrence.

Keywords: Adsorption; Amino acid; Calcium Oxalate.

1. Introduction

Urinary stone is a worldwide disease, and more common in developed countries. Calcium oxalate (CaOx) stones are the most common type. Thus, the study of CaOx stones has great significance for the prevention and treatment of stones. Some scholars showed a great interest in the role of amino acids in CaOx stone formation [1,2] because human urine contains glycine, histidine, asparagine, lysine, serine, 3-methyl histidine, and glutamine etc. [3,4]. The study suggested that the amount of amino acid in urine of healthy and calcium oxalate stone formers has obvious differences. The concentrations of free γ-carboxyglutamic acid in urine of stones formers (50.76±14.48 μmol/L) is significantly higher than normal control group (40.87±8.15 μmol/L), while the amount of protein-bound γ-carboxy glutamic acid (1.44±0.32 μmol/L) is significantly lower than normal control group (1.82±0.42 μmol/L) [5]. This indicates that the increase of free γ-carboxy glutamic acid and the decrease of protein-bound γ-carboxy glutamic acid in urine may be a reason of forming calcium oxalate stones. Moreover, the

* Corresponding author

organic substrates of urinary stones also contain rich amino acid residues. Therefore, amino acids play an important role in the formation of stones.

Amino acid in urine may be adsorbed on the surface of urinary nanocrystallites (such as calcium oxalate), thereby affecting the growth, aggregation and solid phase transformation of urinary nanocrystallites [6]. Because, the study about the adsorption differences of amino acids onto calcium oxalate crystals are very few, we studied the adsorption differences of various amino acids with different pH onto calcium oxalate monohydrate (COM) and calcium oxalate dihydrate (COD) crystals with the size of 50 nm, 100 nm, 1 μm, 3 μm and 10 μm.

2. Materials and Methods

2.1. *Reagents and instruments*

Ninhydrin, glutamic acid, threonine, histidine and other reagents were all analytically pure. Instrument includes Varian Cary 500 UV-visible spectrophotometer (Varian, USA).

COM and COD crystals with the size of 50 nm, 100 nm, 1 μm, 3 μm and 10 μm were prepared according to the our previous work [7,8].

2.2. *Adsorption experiments of amino acids*

The amount of adsorbed amino acids was determined using the depletion method [9]. In brief, 50 mg (*m*) COM or COD crystals were added into a 10 mL (V) amino acid solution with different concentrations (C_0) (pH close to neutral). The suspension was dispersed by ultrasonic processing for 10 min and then placed into a thermostat water bath at 37 °C. The suspension was centrifuged after adsorption for 24 h, and the residual concentration of amino acids was measured (C_{eq}). Three parallel experiments were carried out. The adsorbed amount (Q_{ads}) of AOT on COM or COD was calculated using the formula: $Q_{ads} = V(C_0-C_{eq})/m$.

3. Results and Discussion

3.1. *Standard curve*

Figure 1 shows the molecular structure of acidic L-glutamic acid (L-Glu), neutral L-threonine (L-Thr) and alkaline L-histidine (L-His).

The standard curve of amino acids obtained by spectrophotometrically was shown in Figure 2. (L-glutamic acid as an example). The regression equation is:

$$Y = 9.6629X + 0.0443,\ R^2 = 0.9897.$$

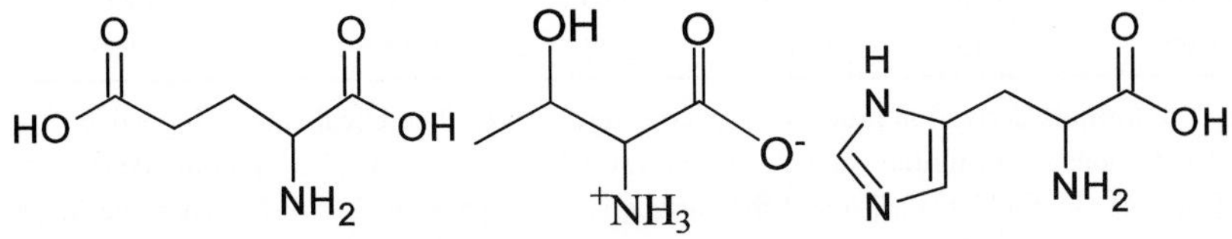

(a) (b) (c)

Fig. 1. The molecular structure of acidic L-glutamic acid (L-Glu) (a), neutral L-threonine (L-Thr) (b) and alkaline L-histidine (L-His) (c).

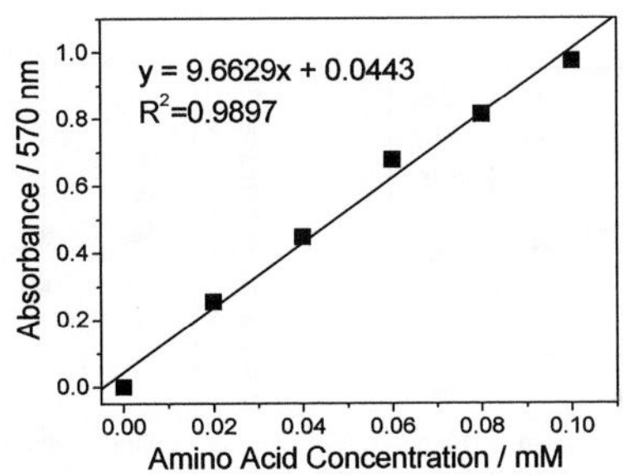

Fig. 2. Standard curve of L - glutamic acid.

3.2. *Adsorption differences of three kinds of amino acids onto COM*

When the concentration of amino acid was 0.2 mM, the adsorption amount differences of acidic L-glutamic acid (L-Glu), neutral L-threonine (L-Thr) and alkaline L-histidine acid onto calcium oxalate monohydrate (COM) with the size of 50 nm, 100 nm, 1 μm, 3 μm and 10 μm was shown in Figure 3A.

(1) Effect of different amino acids: At pH=7, COM with different sizes had the adsorption behavior to L-Glu, while the adsorption amount to L-Thr and L-His were less and substantially has no adsorption.

(2) Effect of COM size: With the increase of COM size from 50 nm to 10 μm, its adsorption amount to three kinds of amino acids gradually decreased. Indicating that the adsorption amount of nano COM is higher than that of micron COM, which might be due to the specific surface area of smaller-sized crystals is larger at the same mass concentration, therefore the adsorption amount is greater. Table 1 shows the specific surface area of COM and COD crystals with different sizes calculated by the surface area density value.

Table 1. Specific surface area of COM and COD crystals with different sizes calculated by the surface area density value / m^2/g.

Size /μm	0.05	0.20	1	3	10
COM	26.28	14.74	3.64	1.53	0.83
COD	21.45	13.64	3.16	1.09	0.80

Note: The formula derivation process of specific surface area theoretical calculating value ($S_{calculated}$) of calcium oxalate: Assuming calcium oxalate crystals are standard sphere and the density is ρ, so the quality is $m = \rho V = \frac{4}{3}\pi R^3\rho$ and the surface area of sphere is $S = 4\pi R^2$. Then the theoretical calculating value of specific surface area is:

$S_{计算} = S/m = 4\pi R^2/(\frac{4}{3}\pi R^3\rho) = \frac{3}{\rho R}$. Putting R= d/2 into the equation obtained $S_{calculated} = 6/(\rho d)$, wherein $\rho_{(COM)} = 2.22\ g/cm^3$, $\rho_{(COD)} = 1.94\ g/cm^3$.

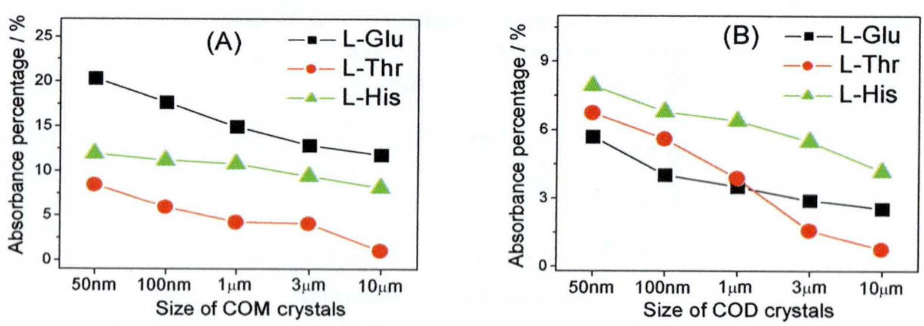

Fig. 3. Adsorption percentage of 0.2 mM L-glutamic acid (Glu), L-threonine (Thr), and L-histidine (His) onto COM, COD crystals with different sizes. (A) COM; (B) COD.

3.3. *Adsorption differences of three kinds of amino acids onto COD*

The comparison of different amino acid: Unlike COM, the adsorption amount of alkaline L-His onto COD is significantly higher than that of L-Glu and L-Thr. (Figure 3B).

Effect of COD sizes: with the sizes of COD increases, its adsorption amount to different amino acid decreased.

3.4. *The adsorption differences amino acids onto COM and COD*

COM crystals mainly adsorbed acidic L-Glu, while COD mainly adsorbed alkaline L-His. This attributed to the difference of surface charge of COM and COD.

COM crystal ($\bar{1}01$) plane with a large positive charge can adsorb COO^- of acidic L-Glu. The COO^- of neutral L-Thr and alkaline L-His exposed less, so the electrostatic interactions with COM is weak, resulting in the adsorption amount is very small, and even have no absorption.

COM is more easily adsorbed substances with negative charge, which is identify with Tunik etc. [9]. According to the literature [9], the adsorption amount of anionic surfactant sodium diisooctyl sulfosuccinate onto COM is also much larger than COD.

Since the crystal water in crystal lattice of COD is more than that of COM, the surface of crystals is easily formed a hydro-layer in aqueous, so that the surface charge of COD is very small in aqueous medium. Therefore, COD has less adsorption to amino acids.

At the same sizes, the absorption amount of amino acid onto COM is greater than that of COD, which is also attributed to the greater specific surface area of COM than COD (Table 1).

3.5. *Effect of three kinds of amino acid concentration on adsorption*

The effect of initial concentration of amino acids (C_0) on the absorption amount of COM and COD was studied, the representative results shown in Figure 4. When the concentration of amino acids was increases, the absolute adsorption amount was also increased, whereas the adsorption percentage was decreased. With the concentration increased to 0.5 mM, because of the adsorption close to saturation, the adsorption percentage increases slowly.

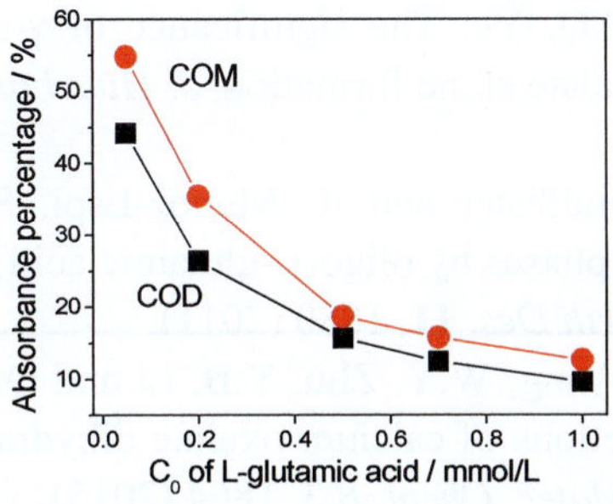

Fig. 4. The adsorption percentage of different concentration of amino acids onto COM, COD.

4. Conclusions

The adsorption of different kinds of amino acid onto COM, COD crystals with the sizes of 50 nm, 100 nm, 1 μm, 3 μm and 10 μm is different. At pH = 7.3, COM mainly adsorbed acidic L-Glu and has less adsorption to neutral L-Thr and alkaline L-His, while COD mainly adsorbed alkaline L-His. The absorption amount of nano crystals is slightly larger than that of micron crystals, this attributed to the larger surface area of nano crystals. At the same sizes, the absorption amount of amino acid onto COM is greater than that of COD, which is attributed to the greater specific surface area of COM than COD.

Acknolwedgements:

This work is supported by grant 21371077 of the National Natural Science Foundation of China.

References

1. T. Lee and Y.C. Lin, Mimicking the initial development of calcium urolithiasis by screening calcium oxalate and calcium phosphate phases in various urine like solutions, time points, and pH values at 37°C, *Cryst. Growth Des.* **11**, 2973 (2011).
2. F. Grases and J.G. March, The crystallization of calcium oxalate in the presence of amino acids, *J. Cryst. Growth* **87**, 299 (1988).
3. S. Atanassova, K. Neykov and I. Gutzow, Solubility and dissolution kinetics of calcium oxalate renal calculi in solutions, containing DL-lysine: in vitro experiments, *J. Cryst. Growth* **212**, 233 (2000).
4. S.P. Dend, D.Z. Chen and J.M. Ouyang, Research progress of the influence of amino acids on the formation of calcium oxalate stones, *Chem. World* **10**, 628 (2005).
5. A.M. Radman, J.H. Liu, S.W. Zhou, S.G. Wang, D. Cai, Q.Y. Zhuang, W.M. Yang and Z.Q. Ye, The significance of γ-carboxyglutamic acid in urinary calcium oxalate stone formation, *J. Huazhong Univ. Sci. Tech.* **4**, 40 (2002).
6. V. Fischer, K. Landfester and R. Munoz-Espí, Stabilization of calcium oxalate metastable phases by oligo(L-glutamic acid): effect of peptide chain length, *Cryst. Growth Des.* **11**, 1880 (2011).
7. X.Y. Sun, J.M. Ouyang, W.Y. Zhu, Y.B. Li and Q.Z. Gan, Size-dependent toxicity and interactions of calcium oxalate dihydrate crystal on Vero renal epithelial cells, *J. Mater. Chem. B* **3**, 1864 (2015).
8. Q.Z. Gan, X.Y. Sun and J.M. Ouyang, Adhesion and internalization differences of COM nanocrystals on Vero cells before and after cell damage, *Mater. Sci. Eng.: C* **59**, 286 (2016).
9. G.M.S. El Shafei and N.A. Moussa, Adsorption of some essential amino acids on hydroxyapatite, *J. Colloid Interface Sci.* **238**, 160 (2001).

Uniform TiO_2/Graphene hybrid nanosheets in a sandwich-like configuration as high-performance anodes for lithium-ion batteries

Guo-Shuai Zhang and Yue-Ping Fang[1, a]

The institute of Biomaterial, College of Science, South China Agricultural University, Guangzhou, Guangdong, China

E-mail: [a]ypfang@scau.edu.cn

The special surface activity and structural characteristics make Titanium dioxide (TiO_2) an ideal candidate for high-rate anode material. However, because of poor electron conductivity, small diffusion coefficient of lithium ions and the tendency of nanoparticle to aggregate, the full electrochemical performance of TiO_2 as Lithium-ion batteries (LIBs) anode have not been reached. Uniform TiO_2 nanosheets on graphene templates in a sandwich-like configuration with variable content of graphene were synthesized by a hydrothermal treatment and heating reduction continuous process. The TiO_2/graphene nanosheets with about 5 wt% graphene content provided a very high first discharge capacity of 315 mAh g-1 at a rate of 0.2C. When graphene content was increased to 11 wt%, the TiO_2/graphene nanosheets delivered a discharge capacity of 93 mAh g-1 at a rate of 50C. We hence proved that TiO_2/GNS is an excellent anode material for high power lithium storage, with high rate capacity and excellent cycle durability. This sandwich-like TiO_2/graphene architecture is a promising anode material with very high lithium storage capacity and desirable cycling stability.

Keywords: TiO_2, Graphene, Nanosheets, Anode, Lithium-ion Batteries.

1. Introduction

Rechargeable lithium-ion batteries (LIBs) are regularly employed in countless electric devices such as small power tools and portable electronics, yet their current performance is insufficient for standalone electric vehicles to viably replace the fossil-fuel engine. Titanium dioxide (TiO_2) is an ideal candidate for high-rate anode material due to its' special surface activity and structural characteristics [1,2]. Unfortunately, because of poor electron conductivity, small diffusion coefficient of lithium ions, and the tendency of nanoparticle to aggregate, the full electrochemical performance of TiO_2 as LIBs anode have not

[1] Corresponding author.

been reached. Effort to improve the lithium performance of TiO_2-based materials is a hot topic of current material sciences research [3-5].

Herein, we report a success in synthesizing uniform TiO_2/graphene nanosheets (TiO_2/GNS) with desired properties by a simple hydrothermal treatment followed by a continuous thermal reduction process, and we proved TiO_2/GNS is an excellent anode material for high power lithium storage, with high rate capacity and excellent cycle durability.

2. Experimental

2.1 *Synthesis of pure TiO_2 and TiO_2/GNS nanocomposites*

Taking appropriate amount of graphene oxide(GO) well dispersed into 36ml deionized water[6], 0.15 ml ammonia and 1.5 ml tetrabutyl titanate were added, the resulting suspension was transferred to a Teflon-lined stainless steel auto-clave and maintained at 90°C for 10 h, cooled, washed repeatedly and drying. The obtained product were calcinated at 500°C under argon or air, which were denoted as TiO_2/GNS or pure TiO_2 (TiO_2/GNS-x, x stands for weight percentage of graphene in the nanocomposites), respectively.

2.2 *Structure and morphology characterization*

XRD, SEM, TEM, HRTEM, and TG analyses were used to characterize pure TiO_2 and TiO_2/GNS nanocomposites

2.3 *Electrochemical measurements*

CR2025 coin cells were assembled in glovebox under the protection of argon, Galvanostatic charge/discharge tests were performed on LAND CT-2001A cell program-control test system at various current densities. Electrochemical impedance spectroscopic (EIS) measurements were carried out on an IM6e electrochemical workstation over the frequency range of 0.01 Hz to 100 kHz while applying an AC signal of 5 mV.

3. Results and discussion

3.1 *Characterization of the TiO_2/GNS nanocomposites*

As shown in Figure 1A, the XRD patterns exhibit characteristic diffraction peaks, which are well indexed to anatase TiO_2 (JCPDS No.21-1272). No other XRD peaks corresponding to graphene in the composite could not be observed in TiO_2/GNS, suggesting the agglomeration of graphene sheets during the reduction process can be effectively prevented [7]. Crystallite size was calculated to be around 5.4 nm by the Scherrer equation based on the parameters of the (101) peaks [8]. Thermogravimetric analyses in Figure 1B indicate that TiO_2/GNS-5, TiO_2/GNS-8 and TiO_2/GNS-11 samples contain about 5 wt%, 8 wt% and 11wt% of graphene, respectively.

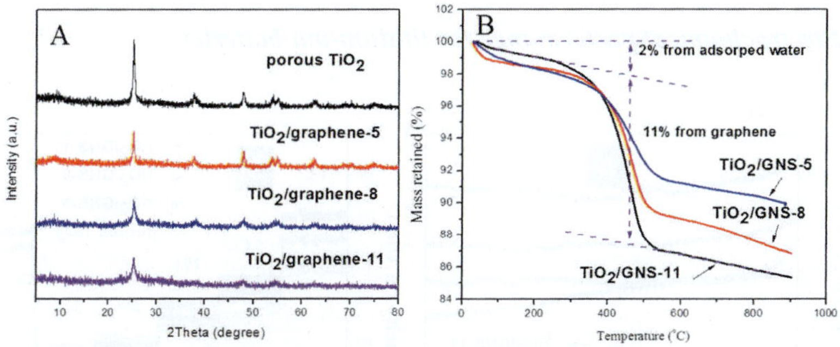

Fig. 1. XRD(A) and TG(B) analyses of samples

Fig. 2. SEM.(A),TEM(B and C) and HRTEM(D) of TiO_2/GNS-5

Figure 2A shows the flexibility of TiO_2/GNS those similar to pure GO nanosheets, suggesting the successful replication from the GO template during the hydrothermal process. The TEM (Figure 2B and C) provide the direct evidence of uniform TiO_2 layer on the GS surface, the TiO_2 nanoparticles interconnected mesoporous network are scattered uniformly on the graphene template. Furthermore, The HRTEM image (inset in Figure 2D) shows clear lattice fringes for the (101) planes anatase TiO_2[8].

3.2 *Electrochemical performance for lithium-ion batteries*

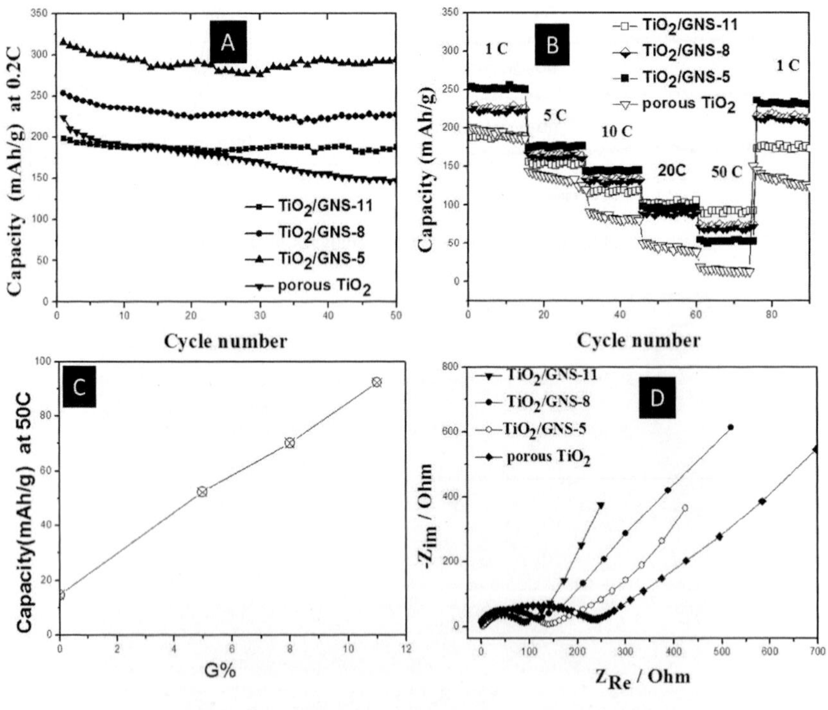

Fig. 3. cycling performances(A) , rate performances(B) and Nyquist plots of pure TiO_2 and TiO_2/GNS-x(D); lithium capacity of TiO_2/GNS-x with different graphene content(G%) at 50C(C).

The cycling performances of TiO_2/GNS and pure TiO_2 samples at a rate of 0.2C were shown in Figure 3A, the reversible capacity for the TiO_2/GNS-5, TiO_2/GNS-8, and TiO_2/GNS-11 samples was determined to be 292, 228, and 187 mA h g^{-1} respectively. More than 90% retention can be retained after 50 cycles at the same current density for three TiO_2/GNS samples. In contrast, the capacity of the pure TiO_2 tends to fade quickly after 50 cycles near a low value of 147 mA h g^{-1} at a rate of 0.2C, demonstrating that the graphene in TiO_2/GNS samples plays an essential role for the improvement of the cycle performance.

The rate performances of the TiO_2/GNS and pure TiO_2 samples at the rates of 1-50C presented in Figure 3B. The TiO_2/GNS-11 exhibit discharge capacities of 190, 153, 119, 104 and 93 mAh g^{-1} for current rates of 1.0C, 5.0C, 10C, 20C and 50C, respectively. The TiO_2/GNS-11 electrode delivers 93 mA h g^{-1} at a rate of 50C, which is 56% of the theoretical capacity. This high-rate performance of the TiO_2/GNS-11 is superior to that of most of the previously reported TiO_2-based nanomaterials[9-11]. Interestingly, as shown in Figure 3C, the discharge capacity at a current rate of 50C increases almost linearly with the increase of G%.

EIS tests of the TiO_2/GNS and pure TiO_2 samples were performed on Figure 3D, which presents analogical depressed semicircles at high frequencies, followed by a vertical tail inclined at about 45° angle toward the real axis corresponds to the lithium diffusion kinetics to the electrode[12]. The TiO_2/GNS electrodes all show a lower resistance [89Ω, 116 Ω and 142 Ω for the TiO_2/GNS-11, TiO_2/GNS-8 and TiO_2/GNS-5, respectively] than that of the pure TiO_2 electrode (236 Ω). This EIS study confirms that the graphene nanosheets sandwiched between mesoporous TiO_2 layer enhance the conductivity of the overall electrode, which is contribute to the improved electrochemical performance of TiO_2/GNS nanocomposites. These results indicate that the introduction of an appropriate graphene component to form a monodisperse and uniform TiO_2/GNS composites increases the utilization efficiency of the TiO_2 and improves the life-cycle of LIBs electrodes by reducing the aggregation problem.

4. Conclusion

The uniform TiO_2/GNS with variable content of graphene have been successfully prepared by a hydrothermal treatment and heating reduction continuous process. Due to its' homogeneous dispersion of ultra-small nanoparticle on graphene layer, the sandwich-like TiO_2/GNS nanocomposites exhibit a high rate capability and excellent cycle performance, holding great potential as a high-rate anode material for lithium storage. This study will inspire better design of graphene-based nanocomposites for high energy density lithium-ion batteries.

Acknowledgments

This research was supported by the NSF of China (21173088)

References

1. M.M. Rahman, J.-Z. Wang, M.F. Hassan, D. Wexler, H.K. Liu, *Advanced Energy Materials*, 2011, 1: 212-220.
2. J. Liu, J.S. Chen, X. Wei, X.W. Lou, X.-W. Liu, *Advanced Materials*, 2011, 23: 998-1002
3. S. Bach, J.P. Pereira-Ramos, P. Willman, *Electrochimica Acta*, 2010, 55: 4952-4959.
4. V. Luca, T.L. Hanley, N.K. Roberts, R.F. Howe, *Chemistry of Materials*, 1999,11: 2089-2102.
5. D.V. Bavykin, J.M. Friedrich, F.C. Walsh, *Advanced Materials*, 2006, 18: 2807-2824.
6. D.C. Marcano, D.V. Kosynkin, J.M. Berlin, A. Sinitskii, Z. Sun, A. Slesarev, L.B. Alemany, W. Lu, J.M. Tour, *ACS Nano*, 2010, 4: 4806-4814.
7. H. Zhang, X. Lv, Y. Li, Y. Wang, J. Li, *ACS Nano*, 2009, 4: 380-386.
8. H.I. Lee, S.H. Joo, J.H. Kim, D.J. You, J.M. Kim, J.-N. Park, H. Chang, C. Pak, *Journal of Materials Chemistry* ,2009, 19: 5934-5939
9. S.-W. Kim, T.H. Han, J. Kim, H. Gwon, H.-S. Moon, S.-W. Kang, S.O. Kim, K. Kang, *ACS Nano*, 2009, 3: 1085-1090
10. Y. Liang, Y. Li, H. Wang, J. Zhou, J. Wang, T. Regier, H. Dai, *Nature Materials*, 2011.10: 780-786
11. Y.S. Hu, L. Kienle, Y.G. Guo, J. Maier, *Advanced Materials*, 2006, 18: 1421-1426
12. S. Yang, X. Feng, K. Müllen, *Advanced Materials*, 2011, 23: 3575-3579.

Green synthesis of silver nanoparticles and their application for catalytic degradation of hazardous reactive dye in aqueous solution

Ping Yao*†, Hui-Ying Wu, Jun Zhang, Mao-Mao Ding, Ran Tao and Lei Xu

Suzhou Institute of Trade and Commerce, 287 Xuefu Road, Suzhou City, Jiangsu Province 215009, P.R. China

†E-mail: szjm_qfx@163.com

Silver nanoparticles (AgNPs) have been recognized in recent years because of their unique properties. In this paper, plant-mediated biosynthesis of silver nanoparticles (AgNPs) was investigated. Aloe extract acted as both reducing and stabilizing agent in the synthesis of silver nanoparticles. The synthesis of nanoparticles by various physical and chemical methods, such as ultrasonic irradiation, laser ablation and evaporative cooling (to name a few) has been studied extensively. However, these methods pose environmental threat. AgNPs will be photosynthesized by aloe extract, and the catalytic activity of green synthesized AgNPs will be discussed by catalytic degradation of reactive orange dye in the presence of NaBH4 in the aqueous solution. The synthesized AgNPs were characterized with UV-visible absorption spectrum, Transmission Electron Microscopy (TEM), and particles size distribution analysis. UV-visible absorption spectrum indicated a significant absorption peak at 463 nm related to the great synthesis of AgNPs in the aqueous solution. TEM images showed the well dispersed AgNPs, and that the average diameter of AgNPs was around 38nm. Furthermore, the AgNPs were utilized for the removal of hazardous reactive dye in the aqueous solution. From the experimental results, the AgNPs showed excellent catalytic activity in the degradation of reactive orange dye in the presence of NaBH4.

Keywords: Silver Nanoparticles; Aloe Extract; Green Synthesis; Catalytic Property; Reactive Dye Catalytic Degradation.

1. Introduction

Silver nanoparticles (AgNPs) have attracted more attention in recent years due to their unique properties[1-3]. The synthesis of nanoparticles by various physical and chemical methods, such as ultrasonic irradiation [4-5], laser ablation [6], evaporative cooling [7], chemical vapor deposition [8], explosion [9], impregnation [10-11], co-precipitation [12-13], sol-gel [14] and deposition-precipitation [15], has been reported extensively. Although these methods have

* Corresponding author.

successfully produced well-defined nanoparticles, there are environmental problems during the reaction processes. At the same time, plant-mediated bioprocess provides an important opportunity for the green synthesis of AgNPs, and many kinds of plant extracts have been used for the synthesis of AgNPs[16]. Aloe extract, as one kind of the plant extract, has already been used for the phytosynthesis of silver nanoparticles[17], but the catalytic application of aloe extract synthesized AgNPs has not been studied. In this paper, AgNPs will be photosynthesized by aloe extract, and the catalytic activity of green synthesized AgNPs will be discussed intensively by catalytic degradation of reactive orange dye in the presence of $NaBH_4$ in the aqueous solution.

2. Materials and Methods

2.1. *Materials*

$AgNO_3$ (99.8%) and $NaBH_4$ (98%) was purchased from Aladdin Industrial Corporation. Aloe extract (98%) was purchased from Run Shanxi Days of Biological Technology Co., Ltd. The reactive orange dye was provided by the Zhejiang Runtu Co., Ltd. All the chemicals and reagents used in the paper were of analytical grade.

2.2. *Synthesis of AgNPs*

10mL silver nitrate (0.01M) was added to 90mL of freshly prepared aloe extract in the ratio of 1:5. Progressively, formation of AgNPs was observed by visual color change from transparent color to darker shades of brown, which was further confirmed by UV-vis spectrophotometer.

2.3. *Characterization*

The reduction of silver ions was recorded by measuring the UV-visible absorption spectra of the liquid solutions in the beginning and after the synthesis reaction. The UV-visible absorption spectrum analysis was done by using UV-visible spectrophotometer (Shimadzu UV-3600) in the wavelength of 200-800nm. The size and morphology of Ag-NPs were determined by a TecnaiG220 (FEI, USA) at an accelerating voltage of 300kv. Samples were prepared by placing a drop of AgNPs solution on the TEM carbon-coated copper grid, followed by solvent evaporation at room temperature. Particle size distribution of bio-reduced Ag-NPs in the aqueous solution was measured by the Zetasizer Nano series (Malven, UK) at 25°C.

2.4. *Catalytic activity of AgNPs*

Catalytic activity of synthesized AgNPs was performed by degrading reactive orange dye. 3mL reactive orange dye (30mg/L) solution was added with 0.1 mL $NaBH_4$ (0.2M) and 0.1ml AgNPs (100μg/mL) solution The reduction process was monitored by recording spectra using UV-vis spectrophotometer at regular time intervals.

3. Results and Discussion

3.1. *Characterization of AgNPs*

When the aloe extract was subjected to the $AgNO_3$ solution, the biosynthesis reaction started within few minutes and the color reaction was observed in which clean $AgNO_3$ solution turned into darker shades of brown which indicated the formation of AgNPs. The UV-visible absorption spectra of original aloe extract and AgNPs were shown in Figure 1. As shown from the UV-visible absorption spectra, a distinct peak of 468 nm was observed which belongs to surface plasmon absorption band of AgNPs.

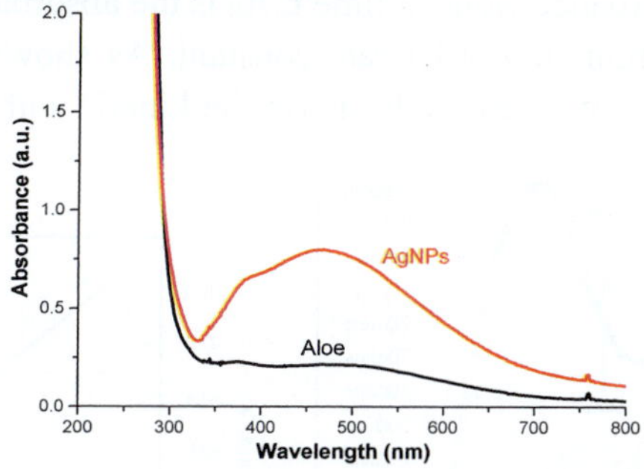

Fig.1. UV-vis spectra of silver nanoparticles by reducing $AgNO_3$ with aloe extract.

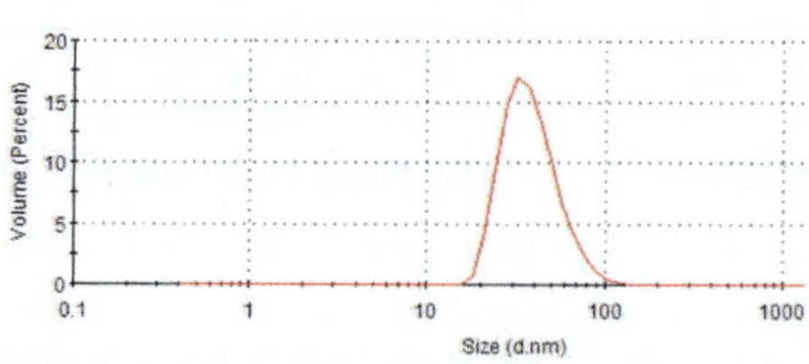

Fig.2. Particle size distribution of synthesized AgNPs.

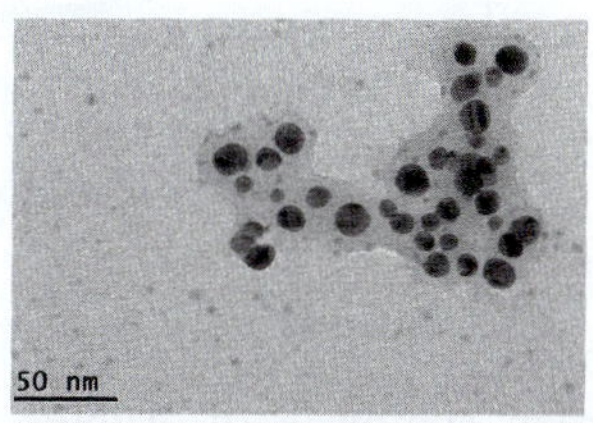

Fig.3. TEM image of synthesized AgNPs.

The particle size distribution and morphology of AgNPs were characterized by Zetasizer Nano series and transmission electron microscope (TEM). Figure 2 and Figure 3 show the particle size distribution and TEM image of AgNPs, respectively. As shown from Figure 2, the average size of AgNPs was around 38nm, which was similar to the TEM images (Figure 3). TEM images also confirmed a typical spherical and an ellipsoidal morphology of AgNPs.

3.2. *Catalytic degradation of reactive orange dye*

AgNPs are well known for their catalytic activity and many reports are available on AgNPs as catalyst for the reduction/degradation of environmental pollutants such as 4-nitrophenol, MB and dyes [1]. Figure 4 shows the degradation of reactive orange dye using $NaBH_4$ as reduction agent in the presence of AgNPs. As soon as the catalyst was added, the absorbance at 477nm decreased significantly in less than 3 mins and growth of a new peak at 386nm was observed. As shown from the Figure 4, most of the reactive orange dye was degraded within 40mins.

In addition, the reaction kinetic of reduction process was calculated using following the equation:

$$\ln (A/A_0) = - kt$$

where A is absorbance value at time t, A_0 is the absorbance value at time t = 0 and k is the apparent first order rate constant. As shown from Figure 5, the reaction process followed the pseudo first-order kinetic with $k = 0.0181\ min^{-1}$.

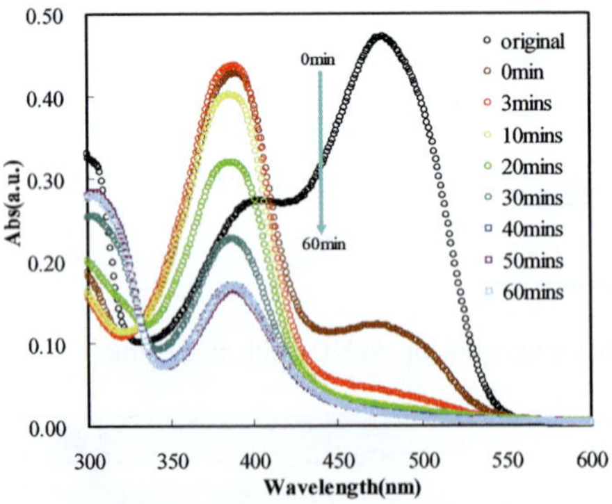

Fig.4. Degradation of reactive orange dye.

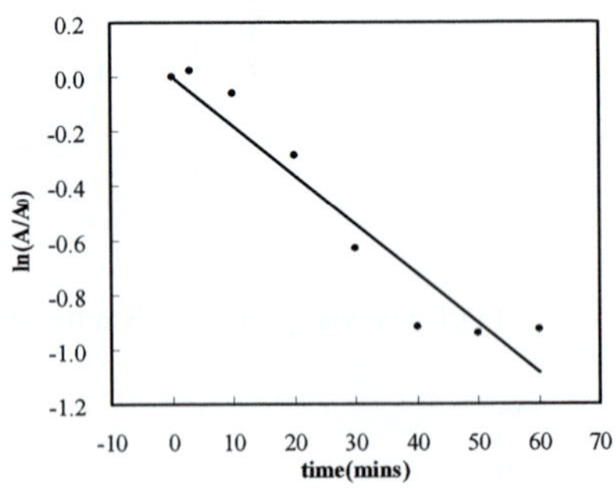

Fig.5. Plot of $\ln(A/A_0)$ against time.

4. Conclusions

Green synthesis of AgNPs was achieved successfully using aloe extract as reducing and capping agent. The average diameter of AgNPs was around 38nm, which was testified by the particle size distribution analysis and TEM image.

The AgNPs was utilized for the degradation of reactive orange dye in the aqueous solution, and most of reactive orange dye was degraded within 40mins.

Acknowledgments

This work was supported by Jiangsu Natural Science Foundation of China (No. BK 20130344), National Special Foundation of Cocoon Silk Development Project 2015 (The Development and Industrialization of Digital Printing Composite Functional Silk Products), Science and Technology Guidance Project of China National Textile & Apparel Council 2015 (No.2015060), and Middle and Young Academic Leaders of Blue Project 2014 (Jiangsu Provincial Education Department).

References

1. Bogireddy, N. K. R., Kiran Kumar, H. A., & Mandal, B. K. (2016). Biofabricated silver nanoparticles as green catalyst in the degradation of different textile dyes. Journal of Environmental Chemical Engineering, 4(1), 56-64.
2. Boroumand, M. N., Montazer, M., Simon, F., Liesiene, J., Šaponjic, Z., & Dutschk, V. (2015). Novel method for synthesis of silver nanoparticles and their application on wool. Applied Surface Science, 346, 477-483.
3. Dhand, V., Soumya, L., Bharadwaj, S., Chakra, S., Bhatt, D., & Sreedhar, B. (2016). Green synthesis of silver nanoparticles using Coffea arabica seed extract and its antibacterial activity. Materials Science and Engineering: C, 58, 36-43.
4. Byeon, J. H., & Kim, Y. W. (2012). A novel polyol method to synthesize colloidal silver nanoparticles by ultrasonic irradiation. Ultrasonics Sonochemistry, 19(1), 209-215.
5. He, C., Liu, L., Fang, Z., Li, J., Guo, J., & Wei, J. (2014). Formation and characterization of silver nanoparticles in aqueous solution via ultrasonic irradiation. Ultrasonics Sonochemistry, 21(2), 542-548.
6. Pathrose, B., Nampoori, V. P. N., Radhakrishnan, P., Sahira, H., & Mujeeb, A. (2016). Effect of femtosecond laser ablated silver nanoparticles in the thermo-optic properties of basic fuchsin dye. Optik - International Journal for Light and Electron Optics, 127(7), 3684-3687.
7. Fisenko, S. P., & Khodyko, J. A. (2009). Low pressure evaporative cooling of micron-sized droplets of solutions and its novel applications. International Journal of Heat and Mass Transfer, 52(15-16), 3842-3849.

8. Zhang, X., Xu, S., Jiang, S., Wang, J., Wei, J., Xu, S., ... Li, H. (2015). Growth graphene on silver-copper nanoparticles by chemical vapor deposition for high-performance surface-enhanced Raman scattering. Applied Surface Science, 353, 63-70.
9. Lerner, M. I., Pervikov, A. V., Glazkova, E. A., Svarovskaya, N. V., Lozhkomoev, A. S., & Psakhie, S. G. (2016). Structures of binary metallic nanoparticles produced by electrical explosion of two wires from immiscible elements. Powder Technology, 288, 371-378.
10. Davidovic, S., Mijkovic, M., Lazic, V., Jovic, D., Jokic, B., Dimitrijevic, S., & Radetic, M. (2015). Impregnation of cotton fabric with silver nanoparticles synthesized by dextran isolated from bacterial species Leuconostoc mesenteroides T3. Carbohydrate Polymers, 131, 331-336.
11. Hassabo, A. G., Nada, A. A., Ibrahim, H. M., & Abou-Zeid, N. Y. (2015). Impregnation of silver nanoparticles into polysaccharide substrates and their properties. Carbohydrate Polymers, 122, 343-350.
12. Lin, C. C., Lin, Y. S., & Ho, J. M. (2016). Adsorption of Reactive Red 2 from aqueous solutions using Fe3O4 nanoparticles prepared by co-precipitation in a rotating packed bed. Journal of Alloys and Compounds, 666, 153-158.
13. Pazhanivelu, V., Selvadurai, A. P. B., Zhao, Y., Thiyagarajan, R., & Murugaraj, R. (2016). Room temperature ferromagnetism in Mn doped ZnO: Co nanoparticles by co-precipitation method. Physica B: Condensed Matter, 481, 91-96.
14. Ivanova, T., Harizanova, A., Koutzarova, T., & Vertruyen, B. (2013). Optical and structural characterization of TiO2 films doped with silver nanoparticles obtained by sol-gel method. Optical Materials, 36(2), 207-213.
15. Zahoranová, T., Mori, T., Yan, P., Ševčíková, K., Václavů, M., Matolín, V., & Nehasil, V. (2015). Study of the character of gold nanoparticles deposited onto sputtered cerium oxide layers by deposition-precipitation method: Influence of the preparation parameters. Vacuum, 114, 86-92.
16. Dhand, V., Soumya, L., Bharadwaj, S., Chakra, S., Bhatt, D., & Sreedhar, B. (2016). Green synthesis of silver nanoparticles using Coffea arabica seed extract and its antibacterial activity. Materials Science and Engineering: C, 58, 36-43.
17. Zhang, Y., Cheng, X., Zhang, Y., Xue, X., & Fu, Y. (2013). Biosynthesis of silver nanoparticles at room temperature using aqueous aloe leaf extract and antibacterial properties. Colloids and Surfaces A: Physicochemical and Engineering Aspects, 423, 63-68.

Chapter 3

Polymer and Concrete Materials

Preparation, characterization and photopatterning of copolymer containing imidazo phenanthroline Langmuir-Blodgett films

Ke-Ke Zou[1,2],Tie-Sheng Li[1,*], Tao Wang[1], Na Zhao[1],Wen-Jian Xu[1]and Yang-Jie Wu[1]
[1]*The College of Chemistry and Molecular Engineering, Zhengzhou University Zhengzhou,450052, P. R China.*
[2]*Pingdingshan Hydrology and Water Resources Survey Bureau of Henan Province, Pingdingshan, 467000, P.R China.*
*E-mail: *13619891880@163.com*

The Langmuir- Blodgett (LB) technique makes it possible to prepare a thin film with a controlled thickness at the molecular level and a well-defined molecular orientation. It has also been studied extensively for photolithography in which it cannot only overcome the weakness of spin-coat films where molecules are distributed randomly, but also be expected to improve the resolution and the sensitivity of the resist effectively. A series of novel copolymer p (AMAc- pIPPMa) has been prepared by reactions of 1H-imidazo [4,5-f] [1,10] phenanthrolin-2-yl)phenyl methacrylate (pIPPMa) and methacrylate (AMAc). They could be formed well-ordered Langmuir monolayer at air/water interface and transferred onto substrates to form Langmuir- Blodgett (LB) films. Preliminary studies of photopatterning of LB films were carried out and the positive patterns were obtained, where a resolution of 0.75μm could be drawn. It confirmed that LB films were a fine material as resist film for photolithography.

Keywords: Copolymer; LB films; Photopatterning; Photolitography;

1. Introduction

Photolithography is extremely important in the production of integrate circuit, because it determines the smallest characteristic size of semiconductor device and the scale of integration of ICs[1,2], in which materials and its applied forms play a important role. Design and synthesis of novel resists have received much attention [3,4] due to controlling the structure of material in the nanometer scale is the key to achieving nonlinear performance improvement[5-7]. The Langmuir- Blodgett (LB) technique makes it possible to prepare a thin film with a controlled thickness at the molecular level and a well-defined molecular orientation [8,9] and has been studied extensively for photolithography [10-14]

* Corresponding author.

in which it cannot only overcome the weakness of spin-coat films[15-16], in which molecules are distributed randomly, but also be expected to improve the resolution and the sensitivity of the resist effectively.

In this paper, a novel class of copolymer p(AMAc-pIPPMa) was prepared by reactions of pIPPMa (1H-imidazo[4,5-f][1,10]phenanthrolin-2-yl)phenyl methacrylate) and AMAc (methacrylate) shown in Scheme 1. Behaviors of the copolymers monolayer at air/water interface and photolithography properties of their LB films were investigated.

2. Experimental

1,10-phenanthroline-5,6-dione was synthesized by the reaction of 1,10-phenanthroline dissolved in concentrated sulfuric acid with potassium bromide and concentrated nitric acid. The product reacted with hydroxyl benzaldehyde to obtain 1H-imidazo[4,5-f][1,10] phenanthrolin -2-yl)phenol.Then, the phenol reacted with methacryloyl chloride toyield 1H-imidazo[4,5-f][1,10]phenanthrolin-2-yl)phenyl methacrylate (pIPPMa). ^{1}H NMR: δ (ppm) DMSO-d6, 1.98 (s, 3H, CH_3), 5.74-5.95 (d, 2H, $=CH_2$), 7.22-7.46 (m, 6H, CH=C), 7.85 (m, 2H, C=CH) 8.91-9.03 (t, 2H, CH =N). IR: ν (cm-1) 3391.15 ($\upsilon_{N\text{-}H}$), 3083.65 ($\upsilon_{=C\text{-}H}$), 1733.52 ($\upsilon_{C=O}$), 1609.86, 1560.50, 1504.26, 1482.36 ($\upsilon_{C=C}$), 1175.37 ($\upsilon_{C\text{-}O\text{-}C}$), 1015.47,946.15 ($\delta_{=C\text{-}H}$), 1451.87,1320.59 ($\delta_{C\text{-}H}$), 808.38, 738.63. MS: 380.8 402.7.

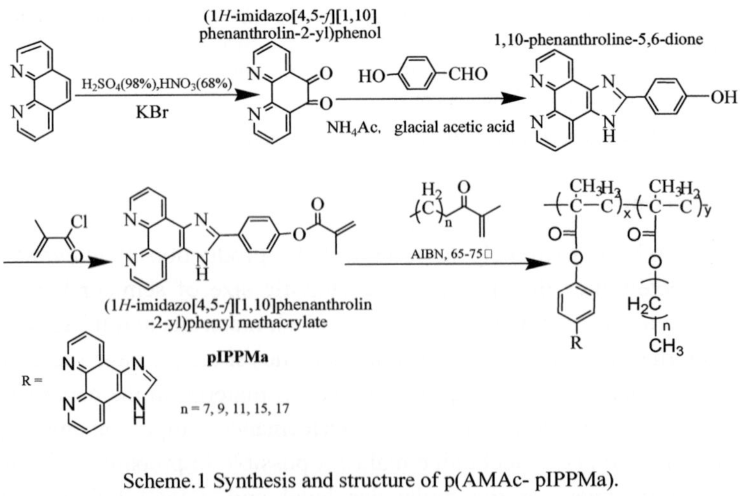

Scheme.1 Synthesis and structure of p(AMAc- pIPPMa).

3. Results and Discussion

3.1. *Behavior of the Langmuir monolayer and LB films*

The Surface pressure (π)-area (A) isotherms of copolymer p(AMAc- pIPPMa)s with different ratio of AMAc were measured at 20°C (Figure 1). The results proved that AMAc with the longer chain had the higher collapse pressure of p (AMAc- pIPPMa) (p7-p17, n=7- 17). Therein, p17 (p (AMAc- pIPPMa) [n=17]) showed palpable solid state and higher collapse pressure. p15 has similar shape to p17, but its collapse pressure is lower. That indicated that the shorter chain of p15 infected the polymer molecular arrangment on the air/ water interface. p7, p9 and p11 did not have palpable solid state and high collapse pressure, because the molecules aggregated with the increasing surface pressure. Then, we took p17 as example to expatiate the films' properties.

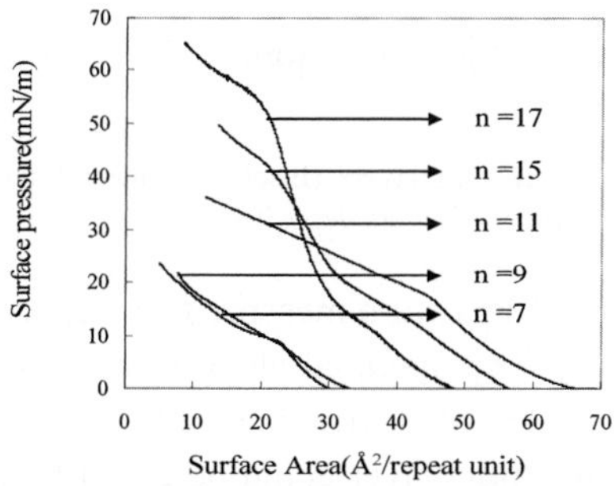

Fig.1. Surface pressure (π)-area (A) isotherms of p (AMAc- pIPPMa).

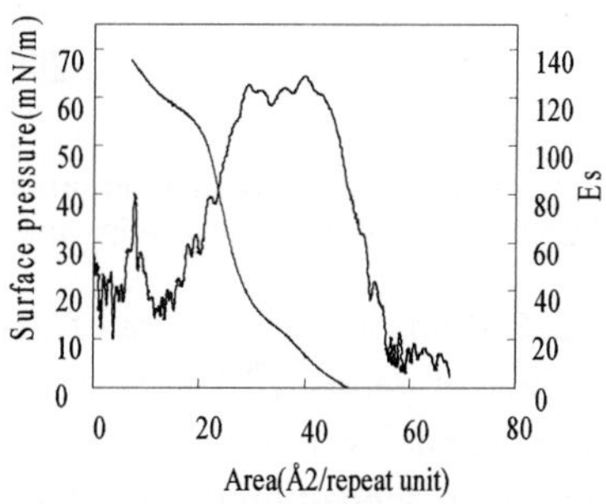

Fig.2. The π-A isotherm and Es-A is otherm of p17 (p (AMAc- pIPPMa),n=17)

In order to achieve optimum conditions for LB films deposition of p (AMAc- pIPPMa) monolayer, the static elasticity (Es)- surface pressure (π) isotherm of copolymer p17 monolayer had been studied (Figure 2). The Es-π curve was better, plainer and more sensitive to describe the monolayer state than π-A isotherm, because elasticity could give more detailed information on molecular interaction and motion. The high Es value and the wide peak implied that p17 could form a stable monolayer in a wider rang of surface pressure. There are different styles during the whole compressing procedure from the two isotherms. In the "gas state", both of them show flattened line without significant transition, because the distance among molecules is too large to interact and increase of surface pressure (π). In the "gas-liquid transition", they show different change, the surface pressure (π) changes smoothly and slightly. The surface pressure increases in "liquid state" smoothly with the decreases of surface area and the elasticity shows flattened line with some small waves. When the monolayer collapse, the surface pressure in π-A isotherm begins to go through unsmooth change, and in the Es-A isotherm, the elasticity decrease severely and even as low as the "gas state", that is because when the monolayer collapse, the close and ordered arrangement of molecules at air/water interface was destructed into disordered arrangement . From the isotherms, we decide that all the p17 LB films were prepared at 30mN/m on the water surface at 20□.

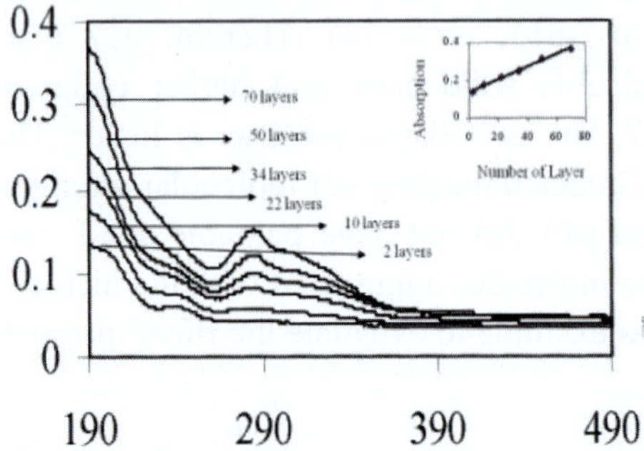

Fig.3. UV spectra of **p17** LB films in different number of layers (Inset is a linear relationship between maximum UV spectra absorption and the number of layers deposited).

The UV absorption spectra of **p17** LB films were measured as a function of the number of deposited layers (Figure 3). The maximal absorbance at 190 nm was proportional to the number of layers deposited. It indicates that the regular homogeneous deposition occurred and **p17** monolayer could be transferred onto solid supports as a Y-type LB films under a given surface pressure with the transfer ratio almost unity.

3.2. ***Morphology of the Langmuir monolayer and LB films***

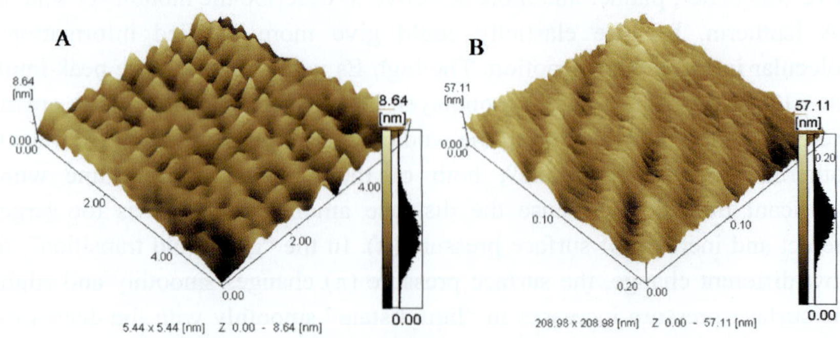

Fig.4 AFM images of **p17** monolayer and LB films deposited on silicon.

The condensed copolymer monolayer at the water surface could be transferred onto solid supports such as mica, quartz, and silicon wafer. Figure 4. showed that monolayer and 10 layers LB films of copolymer **p17** obtained by transferring onto silicon wafers at a surface pressure of 30mN/m, then the LB films were imaged by AFM at different scales. Figure 4A showed the typical AFM images of **p17** LB monolayer. The high region was long AMAc chain in copolymer stretching in the air, and the low region was referred to the pIPPMa molecules which were arranged with regular "pyramid" in uniform line zone. They showed that monolayer had homogeneous topography. Figure 4B showed the 10 layers LB films. The films presented regular "waviness". It indicated that each layer could be transferred uniform and a condensed LB films were formed.

3.3. *Photodecomposition of copolymer LB films*

Photodecomposition mechanisms in copolymer LB films irradiated by deep UV light were investigated. The changes of UV absorption spectra of copolymer **p17** LB films (50 layers) under 248nm UV irradiation were observed (Figure 5). The absorbance at 193 nm and 280 nm were assigned to backbone (carbon chain) and aromatic group of copolymer, respectively, decreased by irradiation time. Apparently, upon irradiation, the **p17** LB films may underwent photodegradation of the polymer backbone or scission of side chain (ester group, C-O-C).

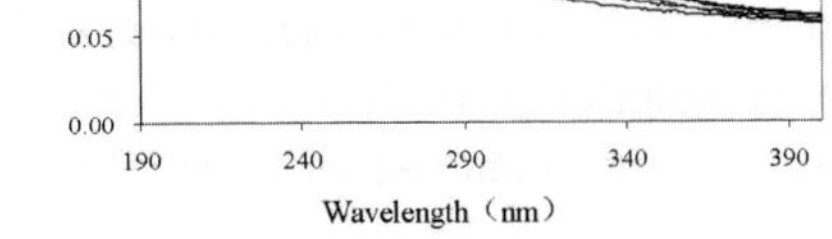

Fig.5 Changes of maximum UV absorption spectra of **p17** LB films by 248 nm UV irradiation at different time.

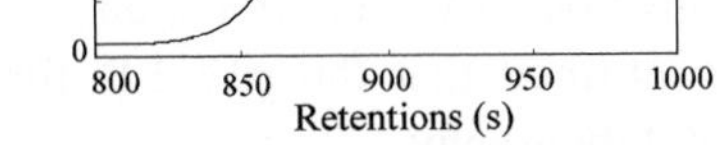

Fig.6 GPC curves of cast film of **p17** before and after UV irradiation for 60 min

The GPC measurements of copolymer **p17** cast films were carried out (Figure 7). It confirmed that the back bone of copolymer really happened during the deep UV irradiation process. As it is clearly shown, the molecular weight decreased with increasing irradiation time and the curve became broader, indicating that the molecular weight distribution is larger than that of the initial state. These findings suggest that the main chain (carbon chain, C-C) scission occurred during deep UV irradiation.

3.4 . *Photopatterning*

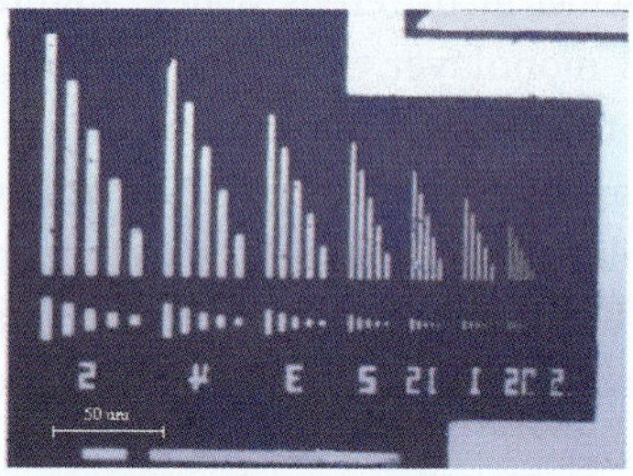

Fig.7. Optical micrograph of positive fine patterns with **p17** LB films (41 layers) deposited on a silicon wafer after UV irradiation for 45 mins , followed development with acetone for 15s.

4. Conclusions

A series of copolymer p(AMAc- pIPPMa) (p17) were prepared and characterized. They could form condensed monolayer at the air/water interface and be transferred onto solid substrate successfully. The maximum absorption of p17 at 190 nm is proportional to the number of layers deposited, indicating that the monolayer were deposited regularly onto substrates to form LB films. Morphology of the p17 monolayer and multilayer were also investigated with AFM method. The images obtained showed that p17 monolayer and multilayer had homogeneous topographic, indicating that p17 monolayer could be transferred uniformly and a condensed LB films were formed. Photodecomposition of copolymer LB films were carried out by GPC. It confirmed that both the main chain (carbon chain, C-C) and the side chain (ester group, C-O-C) scission occurred during deep UV irradiation at same time. Preliminary studies of photopatterning of p17 LB films were carried out and the positive patterns were obtained, , where a resolution of 0.75μm could be drawn. It confirmed that the p17 LB films were a fine material as resist film for photolithography.

Acknowledgements

This work is supported by the Innovation Fund for Young Scholar of Henan Province (074100510015).

References

1. L. R. Harriott, Next Generation Lithography [J]. *Materials Science in Semiconductor Processing*. 1998, 1, 93-97.
2. C. Ngo and C. Rosilio, Lithography for Semiconductor Technology [J]. *Nuclear Instruments and Methods in Physics Research B*. 1997,131, 22-29.
3. H. Gokan, S. Esho and Y. J. Onishi, A gap reduction technique for obtaining submicron geometries utilizing reduction effect, [J]. *Electrochem. Soc*., 1983, 130(1), 143-146
4. Ren et al. 3rd International Symposium on Advanced Materials in Asia-Pacific Rim. *Instruments and Methods in Physics Research B*. 1997,131, 22-29.
5. Kuan et al. Journal of Vacuum Science & Technology, B: Microelectronics and Nanometer Structures, 1988, 6 (6), 2274-2279.
6. Y.Z. Guo, M. Mitsuishi and T. Miyashita, Poly(N-alkylmethacrylamide) LB Films with Short- Branched Alkyl Side Chains for a Self-Developed Positive Photoresist, *Macromolecules*, 2001, 34 (11), 3548- 3551
7. T.S. Li, M. Mitsuishi and T. Miyashita, Photodegradable polymer LB films for nano-lithographic imaging techniques, *Thin Solid Films,* 2001, 389, 267-271
8. K. B. Blodgett, Non-ideal adsorption of divalent cations on a Langmuir monolayer: A theoretical model for predicting the composition of the resulting Langmuir-Blodgett films , *J. Am. Chem. Soc*. 1935, 57,1007-1022.
9. K.B. Blodgett and Langmuir I, Built-Up Films of Barium Stearate and Their Optical Properties, *Phys. Rev*. 1937, 51,964-982.
10. A. Aoki, M. Nakaya and T. Miyashita, Drawing Fine Patterns with Cross-Linkable Polymer Langmuir-Blodgett Film, *Chemistry Letters*, 1996, 8, 667-668.
11. Y.Z. Guo, F. Feng and T. Miyashita, A New Type of Dry-developed Positive Deep UV Resist Using Polymer Langmuir-Blodgett Films, *Chemistry Letters*, 1998, 12, 1269-1270.
12. Y.Z. Guo, F. Feng and T. Miyashita, Preparation of Poly(N-alkylmethacrylamide) Langmuir- Blodgett Films for the Application to a Novel Dry- Developed Positive Deep UV Resist, *Macromolecules*, 1999, 32 (4), 1115-1118
13. A. Barraud, Polymerization in Langmuir-Blodgett films and resist applications, *Thin Solid Films*, 1983, 99,317-321.
14. Sudiwala et al. Polydiacetylene mono- and multilayer Langmuir-Blodgett films as electron beam resists: scanning electron microscope and Raman images, *Thin Solid Films*,1992, 210-211,452-454.

15. Hui et al. Fabrication of ceramic films for solid oxide fuel cells via slurry spin coating technique, *Journal of Power Sources*, 2007, 172, 840-844
16. Jasinski et al. Applications of spin coating of polymer precursor and slurry suspensions for Solid Oxide Fuel Cell fabrication, *Journal of Power Sources* , 2009, 194, 10-15

Mechanical properties of desert sand high strength concrete with different desert sand replacement ratio and fly ash dosage

Hai-Feng Liu*, Qiao-Li Chen and Jie Fu

School of Civil and Hydraulic Engineering, Ningxia University,
Yinchuan, Ningxia 750021, China
†E-mail: liuhaifeng1557@163.com

Experiments on the compressive strength and splitting tensile strength of desert sand high strength concrete with different fly ash dosage and desert sand replacement ratio were carried out. The regulation on the influence of fly ash dosage and desert sand replacement ratio on the compressive strength and splitting tensile strength of high strength desert sand concrete was analyzed. Experiment results show that with the increase of desert sand replacement ratio, the compressive strength and splitting tensile strength of high strength desert sand concrete increases first, followed by a decline. With the enhancement of fly ash dosage, the compressive strength and splitting tensile strength of high strength desert sand concrete also increases first, then declines, which provides guidance for the engineering application of desert sand.

Keywords: Desert Sand High Strength Concrete; Fly Ash; Desert Sand; Compressive Strength; Splitting Tensile Strength.

1. Introduction

Fly ash was additive in concrete mixtures to produce good workability and adhesiveness [1-2]. China is one of districts distributed widely with desert. With the development of western strategy, medium sand resource no longer meets the requirement of engineering. In order to use local desert sand resource fully, it is necessary to research the mechanical properties of desert sand concrete.

Currently, many researches [3-10] had been focused on the mechanical properties of desert sand concrete. There are few documents concentrated on the

* Corresponding author.

influence of fly ash dosage and desert sand replacement ratio on the mechanical properties of desert sand high strength concrete.

In this paper, experiments on the compressive strength and splitting tensile strength of desert sand high strength concrete with different fly ash dosage and desert sand replacement ratio were carried out. The regulation on the influence of fly ash dosage and desert sand replacement ratio on the compressive strength and splitting tensile strength of desert sand high strength concrete was analyzed.

2. Material

Cement was a locally produced ordinary Portland cement P.O 42.5R. Coarse aggregate with the maximum size of 25mm in concrete mixtures was bought from Zhengbeipu town, Yinchuan, Ningxia, China. Medium sand was man-made water washed sand. Desert sand was brought from Mu us sandy land, Yanchi county, China, the finesses of which was 0.194. Fly ash was bought from local electric power plant. Carboxylic acid high performance water reducing agent produced by Beijing Muhu Admixture Company was used.

3. Experiment design

Water-binder, water usage and sand ratio were 0.24, 180kg/m^3 and 30%, respectively. Fly ash dosage was 0%, 15% and 25%. Desert sand replacement ratio was 0%, 20%, 30%, 40%, 60%, 80% and 100%, respectively. The dimension of specimen was 100mm cube. After demoulding, the specimens were wet-cured in the laboratory until testing. The uni-axial compression experiment and splitting tensile test were carried out by Hydraulic universal testing machine.

4. Experimental results and analysis

4.1. *Uni-axial compression experiment*

Uni-axial compression strength of desert sand high strength concrete was listed in Table 1. Compared with the medium value of three compression strength values every group, the max error is less than 15%, which shows that the experimental results are effective[9].

Table 1. Results of uni-axial compressive strength

No	Fly ash dosage /%	Desert sand replacement ratio/%	Uni-axial compression strength/ MPa				
			1	2	3	max error/%	average value
F01	0	0	76.2	81.2	85.8	6.2	81.1
F02		20	76.5	84.8	82.4	7.2	81.2
F03		30	79.0	80.5	81.6	1.8	80.4
F04		40	85.1	81.13	72.7	10.4	79.6
F05		60	74.7	76.5	73.5	1.3	74.9
F06		80	68.9	73.4	76.6	6.2	73.0
F07		100	71.5	67.6	76.8	7.3	72.0
F101	15	0	78.2	88.7	70.2	13.4	82.0
F102		20	93.3	85.7	87.9	6.4	88.9
F103		30	80.5	78.1	84.6	5.1	81.0
F104		40	77.6	81.5	82.3	5.0	80.5
F105		60	75.8	78.7	73.2	3.8	75.9
F106		80	73.1	69.7	82.6	13.1	75.1
F107		100	73.5	74.7	72.6	1.5	73.6
F151	25	0	69.3	80.8	72.3	11.7	74.1
F152		20	81.4	79.9	68.8	13.9	76.7
F153		30	78.8	76.6	73.6	3.8	76.3
F154		40	73.1	71.3	77.8	6.5	74.1
F155		60	66.4	79.2	72.6	9.2	72.7
F156		80	73.2	66.4	63.4	10.2	67.6
F157		100	66.8	65.7	69.4	3.8	67.3

Figure 1 showed the relation between desert sand replacement ratio and compressive strength of desert sand high strength concrete. It can be seem from Table 1 and Fig. 1 that with the enhancement of desert sand replacement ratio, the compression strength of desert sand high strength concrete increases firstly, and then declines. When the desert sand replacement ratio is equal to 20%, the compression strength of desert sand high strength concrete reaches the maximum value.

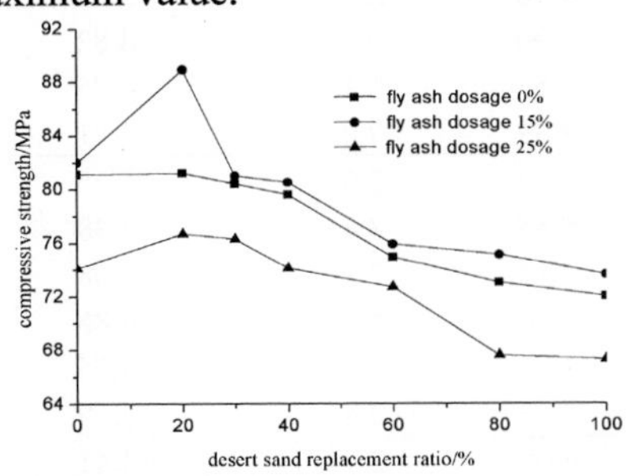

Fig.1. Relation between desert sand replacement ratio and compressive strength

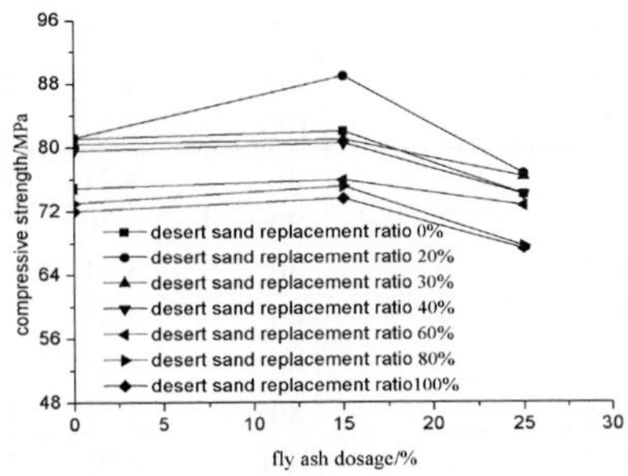

Fig. 2 Relation between fly ash dosage and compressive strength

Figure 2 showed the relation between fly ash dosage and compressive strength of desert sand high strength concrete. As shown in Table 1 and Fig. 2, with the enhancement of fly ash dosage, the compression strength of desert sand high strength concrete increases firstly, and then declines. The compression strength of desert sand high strength concrete with the fly ash dosage of 15% is larger than that of desert sand high strength concrete with the fly ash dosage of 25% and with the fly ash dosage of 0%.

4.2. *Splitting tensile strength experiment*

Splitting tensile strength, max error and average value of desert sand high strength concrete were listed in Table 2. It can be seem from Table 2 and Figure 3 that with the enhancement of desert sand replacement ratio, the splitting tensile strength of desert sand high strength concrete increases firstly, and then declines. When desert sand replacement ratio is equal to 20%, the splitting tensile strength of desert sand replacement ratio reaches the maximum value.

Table 2. Results of splitting tensile strength

No	Fly ash dosage /%	Desert sand replacement ratio/%	Splitting tensile strength /MPa				
			1	2	3	max error /%	average value
F01	0	0	4.42	4.67	4.39	5.6	4.49
F02		20	5.89	5.89	5.71	3.0	5.83
F03		30	5.41	5.50	5.26	2.8	5.39
F04		40	5.28	5.07	5.31	3.9	5.22
F05		60	4.54	4.78	5.08	6.3	4.80
F06		80	4.76	4.63	4.69	9.1	4.69
F07		100	4.47	4.44	1.53	1.5	4.48
F101	15	0	4.35	4.97	4.72	7.8	4.68
F102		20	6.03	6.11	6.05	0.9	6.06
F103		30	5.80	5.57	5.40	2.9	5.59
F104		40	5.08	5.73	4.98	12.7	5.26
F105		60	4.58	4.98	4.90	6.5	4.82
F106		80	4.97	4.77	4.53	5.1	4.76
F107		100	4.72	5.03	4.39	7.0	4.71
F151	25	0	4.21	4.29	4.33	2.0	4.28
F152		20	5.20	5.38	5.54	3.4	5.38
F153		30	5.01	5.13	5.12	2.1	5.08
F154		40	4.92	4.87	4.80	1.3	4.86
F155		60	4.87	4.67	4.72	3.2	4.75
F156		80	4.51	4.64	4.63	2.8	4.59
F157		100	4.01	4.63	4.60	2.7	4.41

Figure 4 showed the relation between fly ash dosage and compressive strength of desert sand high strength concrete. It can be seem from Table 2 and Fig. 4 that with the improvement of fly ash dosage, the splitting tensile strength of desert sand high strength concrete increases firstly, and then declines. The splitting tensile strength of desert sand high strength concrete with the fly ash dosage of 15% is larger than that of desert sand high strength concrete with the fly ash dosage of 25% and with the fly ash dosage of 0%.

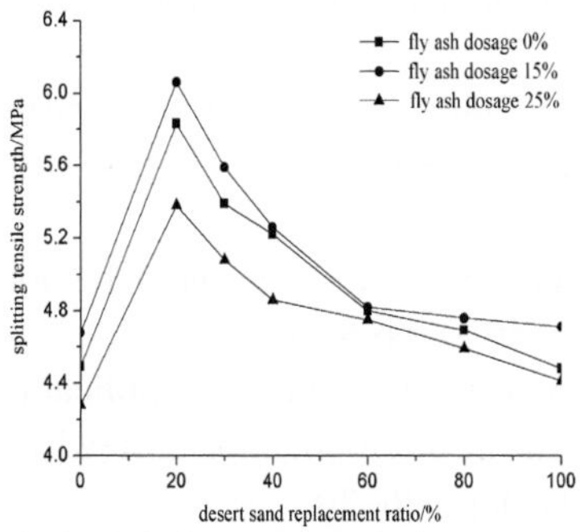

Fig.3 Relation between desert sand replacement ratio and compressive strength

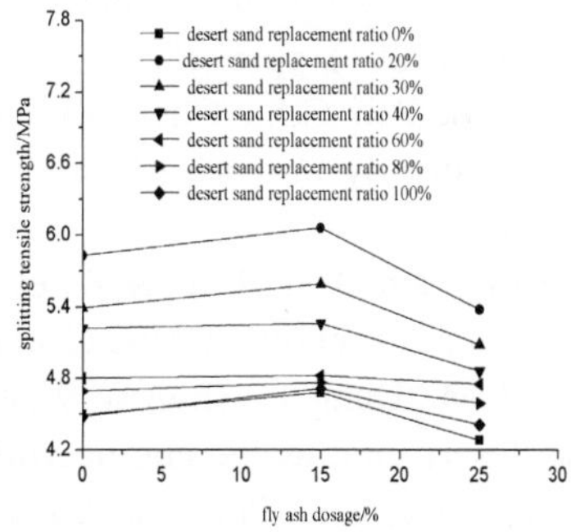

Fig.4 Relation between fly ash dosage and compressive strength

5. Conclusion

Uni-axial compression strength experiment and splitting tensile strength test of desert sand high strength concrete were carried out. The influence of fly ash dosage and desert sand replacement ratio on the 28d compression strength and splitting tensile strength was analyzed. Experimental results show that with the enhancement of desert sand replacement ratio, the compression strength and splitting tensile strength of desert sand high strength concrete increases firstly, and then declines. When desert sand replacement ratio is equal to 20%, the compression strength and splitting tensile strength of desert sand high strength concrete reached the maximum value. With the improvement of fly ash dosage, the compression strength and splitting tensile strength of desert sand high strength concrete increases firstly, and then declines, which provided the guidance to engineering.

Acknowledgements:

The project is supported by National Natural Science Foundation of China (No.51368048 and 11162015) and Postgraduate Innovation Program of Ningxia (GIP2015048).

References

1. Liu A.X. Performance and application of fly ash concrete. *Concrete*, 12(146) (2001), pp.6–15.
2. Lu L.H., Pan G.S., Chen S.L., et al. Strength of concrete with different contents of fly. *Journal of Shenyang University of Technology,* 1(31) (2009), pp.107–11.
3. Liu J.H., Jin D.M., Bao W.Z., et al. Experimental study on the behaviors of desert sand concrete. *China Concrete*, 13(2013), pp.66–68.
4. Zhu T.M., Dang T., Shun W.W, et al. Experimental study and application of the desert sand as the engineering material in the foundation instead of the engineering sand. *Building Science*, 4(1997), pp.26–30.
5. Zhang G.X., Song J.X., Yang J.S., et al. Performance of mortar and concrete made with a fine aggregate of desert sand. *Building and Environment,* 41(11) (2006), pp.1478–1481.
6. Chen M.M., Song J.X., Zhao W.B., et al. Study on the concrete mechanical properties mixed with fly ash and Tenggeli desert sand. *Ningxia Engineering Technology,* 10(1) (2011), pp. 61-63.
7. Chen Y.L., Ma J.R., Han L., et al. Influence of fly ash and desert sand content on the compressive strength of high strength concrete. *Concrete*, 7(2014), pp. 80-84.
8. Yang W.W., Chen Y.L., Ma J.R., et al. Influence of desert sand replacement ratio on the compressive strength of high strength concrete. *Science Technology and Engineering*, 14(19) (2014), pp. 306-309.
9. China Academy of Building Research. Standard for test method of mechanical properties on ordinary concrete (GB/T 50081-2002). Beijing: China Architecture and Building Press, 2002.
10. Wang X.Y. Analysis of effect of superfine fly ash in high performance concrete. *Guangdong Building Material*, 12(2008), pp.32-33.

Experimental research on mechanical behavior of recycled concrete-filled square steel tubular short columns under axial compression

Jing-Feng Wang*,a, Ke-Hui Wang[b] and Xin-Tao Wang[c]

School of Urban construction, Yangtze University
Jingzhou Hubei 434023, P. R. China
[a]625610678@qq.com, [b]s_wu668@163.com, [c]719070934@qq.com

By performing the axial compression experiments of 9 recycled concrete-filled square steel tubular short columns (RCFSST) with different recycled coarse aggregate replacement ratio η of core concrete and width-thickness ratio *D/t* of steel tube, the damage feature of specimens under the action of axial pressure were observed. The paper analyzes the impact of the η, the *D/t* on the loading-longitudinal displacement curves of specimens. The results show that the failure patterns of the recycled concrete-filled steel tube columns are similar to ordinary concrete-filled steel tube columns, and the whole specimens are shear failure pattern. With the recycled coarse aggregate replacement ratio η decreasing, the descent stage of load-displacement curves is steeper, but the difference of the bearing capacity of specimens with different *D/t* is far closer. It shows that the result, which takes the negative influence of recycled aggregate into account, would be more reasonable.

Keywords: Recycled Concrete-filled Square Steel Tube, Width-thickness Ratio *D/T*, Replacement Percentage *H*, Bearing Capacity.

1. Introduction

Compared with ordinary concrete, recycled concrete is lower on strength, resistance deformation ability and durability because of the different internal damages, which hindered the application in practical engineering. If recycled concrete and ordinary concrete mixed in a certain proportion are filled in steel tube, the mechanical properties of recycled concrete will be improved in the three dimensional compression state. So it is an important significance for the research on mechanical behavior of recycled concrete-filled steel tube.

By now, scholars at home and abroad have carried out the research on the mechanical behavior of recycled concrete-filled square steel tube short columns by using ordinary method, and have made much progress [1-2]. [1] conducted

* Corresponding author.

axial compression tests on recycled concrete-filled square and circular steel tubular short columns, then, analyzed the mechanical properties of short column under axial compression in the different stages. [2] simulated the whole process of the load deformation of recycled concrete filled steel tube, and it indicated that this model was correct, which showed that the η of recycled aggregate had influence on ultimate bearing capacity.

In this paper, the influences of the *D/t* and η on the mechanical behavior of recycled concrete filled square steel tube short columns are proposed.

2. Experimental Situation

2.1. *Test Setup and Fabrication*

In order to study the influence of the *D/t* and η on bearing capacity, the paper designed nine recycled concrete filled steel tube short columns, which were 450 millimeter long. The specific parameters of tested specimens were shown in Table 1.

Table 1.Basic parameters of tested specimens

No.	*B*/mm	*t*/mm	*L*/mm	η/%	*B/t*	f_y /MPa	f_{cu}/ MPa	P_{ue}/kN
RCFSST3-00	150	2.6	450	0	57.7	336	51.02	1751
RCFSST4-00	150	3.6	450	0	41.7	342	51.02	1881
RCFSST5-00	150	4.5	450	0	33.3	338	51.02	2004
RCFSST3-30	150	2.6	450	30	57.7	336	50.12	1644
RCFSST5-30	150	4.5	450	30	33.3	338	50.12	1962
RCFSST3-60	150	2.6	450	60	57.7	336	49.71	1545
RCFSST4-60	150	3.6	450	60	41.7	342	49.71	1841
RCFSST5-60	150	4.5	450	60	33.3	338	49.71	1860

Note: RCFSST3-00. Recycled concrete filled short steel tubular columns were abbreviated to RCFSST. The number 3 denoted that the tube thickness was 3mm. and the rest meant that recycled aggregate replacement ratio was 0%.

The materials used in this test mainly included seamless square steel tube with yield strength of 235MPa, Ordinary Portland cement with a 28 day compression strength of 32.5MPa, River sand, natural aggregate, the recycled coarse aggregates, the recycled fine aggregates and High Range Water Reducing Agent FDN-5. The particle diameter of recycled fine aggregates was 5-10mm, and the coarse one was 10-20mm. Firstly, the mud on the surface of the natural aggregates and the recycled aggregates would be washed clearly before concreting. Then, aggregates and screened sand were dried thoroughly. Each concrete specimen had the same mixture proportion, i.e. Cement: Sand:

Aggregate: Water=491:627:1134:147. Mechanical properties index of steel and the 28 day cube compression strength of inner concrete were given in Table 1.

2.2. *Loading Device and Measuring Points Disposition*

Loading device was shown in Figure 1, which could provide 5000kN pressure by controlling micro-computer hydraulic servo tester. The test was carried out in the Building Structure Laboratory of Yangtze University. In order to make steel tube and core concrete pressured at the same time, firstly, the concrete and the upper surface of steel tube should be polished smoothly by burnish machine. Then, the specimens were placed on the loading table.

3. Test Phenomenon

It can be seen from the similar failure mechanism of all specimens that the mechanical characteristics of recycled concrete filled square steel tubular short column is similar to the ordinary one. The failure phenomenon of specimens was shown in Figure 1. Specimen number from left to right in Figure 1. corresponded to that from top to bottom in Table 1.

Fig. 1. Failure phenomena of specimens

With the increase of axial loading, the specimen entered elastic-plastic stage. Some rust was exfoliated from the tube surface where the rust was thick. Meanwhile, diagonal shear slip lines occurred on the both ends of steel tube surface, which became obvious with the increase of load and extended to the middle of the specimen. There was no obvious bulge before the ultimate load, and a local bulking was formed at one quarter of the steel tube height near the ultimate load. By this time the vertical displacement of specimen became large suddenly, and the axial deformation of the specimens was obvious, especially recycled aggregate replacement reached 60%. It related to the internal small void of the recycled aggregate. The deformation of the specimen was accelerated with the increase of the load, and a transverse bulge was formed at the 1/3-1/4 height of the specimen about 1.5mm-4mm. After the peak load, the bearing capacity of the specimen decreased rapidly and the deformation of the steel tube grew quickly. The fracture was formed at the central height of the steel tube in part of the specimens, and the core concrete was crushed. Finally, the longitudinal deformation was over 30mm.

4. Test Results and Analysis

The loading-displacement curves of specimens were shown in Figure 2 and Figure 3. The whole curves included elastic and plastic stage, which is similar with that of the ordinary one. The loading-displacement curves increased straightly before the 0.65-0.85N_{ue}, when the specimens were at the elastic stage. With the increase of axial loading, they entered the plastic stage, and there was small deflection in the curves.

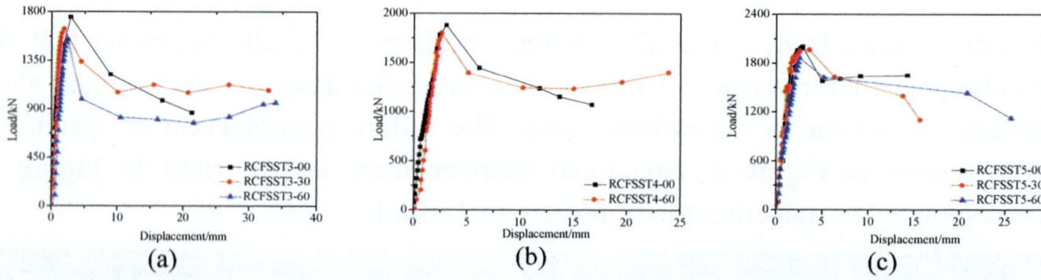

Fig. 2. Loading-longitudinal displacement curves of specimens of different recycled aggregate replacement percentage

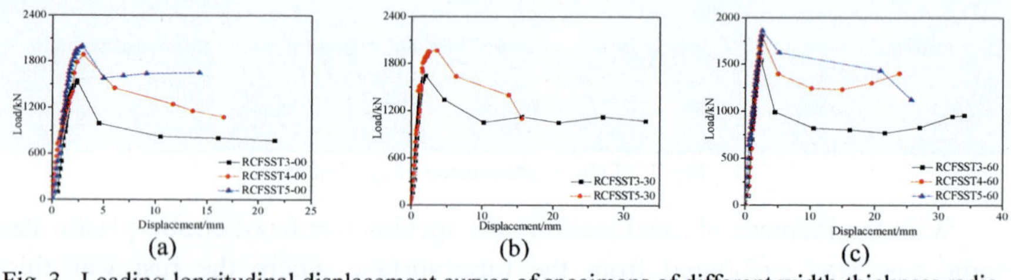

Fig. 3. Loading-longitudinal displacement curves of specimens of different width-thickness radio

The loading-displacement curves of specimens of different recycled aggregate replacement percentage were shown in Figure 2 with the same width-thickness radio of the specimens. It can be observed that the influence of the recycled aggregate replacement percentage on the ultimate bearing capacity of the specimens is obvious among the specimens whose width-thickness radio is smaller. The greater the recycled aggregate replacement percentage is, the lower the ultimate bearing capacity is. However, the recycled aggregate replacement percentage has little effect on the ultimate bearing capacity with the increase of the width-thickness of the specimens. Moreover, the stiffness of the specimens with the same width-thickness is basically similar at the elastic stage. As the recycled aggregate replacement percentage increases, the bearing capacity reduces faster after the peak load.

From Figure 3, it can be noticed that the width-thickness radio has great effect on the ultimate bearing capacity of specimens when the replacement rate of recycled aggregate is less than 30%. As the width-thickness of specimens decreases, the stiffness and ultimate bearing capacity of specimens increase in the context of the same recycled aggregate replacement percentage, which is due to the increase of the binding force of steel tube.

5. Calculation of Bearing Capacity

The ultimate bearing capacity of the recycled concrete filled square steel tubular short columns in this paper is calculated by some specification at home and abroad [3-5], and the results are compared with the experimental results. It can be observed that there is larger error between experiment value and computational results without considering the influence of recycled aggregate replacement percentage.

Based on the analysis above, considering the negative impact of the recycled aggregate replacement percentage is more reasonable. Through the limit equilibrium method, the calculation formula of the ultimate bearing capacity can be presented as follows.

$$P_u = \varphi A_c f_{ck}(1+1.96\xi)$$

Where $\varphi = 1/(0.095\eta^2+0.09\eta+1.25)$, $\xi = f_y A_s / f_{ck} A_c$, $f_{ck} = 0.67 f_{cu}$. Where φ is the reduction coefficient and η is the recycled aggregate replacement percentage.

Applying the fitting formula for calculating the bearing capacity of recycled concrete filled square steel tubular short columns in [6] and circular steel tubular short columns in [7], It is found that when the recycled aggregate replacement percentage is more than 20%, the calculated results are close to the test value, meanwhile, the error range is within 5%, which is within engineering allowance. It is quite clear that the calculation values fit well with experimental data.

6. Conclusion

The mechanic characteristic of the recycled concrete-filled square steel tubular short column under axial compression is similar to the ordinary concrete filled steel tubular short column, and they emerge shearing failure.

During the loading process, when the width-thickness ratio of specimens is the same, the stiffness of them in the elastic and elastic-plastic stage is similar, respectively. As the recycled coarse aggregate replacement percentage increases, the descent stage of load-displacement curves is steeper after the peak load.

As the increase of the width-thickness ratio, the ultimate bearing capacity of specimens increases. However, the ultimate bearing capacity increases slightly when the recycled aggregate replacement percentage is over 30%. When the recycled aggregate replacement percentage reaches 60%, the specimens with different width-thickness have the same ultimate bearing capacity.

The proposed formula can be used to calculate the bearing capacity of recycled concrete filled square steel tubular short columns as well as recycled concrete filled circular steel tubular short columns.

Acknowledgements:

This work gets its funding from Projected (D20151304) supported by Science and Technology Research Project of Hubei Province's Education Department of China; Projected (104892014001) supported by College Students' Innovative Training Program project of Hubei Province of China.

References

1. Zong-ping Chen, XU Jinjun, XUE Jianyang. Influnence of replacement ratio on axial compression performance degradation of recycled aggregate concrete-filled steel tube short column. Journal of experimental mechanics,1(2014), (in Chinese).
2. Dong He, Qing-yuan Wang, Cichang Qiu, et al. The theoretical and experimental analysis of load-deformation of recycled concrete filled steel tubes under axial compression. Journal of Sichuan University(engineering science edition) ,42, S1(2010), pp.49-53, (in Chinese).
3. Lin-hai Han, You-fu Yang, Modern steel tube concrete structure technology. China Building Industry Press, 2004, (in Chinese).
4. British Standards Institutions, 1979. BS5400 Steel,concrete and composite bridges, Part 5; Code of practice for design of bridges. London,U K.
5. National military standard of the people's Republic of China GJB 4142-2000,Wartime naval repair technical code for early strength composite structure. Beijing: Chinese people's Liberation Army General Logistics Department, (in Chinese).

6. Xiao-jun Ke, Zong-ping Chen, Jian-yang Xue , et al. Experimental study on the bearing capacity of recycled aggregate concrete-filled square steel tube short columns under axial compression. Engineering mechanics,30,8(2013) , pp.35-41, (in Chinese).
7. Jun-tao Li, Zong-ping Chen, Feng Liu. Experimental study on the axial compress mechanical behavior of recycled aggregate concrete-filled steel tube. Concrete,7 (2013) , pp.13-15, (in Chinese).

CFD simulation of liquid Mg drop impact on an Al substrate for compound casting

Yong-Ning Liu[1*], Yi-Qing Chen[1,a] and Chun-Hui Yang[2,b]
[1]*School of Material and Engineering, Hefei University of Technology 230009, Tunxi Road 193, Hefei City, China.*

[2]*School of Computing, Engineering and Mathematics, Western Sydney University Locked Bag 1797, Penrith, NSW 2751, Australia.*
E-mail: chenyq63@126.com[a], r.yang@westernsydney.edu.au[b]

In this paper, to investigate a compound casting process for creating Al coating layer for Mg alloy, a computational fluid dynamics (CFD) model is developed using the VOF method to simulate the molten Mg drop impact on the Al substrate at solid state. A spreading factor, which is the ratio to the varied diameter of the Mg droplet after impacting on the Al substrate and its initial diameter before the impact, is used to describe and measure the impact process generated by the Mg drop, which could be divided into several steps: spreading, retracting and oscillating, and equilibrium as final step. Some key parameters are investigated via parametric studies: impact velocities, initial diameters of drop, the temperature of the Al substrate, which have important influences on the quality of compound casting products. The spread diameter of the Mg drop increases when the impact velocity and initial diameter of drop increase. While the impact velocities and initial diameters of the drop are larger than critical values, the liquefied Mg may partially have bubbles. The higher the temperature of the Al substrate is, the larger the spread diameters could be obtained, though oscillating and equilibrium steps take more time to finish in the CFD simulations developed.

Keywords: Mg, Al, Compound Casting, CFD, VOF, and Spread Factor.

1. Introduction

Compound casting process was developed to produce bi-metallic material products via casting a metal onto or around the other metal at solid state. Recently the compound casting was attempted to produce a bi-metal structure using Aluminium to protect Magnesium as coatings [1-2].With the aid of computer aided engineering, its processing can be optimally controlled via numerical modelling and simulations using powerful engineering analysis means, i.e., including finite element analysis (FEA)[3] and computational fluid

---

* Corresponding author.

dynamics (CFD)[4-11]. To simplify the complex casting process, a single liquid drop-solid substrate CFD model could be applied which has been developed and employed to study coating problems using plasma spaying process. In fact, the thermal spraying process for producing coatings on metals is very similar to the compound casting process in some ways, and the flow of droplet in the plasma thermal spaying process using CFD with the Volume of Fluid (VOF) method was studied [4-11]. In such a model, several important parameters, i.e., the size, impact velocity, temperature of droplet, and the substrate material and its temperature, play significant roles in the whole process, especially at both impact and solidification stages. These parameters affect the quality of final product. A. Abdellah et al. [4] found that when the ambient temperature was above the melting temperature of the particle, a delay of the solidification of splat was observed allowing the shrinkage of the splat. On the other hand, the high gas temperature could lead to a good adhesion of the splat with the substrate due to relative improvement of the contact pressure compared to low gas temperature case. The effect of the impact velocity and the size of droplet were investigated [6-9] and it was found that the spreading factor obviously increases when impact speed increases; however, splashing will appear in the status when the speed is high enough. For the effect of drop size, the smaller the droplet was, the earlier the drop reached its maximum spread factor. The influences of drop temperature and substrate temperature on the evaporation process were also studied [9]. It was indicated that when the drop temperature is attained the saturation point after had impinged on the heat substrate, it evaporates much faster.

In this study, a 2-D single liquid drop-solid substrate CFD model using the VOF method is developed to simulate the compound casting process for Al-Mg bi-metallic materials, which are Magnesium alloy AM60 and aluminum alloy A390. To investigate the influences of the main parameters on the spread of drop in such a process, parametric studies are conducted to look through different cases with different size, velocity, initial temperature of the liquid Mg drop, and the initial temperature of the solid Al substrate.

2. 2-D axisymmetric CFD model

The 2-D single liquid Mg drop-solid Al substrate CFD model using the VOF method was devised using a commercial numerical modelling and simulation package – ANSYS/FLUENT to simulate the flow of the single drop and its spread in the substrate in the compound casting process, as shown in Figure 1.The casting was assumed to be conducted in an Argon environment to avoid oxidation. The drop was assumed spherical with a diameter of 4mm and thus the2-D axisymmetric model was used. The material properties of the magnesium alloy AM60 and aluminum alloy A390are shown in Tables 1 and 2, respectively. The latent heat, liquid-solid temperature and density of AM60 are 373 kJ/kg, 610-550°C, and 1800 kg/m^3, respectively. The Argon and Al liquid were both assumed as incompressible, Newtonian fluids. The flow was assumed to be laminar, and the internal circulation of fluid within the drop was neglected. In the parametric studies, the velocity, size and initial temperature of the liquid Mg drop, and the initial temperature of the solid Al substrate were varied as 0.5 m/s, 1 m/s, 1.5 m/s, and 2.0 m/s;3 mm, 4 mm, 5 mm, and 6 mm; and 100°C, 300 °C, and 500 °C, respectively.

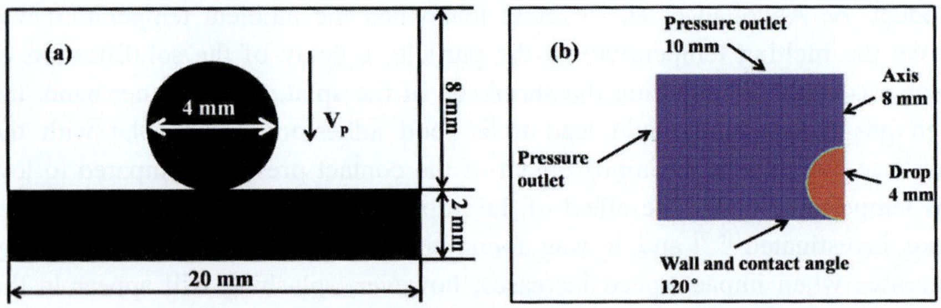

Fig. 1 The CFD model: (a) geometrical model and (b)2-D axisymmetric model

Table 1. The thermal propertied of AM60 at different temperatures

AM60 Temperature(°C)	Thermal conductivity (W/m K)	Specific heat (J/Kg K)	Viscosity (Pa·s ×10^{-3})
20	68.8	1000	-
200	78.2	-	-
300	85.6	-	-
400	91.3	1000	-
550	-	1068	2.22477
615	-	1150	-
640	-	-	1.46478
700	95	-	-
780	-	-	0.858717
900	-	1150	0.601331

Table 2.The thermal propertied of A390 at different temperatures

A390 Temperature(°C)	Thermal conductivity (W/m K)	Specific heat (J/Kg K)	Density, (kg/m^3)
20	116.7	851	
100	126.7	904	
200	135.3	945	
300	-	981	2730
400	152.2	-	
492	161.1	1163	
636	100	1163	
1000	100	1163	

3. Results and Discussion

To descript the evolution history of the liquid Mg, the drop spread factor (D/D_0) was an effective parameter, in which D is the diameter of the Mg drops on the Al substrate after impact and $D_0$are its initial diameter, respectively. The temperatures of Mg, Al, and Argon were set as 700 °C, 500 °C, 26 °C, respectively. As shown in Figure 2, the spread factor reached its largest value at *t=3ms*. In other words, the drop will begin to retract after this time. Based on the spreading factor's variation, this impact process can be divided into several steps: (a) spreading step,(b)retracting step, (c)oscillating step, and (d)equilibrium step.

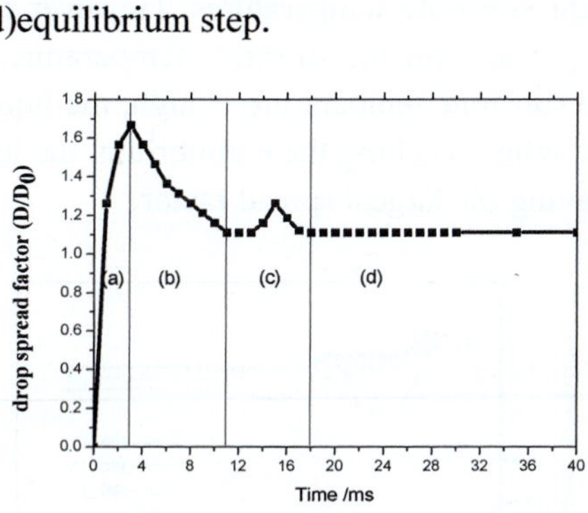

Fig. 2. The variation of the spread factor of the liquid Mg.

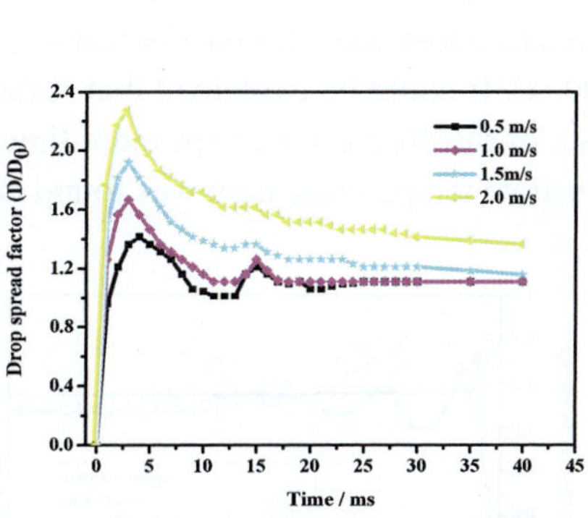

Fig.3. Variation of spread factors with time different velocities.

Figure 3 showed the time evolution of spread factors of the drops with four impact velocities. The four drops reached their maximum spread diameters at 3 ms, and then started to retract. The spread factors were found fluctuated for the drops with lower velocities of 0.5 and 1.0 m/sat 15 ms. it is clear to find that the drop spread speed and spread factor increased correspondingly when the impact velocity increases. The low impact velocities had less influence on the spread factor when the equilibrium is reached, which are kept the same to all two cases with lower impact velocity.

Four different drop diameters, 3.0 mm, 4.0 mm, 5.0mm and 6.0 mm were chosen to investigate the effect of drop size on the impact process, with the impact velocity 0.5 m/s and the temperatures of Mg, Al, and Argon were 700 °C, 500 °C, 26 °C respectively. The time evolution of the spread factors of the four drops with different diameters but at the same impact and temperature conditions was shown in Figure 4.It can be seen that although the drop diameters were different, the drops spread at the same rate in their spread step. However, with the increase of the diameters, the maximum spread factor increases, and the smallest drop reaches its maximum spread first. Besides that, the receding speeds were almost the same at the receding step. Based these results, the drop diameter has almost no influence on its spreading speed and receding speed. In this impact process, the spreading of the drop is accompanied with heat transfer. Thus it is important to investigate the effect of temperatures on the spreading. The spread factors of the Mg drop at the different Al substrate temperatures were investigated, as shown in Figure 5. The temperatures of the substrate were chosen as 100 °C, 300 °C, and 500 °C. The impact velocity of the drop was still set as 0.5 m/s. For the spread step of the drops, the variation trends of spread factors were found the same with different drop diameter, in other words, the spread rates were the same for different substrate temperatures. However the spread factors were found fluctuated especially for the substrate temperature at 500 °C. It could be explained that if the substrate temperature is high, the liquid state of the drop can be kept more time. When reaching the equilibrium, the low substrate temperature case was found having the largest spread factor.

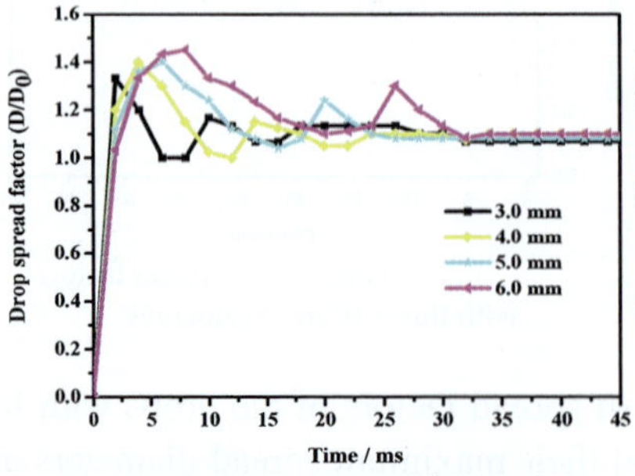

Fig. 4. Variation of spread factors for drops with different diameter

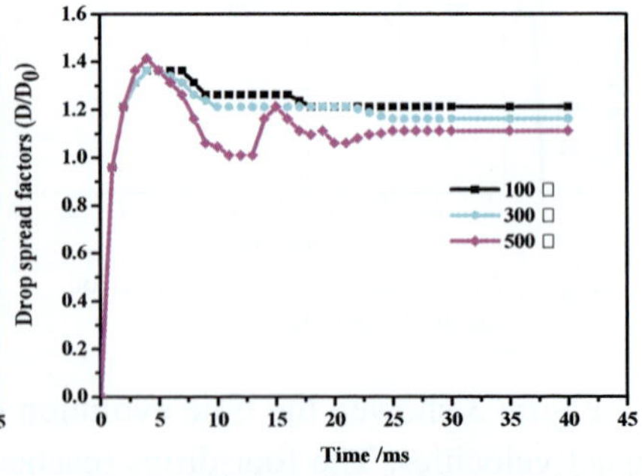

Fig. 5. Variation of spread factors with different substrate temperature

4. Conclusions

In this study, the 2-D single liquid Mg drop-solid Al substrate CFD model using the VOF method has been devised and employed to trace the free surface deformation of the liquid Mg drop into the solid Al substrate, which was extensively used in the impact process problems. According to the spreading factor's variation, this impact process can be divided into several steps: spreading step, retracting step, and oscillating step, equilibrium step. As for the influence of the drop impact speed, the drop spread speed and the spread diameter on the substrate increased correspondingly when the impact speed increased. The low impact speeds have almost no influence at the equilibrium step due to almost the same spread factors. The diameters of drop have an obvious influence on the retracting step, and oscillating step, while the spread factors were almost same for the spread and equilibrium steps. There are the same spread rates in the spreading step for the cases of the substrate with different temperature. The higher the temperature of the substrate is, the larger the spread diameter can be obtained, while they may take more time for the oscillating step and equilibrium step. These research findings can be further used to guide the optimal control of such a compound casting.

Acknowledgments

The authors would like to show their gratitude to Intersect Australia Limited (INTERSECT) and National Computational Infrastructure (NCI), Australia for their high-performance computing resources.

References

1. G.C. Xu, A. Luo, Y.Q. Chen, and A.K. Sachdev, Interfacial phenomena in magnesium/aluminum bi-metallic castings, *Materials Science & Engineering A***595**, 154 (2014).
2. Y.N. Liu, Y.Q. Chen, and C. Yang, A study on atomic diffusion behaviours in an Al-Mg compound casting process, *AIP Advances***5(8)**,087147 (2015).
3. M. Zirari, A.A. El-Hadj, and N. Bacha, Numerical analysis of partially molten splat during thermal spray process using the finite element method,*Applied Surface Science***256(11)**, 3581-3585(2010).
4. H. Tabbara, and S. Gua, Modelling of impingement phenomena for molten metallic droplets with low to high velocities, *International Journal of Heat and Mass Transfer***55(7-8)**, 2081–2086(2012).

5. A.A. El-Hadj, M. Zirari, and N. Bacha, Numerical analysis of the effect of the gas temperature on splat formation during thermal spray process, *Applied Surface Science* **257(5)**, 1643-1648(2010).
6. Z.M. Liu, H.M. Liu, and X. Liu, Dynamical analysis of droplet impact spreading on solid substrate, *Front. Mech. Eng. China***5(3)**, 308-315(2010).
7. A. Kumar, S.Sahoo, S. Ghosh, and B.K.Dhindaw, Effect of Process Parameters on Splat Formation during Impingement of Liquid Metal Droplets over a Cold Substrate, *Materials Science Forum***710**, 186-191(2012).
8. J.Y. Li, Q. Han, Y.J. Zhao, and X.F. Yuan, Deposition Process of Droplets Impacting on a Horizontal Surface, *Advanced Materials Research***354-355(1)**, 579-584(2012).
9. M.Toivakka, Numerical Investigation of Droplet Impact Spreading in Spray Coating of Paper,*TAPPI Advanced Coating Fundamentals Symposium*(2003).
10. Y.L. Guo, S.Q. Shen, andS. Quan, Numerical simulation of dynamics of droplet impact on heated flat solid surface, I*nternational Journal of Low-Carbon Technologies Advance Access* **8(2)**,134-139(2012).
11. S. Afkhami, and M. Bussmann, Height functions for applying contact angles to 3D VOF simulations, *Int. J. Numer. Meth. Fluids***61**, 827–847(2009)

Chapter 4

Structural Materials

The model of the total state function for structural materials

Zhen-Hai Wei
School of Rear Professional Service, Logistics Academy,
Beijing, China 100858
E-mail: wzhai@vip.sina.com

Meng-Shu Wang and Ding-Li Zhang
School of Civil Engineering, Beijing Jiaotong University,
Beijing, China 100044

This paper applies the theoretical method of the structural total state function to obtain control partial differential equations, path equations, and constitutive relations of the total state function of structural materials, as well as three analytical solutions for special states, namely the plane-form solution, cylindrical-form solution, and spherical-form solution. According to the research, the total state function of elastic-perfectly plastic materials is completely consistent in form with the strength theory of classic elastic-plastic theory; moreover, the path equation is consistent with the flow rule, and the constitutive relation conforms completely to the form in classic elastic-plastic theory as well. So far, these equations have been able to obtain conclusions and results that completely agree with classic elastic-plastic theory by means of applying the theories of the structural total state function to elastic-perfectly plastic materials. More importantly, the concepts, theories and methods of the total state function unify the strength theory, flow rule, and constitutive relation in classic elastic-plastic theory under a theoretical framework, endowing them with a unified theoretical basis.

Keywords: Structural Material; Total State Function; Path Equation; Constitutive Relation; Elastic-perfectly Plastic Materials

1. Introduction

The strength theory, flow rule, and constitutive relations between materials form the basis of the classic elastic-plastic theory. A great deal of earnest and careful research has demonstrated and clarified theories related to these concepts, including the Trasica and Misses theories that resulted in the intensity rule [1-5]. Although other researchers have proposed new rules, the basic concepts of the strength theory remain nearly the same. Many new ideas have been proposed for the flow rule based on the concept of plastic potential, but they have not yet produced any significant breakthroughs [6-9]. As for discussions and research around constitutive relations, most new achievements depend on the concept of the stress-strain relation [7-13]. The strength theory, flow rule, and constitutive relations function basically independently in a conceptual system, requiring a corresponding unified theoretical basis.

According to existing research, the conceptual system of elastic theory stills constitutes the essence of material research. The study of plastic materials resulted in a partial new hypothesis related to material plasticity, which supplements elastic theory in order to account for the characteristics of plastic materials. The theory of a structural total state function, which treats plasticity problems from a new viewpoint [14-18], suggests that plasticity is an outward manifestation of geometrical shape changes in a material. Changes in the material structure represent the essence of plasticity. All materials have structures, and a material's structural properties are universal. Moreover, a structural total state function describes a material's characteristics and structural properties; it serves as a comprehensive reflection of the mechanical characteristics of a material with a structural property.

An elastic-perfectly plastic material can be regarded as a special type of structural material; that is to say, the material structure tends toward being a homogeneous structure. The present study analyzes this special structural material using the theory and method of a structural total state function. Being more universal than the classic elastic-plastic theory, the theory of a structural total state function can both determine and effectively analyze the special case of elastic-perfectly plastic materials with homogeneous structures. The results of this study support this assertion.

2. Basic Hypothesis and Total State Function Model

2.1. *Basic hypothesis*

The following hypotheses are essential for deriving an elastic-perfectly plastic material with a homogeneous structure:

(1) The material structure distributes uniformly, and it can be regarded as continuous and approximately isotropic;

(2) The energy density of the material structure is a constant or finite quantity that can be described;

(3) In the stress-modulus of the elasticity-strain space, the intensity-rigidity-structural geometric strain of the material's structural state constitutes a structural state parameter, and the set of all possible structural states constitutes a structural state function, which is also referred to as a total state function;

(4) The structural total state function of an elastic-perfectly plastic material is not relevant to the geometric strain components, and the structural rigidity is a constant;
(5) The structure changing process conforms to the state conservation principle. That is to say, the material structure can take on new characteristics while maintaining its original state. Even when the structure must change, it has a tendency to keep as few changes as possible;
(6) The intensity-structural strain path conforms to the least-energy principle.

2.2. *Model and equations of total state function*

In compliance with the fifth hypothesis, the material structure attempts to maintain its elastic state for as long as possible during its changes. Furthermore, structural materials always try to maintain their original states when changing under an environmental or load effect. In this way, the total state function surface of a material structure can be regarded as the limit surface for maintaining its elastic structure. The material maintains its elastic state within the total state function surface, but the material structure is destroyed beyond the total state function surface. The material structure changes on the total state functional surface, transforming to another structural state.

The structural total state function of a material with parameter forms is expressed as follows:

$$\begin{cases} \sigma_1 = \sigma_1(u,v) \\ \sigma_2 = \sigma_2(u,v) \\ \sigma_3 = \sigma_3(u,v) \end{cases} \tag{1}$$

where σ_1, σ_2, and σ_3 are the three principal stresses, and u and v are the parameters.

Therefore, an function exists and is as follows:

$$\Pi = \int_\Omega (\sqrt{EG - F^2} + \lambda U)\ d\Omega \tag{2}$$

In this formula,

$$E=(\frac{\partial\sigma_1}{\partial u})^2+(\frac{\partial\sigma_2}{\partial u})^2+(\frac{\partial\sigma_3}{\partial u})^2 \text{ and } G=(\frac{\partial\sigma_1}{\partial v})^2+(\frac{\partial\sigma_2}{\partial v})^2+(\frac{\partial\sigma_3}{\partial v})^2,$$

$$F=(\frac{\partial\sigma_1}{\partial u})(\frac{\partial\sigma_1}{\partial v})+(\frac{\partial\sigma_2}{\partial u})(\frac{\partial\sigma_2}{\partial v})+(\frac{\partial\sigma_3}{\partial u})(\frac{\partial\sigma_3}{\partial v}), \text{ and } \int_\Omega U d\Omega = C_0,$$

where U is the intensity energy density function of the material structure.

Thereby, the following Euler equations can be obtained:

$$\begin{cases} \frac{\partial}{\partial u}\left(\frac{\frac{\partial\sigma_1}{\partial u}G-\frac{\partial\sigma_1}{\partial v}F}{\sqrt{EG-F^2}}\right)+\frac{\partial}{\partial v}\left(\frac{\frac{\partial\sigma_1}{\partial v}G-\frac{\partial\sigma_1}{\partial u}F}{\sqrt{EG-F^2}}\right)=\lambda\frac{\partial U}{\partial\sigma_1} \\ \frac{\partial}{\partial u}\left(\frac{\frac{\partial\sigma_2}{\partial u}G-\frac{\partial\sigma_2}{\partial v}F}{\sqrt{EG-F^2}}\right)+\frac{\partial}{\partial v}\left(\frac{\frac{\partial\sigma_2}{\partial v}G-\frac{\partial\sigma_2}{\partial u}F}{\sqrt{EG-F^2}}\right)=\lambda\frac{\partial U}{\partial\sigma_2} \\ \frac{\partial}{\partial u}\left(\frac{\frac{\partial\sigma_3}{\partial u}G-\frac{\partial\sigma_3}{\partial v}F}{\sqrt{EG-F^2}}\right)+\frac{\partial}{\partial v}\left(\frac{\frac{\partial\sigma_3}{\partial v}G-\frac{\partial\sigma_3}{\partial u}F}{\sqrt{EG-F^2}}\right)=\lambda\frac{\partial U}{\partial\sigma_3} \end{cases} \tag{3}$$

These are the control partial differential equations for the structural total state function of the material with parameter forms. With the appropriate boundary conditions, these equations can determine the specific surface equation of the total state function. However, since this equation set is nonlinear, it is much more difficult to calculate its analytical solutions. In fact, it is only possible to calculate an analytical solution under special conditions.

2.3. *State path equation*

According to the hypothesis of the structural total state function, this function reflects the overall performance of a material. Specific state changes that take place because of environmental factors occur along the total state function surface, as long as they meet certain constraint conditions. Moreover, inherent constraint conditions can force the material state to change in order to minimize energy state at all times.

Assuming there is a total state function,

$$\Phi=\Phi(\sigma_1,\sigma_2,\sigma_3) \tag{4}$$

Because of the least-energy principle and the Lagrange method, the following function exists:

$$\Gamma = \int_s (\sigma_1 d\varepsilon_1 + \sigma_2 d\varepsilon_2 + \sigma_3 d\varepsilon_3) + \alpha\Phi(\sigma_1, \sigma_2, \sigma_3) \quad (5)$$

where α is an undetermined constant. Assuming the path equation can be expressed in a parameter form, then

$$\begin{cases} \sigma_1 = \sigma_1(t) \\ \sigma_2 = \sigma_2(t) \\ \sigma_3 = \sigma_3(t) \end{cases} \text{ and } \begin{cases} \varepsilon_1 = \varepsilon_1(t) \\ \varepsilon_2 = \varepsilon_2(t) \\ \varepsilon_3 = \varepsilon_3(t) \end{cases} \quad (6)$$

$$\Gamma = \int_s (\sigma_1 \dot{\varepsilon}_1 + \sigma_2 \dot{\varepsilon}_2 + \sigma_3 \dot{\varepsilon}_3) dt + \alpha\Phi(\sigma_1(t), \sigma_2(t), \sigma_3(t)) \quad (7)$$

Therefore, the following Euler equation can be obtained:

$$\begin{cases} \dot{\varepsilon}_i + \alpha \dfrac{\partial \Phi}{\partial \sigma_i} = 0 \\ \alpha \dfrac{\partial \Phi}{\partial \varepsilon_i} - \dot{\sigma}_i = 0 \end{cases} \quad (i = 1, 2, 3) \quad (8)$$

Clearly, as $\frac{\partial \Phi}{\partial \varepsilon_i} = 0$, then $\dot{\sigma}_i = 0$. In other words, the increment of stress intensity is zero. In fact, the intensity of an elastic-perfectly plastic material is a constant, so it is unchangeable and the increment is zero. Therefore, the following can be obtained:

$$\dot{\varepsilon}_i = -\alpha \frac{\partial \Phi}{\partial \sigma_i} \quad (i = 1, 2, 3) \quad (9)$$

This represents the relationship between the strain and the total state function, and its form is completely consistent with the flow rule in classic elastic-plastic theory.

2.4. *Constitutive relation*

The path equation (8) is able to produce the tangential equation $\frac{\varepsilon_i - \varepsilon_i^0}{\dot{\varepsilon}_i} = k$.

Assuming that $d\varepsilon_i = \varepsilon_i - \varepsilon_i^0$, then $d\varepsilon_i = k\dot{\varepsilon}_i$. By substituting Formula (9) in, we can obtain the following:

$$d\varepsilon_i = -\alpha k \frac{\partial \Phi}{\partial \sigma_i} \quad (i = 1, 2, 3) \quad (10)$$

Let $\bar{\lambda} = -\alpha k$; then the structural strain can be granted as the plastic strain in classical theory, yielding $d\varepsilon_i^p = \bar{\lambda}\frac{\partial \Phi}{\partial \sigma_i}$. Thus, the expression form that is completely consistent with the flow rule in classic elastic-plastic theory is obtained.

In addition, in view of the tangent plane equation of the total state function, the following formula can be obtained:

$$\frac{\partial \Phi}{\partial \sigma_i} d\sigma_i = 0 \tag{11}$$

With a view to the total strain of the material, the sum of the elastic and plastic strains is as follows:

$$d\bar{\varepsilon}_i = d\varepsilon_i^p + d\varepsilon_i^e \qquad (i = 1,2,3)$$

where $d\bar{\varepsilon}_i$ represents the total strain of the material; $d\varepsilon_i^p$ represents the plastic strain or structural strain of the material; and $d\varepsilon_i^e$ represents the elastic strain of the material.

Therefore,

$$\begin{aligned} d\sigma_i &= D_{ij} d\varepsilon_i^e = D_{ij}(d\bar{\varepsilon}_i - d\varepsilon_i^p) \\ &= D_{ij} d\bar{\varepsilon}_i - D_{ij}\bar{\lambda}\frac{\partial \Phi}{\partial \sigma_i} \end{aligned} \tag{12}$$

Substituting Formula (12) into Formula (11), we can obtain:

$$\bar{\lambda} = \frac{\frac{\partial \Phi}{\partial \sigma_i} D_{ij}}{\frac{\partial \Phi}{\partial \sigma_i} D_{ij} \frac{\partial \Phi}{\partial \sigma_j}} d\bar{\varepsilon}_i \tag{13}$$

Substituting Formula (13) into Formula (12), we can obtain:

$$d\sigma_i = [D_{il} - \frac{D_{ij}\frac{\partial \Phi}{\partial \sigma_j}\frac{\partial \Phi}{\partial \sigma_k} D_{kl}}{\frac{\partial \Phi}{\partial \sigma_m} D_{mn} \frac{\partial \Phi}{\partial \sigma_n}}] d\bar{\varepsilon}_l \tag{14}$$

This is the constitutive relation of elastic-perfectly plastic materials, and D_{ij} represents the material's elastic modulus.

The total state function, state path equation, and constitutive relation form the complete solution and solving process for the problems.

3. Solutions of Special Problems

The control equations for the structural total state function are a group of nonlinear partial differential equations. Generally speaking, it is very difficult to calculate analytical solutions for these equations, especially under certain boundary conditions. Based on the previous discussion, we know that the specific forms of the path equation and constitutive relation can only be solved when the specific form of the total state function is determined. Therefore, the solution of a specific form of the total state function is the key to solving this type of problem.

In this study, problems are solved according to the following three conditions.

3.1. *Plane-form solution*

Assume that the form of the intensity energy density is as follows:

$$U = C_0 \tag{15}$$

where C_0 is a constant. Further assume that the solution form is as follows:

$$\sigma_i = a_i u + b_i v + c_i \qquad (i = 1, 2, 3) \tag{16}$$

As a result,

$$\frac{\partial \sigma_i}{\partial u} = a_i, \quad \frac{\partial \sigma_i}{\partial v} = b_i \qquad (i = 1, 2, 3) \tag{17}$$

$$E = \sum_{i=1}^{3} a_i^2, \qquad G = \sum_{i=1}^{3} b_i^2, \qquad F = \sum_{i=1}^{3} a_i b_i \tag{18}$$

Substitute Formulas (15), (17), and (18) into Equation (3), which will automatically conform. However, there remain a total of nine undetermined constants, such as a_i, b_i and c_i, which need to be determined according to boundary conditions.

Some simple boundary conditions are discussed below.

Assuming that $a_i = k_1$, $b_i = k_2$, and $i = 1, 2, 3$, the solution of the equation is as follows:

$$\sigma_i = k_1 u + k_2 v + c_i \qquad (i = 1, 2, 3) \tag{19}$$

Consequently, we obtain:

$$\begin{cases} \sigma_1 - \sigma_2 = c_1 - c_2 = \overline{C}_1 \\ \sigma_2 - \sigma_3 = c_2 - c_3 = \overline{C}_2 \\ \sigma_3 - \sigma_1 = c_3 - c_1 = \overline{C}_3 \end{cases} \tag{20}$$

If symmetry is considered, the graph of the total state function can be obtained on an isoclinic surface in a stress space. When $\overline{C}_1 = \overline{C}_2 = \overline{C}_3$, we can obtain the Trasica strength theory using classical theory.

Assuming that $a_i = b_i = 0$, then the solution of the equation is as follows:

$$\sigma_i = c_i \qquad (i = 1, 2, 3) \tag{21}$$

Considering symmetry allows us to obtain the function graph on an isoclinic surface in a stress intensity space, which is in fact the maximum tensile-stress criterion (Rankine rule) in classical theory. When C_i takes different values, the graph encloses different areas, yielding the strength criterion in different forms.

We can now obtain different graphs of the total state function with different boundary conditions, which can also be regarded as different strength criteria.

3.2. *Cylindrical-form solution*

Assume that the intensity energy density function is as follows:

$$U = \frac{\sigma_1^2 + \sigma_2^2}{D} \tag{22}$$

where D represents the elastic modulus of the materials.

Assume that the solution form is as follows:

$$\begin{cases} \sigma_1 = R \sin v \\ \sigma_2 = R \cos v \\ \sigma_3 = u \end{cases} \tag{23}$$

Therefore,

$$\begin{cases} \dfrac{\partial \sigma_1}{\partial u} = 0 \\ \dfrac{\partial \sigma_2}{\partial u} = 0, \\ \dfrac{\partial \sigma_3}{\partial u} = 1 \end{cases} \begin{cases} \dfrac{\partial \sigma_1}{\partial v} = R\cos v \\ \dfrac{\partial \sigma_2}{\partial v} = -R\sin v \\ \dfrac{\partial \sigma_3}{\partial v} = 0 \end{cases}, \begin{cases} E = 1 \\ G = R^2 \\ F = 0 \end{cases}, \quad \sqrt{EG - F^2} = R \tag{24}$$

Substituting this into Equation (3), we obtain

$$\begin{cases} \frac{\partial}{\partial v}\left(\frac{R\cos v}{R}\right) = \frac{2\lambda}{D}\sin v \\ \frac{\partial}{\partial v}\left(\frac{R\sin v}{R}\right) = \frac{2\lambda}{D}\cos v \\ \frac{\partial}{\partial v}(R) = 0 \end{cases} \tag{25}$$

If $\lambda = -\frac{D}{2R}$, then the two sides of the equation are equal, and the solution meets Equation (3). In actuality, all cylindrical equations with the center passing through the origin could potentially meet partial differential Equation (3), but the form of the intensity energy density function changes correspondingly. The total state function with the same form as the Misses rule in classical theory will be derived below.

Assume that the form of the total state function is as follows:

$$\begin{bmatrix} \sigma_1 \\ \sigma_2 \\ \sigma_3 \end{bmatrix} = \begin{bmatrix} k_{11} & k_{12} & k_{13} \\ k_{21} & k_{22} & k_{23} \\ k_{31} & k_{32} & k_{33} \end{bmatrix} \begin{bmatrix} R\sin v \\ R\cos v \\ u \end{bmatrix} \tag{26}$$

Assume that the intensity energy density function is as follows:

$$U = \frac{(\sigma_1 - \sigma_2)^2 + (\sigma_2 - \sigma_3)^2 + (\sigma_3 - \sigma_1)^2}{D} \tag{27}$$

We can then obtain

$$\begin{cases} \frac{\partial \sigma_1}{\partial u} = k_{13} \\ \frac{\partial \sigma_2}{\partial u} = k_{23} \\ \frac{\partial \sigma_3}{\partial u} = k_{33} \end{cases}, \quad \begin{cases} \frac{\partial \sigma_1}{\partial v} = R(k_{11}\cos v - k_{12}\sin v) \\ \frac{\partial \sigma_2}{\partial v} = R(k_{21}\cos v - k_{22}\sin v) \\ \frac{\partial \sigma_3}{\partial v} = R(k_{31}\cos v - k_{32}\sin v) \end{cases} \tag{28}$$

Assuming that the three column vectors in Matrix [K] are unit vectors, and that they are mutually vertical to each other, then we can obtain the following:

$$\begin{gathered} k_{11}^2 + k_{21}^2 + k_{31}^2 = 1,\ k_{12}^2 + k_{22}^2 + k_{32}^2 = 1,\ k_{13}^2 + k_{23}^2 + k_{33}^2 = 1 \\ k_{11}k_{12} + k_{21}k_{22} + k_{31}k_{32} = 0,\ k_{11}k_{13} + k_{21}k_{23} + k_{31}k_{33} = 0, \\ k_{12}k_{13} + k_{22}k_{23} + k_{32}k_{33} = 0 \end{gathered} \tag{29}$$

For convenience, this study assumes $k_{31}=\sqrt{\frac{2}{3}}$ and can therefore obtain the following formula:

$$[k_{ij}]=\begin{bmatrix} -\frac{1}{\sqrt{6}} & -\frac{1}{\sqrt{2}} & \frac{1}{\sqrt{3}} \\ -\frac{1}{\sqrt{6}} & \frac{1}{\sqrt{2}} & \frac{1}{\sqrt{3}} \\ \sqrt{\frac{2}{3}} & 0 & \frac{1}{\sqrt{3}} \end{bmatrix} \tag{30}$$

In practice, this is a rotation matrix. That is to say, it is possible to convert the cylinder surface by means of rotating the coordinate axis through superposition between the cylinder axe and σ_3 axis to the isoclinic axis, in other words, between the axis of the cylindrical surface and the isoclinic axis.

No matter whether or not the axis of the cylindrical surface passes the origin in any direction, the surface is a solution for Equation (3) as long as the projection of the intensity energy density function on the plane vertical to the cylinder axis is a circle. This demonstrates that the total state function of an elastic-perfectly plastic material is a cylindrical surface, which changes its material characteristics under complicated stress conditions.

The path equation of an elastic-perfectly plastic material is as follows:

$$\begin{cases} \dot{\varepsilon}_i+\alpha\dfrac{\partial\Phi}{\partial\sigma_i}=0 \\ \alpha\dfrac{\partial\Phi}{\partial\varepsilon_i}-\dot{\sigma}_i=0 \end{cases} \quad (i=1,2,3)$$

where $\frac{\partial\Phi}{\partial\varepsilon_i}=0$, $\dot{\sigma}_i=0$, and $\sigma_i=t+C_i \quad (i=1,2,3)$

$$\begin{cases} \dot{\varepsilon}_1=-2\alpha K(2\sigma_1-\sigma_2-\sigma_3)=B_1 \\ \dot{\varepsilon}_2=-2\alpha K(2\sigma_2-\sigma_1-\sigma_3)=B_2 \\ \dot{\varepsilon}_3=-2\alpha K(2\sigma_3-\sigma_2-\sigma_1)=B_3 \end{cases} \tag{31}$$

Therefore,

$$\varepsilon_i=B_i t+\overline{C}_i \quad (i=1,2,3) \tag{32}$$

In this formula, B_1, B_2, and B_3 are constants.

Both the intensity and the structural strain take on linear changes.

The total state function has the same form as the Misses theory in classical theory.

3.3. *Spherical-form solution*

Assume that the solution form is as follows:

$$\begin{cases} \sigma_1 = r\sin\theta\cos\varphi \\ \sigma_2 = r\sin\theta\sin\varphi \\ \sigma_3 = r\cos\theta \end{cases} \tag{33}$$

Therefore, the intensity energy density function is $U = \dfrac{\sigma_1^2 + \sigma_2^2 + \sigma_3^2}{D}\sin\theta$.

Correspondingly,

$$\frac{\partial\sigma_1}{\partial\theta} = r\cos\theta\cos\varphi, \quad \frac{\partial\sigma_2}{\partial\theta} = r\cos\theta\sin\varphi, \quad \frac{\partial\sigma_3}{\partial\theta} = -r\sin\theta$$

$$\frac{\partial\sigma_1}{\partial\varphi} = -r\sin\theta\sin\varphi, \quad \frac{\partial\sigma_2}{\partial\varphi} = r\sin\theta\cos\varphi, \quad \frac{\partial\sigma_3}{\partial\varphi} = 0$$

$$E = (r\cos\theta\cos\varphi)^2 + (r\cos\theta\sin\varphi)^2 + (-r\sin\theta)^2 = r^2$$

$$G = (-r\sin\theta\sin\varphi)^2 + (r\sin\theta\cos\varphi)^2 = r^2\sin^2\theta$$

$$F = -r^2\cos\theta\cos\varphi\sin\theta\sin\varphi + r^2\cos\theta\cos\varphi\sin\theta\sin\varphi = 0$$

$$\frac{\partial U}{\partial\sigma_1} = \frac{2\lambda}{D}\sigma_1\sin\theta, \quad \frac{\partial U}{\partial\sigma_2} = \frac{2\lambda}{D}\sigma_2\sin\theta, \quad \frac{\partial U}{\partial\sigma_3} = \frac{2\lambda}{D}\sigma_3\sin\theta$$

Substituting the above formulas into Equation (3), we can obtain the following:

$$\begin{cases} \dfrac{\partial}{\partial\theta}(r\sin\theta\cos\theta\cos\varphi) + \dfrac{\partial}{\partial\varphi}(-r\sin\varphi) = \dfrac{2\lambda}{D} r\sin^2\theta\cos\varphi \\ \dfrac{\partial}{\partial\theta}(r\sin\theta\cos\theta\sin\varphi) + \dfrac{\partial}{\partial\varphi}(r\cos\varphi) = \dfrac{2\lambda}{D} r\sin^2\theta\sin\varphi \\ \dfrac{\partial}{\partial\theta}(-r\sin^2\theta) = \dfrac{2\lambda}{D} r\sin\theta\cos\theta \end{cases} \tag{34}$$

After reorganization, it becomes apparent that if $\lambda = -D$, then the two sides of the equation are equal. Therefore, the assumed solution is a solution of the equation.

As Equation (3) is a nonlinear partial differential equation, it is very difficult to calculate analytical solutions. However, a small number of particular solutions can be calculated under some special conditions. Moreover, these particular solutions can reflect the characteristics of the structural total state function, especially the characteristics of the total state function of elastic-perfectly plastic materials according to plane-form, cylindrical-form, and spherical-form solutions. In addition, the boundary condition of the solution represents an important factor that influences the form of the solution. Another important factor is the function of the material intensity energy density, which reflects the material features and influences the solution's characteristics. The material intensity energy density reflects the important features of the materials and is closely related to the microcosmic or macroscopic structure, linking form and connection performance.

4. Conclusions

Through this study, we have acquired a better understanding of the total state function theory of the material structure. Based on the above analysis of the elastic-perfectly plastic material, we are able to draw the following conclusions:

(1) As a method for solving the material structure problem (a theoretical system for solving the problem of the complicated material), the total state function theory is fully effective and feasible for solving simple materials such as the elastic-perfectly plastic material. Therefore, we can obtain conclusions and results that are almost completely consistent with the typical elastic-plastic theory. However, their substantial concepts are completely different.

(2) The total state function theory establishes a set of complete conceptual systems and methods for solving the problem of the structural material. This is mainly reflected by the governing, path, and constitutive relation equations, which form the core content and method of the total state function theory of the material structure.

(3) The total state function theory integrates the strength theory, flow rule, and constitutive relation with the typical elastic-plastic theory by using a uniform concept and method. It gives new meaning to the three concepts, as follows: The strength theory with a complicated stress condition is only the reflection of the total state function in the stress subspace; the flow rule is the specific path of the material on the total state function surface with a specific environmental effect; and the constitutive relation is the incremental relationship of the points on the path curve.

(4) The total state function theory provides a new concept for analyzing and considering problems from the whole to the part.

In brief, as a method for solving the problem of the material structure, the total state function theory of the material structure demonstrates its unique aptitude and feasibility through the analysis of the elastic-perfectly plastic material. Further research much extend the analysis to more complicated materials or circumstances, and it is hoped that more researchers will participate in these studies in the future.

References

1. YU Maohong, LINEAR AND NONLINEAR UNIFIED STRESCAENGTH THEORY[J]. Chinese Journal of Rock Mechanics and Engineering,2007 , (26)4: 662-669(in Chinese).
2. X.-L. Gao, M.Q. Liu. Strain gradient solution for the Eshelby-type polyhedral inclusion problem[J]. Journal of the Mechanics and Physics of Solids. 2012,60, 261–276.
3. Xuanhe Zhao. A theory for large deformation and damage of interpenetrating polymer networks[J]. Journal of the Mechanics and Physics of Solids. 2012, 60, 319–332.
4. LIU De-wen,XU Chun-mei ,DENG Yong-1iang ,GUO Chun-li, Discussion on "Several shear spatially mobilized planes and anisotropic strength criteria of soils"[J], Chinese Journal of Geotechnical Engineering,Vol.32 No.10 Oct, 2013:1947-1949.
5. NIU Jie, LIU Guang-lian,TIAN Jie,ZHANG Ya-xin,MENG Li-ping,COMPARISON F YIELD STNGTH THEREORIES WITH EXPERIMENTAL RESULTLTS[J],ENGINEERING MECHANICS, 2014, (31)1: 181-187(in Chinese).
6. KONG Wei-xue, RUI Yong-qin, DONG Bao-di, Determination of dilatancy angle for geomaterials under Non-associated flow rule[J],Rock and Soil Mechanics, 2009,(30)11: 3278-3282(in Chinese).
7. Alexander E.Ehre, Markus B., MikhailItskov. A continuum constitutive model for the active behavior of skeletal muscle. Journal of the Mechanics and Physics of Solids 59 (2011) 625–636.
8. ZHAO Lianheng, SUN Qiuhong, HUANG Fu, ZOU Jinfen, Influences of Non associated Flow Rule on Shallow Tunnel Stability Based on Upper Bound Limit Analysis[J]. Journal of Highway and Transportation Research and Development, 2012, (29)12: 101-106(in Chinese).
9. HU Ya-yuan, Mechanism of non-associated flow rule based on Hill'S principle[J]. Journal of Zhejing University(Engineering Science), 2013, (47)7: 1192-1198(in Chinese).

10. Ercan Gurses, Tamer El Sayed. A variational multiscale constitutive model for nanocrystalline materials. Journal of the Mechanics and Physics of Solids 59 (2011) 732–749.
11. NIU Bin, YANG Hai-tian, Homogenization-based investigation on constitutive relationship of oblique joint rock[J], Chinese Journal of Geotechnical Engineering, 2012, (18)4: P390-395(in Chinese).
12. Victor L. Berdichevsky. On the origin of microstructure self-similarity in severe plastic deformation. Journal of the Mechanics and Physics of Solids ,2012,60, 1–13.
13. XIANG Biao,ZHANG Zong-liang,CHI Shi-chun1,LIN Gao,Study of constitutive relations of rock fill under complex stress path[J], Rock and Soil Mechanics,(in Chinese).
14. Wei ZhenHai, Wang MengShu, Zhang DingLi, THE CONSTITUTIVE RELATIONSHIP MODEL OF BASED ON THE FULL STATE FUNCTION, PROCEEDINGS OF THE TWENTY-FIRST NATIONAL CONFERENCE ON STRUCTURAL ENGINEERINGTHE,2012.10(in Chinese).
15. Wei ZhenHai, Wang MengShu, Zhang DingLi, A research on the full state function constitutive relation model, The Second International Symposium on Constitutive Modeling of Geomaterials: Advances and New Applications 2012.10.
16. Wei ZhenHai Research on the Geomaterials Structure[D] Beijing Jiaotong University 2012(in Chinese).
17. Wei ZhenHai, Wang MengShu, Zhang DingLi, Study on the Full State Function of Structure Parameters For Materials, Mechanical Engineering Materials Science and Civil Engineering, Dct.31,2013.
18. Wei ZhenHai, Wang MengShu, Zhang DingLi, A Reserch on Strength Criteria Based on Material Structure, Experimental and Applied Mechanics,(EAM2014) 2014,Miami,USA.

PNIPAM/PDAMEMA network-grafted hydrogels: thermal response and drug delivery on network/graft architecture

Sheng-Qi Chen*, Wei-Dong He† and Ting-Ting Pan

Department of Polymer Science and Engineering, CAS Key Laboratory of Soft Matter Chemistry, University of Science and Technology of China,
Hefei, Anhui 230026, China
*E-mail: † wdhe@ustc.edu.cn, *sqchen@mail.ustc.edu.cn*

Hydrogels have drawn wide-spread research attention. Intelligent hydrogels are capable of responding to pH, temperature, electric and magnetic fields, through the change in their volume and properties. Therefore, they have broad application prospects in cell isolation and culture, immobilized enzyme, biosensors and tissue engineering, especially in drug control release Poly (N-isopropylacrylamide) (PNIPAM) is a well-known thermal responsive polymer. PNIPAM and poly(2-dimethylaminoethyl methacrylate) (PDMAEMA) network-grafted hydrogels with reversed crosslinkage/graft architecture were obtained through click reaction of alkynyl-pending PNIPAM (or PDMAEMA) as network backbone and PDMAEMA (or PNIPAM) with two azido groups at one chain end as macro-crosslinker. Both series of hydrogels have the high swelling degree in water, exhibited thermal response and have the ability of controllable drug release. Moreover, those properties were found to be dependent on the crosslinkage/graft architecture.

Keywords: Hydrogel; Thermal Responses; Architecture; Drug Release; PNIPAM;

1. Introduction

Hydrogels, as hydrophilic polymer network to retain large amount of water, have drawn wide-spread research attention. Intelligent hydrogels are capable to response to pH, temperature, electric and magnetic fields, through the change in their volume and properties [1-2]. Therefore, they have broad application prospects in cell isolation and culture, immobilized enzyme, biosensors and tissue engineering, especially in drug control release [3-4]. Poly (*N*-isopropylacrylamide) (PNIPAM) is a well-known thermal responsive polymer[5-6].

To improve the performances of hydrogels, some side-chains are grafted onto the cross-linked network. Upon proposing this strategy, Okano et. al.

* Corresponding author.

prepared this network-grafted PNIPAM hydrogels through grafting polymerization of NIPAM onto crosslinked PNIPAM and attributed their rapid swelling to the spontaneous collapse of grafting chains[7]. Zhang et. al. [8] found that PAA/PNIPAM network-grafted hydrogels exhibited greater swelling ratio than ordinary hydrogel. Lee et. al. grafted PNIPAM chains onto the alginate or chitosan network, to obtain different hydrogels with fast responsiveness[9-10]. However, the influence of network-graft architecture on the properties of dual-responsive hydrogels has gained little concern.

In this paper, we synthesized two kinds of PNIPAM/PDMAEMA hybrid hydrogels. One is composed of PNIPAM network and PDMAEMA grafting-chain and the other has the opposite allocation of network and grafting-chain. Those hydrogels exhibited different responsive behavior and drug release behavior.

2. Experiment

2.1. *Synthesis of PNIPAM-(N_3)$_2$ and PDMAEMA-(N_3)$_2$ via ATRP*

PDMAEMA with two azido groups at one chain end [PDMAEMA-(N_3)$_2$] was synthesized through ATRP with *N,N*-bis(2-azidoethyl)-2-bromoisobutyrylamide (AEBIA) and pentamethyldiethylenetriamine (PMDETA) as the initiator and ligand in isopropyl alcohol at 25°C for 4 h. After the polymerization, the mixture was dialyzed against deionized water for 48 h and lyophilized. The block length of PDMAEMA-(N_3)2 was determined to be 60 based on ^{1}H-NMR analysis, and molecular weight was also characterized with GPC measurement.

PNIPAM-(N_3)$_2$ was in the similar way but *N,N*-bis(2-azidoethyl)-2-chloropropionylamide and tris[2-(dimethylamino)ethyl]amine was used as the initiator and ligand while the mixture of DMF and H_2O (3:2 in volume) was used as the polymerization medium.

2.2. *Synthesis of Alkynyl-pending Copolymers via RAFT Polymerization*

Alkynyl-pending poly(NIPAM-*co*-propargyl acrylate) [P(NIPAM-*co*-ProA)] with different compositions were synthesized via RAFT copolymerization of NIPAM and ProA using dibenzyltrithiocarbonate (DBTTC) as chain transfer agent. The copolymerizations were carried out in THF at 80°C for 24h. The copolymer composition was determined with 1H-NMR spectroscopy and the molecular weight distribution was characterized by GPC.

Similarly, alkynyl-pending poly(DMAEMA-*co*-propargyl acrylate) [P(DMAEMA-*co*-ProA)] with different compositions were synthesized.

2.3. *Synthesis of Network-Grafted Hydrogels through Click Chemistry*

Network-grafted hydrogels with PNIPAM network and PDMAEMA grafting chains (N-PNIAPM/G-PDMAEMA) were synthesized through click reaction between P(NIPAM-co-ProA) and PDMAEMA-$(N_3)_2$. The click reactions were performed at 50°C for 24 h in THF with CuBr and 2,2'-bipyridine as the catalyst and ligand. The hydrogels was purified by dialysis against THF/water (1/1) for 2 d and water for 3 d. The final dry hydrogels were obtained by the lyophilization.

In the same way, different hydrogels with PDMAEMA network and PNIPAM grafting chains (N-PDMAEMA/G-PNIAPM) were obtained.

2.4. *Swelling and De-swelling of Hydrogels under Different Temperatures*

The dried hydrogel samples were immersed into distilled water at 20°C. The swollen hydrogels were taken out at regular intervals. After wiping off the water on the sample, the weights of the swollen hydrogels were recorded. The data presented are an average of three sampling. The swelling ratio (R_s) was defined as followed.

$$R_s = \frac{W_s - W_d}{W_d} \times 100\% \qquad (1)$$

where W_s and W_d is the weight of swollen hydrogel and dried hydrogel.

After saturated swelling, the hydrogels were kept in distilled water at 40°C. The swelling ratio was determined in the same way.

2.5. *Drug Loading and Release of Ceftriaxone Sodium by Hydrogels*

The dried hydrogels (50.0 mg) were swollen in aqueous solution of ceftriaxone sodium (10.0 mg/g, 10.0 g) at 20ºC for 12 h. After the excess water on the surfaces was wiped off, the loading content of ceftriaxone sodium was estimated based on the weight difference of swollen gel and dry gel. After the drug loading, the drug-loaded hydrogels were immersed into the distilled water at 20 or 40°C, 2.0 mL solution was taken out for UV-vis measurement at different intervals. The concentration of ceftriaxone sodium was determined at $\lambda = 292$ nm. At each sampling, distilled water (2.0 mL) was added.

3. Results and Discussion

Network-grafted hydrogels with reversed network/graft architectures were synthesized through click reaction as shown in Scheme 1. Herein, macro-crosslinkers with two azido groups at one chain-end was used and crosslinking degree was modulated by changing the amount of alkynyl groups in ProA copolymers as the network backbone. The thermal response and drug release of those hydrogels were comparatively investigated to disclose the influence of hydrogel architecture on its properties.

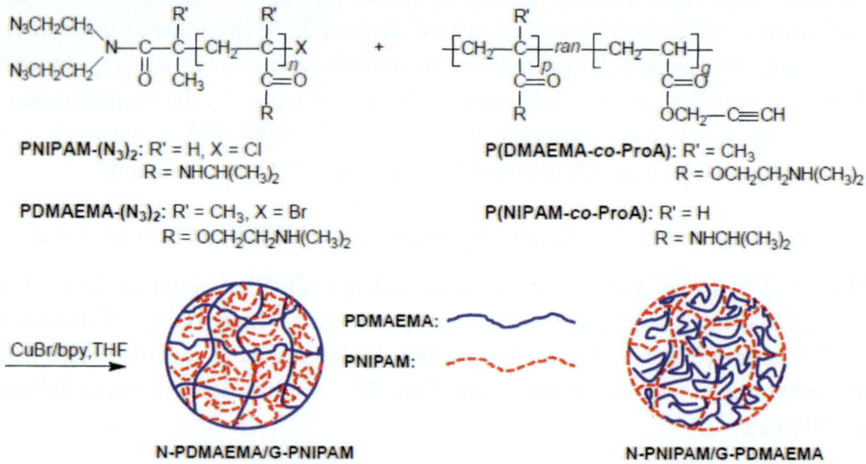

Scheme. 1. Illustration of preparing network-grafted hydrogels with reversed architectures

3.1. *Synthesis of Network-Graft Hydrogels*

Figure 1 shows the typical ^{1}H-NMR spectra of macro-crosslinkers and network backbone, as well as the assignment of characteristic signals. All the spectra show the appearance of characteristic signal from different units, such as $COOC\boldsymbol{H}_2CH_2$ (4.07 ppm) from DMAEMA, $NHC\boldsymbol{H}(CH_3)_2$ (4.01 ppm) from NIPAM, $COOC\boldsymbol{H}_2C{\equiv}CH$ (4.70 ppm) from ProA and $N(C\boldsymbol{H}_2C\boldsymbol{H}_2N_3)_2$ (about 3.5 ppm) from ATRP initiator. Based on the integration ratio of those signals, the polymerization degrees of macro-crosslinkers and the *p/q* ratio of ProA-containing copolymers were obtained. GPC traces of all polymers in DMF reveal the symmetric and narrow distribution of molecular weight. The detail of structural characterization is summarized in Table 1.

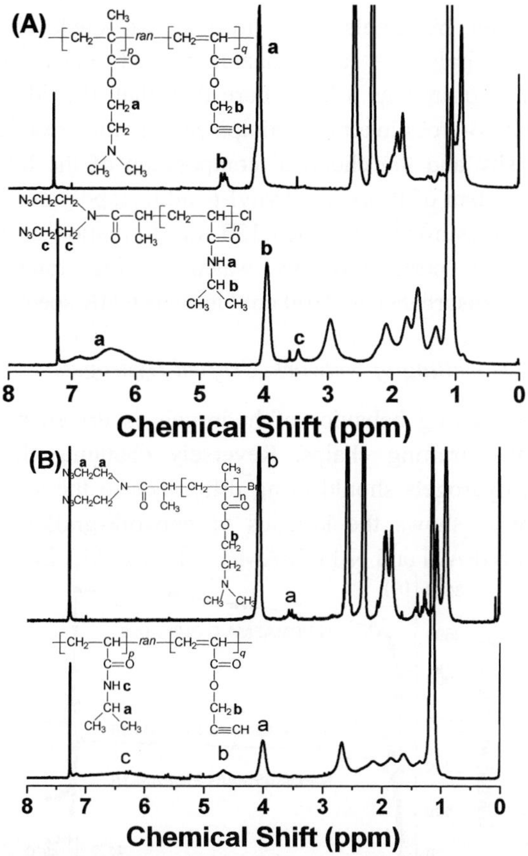

Fig. 1. ^{1}H-NMR spectra of PNIPAM-$(N_3)_2$ (A, lower), P(DMAEMA$_{100}$-*co*-ProA$_{15}$) (A, upper), P(NIPAM-*co*-ProA) (B, lower) and PDMAEMA-$(N_3)_2$ (B, upper) in $CDCl_3$

Table 1.Characterization of alkynyl-pending copolymers and diazido-terminated polymers.

polymer	NMR result		GPC result	
	NIPAM (or DMAEMA):ProA (mole)	$M_{n,NMR}$	$M_{n,GPC}$	M_w/M_n
PNIPAM$_{62}$-$(N_3)_2$	-	7100	7300	1.12
P(DMAEMA$_{100}$-*co*-ProA$_5$)	5:100	32100	35700	1.44
P(DMAEMA$_{100}$-*co*-ProA$_{10}$)	10:100	33800	36200	1.52
P(DMAEMA$_{100}$-*co*-ProA$_{15}$)	15:100	35200	34700	1.42
PDMAEMA$_{60}$-$(N_3)_2$	-	9340	10100	1.15
P(NIPAM$_{100}$-*co*-ProA$_5$)	100:5	12100	13400	1.34
P(NIPAM$_{100}$-*co*-ProA$_{10}$)	100:10	12600	15200	1.33
PNIPAM$_{100}$-*co*-ProA$_{15}$)	100:15	13100	16800	1.39

Network-grafted hydrogels were formed via click chemistry between alkynyl-pending copolymers and diazido-capped macro-crosslinkers. The amount of alkynyl group was a little more than that of azido groups to ensure that all the macro-crosslinkers participated in the reaction. Herein, the crosslinking density and the chemical composition of the hydrogels could be adjusted by the content of ProA in alkynyl-pendingcopolymers. When the chain length of macro-crosslinkers exceeded 120, the formation of hydrogel was not observed. After the purification by washing with water thoroughly, the hydrogels have no absorbance at 2100 cm^{-1} in their FTIR spectra.

3.2. *Swelling/De-swelling of Network-Grafted Hydrogels*

The swelling/de-swelling behavior of hydrogels is driven by the crosslinked backbone and the grafting chains. Reversely changing the architecture of network grafted hydrogels should bring difference in the swelling/de-swelling behavior. Figure 2. shows the kinetics of network-grafted hydrogels of N-PDMAEMA/G-PNIPAM and N-PNIPAM/G-PDMAEMA in neutral water.

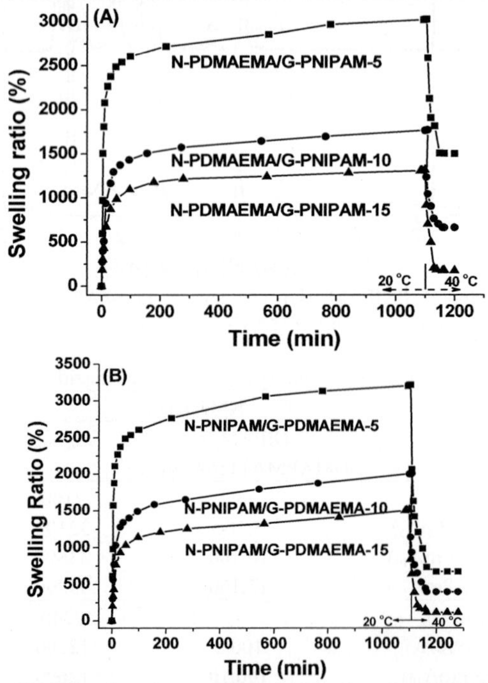

Fig. 2. Swelling/de-swelling kinetics of network-grafted hydrogels of N-PDMAEMA/G-PNIPAM (A) and N-PNIPAM/G-PDMAEMA (B) from different ProA-containing copolymers (the number indicates the content of ProA as shown in Table 1) in neutral water at 20 °C and 40 °C.

When the dry hydrogels were immersed in water at 20 °C and pH = 7.0, all of them exhibited excellent ability to uptake water, with saturated swelling ratio of from 1300 % to 3700 %.N-PNIPAM/G-PDMAEMA hydrogels have higher swelling ratios than N-PDMAEMA/G-PNIPAM hydrogels. The saturated swelling ratio decreases obviously with the increase of crosslinking density.

When positioned in neutral water at 40 °C, the swollen hydrogels shrank to the equilibrium state immediately in about 30 min. It should be assigned to the response of PNIPAM segment in hydrogel network. The percent of excluded water increases with crosslinking density, indicating that the crosslinking density is the most important factor for water-holding capacity of the comb-type grafted hydrogels. Comparing N-PDMAEMA/G-PNIPAM with reversed network-grafting architecture, N-PNIPAM/G-PDMAEMA excluded much more water under the same condition, suggesting the influence of hydrogel architecture.

3.3. *Loading and in vitro Release of Ceftriaxone Sodium*

Ceftriaxone sodium was loaded into hydrogels by incubating the freeze-dried hydrogels (50.0 mg) in the aqueous solution of drug (10 mg/g) at 20 ºC and pH = 7.0 for 12 hr. The amount of loaded drug is 16.8 mg and 15.6 mg for N-PNIPAM/G-PDMAEMA-5 and N-PDMAEMA/G-PNIPAM-5. It is found that the amount of loaded drug decreases with the increase of crosslinking degree.

Figure 3 shows the drug release from two hydrogels at different temperatures. Ceftriaxone sodium releases at 40 ºC much faster than at 20 ºC, considering the time needed for the maximum release. At the same time, the maximum release amount also rises when the temperature goes up from 20 ºC to 40 ºC. As well, the hydrogel with PNIPAM as grafting chains has better ability of drug release than that with PNIPAM as network.

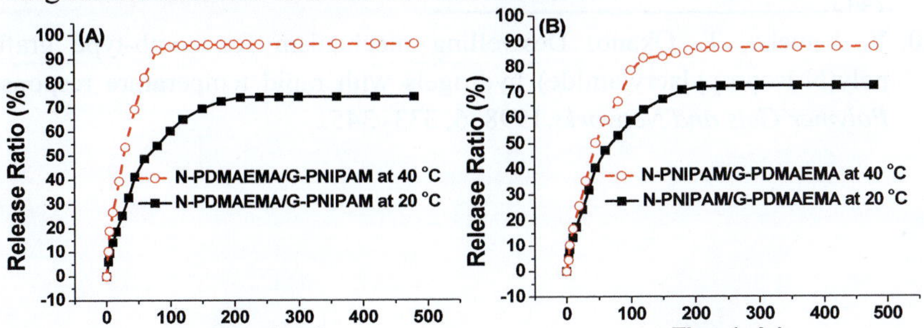

Fig. 3. Drug release behavior of network-grafted hydrogels from P($NIPAM_{100}$-*co*-$ProA_5$) and P($DMAEMA_{100}$-*co*-$ProA_5$)

Acknowledgments

The financial support of the National Natural Scientific Foundation of China Projects(20934005 and 21274136) are greatly acknowledged.

References

1. A.S. Hoffma, Hydrogels for biomedical applications, *Adv Drug Deliver Rev*, 2002, 43, 3-12.
2. S.J. Buwalda, Hydrogels in a historical perspective: From simple networks to smart materials, *J Control Release*, 2014, SI, 254-273.
3. N. Annabi, Rational Design and Applications of Hydrogels in Regenerative Medicine, *Adv Mater*, 2014, 26, 85-124.
4. A. Kumar, Smart polymers: physical forms and bioengineering applications, *Prog Polym Sci*, 2007, 32, 1205-1237.
5. Y.Y. Zhang, Study on the room temperature synthesis of highly photoluminescent and temperature-sensitive CDs/PNIPAM hybrid hydrogels and their properties, *RSC Advances*, 2015, 5, 71030-71034.
6. J.J. Cheng, Temperature and pH-dependent swelling and copper (II) adsorption of poly(N-isopropylacrylamide) copolymer hydrogel, *RSC Advances,* 2015, 5, 62091-62100.
7. T. Okano, Thermally on-off switching polymers for drug permeation and release, *J Control Release*, 1990, 11, 255-265.
8. J. Zhang, Dual thermo- and pH-sensitive poly(N-isopropylacrylamide-co-acrylic acid) hydrogels with rapid response behaviors, *Polymer*, 2007, 48, 1718-1728.
9. Z.L. Tang, Comb-type grafted poly(N-isopropylacrylamide) gel modified surfaces for rapid detachment of cell sheet, *Biomaterials*, 2010, 31,7435-7443.
10. Y. Kaneko, T. Okano, Deswelling mechanism for comb-type grafted poly(N-isopropylacrylamide) hydrogels with rapid temperature responses, *Polymer Gels and Networks*,1998, 6, 333- 345.

Scan trajectory optimization for laser rapid prototyping based on tangent arc transition

Bo Qian[a*] and Ming-Yi Wang[b]

School of Mechanical and Power Engineering, East China University of Science and Technology,
Shanghai,200237, China
E-mail: [a]qianbo@ecust.edu.cn, [b]wangmingyi1892@126.com

In the laser rapid prototyping machine, the parts are manufactured through layers. Every layer is formed by realizing scan trajectory to solidify or sinter materials by laser machine and galvanometer. The scanning trajectories of every layer comprise of filling path and hollow path. In order to improve the scanning efficiency and quality and reduce scanning time of every layer, this paper deals with the connection of filling path in scanning trajectory based on tangent arc transition way, to ensure the shortest transition distance among the filling paths and ensure all their connections lengths are short. Meanwhile, the processing orders of the hollow paths are sorted by using the principle of cheaper algorithm again ensuring the sum of all hollow paths lengths are short and time is saved while processing. Finally, when the algorithm is applied into the laser rapid prototyping machine, it can not only product parts, but also improve quality and efficiency of the actual scanning trajectories.

Keywords: Laser scan; Rapid Prototyping; Tangent Arc Transition; Trajectory Optimization; Cheap Algorithm.

Introduction

The laser rapid prototyping is a kind of manufacturing technology which is driven by CAD model directly and can quickly generate the three dimensional entity in arbitrary complex shapes. Based on the discrete layered manufacturing principle, the three dimensional CAD model is divided into a series of orderly two-dimensional layers (slices) along a certain direction (usually along Z direction). And then the slices will be manufactured in sequence by the rapid prototyping machine and be connected automatically, thus, the three-dimensional physical entity is generated [1, 2].

* Corresponding author

The rapid prototyping scanning system usually consists of laser, two-dimensional or three-dimensional galvanometer, and optical path system and so on. Based on the discrete layered manufacturing principle, before any parts is manufactured by the rapid prototyping system, firstly, the three-dimensional parts will be sliced into layers and layers of two-dimensional graphics through slicing software or algorithm and then will be saved in the computer memory[3], secondly, the padding trajectory and contours trajectory of two-dimensional graphics is computed by the path generation algorithm after reading two-dimensional graphics of every layer. Thirdly, the resin will participate in the curing photochemical reaction and powder sintering after scanning photosensitive resin or polymer powder by the scanning system according to the above mentioned paths, finally, the three-dimensional parts are formed by the accumulation layer by layer.

The trajectory can be divided into filling path and vacant path when processing according to the above trajectory by the scanning system. The filling path is scanning trajectory, while vacant path is the path of laser shutdown between two scanning trajectories. As shown in figure 1, the traditional vacant path makes the transition by connecting the adjacent endpoints of the filling path. However, the deflection lenses of the galvanometer for controlling the laser path have a certain moment of inertia, thus this kind of transitional form on the scanning process will come about the overshot of the end and the lateral displacement of the beginning especially on scanning the fine patterns. The vacant path will not directly affect the final effect of the laser scanning, so it is feasible to run the vacant path in any way. But the routine of the vacant path will affect the factual scanning efficiency, thus the optimization of the vacant path is a very important step to improve the scanning efficiency of the actual graphics and the quality of the graphics after scanning. According to the statistics, if the graphics are processed directly before the optimization of the vacant path, it is just from 30 percent to 40 percent. If the total length of the vacant path can be achieved the shortest after adjusting he routine order of the scanning trajectory, it will obviously reduce the scanning time of the vacant path. Therefore, the optimization of the vacant path will adopt the cheap algorithm. Although the cheap algorithm cannot attain the optimal solution, its solving speed is fast enough and can meet the requirement of the real-time solution; meanwhile, it will not appear that the time after the optimization is needed longer than the saved time after the optimization 0.

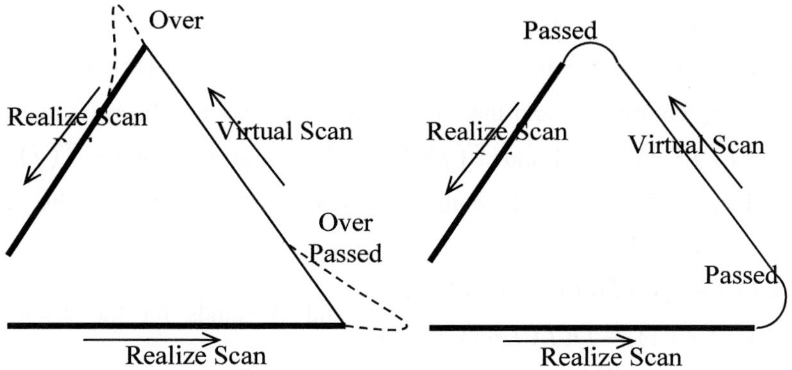

Fig.1. Tangent arc transition in hollow path and filling path

Furthermore, the vacant path will not directly affect the marking effect, so the galvanometer should be run in the critical state of the maximum acceleration and speed. On the situation, although there will be some deviations between the moving path and the default path on the swing of the galvanometer lenses, the scanning time of the vacant path especially the short vacant path will decrease sharply because of the large acceleration. However, scanning the vacant path by the critical acceleration will make the deviation towards the direction of the vacant path between the starting and ending points of the scanning trajectory. The solution is to find the curve tangent to the actual scanning path rather than walking the straight line on the starting point and the ending point.

1. The order optimization of the vacant path with cheap algorithm

The generation of two-dimensional graphics is complex [5]. In order to improve the whole prototyping efficiency, it not only accomplishes the generation process of the trajectory, but also reduces the vacant path on the scanning process[6]; therefore, the vacant path is computed by the cheap algorithm7 as shown in the figure. The computation processes are shown as follows:

Step 1:defining a construe, and its data members including the starting point and the ending point, stipulating the direction of the vacant path, that is to say, the vacant path has the order at this time. Step 2:creating a contractual array S(initial array), putting all the information of the vacant path in the scanning graphics into it, creating another contractual array T(new array), putting the information closest to the origin on array S of the laser beam into array T, and deleting the information of the vacant path on array S.

Step 3:finding out the shortest distances $T(j)$ and $S(i)$ among the points on array T and S. If there is only one element on array T, $S(i)$ will be the second element on array T; and then delete $S(i)$ on array S. Otherwise, as for the edges $(T(j-1),T(j))$ and $(T(j),T(j+1))$ on array T, when $T(j)$ is the first element, $T(j-1)$ will be the last element; when $T(j)$ is the last element, $T(j+1)$ will be the first element. If $D(S(i),T(j-1))-D(T(j-1),T(j)) \leq D(S(i),T(j+1))-D(T(j+1),T(j))$, and D stands for the distance between two points, $S(i)$ should be inserted between $T(j-1)$ and $T(j)$, or $S(i)$ will be inserted between $T(j)$ and $T(j+1)$, and delete the element i from array S. If S=0, it will conclude. Otherwise, it will return to step 3.

On the computation, array T is always an ever-expanding primary circuit, and the initial is a self-loop. The position where array T should be inserted into based on the increment size of the length on circuit T after adding the new elements. Practices have proven that using cheap algorithm principle on the optimization of the scanning trajectory can improve the efficiency. As shown in the scanning graphics, the time taken is 3.7 seconds before the optimization, while the time taken is 2.5 seconds after the optimization which is shorten by 10 percent than the past scanning way, and reduce the overall marking time as shown in the figure.

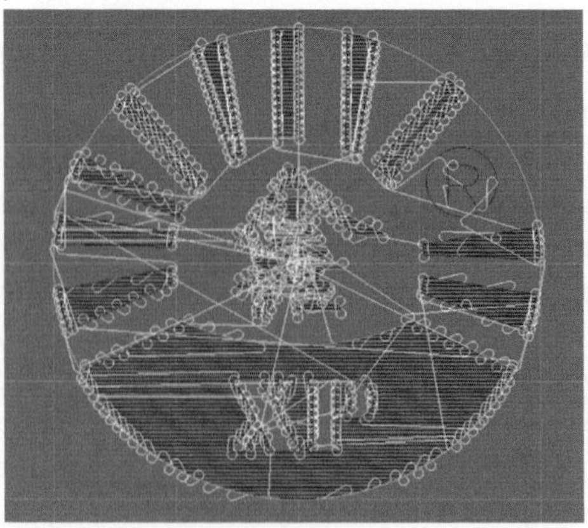

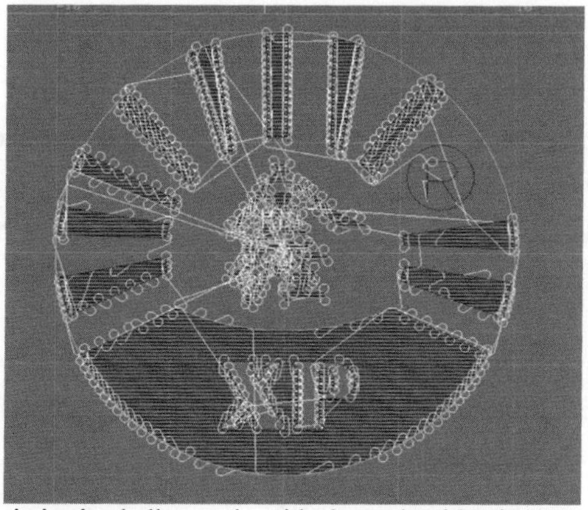

Fig. 2 Optimization hollow paths with cheap algorithm in laser scanning

2. Arc transition connecting the scanning trajectory in the filling path

In the paper, the defined connection curve is described by a circular arc with a fixed radius R and a straight line. As for circle is a kind of the simplest two-dimensional curve, it is simple on the mathematical definition, and it is easiest to implement on the piratical curve segmentation, therefore, it is feasible to realize the curve transition8 by the above methods.

In theory, there are two circles tangent to every filling path as shown in the figure 3. As for the previous marked path, it is bound by the direction of the walking vector, so the circle tangent to the path should be on the ending point, as shown in the figure; the radius of the point B is the fixed value R. An important principle to solve the problem is that the transition curve should be short enough after connection to save the walking time of the vacant path. On the actual computation, two tangent circles of the shortest distance between centers are usually selected for the sake of simplicity. After selecting such fixed tangent circles, there will be four tangent lines in mathematics (as shown in figure 4), and the computation processes of four tangent lines are as follows:

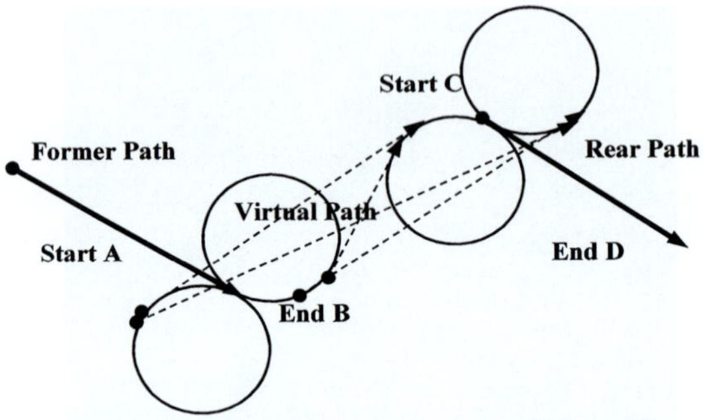

Fig.3. Four tangent arc transition styles in two filling paths

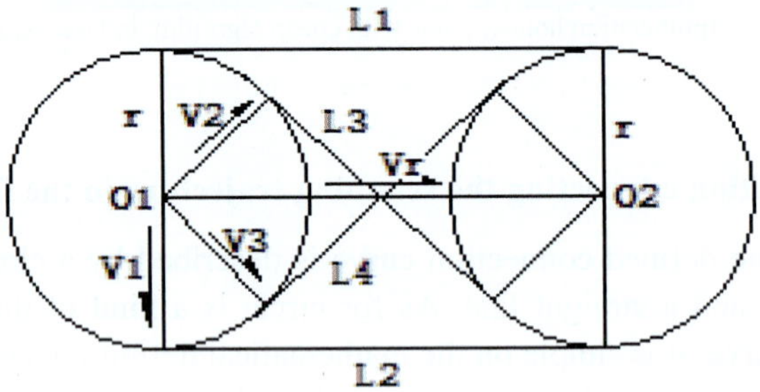

Fig.4 Calculate four tangent lines in two circles

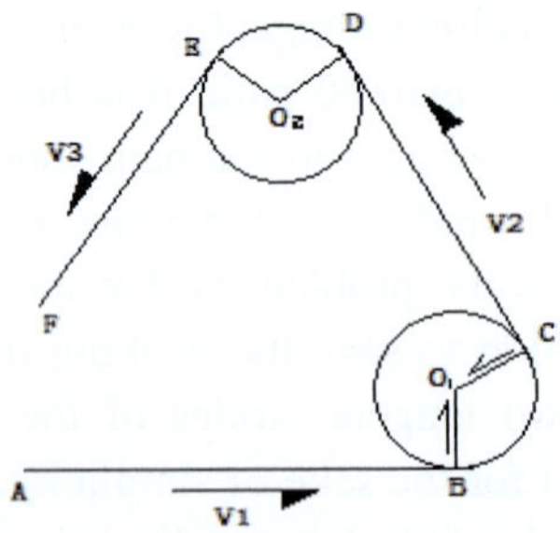

Fig.5 Only one tangent line according to the principle of same rotation direction

On calculating the tangent lines of two circles 9, the related definition in the tangent circles are shown as follows:

r is the radius of two circles, O_1, O_2 are respectively the centers of two circles, and V_r is the vector between two centers, that is: $V_r = O_2 - O_1$.

From the above definition, the calculation formula of the four tangent lines is: $L1:(O_1 - rV_1, O_2 - rV_1)$, from the formula, V_1 has two vector conditions, which are $V_1 \cdot V_r = 0$, and $|V_1| = 1$, and it can limit the direction of V_1 by the condition $V_1 \times V_r \leq 0$. Similarly, it is on the formula $L2:(O_1 + rV_1, O_2 + rV_1)$.On the formula $L3:(O_1 + rV_2, O_2 - rV_2)$, V_2 has also two vector conditions, which are $V_2 \cdot V_r = 2r$ and $|V_2| = 1$, and it can limit the direction of V_2 by the condition $V_2 \times V_r \geq 0$.Similarly, it is on the formula $L4:(O_1 + rV_3, O_2 - rV_3)$, and $V_3 \cdot V_r = 2r$, $|V_3| = 1$, $V_3 \times V_r \leq 0$.

The direction of above four tangent lines is all from circle O_1 to circle O_2, but it might be calculated the opposite direction on the actual computation which will not affect subsequent calculations. The adjacent filling paths have the vector direction, so the path of every tangent line should be in accordance with the path of connecting circular arc, and filling path is a tangent line, therefore, it can be described as: the rotation direction of two tangent lines is the same with the center of the tangent circle. There is a formula in mathematics as follows: $W_1 = V_1 \times (O_1 - B) = V_2 \times (O_1 - C)$, and W_1 is the rotation direction which is also cross multiply product of two vectors, and V_1 is the vector of the previous filling path, O_1 is the center of the first tangent circle, B is the ending point of the previous filling path, C is the starting point of the tangent line, and V_2 is the vector direction of the tangent line.

On the formula $W_2 = V_2 \times (O_2 - D) = V_3 \times (O_2 - E)$, W_2 is the rotation direction, V_3 is the vector of another filling path, O_2 is the center of the second tangent circle, E is the starting point of another marked path, D is the ending point of the tangent line, and V_2 is the vector direction of the tangent line.

In theory, there is only one tangent which meets the condition of the above two rotation direction, because there are two tangent lines among four tangent lines of the previous filling path whose direction is opposite to the rotation direction of W_1. On the left two tangent lines of another filling path, only one rotation direction is the same with W_2, while another one is opposite to W_2. And then except to the rotation direction of the tangent line opposite to W_2, there is only one tangent line meeting the above mentioned conditions. The tangent line will finally be selected.

3. Conclusion

The optimization computation of tangent arc transition scanning on the laser rapid prototyping scanning process software PowerRP(as shown in figure 6) has been applied into HRPS series laser rapid prototyping machine(as shown in figure 7) on Huazhong University of Science and Technology.

Compared to the past laser scanning methods, its speed is improved by 15 percent on the high speed scanning. Moreover, on the starting point, ending point and sharp point on the scanning process, it eliminates the defects such as the contour distortion, over-burnt and so on; meanwhile, scanning efficiency and prototyping quality also get improvement.

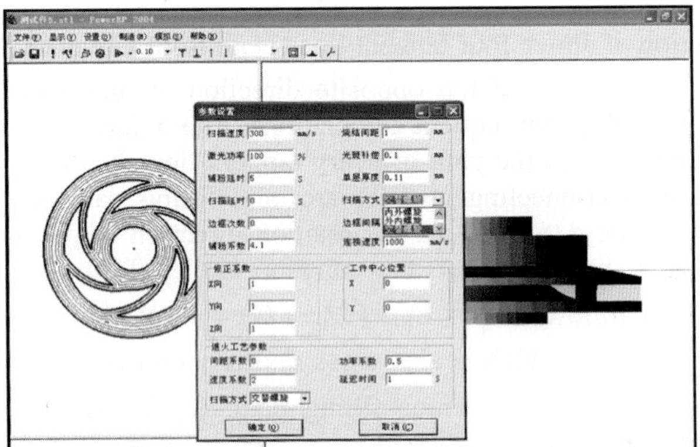

Fig.6 Running interface of laser scanning technology software

Fig.7 The actual process of laser scanning and formed part after optimization

Acknowledgement

This work is supported by Beijing Natural Science Foundation, No. 3132007 and National Natural Science Foundation of China (Grant No. 21206152).

Reference

1. S.H Huang, Y.J Xiao, J.H Mo et al. Prospects of Rapid Prototyping Technology [J].China Mechanical Engineering, 2000, 11 (1):195-200.
2. L.C Zhang, M Han, S.H Huang. Fault tolerant slicing of STL file based on crack tracking technology. Forging machinery[J], 2002, 2 (1): 52-54.
3. J. P. Kruth., Material Increase Manufacturing by Rapid Prototyping Techniques[J]. Annual of the CIRP, 1991, 40(2): 603-614
4. N.C. Bruno, J.L. Lage-Marques, S. M. Rode. Pulpal temperature increases with Er:YAG laser and high-speed handpieces [J]. The Journal of Prosthetic Dentistry, 2003, 90(5): 447-451.
5. T Hothorn, B Lausen. Bagging tree classifiers for laser scanning images: a data- and simulation-based strategy [J]. Artificial Intelligence in Medicine, 2003, 27(1): 65-79.
6. L.C Zhang, B Qian, S.H Huang. Software Architecture for High-performance Laser Marking Machi [J]. Modern machinery, 2006,2(1) :1-2.
7. Y.Q Dai, G.Z Hu. Graph theory and algebra structure [M].Beijing: Tsinghua University press, 1995, 12-46.

8. Haiyang Sun, Dapeng Fan. A new algorithm for the compensation of cutting tool arc [J]. China Mechanical Engineering,2007, 18(3):276-281.
9. J.G Sun, C.G Yang. Computer graphics (New Version) [M]. Beijing: Tsinghua University press, 1997, 368-458.

Microstructure and compound developed from the Fe-Ce-As at 1323K

Wei-Yu Yang[1], Jin-Zhu Zhang[2] and Shuie-Li[3]
College of Materials and Metallurgy, Guizhou University
Guiyang 550025, China
E-mail: [1] 404333078@qq.com, [2] zjz-yjx@163.com,
[3] longlong9812@163.com

The Arsenic in steel commonly comes from a certain amount of iron ore but it is difficult to remove them in the iron and steelmaking process because its oxidation potential is lower than that of Iron. In this study, a certain amount of rare earth metal, Cerium and a metal with low melting point, Arsenic were closed in the barrel-shaped cylinder machine by H08 steel, heated to 1323K for 50 hours. Interaction among the cerium, arsenic and iron in the barrel-shaped cylinder were studied by X-ray diffraction, optical microscopy, and electronic probe microscopy analysis. The result shows that the ternary compound Ce12Fe57.5As41 can be developed at 1323K when the atomic ratio of Cerium to Arsenic is 1:3. The binary compound CeAs is the main product at 1323K in the Fe-As-Ce system. The eutectic compound Fe2As can be precipitated from ferrite with the temperature decreasing.

Keywords: Arsenic; Microstructure; Compound; X-Ray Diffraction; Optical Microscopy; Electron Probe Microscopy

1. Introduction

Arsenic in steel mainly segregated on grain boundary and it is probably that the impact toughness, the cold brittleness and the hot-working performance of steel would be deteriorated [1-4]. The Arsenic in steel commonly come from a certain amount of iron ores but it is difficult to removal in the conventional ironmaking and steelmaking process because its oxidation potential is lower than that of Iron. The rare earth metals, which have a stronger affinity with the deleterious impurities such as oxygen, sulfur, phosphorus and some low melting point metals contained in steel, are often used to purify the liquid steel [5]. According to the thermodynamic property in binary phase diagram, the compounds CeAs, Ce_4As_3 and $CeAs_2$ can be developed in Cerium Arsenic system [6-8]. With a higher content of Cerium in steel, the compounds Fe_2Ce and $Fe_{17}Ce_2$ can be developed by heating the cerium and iron at 1180°C for 30 hours, and slowly

[2] Corresponding author.

cooling to 900°C, then annealing[6]. Moreover, it was reported if the Ce pieces, Fe powder, and As lumps in a molar ratio of 1:5:3 was placed in an evacuated fused-silica tube together with a 6-fold (by weight) excess of Sn, stepwise heated to 900 °C over 3 days, held at this temperature for 4 days, and slowly cooled (4 °C /h) to 600 °C, the ternary arsenide $RE_{12}Fe_{57.5}As_{41}$ and FeAs were obtained [7-9]. In this study, a certain amount of rare earth metal Cerium and low melting point Arsenic were closed in the barrel-shaped cylinder machined by H08 steel, heated to 1323K for 50h, and the interaction among the cerium and arsenic in steel was studied by X-ray diffraction, optical microscopy, and electronic probe microscopy analysis.

2. Experiment

The sample preparation device like a pressure-tight cylinder is shown in Figure 1, and machined by H08 steel of which the main composition is as follows: 0.35% (mass percent) Manganese, 0.03% Silicon, 0.082% Carbon, 0.016% Phosphorus, 0.014% Sulphur, and 99.3% Iron. The metal Cerium pieces (purity 99.0%) and the non-metal Arsenic particle (purity 99.99%) with a size below 5 mm are weight 3.835g and 6.164g respectively with an atomic ratio of Ce to As 1:3, put into the cylinder, then screwed down and welded the cylinder.

The cylinder was put into SRJK-2-9-tube vacuum resistance, and heated slowly in the reduction atmosphere as H_2:Ar=1:10. The temperature was controlled based on the vapor pressure of Arsenic as follows: taking 2h for room temperature to 773K, from 773K to 923K, taking 10min for increasing every ten degrees, from 923K to 983K, taking 20min for increasing every ten degrees, from 983K to 1023K, taking 30min for increasing every ten degrees, from 1023K to 1073K, taking 2h for increasing every ten degrees, from 1073K to 1123K, taking 4h for increasing every ten degrees, from 1123K to 1173K, taking 5h for increasing every ten degrees, from 1173K to 1223K, taking 6h for increasing every ten degrees, from 1223K to 1273K, taking 7h for increasing every ten degrees, from 1273K to 1323K, taking 8h for increasing every ten degrees, finally, at 1323K, keeping for 50h. Then the furnace was turned off, cooled to room temperature and switched off the protective hydrogen and argon.

The sample was sawed off laterally about 16mm away from the bottom by hand. One part of the samples was prepared for metallographic observation by the conventional methods. The other part was scratched, and taken the central substance that was mainly developed from Cerium, Arsenic, and Iron in the high temperature off the cylinder. Then the substance from the internal infiltration layer scratched off was crushed, ground, and used for phases or compounds determination by X-ray diffraction.

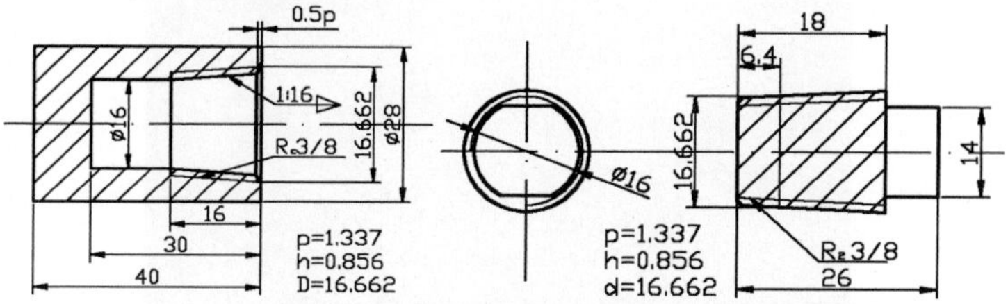

Fig.1 The diagram of the barrel-shaped cylinder and the screw plug

3. Results and Discussions

There are at least three different microstructures seen on the metallograph marked by p_a, p_b, p_c (see Figure 2) or the backscattered electron image marked by P_{ai}, P_{bi} *and* P_{ci} (see Figure 3) in the sample. The microstructures marked by P_{a1}, P_{a2} and P_{a3} show in Platinum grey on the backscattered electron image and in dark grey marked by p_a on the metallograph. The phases labeled by P_{b1}, P_{b2} and P_{b3} show in deep grey on the backscattered electron image and in light grey marked by p_b on the metallograph. The microstructures indicated by P_{c1}, P_{c2} and P_{c3} show in ash grey on the backscattered electron image and in grey indicated by p_c on the metallograph.

The main chemical compositions on various microstructures determined by means of electron microprobe are shown in table 1. The X-ray diffraction pattern of the sample is shown in Figure 4.

The phases marked by P_{a1}, P_{a2} and P_{a3} are mainly composed by cerium and arsenic with an atomic ratio approximately 1:1 (see Figure 3 and table 1), based on the results determined by X-ray diffraction and the Cerium Arsenic system diagram (see Figure 4) [3-5], this phase has a CeAs crystal structure.

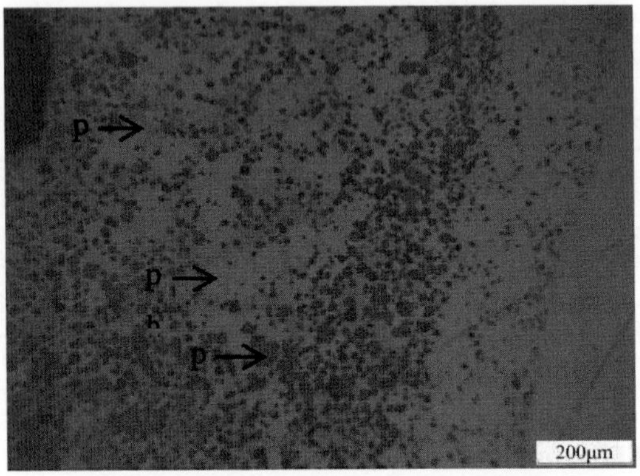

Fig.2 Metallograph of the area closes to the cylinder of the sample (1323K, 50h)

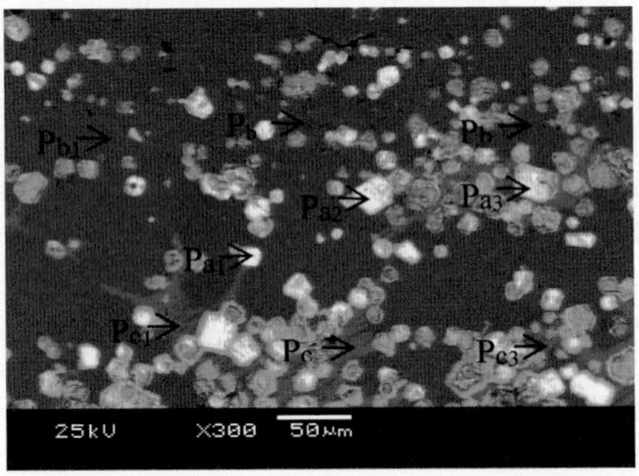

Fig.3 Backscattered electron image of the central area of the sample (1323K, 50h)

The microstructures labeled by P_{b1}, P_{b2} and P_{b3} on Figure 3 shown in dark grey are mainly composed by Iron and Arsenic besides a little amount of Cerium (see table.1 and Figure 3.). Based on both the phase diagram of Fe-As system and the results determined by X-ray diffraction (see Figure 4.), the content of arsenic at the area marked by P_{b1} or P_{b3} is bigger than the bigest saturated solubility of arsenic in ferrite of iron (9%As, at 1113K) [4], it can be seen that the compound Fe_2As may be precent at 1323K; moreover, the compound Fe_2As will be precipitated from ferrite with the temperature decreasing. Therefore, the microstructure labeled by P_{b1}, P_{b2} and P_{b3} should be a mixture composed by ferrite and a certain amount of Fe_2As.

The phases labeled by P_{c1}, P_{c2} and P_{c3} on Figure 3. shown in ash grey are mainly composed by iron, arsenic and cerium (see table.1). There are small differences among the molar ratios of Ce, Fe and As. Comparing with the ternary arsenide $Ce_{12}Fe_{57.5}As_{41}$[7, 9], it can be written as $Ce_{11.69}Fe_{53.78}As_{41}$, $Ce_{11.49}Fe_{53.68}As_{41}$ and $Ce_{12.45}Fe_{55.79}As_{41}$ respectively by the compositions of the phases labeled by P_{c1}, P_{c2} and P_{c3} according to the stoichiometric (see table.1 and Figure 3). Based on the results determined by X-ray diffraction, this phase is the ternary compound $Ce_{12}Fe_{57.5}As_{41}$ (see Figure 4).

Table 1 Atomic ratio of element for different microstructures analyzed by electron probe (At %)

area	Ce	Fe	As	area	Ce	Fe	As
P_{a1}	48.45	2.36	49.19	P_{b3}	0.97	90.71	10.32
P_{a2}	47.30	3.19	49.51	P_{c1}	10.98	50.51	38.51
P_{a3}	47.41	4.33	48.26	P_{c2}	10.82	50.56	38.62
P_{b1}	0.17	90.53	9.29	P_{c3}	11.40	51.07	37.53
P_{b2}	0.22	90.87	8.91				

Based on both the weaker diffraction intensity in the X-ray diffraction pattern (see Figure 4) and the less proportion on the metallograph (see Figure 2) or the backscattered electron image (see Figure 3), the amount of the ternary compound $Ce_{12}Fe_{57.5}As_{41}$ developed in the sample with an atomic ratio of Ce to As 1:3 at 1323K keeping for 50h is reletively small. The binary compound CeAs is the main product basaed on the highest diffraction intensity in the X-ray diffraction pattern of the sample (see Figure 4).

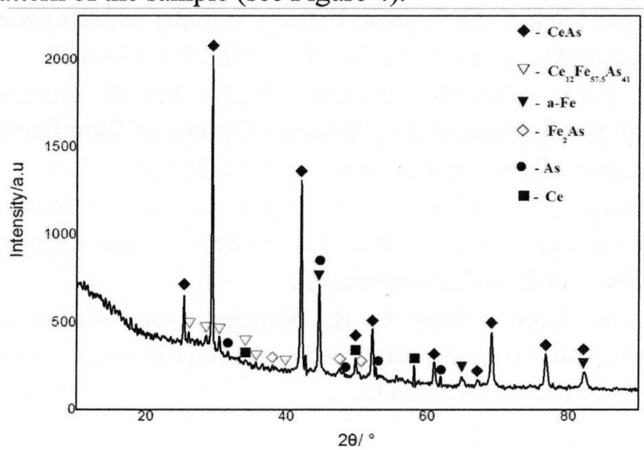

Fig.4 X-ray diffraction pattern of the sample (1323K, 50h)

4. Conclusions

The binary compound CeAs is the main product at 1323K in the Fe-As-Ce system when the atomic ratio of Cerium to Arsenic is 1:3.

The ternary compound $Ce_{12}Fe_{57.5}As_{41}$ can be developed at 1323K in the Fe-As-Ce system when the atomic ratio of Cerium to Arsenic is 1:3, and shown in ash grey on the backscattered electron image and in grey on the metallograph.

The eutectic compound Fe_2As can be precipitated from ferrite in which there is a high content Arsenic dissolved at high temperature with the temperature decreasing.

Acknowledgment

The authors gratefully acknowledge the financial support provided by National Natural Science Foundation of China (50964002) and Graduate innovation fund of Guizhou University (Graduate student of Science and engineering 2015006)

References

1. W. F. Gale&T. C. Totemeir. Smithells Metals Reference Book[M]. Boston: Butterworth-Heinemann, 2004: 497.
2. Yin Guojin. The Distribution of Arsenic in Steel [J]. Iron & Steel, 1981, 16(2): 20-28(in Chinese).
3. K.A. Gschneidner, Jr. and F.W. Calderwood. The As-Ce (Arsenic-Cerium) System [J]. Bulletin of Alloy Phase Diagrams. 1986, 7(3): 276-277.
4. V. Babizhets'ky, C. Le Sénéchal, J. Bauer. Interaction of Lanthanum or Cerium with Nickel and Arsenic Phase Diagrams and Structural Chemistry [J]. Journal of Alloys and Compounds. 1999, 287: 174–180
5. Riccardo Ferro, Gabriella Borzone, Nadia Parodi. Comments on the Formation Thermodynamics of Selected Groups of Rare Earth Compounds [J]. Journal of Alloys and Compounds, 2001, 321: 248–260
6. Li Daizhong, Gao Shuqin and Zhang Liehuo. Properties of Ce-Fe Intermetallic Compound in Steel [J]. Journal of the Chinese Rare Earth Society, 1985, 3(2):71-72(in Chinese).
7. Stanislav S. Stoyko, Peter E. R. Blanchard, and Arthur Mar. Ternary Rare-Earth Iron Arsenides RE12Fe57.5As41 (RE = La, Ce) [J]. Inorganic Chemistry, 2010, 49: 2325-2333.

8. Xiaodong Liu, Jinzhu Zhang, Sisi Zhu. Microstructure and Compound Developed from La-As-Fe System at 1223K [J]. Advanced Materials Research, 2013, 702: 145-148.
9. Jinzhu Zhang, Shengtao Dou. Study on Interaction between Cerium and Arsenic. [J]. Advanced Materials Research, 2011, 194: 1231-1234.

Numerical investigation of sediment wear in a centrifugal pump impeller

Hong-Ming Zhang †, Li-Xiang Zhang and Ji Yao
Department of Engineering Mechanics, Kunming University of Science and Technology, Kunming, Yunnan, China
†E-mail: hmzhang8888@126.com

OpenFOAM is an object-oriented C++ library of classes and routines of use for writing CFD codes. It has a set of basic features similar to any commercial CFD solver, such as turbulence models and discretization schemes. The paper presents the numerical studies of sediment wear in a centrifugal pump impeller using OpenFOAM code, which is an Open Source CFD Package. The 3-D turbulent particulate-liquid two-phase flow equations are employed in this study. Hashish erosion model was implemented in this code. The sand volume fraction distribution, sand erosion rate distribution, wall shear strain rate distribution and wall stress distribution in the impeller were analyzed. Simulation results have shown that the main sediment wear of impeller is at the suction side of the inlet and the pressure side of the outlet.

Keywords: CFD; Sediment Wear; Centrifugal Pump; OpenFOAM.

1. Introduction

Centrifugal pumps are widely used in agriculture, mining, electricity, sewage treatment and chemical industries. Since the transmission medium containing sediment and other solid particles, centrifugal flow components will be severe wear and results in reduced efficiency. The sediment wear is due to the dynamic action of sediment or solid particles flowing along with water impacting against a solid surface of hydraulic components. In general, a number of factors influence the development of sediment wear process of hydraulic machinery. These factors include mean velocity of particles, mass of the particle, concentration of the wear particles in a liquid flow, grain size and shape of the particles and angle of attack at which the particles collide with the surface etc. With the aid of CFD, numerical solutions for the equations governing fluid flows can be successfully obtained in order to accurately predict the different outcomes of fluid-surface interactions. Bross and Addie[1] developed a simple

† Corresponding author.

model to predict the influence of different impeller design parameters on the wear behaviour of the impeller suction sealing in a slurry pump. They calculated the velocity field in the impeller suction side and sealing gap and the local wear intensity is predicted with respect to the acting velocities and fluid materials properties. The influence of different design parameters on the wear behaviour were analysed by this model. [2] sudied the erosion evaluation for the new type of jet pump by using a CFD-based particle erosion model. Finnie's model is used as the erosion prediction model and a particle-tracking model of the sand particles is used to track the trajectory of sand particles. The results demonstrate the erosion intensity and distribution in the jet pump. Ma et al.[3] analyzed the wear behavior of swash plate/slipper pair using a method on the basis of Elasto Hydrodynamic Lubrication(EHL) model. They compared the wear masses of theoretical analysis and wear tests and analyzed the relationships between wear rate of the friction pair and associated parameters. Qian et al.[4] analyzed the influence of inlet and outlet shapes on silt abrasion and predicted the blade erosion rates in a double-suction centrifugal pump using a particle erosion model. The results show the blade inlet and outlet are severely silt abraded and the average erosion rate is always larger on the suction side than on the pressure side. Lopez et al. [5] analyzed the erosion rates of a single channel of a centrifugal pump subjected to solid particle impingement using four different methods with OpenFoam. Results yielded some disparities between models due to the different factors taken into consideration.

This paper presents using OpenFOAM code only to make a qualitative analysis of the sediment wear in a centrifugal pump impeller. OpenFOAM is an object-oriented C++ library of classes and routines of use for writing CFD codes. It has a set of basic features similar to any commercial CFD solver, such as turbulence models and discretization schemes. With OpenFOAM it is easy to add any modification to any part of the implementation. Hashish erosion model was implemented here with C++.

2. Numerical Method

2.1. *Governing equations*

The flow model used for the numerical simulations is based on the generalized homogeneous multiphase flow model with the additional sources of momentum for the effects of the Coriolis and centrifugal accelerations in a steady rotating frame of reference. The governing equations are described below[6]

Continuity equation of liquid:

$$\frac{\partial}{\partial t}(\rho)+\frac{\partial}{\partial x_j}(\rho \boldsymbol{u}_j)=S=-\sum n_p \dot{m}_p \tag{1}$$

Momentum equations of liquid:

$$\begin{aligned}\frac{\partial(\rho \boldsymbol{u}_i)}{\partial t}+\frac{\partial}{\partial x_j}(\rho \boldsymbol{u}_j \boldsymbol{u}_i) &= -\frac{\partial p}{\partial x_i}+\frac{\partial}{\partial x_j}[\mu_e(\frac{\partial u_i}{\partial x_j}+\frac{\partial u_j}{\partial x_i})] \\ &+\sum \rho_p C_D \frac{3}{4}\frac{R_{ep}\mu}{\bar{\rho}_p d_p^2}(\boldsymbol{u}_{pi}-\boldsymbol{u}_i) \\ &+\sum C_D \frac{3}{4}\frac{R_{ep}\mu}{\bar{\rho}_p d_p^2}(\frac{v_p}{\sigma_p}\frac{\partial \rho_p}{\partial x_j})+F_{cj}\end{aligned} \tag{2}$$

where F_{cj} represents the Coriolis force. $P=P^*+0.5\rho(\omega r)^2$ is total pressure, P^* is static pressure, ρ_p is the volume density of the particle phase, n_p is the number of particles per unit volume, m_p is the particle mass, C_D is the drag coefficient of unsteady flow.

Continuity equation of particle:

$$\frac{\partial}{\partial t}(\rho)+\frac{\partial}{\partial x_j}(\rho \boldsymbol{u}_j)=\frac{\partial}{\partial x_j}(\frac{v_p}{\sigma_p}\cdot\frac{\partial \rho_p}{\partial x_j}) \tag{3}$$

Momentum equations of particle:

$$\begin{aligned}\frac{\partial(\rho_p \boldsymbol{u}_{pi})}{\partial t}+\frac{\partial}{\partial x_j}(\rho_p \boldsymbol{u}_{pj}\boldsymbol{u}_{pi}) &= \frac{\partial}{\partial x_j}[\rho_p \boldsymbol{u}_p(\frac{\partial u_{pi}}{\partial x_j}+\frac{\partial u_{pj}}{\partial x_i})] \\ &+\sum \frac{m_p C_D \frac{3}{4}\frac{R_{ep}\mu}{\bar{\rho}_p d_p^2}+m_p}{m_p}\frac{v_p}{\sigma_p}\frac{\partial \rho_p}{\partial x_i} \\ &+\frac{\partial}{\partial t}(\frac{v_p}{\sigma_p}\frac{\partial \rho_p}{\partial x_i})+F_{pci}+F_{pli}+F_{ppi}\end{aligned} \tag{4}$$

Where F_{pci} is the Coriolis force, F_{pli} is the particles centrifugal force, F_{ppi} is the particle phase pressure.

2.2. *Erosion model*

Hashish erosion model is used here [7]. It modified Finnie's model for erosion by solid particle impingement in ductile materials taking into account particle shape and including no empirical constants. The equation, taken from [5], and used for the simulation is equation (5).

$$W = \frac{7}{\pi}\frac{M}{\rho_p}\left(\frac{V}{C_k}\right)^{2.5}\sin(2\alpha)\sqrt{\sin\alpha} \tag{5}$$

Where M is the mass of the abrasive particles, ρ_p is the density of the particles, V is the velocity magnitude of the particles, α is the angle of impingement, C_k is a coefficient which can be obtained from equation (6).

$$C_K = \sqrt{\frac{3\sigma_f R_f^{\frac{2}{5}}}{\rho_p}} \tag{6}$$

Where R_f is the roundness factor of the particulate phase.

2.3. *Numerical algorithm*

We applied OpenFOAM code to a centrifugal pump. In the OpenFOAM, the spatial discretization is performed using a cell centered co-located finite volume method for unstructured meshes with arbitrary cell-shapes, and a multi-step scheme are used for the time derivatives. For the simulations presented in this paper, a second order implicit time scheme is used combined with second order linear interpolation in space, except for the convective terms. The time step is equal to 1 degrees of runner revolution. The iterative solvers are considered converged when the residuals have been reduced by a factor of 10-5. The pressure-velocity coupling is handled via a Pressure Implicit with Splitting of Operators (PISO) procedure. The turbulence model is adopted for simulation of the unsteady flow through a centrifugal pump. It has been implemented in the OpenFOAM libraries.

3. Calculation Conditions

Figure 1 shows the calculation domain of the centrifugal pump. The parameters are listed in Table 1.

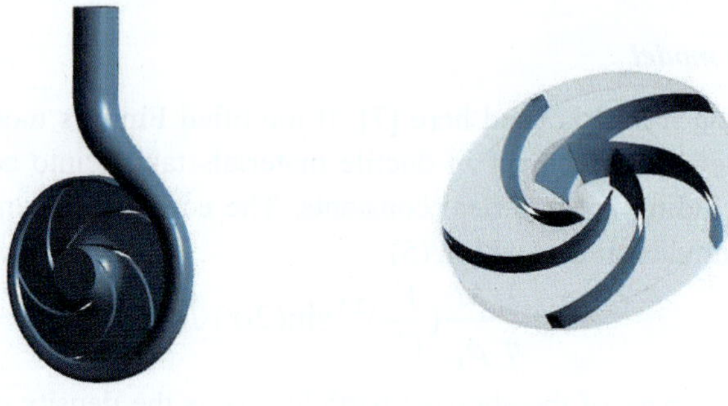

Fig. 1.Centrifugal pump and impeller

Table 1. Parameter of the centrifugal pump

Diameter of impeller (m)	0.4
Number of impeller blade	5
head (m)	68
Rotational speed (rpm)	1480
Rated flow rate (m^3/s)	0.035

The computational grid of the centrifugal pump is generated using ICEM CFD, including impeller and volute-type casing with 20,800,000 elements was used. After importing the meshed geometry in OpenFOAM, it is divided into two domains namely impeller and casing, in which impeller is rotating domain and casing is stationary domains. The element number of impeller domain is 12,600,000.

In the algorithm, the length of the time step is equal to one of the impeller revolution. Rotation speed of the impeller is 1480 r/min. Thus, the time step was 0.0001126s. Inlet bulk mass flow rate and fixed outlet pressure boundary conditions are applied. The calculation is done on a SUGON high-performance computers using 240 CPU core.

4. Analysis of numerical calculation results

The collision of sand particle to the wall can be seen from the sand particle track. Figures 2 show the sand particle track with different sand concentration. It can be seen that the movement of sand particles at the inlet and the tongue is more complex, which shows that there are more opportunities for the collision of sand particle to the wall. The sand volume fractions and sand erosion rate is the important feature for sediment wear in hydraulic turbine. The region of higher sand volume fraction and sand erosion rate with the region of sediment wear is consistent qualitatively. Figures 3 show the volume fraction of sand distribution for the impeller on the pressure side and the section side. It is clearly seen that the sand volume fraction at the suction side of the blade inlet and the pressure side of the blade outlet is higher than others. Compared with the sand erosion rate distribution of Figure 4, the maximum of sand volume fraction is at same location with the maximum of sand erosion rate. Figures 5 show the wall shear strain rate distribution for the impeller on the pressure side and section side. The wall shear strain rate is higher on the pressure side of the blade outlet. Figures 6 show the wall stress distribution for the impeller on the pressure side and section side. Its maximum position is also consistent with the maximum position of the sand erosion rate and sand volume fraction. It can be seen that the main sediment wear of impeller is at the suction side of the inlet and the pressure side of the outlet.

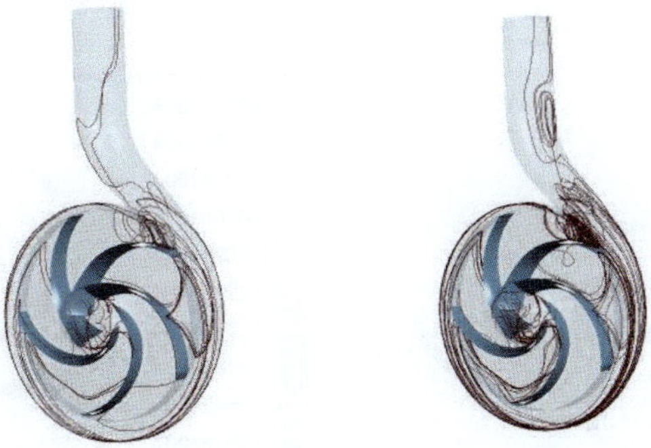

Fig. 2. The sand particle track with different sand concentration

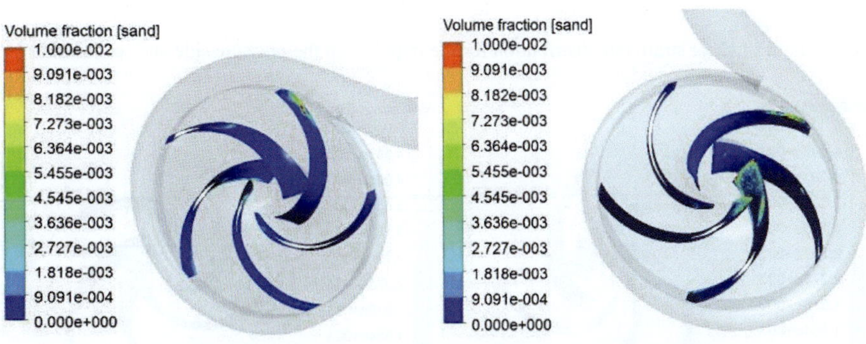

Fig. 3. The sand volume fraction distribution for the impeller on the pressure side and section side

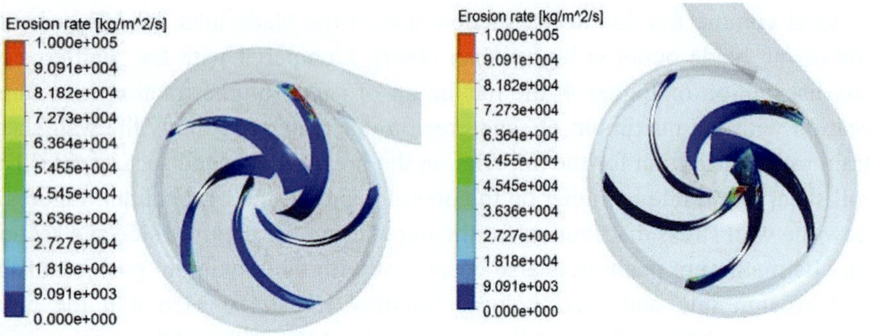

Fig. 4. The sand erosion rate distribution for the impeller on the pressure side and section side

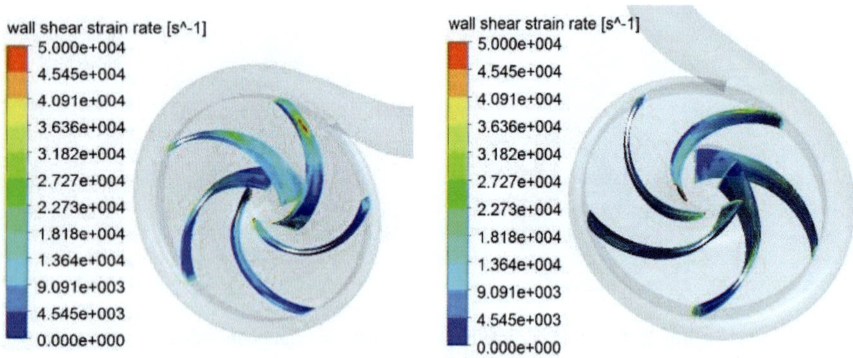

Fig. 5. The wall shear strain rate distribution for the impeller on the pressure side and section side

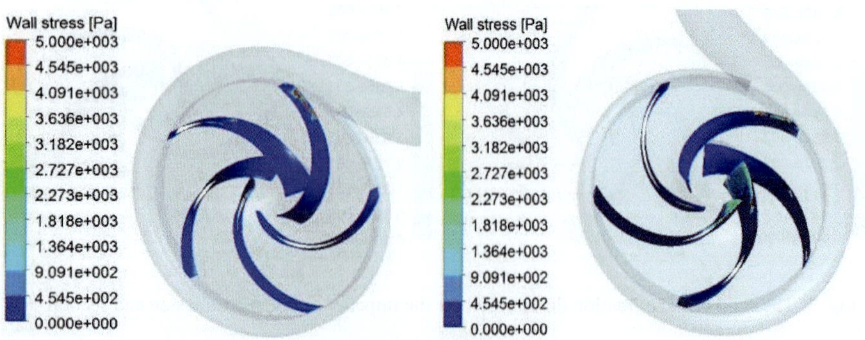

Fig. 6. The wall stress distribution for the impeller on the pressure side and section side section

5. Conclusion

In this paper, the numerical studies of sediment wear in a centrifugal pump has been carried. The sand volume fraction distribution, sand erosion rate distribution, wall shear strain rate distribution and wall stress distribution in the impeller were analyzed. Simulation results have shown that the main sediment wear of impeller is at the suction side of the inlet and the pressure side of the outlet. Hashish erosion model implemented in OpenFoam is able to simulate the sediment wear problem well.

Acknowledgements

This work was financially supported by the National Natural Science Foundation of china (No. 51279071) and the Natural Science Foundation of Kunming University of Science and Technology (No.20149x16).

References

1. Bross S, Addie G. Prediction of impeller nose wear behaviour in centrifugal slurry pumps[J]. Experimental thermal and fluid science, 2002, 26(6): 841-849.
2. Song X G, Park J H, Kim S G, et al. Performance comparison and erosion prediction of jet pumps by using a numerical method[J]. Mathematical and Computer Modelling, 2013, 57(1): 245-253.
3. Ma J, Chen J, Li J, et al. Wear analysis of swash plate/slipper pair of axis piston Hydraulic pump[J]. Tribology International, 2015.
4. Qian Z, Wang Z, Zhang K, et al. Analysis of silt abrasion and blade shape optimization in a centrifugal pump[J]. Proceedings of the Institution of Mechanical Engineers, Part A: Journal of Power and Energy, 2014, 228(5): 585-591.
5. Lopez A, Stickland M, Dempster W. Modeling erosion in a centrifugal pump in an Eulerian-Lagrangian frame using OpenFOAM[J]. Zeszyty Naukowe. Cieplne Maszyny Przepływowe-Turbomachinery/Politechnika Łódzka, 2014 (145): 85-88.
6. Tang Xue-Lin, TANG Hong-Fen,WU Yu-Lin. Simulating and erosion prediction of 3D two phase flow through a turbine runner,JOURNAL OF ENGINEERING THERMOPHYSICS, Vol.22, N0.1, jan.,2001.
7. Hashish M. Modified model for erosion. Proceedings of the 7th International Conference on Erosion by liquid and solid impact, 1987, Cambridge, UK, pp. 461–480.

Improvement on Polymer Blending with Nano-Technology: self-assembled Poly(γ-glutamic acid)/chitosan (γ-PGA/CH)/blends

Ge Yang*, Li-Qun Ma and Xue-Yan Su

College of Life Science,Qufu Normal University,
Qufu, Shandong, 273165 , china
yangge100@126.com

Key potential of polymer blending is to produce a range of materials with characteristics that are superior to that of each individual component polymers via upgrading disadvantageous properties. A new type of self-assembled Poly(γ-glutamic acid)/ chitosan (γ-PGA/CH) nano-blend was derived by a molecular self-assembly process. Exclusive nano-scale morphology generated in perfectly miscible γ-PGA/CH blends was studied using FESEM. Moreover, a range of physiologically relevant concentrations of the protein angiopoietin-2, a candidate cancer biomarker, are incubated with quantum dots conjugated with a polyclonal mixture of anti-angiopoietin-2 antibodies. Antigen concentration modulates the slopes and inflection times of the sigmoidal kinetics curves. An understanding of self-assembly kinetics in this system may lead to improvements in sensitivity and specificity of this novel proteomic biomarker detection technique and improve the screening, diagnostics, and therapy response monitoring for cancers and other diseases. This approach to studying the kinetics of nanoparticle self-assembly may also provide a valuable tool for understanding the fundamental characteristics of nanoscale particle aggregation.

Keywords: Poly(γ-glutamic acid); Self-assembly; Morphology; Kinetics; Diagnostics.

1. Introduction

Key potential of polymer blending is to produce a range of materials with characteristics that are superior to that of each individual component polymers via upgrading disadvantageous properties [1]. Through judicious choice of the constituent polymers and their relative mole fractions, a variety of properties can be realized in the final material [2]. In consequence, design and control of the structural and physical properties of polymer mixtures are crucial for a number of promising applications including microelectronics, membrane technologies, microfluid devices and many others [3 - 5]. Anthology of commercially available polymers thus can be combined in a straightforward and rapid fashion leading to extremely versatile self assembled nano-structures [6]. This approach has a number of advantages over the tedious synthesis of various copolymers or block copolymers because no additional chemical synthesis is required. Directed self-assembly of nanostructures into microstructures through intermolecular interactions is an important phenomenon in many biological systems. Assembly of virus coat proteins into capsids [7], of microtubulin into microtubules [8].

* Corresponding author.

Programmed self-assembly using biomolecular interactions as a route to synthesis of novel nanostructured materials has also been an area of active investigation in the recent past [9-12].Techniques aimed at molecular diagnostics have also been demonstrated using nanoparticle self-assembly mediated by molecular interactions, including polynucleotide interactions and antibody-antigen interactions.

2. Materials and Methods

2.1. *Materials*

Poly(γ-glutamic acid)s (γ-PGA) with molecular weight 100,000 was obtained from BDH. Anhydrous N,N-dimethylacetamide (DMAc) was supplied by Aldrich. 4,40-oxydianiline (ODA) and isophthaloyl chloride (IPC) were procured from Fluka and used as received.

2.2. *Measurements*

Fourier transform infrared (FTIR) spectra of thin blend films were taken at room temperature with a resolution of 4 cm-1 using Excalibur Series FTIR Spectrometer, Polym. Bull. (2014) 71:227 - 242 229 Model No. FTSW 300 MX manufactured by BIO-RAD. Weight-average (Mw) and polydispersity index of aromatic polyamide was determined through gel-permeation chromatography (GPC) with a refractive index (RI) detector using dimethyl formamide (DMF) as an eluent. Stress - strain response of the blend samples dimensions was monitored according to DIN procedure 53,455 having crosshead speed of 5 mm/min using Universal Testing Machine INSTRON 4206. Standard procedures and formulae were used to calculate various mechanical properties including stress, strain, young's modulus and toughness. For phase morphological studies, films were cryogenically fractured in liquid nitrogen and the micrographs were taken by FEI Nova 230 field emission scanning electron microscope.

2.3. *Blend preparation*

γ-PGA/CH blends were prepared by the solution blending technique using DMAc as a solvent. Stock solutions of the component polymers were prepared by dissolving their required amounts in the solvent. For a desired composition, known amounts of the two components were blended with a constant stirring for 24 h. Thin films of the resulting clear blend solutions were cast by evaporating the solvent at 70℃ for 36 h. Prior to characterization, films were vacuum dried at 70℃ to a constant weight.

2.4. *Kinetics experimental*

Streptavidin-coated QDs with 705 nm fluorescence emission were purchased from Invitrogen and used as received. Biotin conjugated anti-angiopoietin-2 polyclonal antibody , and recombinant human angiopoietin-2 were purchased from R&D Systems, reconstituted in Tris-buffered saline (TBS) containing 0.1% bovine serum albumin (BSA) and stored at 20 °C. Appropriate dilutions of all antibodies and antigens were prepared in phosphate buffered saline (PBS) with 0.1% BSA immediately prior to use. Deionized water with resistance of 18 MΩ/cm was used for preparing buffers. All buffers were filtered through 0.2 μm filters. All other reagents were ACS reagent grade. Flow cytometric measurements were carried out with an unmodified Becton Dickinson (BD) FACSCalibur, and the associated software.

3. Results and Discussion

3.1. *Spectroscopy*

FTIR spectra of pure γ-PGA, pure aramid and γ-PGA/CH blend are shown in Figure 1. For pure γ-PGA two bands appeared at 2,978 and 2,967 cm^{-1} owing to asymmetrical and symmetrical stretching of methyl group, whereas bands at 1,456 and 1,387 cm^{-1} emerged due to bending vibrations of CH_3. A band at 1,732 cm^{-1} was attributable to carbonyl groups (C=O) stretching. Broader and stronger bands attributable to C-O stretching vibrations were experiential in the region 1,300-1,000 cm^{-1}. Out of plane C-H bending appeared at 748 cm^{-1}. For the better interpretation of blend structure, pure aramid spectrum is also presented. Major vibrations around 3,269 and 1,605 cm^{-1} were assigned to the N－H stretching and bending, respectively. The bands due to the aromatic C-H stretching and amide C=O appeared at the 3,058 and 1,655 cm^{-1}, respectively. Theγ-PGA/CH (60/40) blend spectrum showed a shift of the carbonyl band of γ-PGA to lower wavenumber (1,723 cm^{-1}). This shift may be due to the hydrogen bonding between C=O of methacrylate group and hydrogen from amide linkage. The broadening of peak in the region of 3,000-3,300 cm^{-1} may also be attributed to H-bonded amide N-H stretching to engender compatible γ-PGA/CH blend. Stretching (2,975 and 2,964 cm^{-1}) and bending vibrations (1,434 and 1,390 cm^{-1}) of CH_3 were also evident from blend spectrum. Other characteristic bands observed were around 3,183 and 1,534 cm^{-1} for aromatic C-H and C=C stretching, respectively, while the spectrum of pure γ-PGA showed no band for aromatic moieties. Nevertheless, in the blend both aromatic and aliphatic C-H peaks were present, signifying the integration of CH chains with the matrix. Out of plane C-H bending vibrations in the range of 750-840 cm^{-1} were also observed for the blend. Thus, in situ generation of nano-blends through solution mixing of γ-PGA/CH was efficiently studied using spectroscopic analysis .

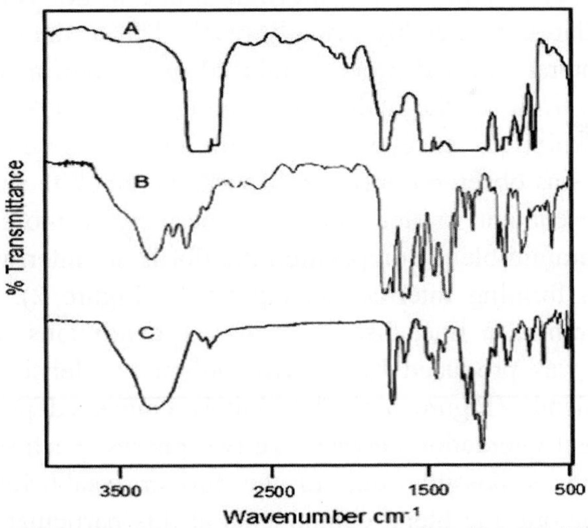

Fig. 1 FTIR spectra *A* pure PGA; *B* pure CH;*C* PGA/CH blend

3.2. *Morphological investigation*

In order to understand the effect of aramid loading in the blend, morphological studies were carried out using FESEM. Figure 2 represents SEM micrographs of cryofractured samples of blends with varying aramid content (10-90 wt%). The blend system exhibited self-assembled nano-phase morphologies ranging from finely dispersed granular matrix to well-patterned knitted nano-fibrillar network. There are a number of factors affecting the structure and morphology such as glass transition temperature, crystallinity as well as specific interactions between the blend components. Micrographs show that PMMA, which is a brittle polymer, acts as a matrix dark portion. While aramid is of white color, present as a second phase spread as miniature white trickles delicately trapped between the dark portions at low polyamide content. Figure 5A showed that the aramid was found evenly distributed in the form of small beads interlocking with PMMA. Size of these beads slightly increased with the increase in weight percent of second phase due to coalescence of the dispersed phase with PMMA (Fig. 5B). With the addition of 30 wt% aramid (Figure 5C) some rearrangement of both the phases begins. Afterwards, in 40 wt% aramid blend (Figure 5D), drops of polyamide along with matrix were stretched in form of fibrils forming a well-knitted self-assembled nanofibrillar frame-work. This exceptional knitted-fibrillar self-assembly actually resulted from the combination of interfacial interaction and process induced orientation at this particular blend ratio. Owing to the hydrogen bonding in 40 wt% blend, the two polymers exhibited excellent compatibility. This change in the morphology was observed only for 40 wt% blend. With further increase in polyamide content, knitted-fibrils became merged into the matrix and for 60 wt% aramid blend composition the fibrils are intertwined with the matrix again forming inter-connecting beads (Figure 2). Upon further loading of aramid, a fine, distinctive and co-continuous self-assembled morphology was produced by the two polymers. Hence, the fractured surface of blends (Figure 3) had infinitely connected periodic pattern owing to finest interaction between the two phases. Such self-assembled nano-structure is possibly due to the further establishment of self-association among the blend components at this particular composition. By blending PMMA with further higher aramid content, polyamide was rather profusely dispersed phase in PMMA and a slight phase separation is apparent in Fig. 4. The interfacial properties of these blends are important for future applications in nano-technology.

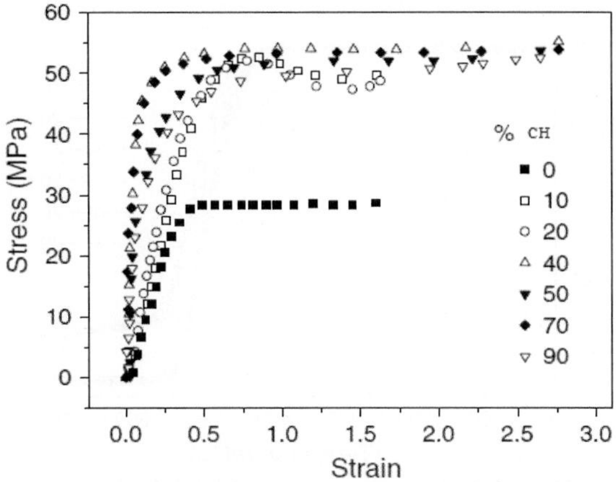

Fig. 2 Stress-strain curves of PGA/CH blends

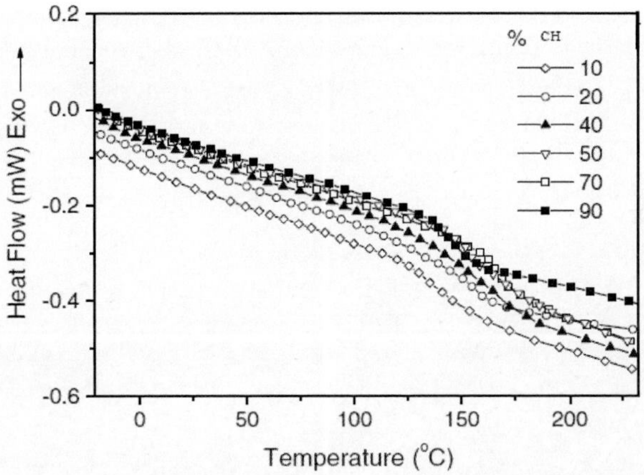

Fig. 3 DSC thermograms of PGA/CH blends

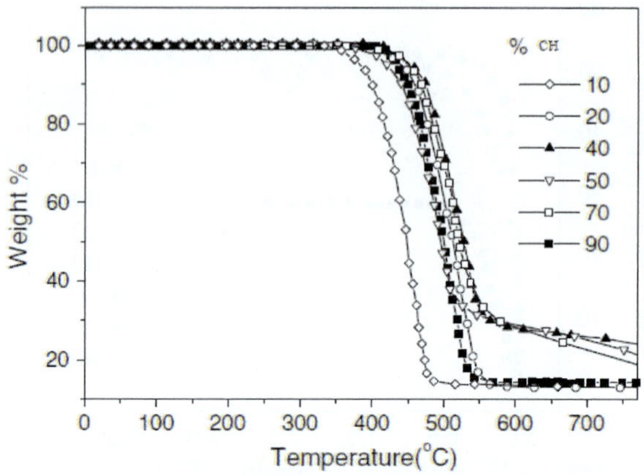

Fig. 4 TGA curves of PGA/CH blends

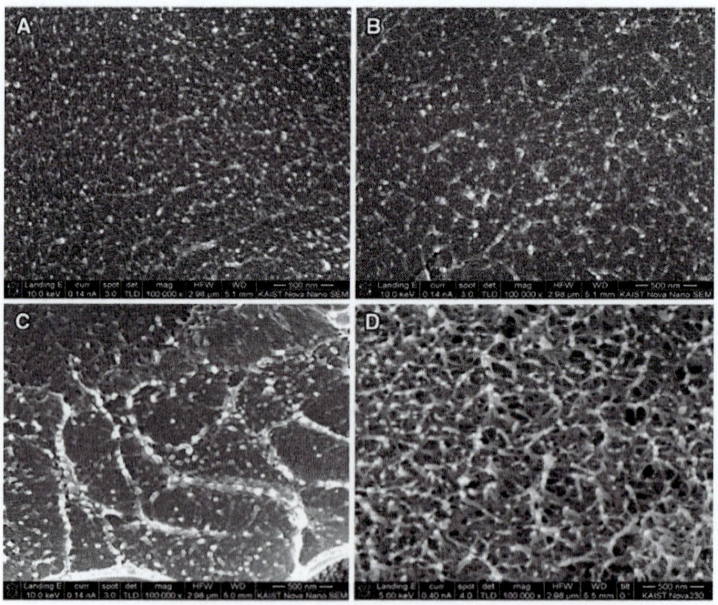

Fig. 5 FESEM micrographs of PGA/CH blends

4. Conclusion

The preparation of the new pH responsive hydrogels based on PGA/CH for oral drug delivery was carried out using a radical polymerization technique initiated by new γ-irradiation method. Varying the composition of PGA, CH, and/or radiation dose generated gels with different gelation and crosslinking degree. The degree of gelation increased gradually due to the crosslinking of PGA/CH mixtures by γ-ray irradiation. Over 80% crosslinked PGA/CH hydrogels were obtained even at only a 10 kGy radiation dose. The gelation increased linearly in proportion to the PGA content in the gel mixture. The equilibrium swelling measurements of these hydrogels clearly show the pH responsive nature of these hydrogels. The swelling percent was low at low pH, while the swelling percent of gels increased with increasing pH. Continuous reduction of the blood glucose was observed without the addition of additives such as absorption enhancers and/orprotease inhibitors. The reduction was found to be strongly dependent on the administered dose. Our results indicate that the PGA/CH (1:1) hydrogels irradiated at 30 kGy are highly suitable as an oral carrier for peptide drugs, specifically insulin.

Acknowledgments

This work was supported by Foundation item: Shandong Provincial Natural Science Foundation, China (ZR2014CM020)
Li-qun Ma and Ge Yang contributed equally to this work.

References

1. Munoz-Bonilla A, Ibarboure E, Bordege V, Fernandez-Garcia M, odriguez-Hernandez J (2010) Fabrication of honeycomb-structured porous surfaces decorated with glycopolymers. Langmuir 26:8552–8558.
2. Lehn J-M (2002) Toward self-organization and complex matter. Science 295:2400–2403.
3. Ikkala O, Ten Brinke G (2002) Functional materials based on self-assembly of polymeric supramolecules.Science 295:2407–2409.
4. Reinhoudt DN, Crego-Calama M (2010) Synthesis beyond the molecule. Science 295:2403–2407.
5. Shelat KJ, Dutta NK, Choudhury NR (2008) Interfacial interaction and morphology of EVOH and ionomer blends by scanning thermal microscopy and its correlation with barrier characteristics.Langmuir 24:5464–5473.

6. Diaz MF, Barbosa SE, Capiati NJ (2007) Reactive compatibilization of PE/PS blends. Effect of copolymer chain length on interfacial adhesion and mechanical behavior. Polymer 48:1058–1065.
7. Boztug A, Coskun A, Bolayır G, Zengin B, Ozdemir KA (2006) Preparation of compatible blends of poly(methyl methacrylate) used in dentistry with a reactive terpolymer containing maleic anhydride and their thermomechanical characterization. J Appl Polym Sci 100:363–367.
8. Urbas A, Sharp R, Fink Y, Thomas EL, Xenidou M, Fetters LJ (2000) Tunable block copolymer/homopolymer photonic crystals. Adv Mater 12:812–814.
9. Black CT, Guarini KW, Milkove KR (2001) Integration of self-assembled diblock copolymers for semiconductor capacitor fabrication. Appl Phys Lett 79:409–411.
10. Passy S (2003) PMMA implants : a new way to correct body defects without surgery. Int J Cosm Surg Aesth Dermatol 5:193–199.
11. Zulfiqar S, Ahmad Z, Sarwar MI (2007) Soluble aromatic polyamide bearing ether linkages: synthesis and characterization. Colloid Polym Sci 285:1749–1754.
12. Dreezena G, Kochb MHJ, Reynaersa H, Groeninckxa G (1999) Polymer 40:6451–6463.

Structural stabilities of $AMgH_3$ hydrides with perovskite structure

Zhen-Zhen Wan[1,2], Zhong-Min Wang[1,2†*], Dian-Hui Wang[1,2], Huai-Ying Zhou[1,2] and Chao-Hao Hu[1,2]

[1]School of Materials Science and Engineering, Guilin University of Electronic Technology, Guilin, 541004, P. R. China.

[2] Guangxi Key Laboratory of Information Materials, Guilin University of Electronic Technology,
Guilin, 541004, P.R. China
[†]E-mail: zmwang@guet.edu.cn

AMgH3 hydride (A=Li, Na, K) is one of the promising candidates for sold-state hydrogen storage. For Na1-xLixMgH3 system, trigonal R-3c structure becomes more stable than Pnma orthorhombic structure when x reaches 0.375. The tendency of spontaneous decomposition will happen in the case of Na0.5Li0.5MgH3, hence, a suitable range of x would be 0 ~ 0.375. For the K1-xLixMgH3 system, the volume of atoms and bond length of Li ions are much smaller than that of K ions. Therefore, an obvious crystal lattice distortion will occur when Li ions take positions of K ions. Cubic Pm-3m phase of KMgH3 will transform to trigonal R-3c phase when x reaches 0.25 and the reasonable range of x is 0 ~ 0.375 to keep its structure stability. All calculations were performed using the Vienna Ab-Initio Simulation Package (VASP). The interaction between core and electrons are described with projector augmented wave (PAW) method. Above studies support that structure stabilities of AMgH3 hydrides (A=Na, K) will be reduced by replacing of A with Li, and possible structure transformation followed, which may be beneficial for the improvement of kinetic performance of hydrogen absorption and desorption.

Keywords: Magnesium Based; First-principles; Doping; Hydrogen Storage Materials; Density Functional Theory

1. Introduction

Solid state hydrogen storage offers several benefits over other means of storing hydrogen such as compressed hydrogen and liquid hydrogen storage, particularly in terms of safety, cost, and high volumetric and gravimetric densities [1, 2]. Recently, Mg-based perovskite-type hydrides, $AMgH_3$ (A = Na,

---

[*] Corresponding author.

K, Rb), have received considerable attention [3-8]. For instance, $NaMgH_3$[9] has an orthorhombic perovskite structure analogous to the $GdFeO_3$ type (space group Pnma) which is common in low tolerance factor oxide perovskites, where the singly charged Na cation occupies eight-fold coordinated voids. $KMgH_3$ is reported to occur in the ideal cubic perovskite structure[10], while $RbMgH_3$ has a non-ferroelectric hexagonal P63/mmc structure, of the type seen in some high tolerance factor oxides and fluorides [11]. Among them, $NaMgH_3$ was studied as a potential candidate for on-board applications because it has high gravimetric and volumetric H densities (6 wt. % and 88 kg / m^3) as well as reversible hydrating and dehydrating reactions [6]. Also, $NaMgH_3$ was pointed out as a potential material for future generation of electronic devices due to the superior hydrogen mobility of the perovskite structure [3]. The desorption process of $NaMgH_3$ shows a two-step reaction: $NaMgH_3 \rightarrow NaH + Mg + H_2 \rightarrow Na + Mg + 3/2H_2$. For the first step desorption, Sheppard et al. measured the kinetic and thermodynamic data for the decomposition of $NaMgH_3$ into NaH and Mg. The desorption enthalpy and entropy are 86.6 ± 1.0 kJ / (mol H_2) and 132.2 ± 1.3 kJ / (mol H_2) respectively, indicating that $NaMgH_3$ is thermodynamically more stable than MgH_2 [12].

The thermodynamic properties of $NaMgH_3$ can be tuned by incorporating new elements [13-18]. For example, Anna-Lisa Chaudhary et al. found that the onset temperature of decomposition with the additions of Si element doped into $NaMgH_3$ by mechanical milling successfully decreased [14]. Martínez-Coronado et al. [15] prepared $Na_{1-x}Li_xMgH_3$ (X=0, 0.25, 0.5) ternary hydrides using a high-pressure technique, and found that as Li is introduced in the system, the unit-cell parameters decrease, the tilting angle increases and the MgH_6 octahedral become more distorted due to the smaller ionic size of Li+ vs Na+, the hydrogen desorption temperature decreases due to the lower structural stability of the Li-containing perovskites. Xiao et al. also predicted the same results based on density functional theory calculations [16]. However, Material functions and also fundamental properties of these doped perovskite hydrides have not been widely investigated and clarified yet.

In this paper, a theoretical investigation of $AMgH_3$ (A=K, Na, Li) hydrides were carried out. To reveal the relationships among Li doping, crystal structure, electronic behaviors and hydrogen storage properties of $AMgH_3$ hydrides, the crystal models of $KMgH_3$ (Pm-3m), $NaMgH_3$ (Pnma) and $LiMgH_3$ (R-3c) were established, and its total energy, formation enthalpy and electronic structure, have been calculated in the frame work of density functional theory (DFT).

2. Computational details

All calculations are performed using the Vienna Ab-Initio Simulation Package (VASP) [19], an implement of density functional theory. The interaction between core and electrons are described with projector augmented wave (PAW) method [20], and Perdew-Burke-Ernzerhof (PBE) implemented Generalized Gradient Approximation (GGA) [21] is used to treat the exchange and correlation energies. The plane wave cutoff energy is set to 550eV and the k-point mesh in brillouin zone [22] is set to $0.03 * 2\pi$ $Å^{-1}$ in all calculation. The lattice parameters, volume and atom positions are allowed to change within symmetry restrictions in structure optimization calculation. And a criterion of $1 \cdot 10^{-5}$ $eV \cdot atom^{-1}$ is placed to obtain accurate total energy in self-consistent convergence. In the calculation of the unit cell, the optimal solution is completely relaxed, which allows the lattice constant, the atomic position and the unit cell volume to change.

In general, the stability of hydride can be represented by its formation enthalpy, the lower the formation enthalpy is, the more stable the hydride will be. For the decomposition reaction of $AMgH_3$ hydride can be described as:

$$A_{1-x}Li_xMgH_3 \longrightarrow (1-x)AH + (x)LiH + MgH_2 \qquad [23]$$

Therefore, the calculation of formation enthalpy is:

$$E_F = E_{A_{1-x}Li_xMgH_3} - ((1-x)E_{AH} + xE_{LiH} + E_{MgH_2}) \qquad [24]$$

Where $E_{A_{1-x}Li_xMgH_3}$, E_{MgH_2} , E_{AH} and E_{LiH} are the energies obtained from their self-consistent calculations, normalized to the number of metal atoms present in the unit cell.

In order to obtain smooth curves showing the relationships between the formation enthalpy and the composition. Stable crystal structures of $AMgH_3$ (A=Na, K, Li) hydrides are constructed and shown in Figure 1,where the white balls represent the H atoms, purple, yellow and green balls represent K, Na and Li atoms, respectively. The arrangement of the metal atoms in the three structures are similar, Mg atoms are placed in the center of octahedral sites formed by H atoms. But obviously the crystal structures of $NaMgH_3$ and $LiMgH_3$ are distorted perovskite-type structures.

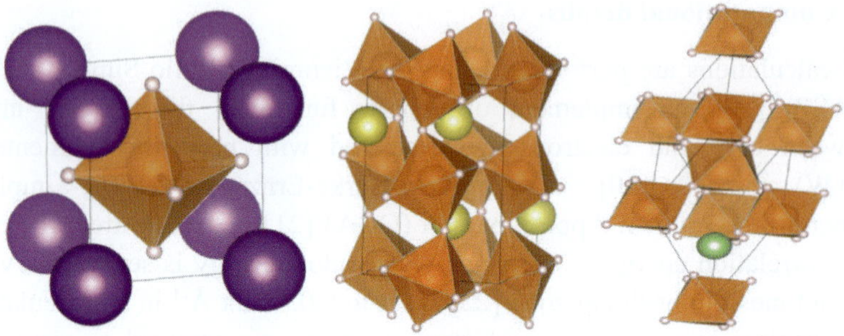

KMgH3 NaMgH3 LiMgH3

Fig. 1. Crystal models of $KMgH_3$ (Pm-3m), $NaMgH_3$ (Pnma) and $LiMgH_3$ (R-3c)

3. Results and discussion

3.1. *Thermodynamic stability of Li doped $AMgH_3$*

We first investigate thermodynamic properties of Li doped $AMgH_3$(A=Na, K). The formation enthalpy is calculated by taking the difference between total energy of the products and the reactants. Calculation results that Li doped $NaMgH_3$ is shown in Figure 2. With no Na atoms doping, the Pnma structure of $NaMgH_3$ has the lowest formation enthalpy. The trigonal R-3c phase becomes the most stable phase when Li content reaches 37.5 at.%. If Li atoms replace 50% more of the Na atoms, the formation enthalpy of will goes exceed 0 and the system becomes unstable. The calculated formation enthalpy of $Li_{0.5}Na_{0.5}MgH_3$ compound is 0.313eV, which indicate that $Li_{0.5}Na_{0.5}MgH_3$ has the tendency of spontaneous decomposition. At the same time, it is found that the formation enthalpies of $LiMgH_3$ in all 3 structures are more than 0, which means that the $LiMgH_3$ is not stable in the ground state.

Calculation results that Li doped $KMgH_3$ is shown in Figure 3. The difference of formation enthalpies between these three structures of $KMgH_3$ is very small, but the cubic Pm-3m phase of $KMgH_3$ is the most stable crystal structure. when the Li contents in the range of 0~37.5 at.%, all the $KMgH_3$ phases are energetically favorable with negative formation enthalpy values, which indicates that the Li doped $KMgH_3$ can be stable. With increasing proportion of Li atoms, the stability of the system gradually decreases. When more than 37.5% of the Li atoms are replaced by K atom, $K_{1-x}Li_xMgH_3$ begins to be unstable and trends to spontaneously decompose. The above calculation indicates that the stability of $KMgH_3$ can be reduced by Li doping, this will be beneficial to the improvement of kinetic performance of hydrogen absorption and desorption.

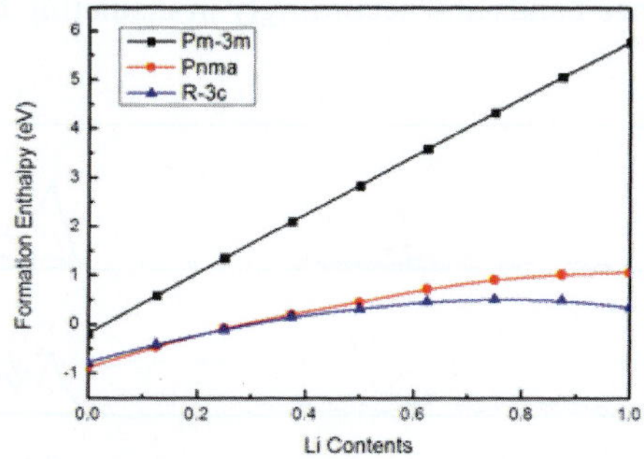

Fig. 2. The formation enthalpy of different crystal structures of $Na_{1-x}Li_xMgH_3$

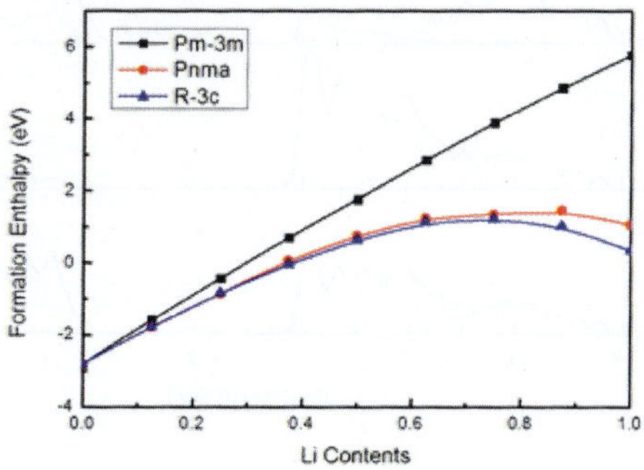

Fig. 3. The formation enthalpy of different crystal structures of $K_{1-x}Li_xMgH_3$

3.2. *Electronic structures of Li doped $AMgH_3$*

The electronic density of states(DOS) at ground-state of $Na_{0.625}Li_{0.375}MgH_3$ is displayed in Figure.4. The calculations show that $Na_{0.625}Li_{0.375}MgH_3$ has a finite energy gap (Eg=3.6eV) between the valence band and the conduction band. Therefore, $Na_{0.625}Li_{0.375}MgH_3$ can be considered as an semi-conductor. From Figure 4, it can be seen that the Li and Na-s and Na-p states are primary contributors to the conduction bands, and the Mg-s and Mg-p states have apparent contribution to both the valence bands and conduction bands. The H-s states are essential in the valence bands. What is more, there is obvious hybridization between Li-p, Mg-p and H-s states in the region from -5.8eV to 0.25eV. This hybridization indicates that covalent bonds are formed between Li atoms and H atoms or Mg atoms and H atoms. The insulating behavior of $Na_{0.625}Li_{0.375}MgH_3$ can be explained as follows: per formula unit, one electron from metal atoms fill one of the three originally half-filled H-s orbitals and the other two of these orbitals are filled by electrons from Mg, resulting in complete filling of the valence bands and accordingly in insulating behavior.

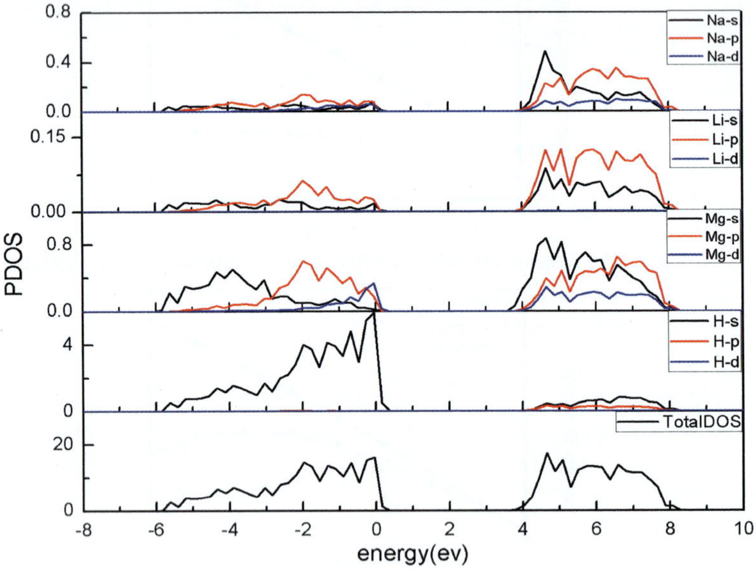

Fig. 4. A $Na_{0.675}$ $Li_{0.325}MgH_3$ of the partial density of States

As shown in Figure 5 is electron localization function(ELF) section close to thermodynamic stability limit of $Na_{0.625}Li_{0.375}MgH_3$. Li atom, Na atoms and Mg atoms ELF values are close to 0.2, 0.25 and 0, respectively, electronics are highly electron delocalization; H atoms ELF values are close to and approximately 0.95, electronics are highly localized. The ELF values between the metal atoms and the H atoms are 0. Therefore, the interaction between Mg(Li) atoms and H atoms is an ionic bond. Mg atoms octahedral interstice of the H atoms in the center to a certain extent can be viewed as a whole. As it is clear that Li atoms the volume and the bond lengths compared with Na atoms are smaller, after Li substitution of Na atoms, crystal lattice had obvious distortion, Li atoms and no longer in equilibrium position between the Mg-H octahedral, leading to decline in the stability of the crystal. This may be beneficial to the improvement of kinetic performance of hydrogen absorption and desorption.

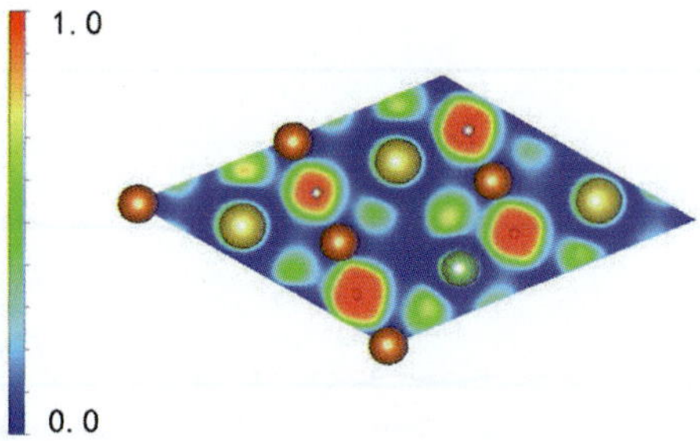

Fig. 5. $Na_{0.625}Li_{0.375}MgH_3$ electron localization function on the surface of the 2D section

As shown in Figure 6, in order to further elaborate the stability of three kinds of alloy compounds. The total density of states (DOS) at the optimized parameters for the ground-state structure of the $K_{0.625}Li_{0.375}MgH_3$ compounds are displayed in Figure 6. The site projected density of states (PDOS) is also illustrated in Figure 6. We calculated the K atom doping density of states of $NaMgH_3$. Obviously, between the conduction bands and valence bands have a wide band gap(Eg=3.0eV), indicating that compounds $AMgH_3$ system can be considered as an insulator. It can be seen that the Li-s states, Mg-s states and Mg-p states have primary contribution to the conduction bands, and the Mg-s states and Mg-p states have apparent contribution to both the valence bands and conduction bands. For the total density of states(TDOS), the top of the valence band is composed of H-s orbital electronic states near the Fermi level, the width is 6.0eV. The bottom of the conduction band is composed of the electronic states of the p s orbital of the Li atom and Mg atom, and the width is 4.5eV. Thus it can be seen the above below the top of the valence band and the conduction band bottom, a hybridization between Li(Mg) electronic states and H electronic states.

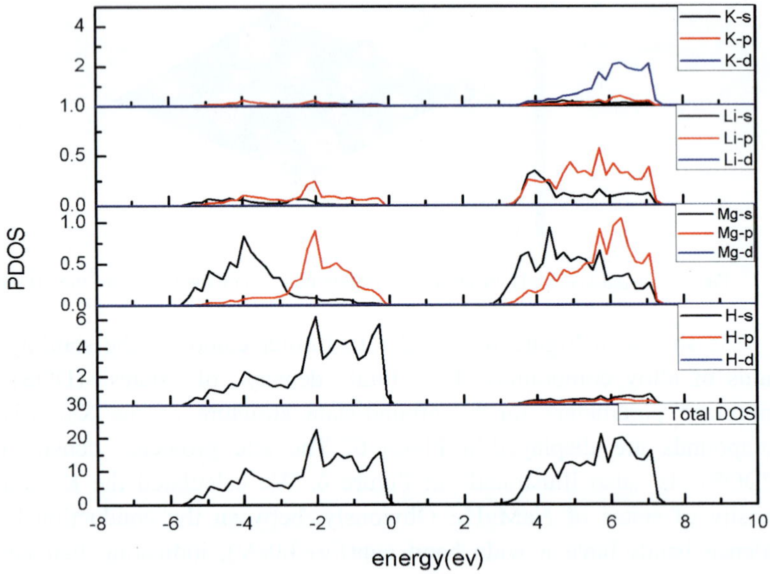

Fig. 6. a $K_{0.675}$ $Li_{0.325}MgH_3$ of the partial density of States

As shown in Figure 7 is electron localization function(ELF) section close to thermodynamic stability limit of $K_{0.625}Li_{0.375}MgH_3$. Mg atoms ELF values close to 0, electronics are highly electron delocalization; Li atoms, K atoms and H atoms ELF values were close to 1, about 0.85 and approximately 0.95, electronics are highly localized; ELF the metal atoms or between metal atoms and H atoms value is 0. This show that the metal atoms between the metal atom or the metal atoms between H atoms are an ionic bond. While the H between the atoms of ELF value is 0.2 ~ 0.3, that charge between H atoms has an itinerant, Mg atoms octahedral interstice of the H atoms in the center to a certain extent can be viewed as a whole. Also from Figure visible, as the basis of K atoms and Mg-H octahedral site is evenly arranged, and Li atoms due to the volume and the bond lengths compared with K atoms are different, after Li substitution of K atoms, crystal lattice had obvious distortion, Li atoms and no longer in equilibrium position between the Mg-H octahedral, leading to decline in the stability of the crystal.

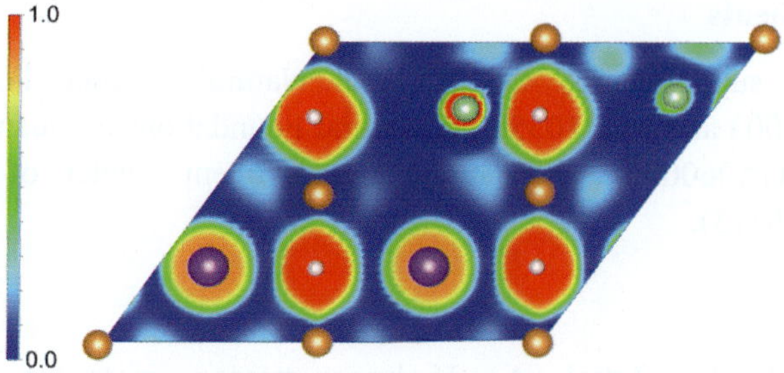

Fig. 7. $K_{0.625}Li_{0.375}MgH_3$ electron localization function on the surface of the 2D section

4. Conclusions

Using first-principles calculations based on density functional theory, we have studied doping effects of $AMgH_3$ (A=Na, Li, Mg) systems. The formation enthalpies and electronic structures, including density of States and electron localization function, are analyzed through calculations. It is found that for Na1-xLixMgH$_3$ system, the value of X should be between 0 - 0.375 to keep the compound stable. Due to the enthalpy values at the critical point, the system can hydrogen desorption at low pressure and temperature. In the x=0.2, the crystal structure of the system has the possibility to change, and the transformation is the trigonal r-3c crystal structure, which may have a good effect on the dynamic performance of the system. For the $K_{1-x}Li_xMgH_3$ system, the value of X should be 0 ~ 0.375, The system of formation enthalpy in the vicinity of the critical point, compounds can exist stable; While x=0.25, the crystal structure of the system will be transformed from the cubic phase to the triangular phase. When the Li doping is more than 40%, Li atoms replace K atoms can make the cubic crystal rapidly instability to transform trigonal r-3c crystal structure and the system value of the formation enthalpy is too high, the compounds cannot exist stable. In addition, For the Li atom doped $AMgH_3$ system, when Li doping to a certain extent, crystal structure are will to transform the more stable trigonal R-3c crystal structure. Above studies support that structure stabilities of $AMgH_3$ hydrides (A=Na, K) will be reduced by replacing of A with Li, and possible structure transformation followed, this may be beneficial to the improvement of kinetic performance of hydrogen absorption and desorption.

Acknowledgments

This work is supported by the National Natural Science Foundation of China (51261003 and 51471055), the Natural Foundations of Guangxi Province (2012GXNSFGA060002), and Guangxi Experiment Center of Information Science (20130113).

References

1. Schlapbach L, Zuttel A. Hydrogen-storage materials for mobile applications. Nature 2001;414:353-8.
2. Ritter JA, Ebner AD, Wang J, Zidan R. Implementing a hydrogen economy. Mater Today 2003;6:18-23.
3. Pottmaier D, Pinatel ER, Vitillo JG, Garroni S, Orlova M, Baro MD, et al. Structure and thermodynamic properties of the NaMgH3 perovskite: a comprehensive study. Chem Mater 2011;23:2317-2326.
4. Bouhadda Y, Fenineche N, Boudouma Y. Hydrogen storage: lattice dynamics of orthorhombic $NaMgH_3$. Physica B 2011; 406:1000-1003.
5. Vajeeston P, Ravindran P, Kjekshus A, Fjellvag H. First principles investigations of the $MMgH_3$ (M=Li, Na, K, Rb, Cs) series. J Alloys Compd 2008;450(1-2):327-337.
6. Bouhadda Y, Boudouma Y, Fenineche N, Bentabet A. Ab initio calculations study of the electronic, optical and thermodynamic properties of $NaMgH_3$, for hydrogen storage. J Phys Chem Solids 2010;71:1264-1268.
7. Wu H, Zhou W, Udovic TJ, Rush JJ, Yildirim T. Crystal chemistry of perovskite-type hydride NaMgH3: implications for hydrogen storage. Chem Mater 2008;20(6):2335-2342.
8. Y. Bouhadda, M. Bououdina, N. Fenineche, Y. Boudouma, Elastic properties of perovskite-type hydride NaMgH3 for hydrogen storage, Int J Hydrogen Energy 38(2013)1484-1489.
9. A. Bouamrane, J.P. Laval, J.P. Soulie, J.P. Bastide, Mater. Res. Bull. 35 (2000) 545.
10. X.-B. Xiao, B.-Y. Tang, S.-Q. Liao, L.-M. Peng, W.-J. Ding, J. Alloys Compd. 474 (1–2) (2009) 522–526.
11. M. Fornari, A. Subedi, D.J. Singh, Phys. Rev. B 76 (2007) 214118.
12. D.A. Sheppard, M. Paskevicius, C.E. Buckley, Chem. Mater. 23 (2011) 4298–4300.
13. Yongtao Li, Luxing Zhang, Qingan Zhang, Fang Fang, Dalin Sun, Kongzhai Li, Hua Wang, Liuzhang Ouyang, Min Zhu, Int. J. Hydrogen Energy, J. Phys. Chem. C 118 (2014). 23635-23644.

14. Anna-Lisa Chaudhary, Mark Paskevicius, Drew A. Sheppard, Craig E. Buckley, J. Alloys Compd. 623 (2015). 109–116.
15. R. Martínez-Coronado, J. Sánchez-Benítez, M. Retuerto, M.T. Fernández-Díaz, J.A. Alonso, Journal of Alloys and Compounds 522 (2012). 101– 105.
16. Xiao-Bing Xiao, Bi-Yu Tang, Sun-Qi Liao, Li-Ming Peng, Wen-jiang Ding, J. Alloys Compd. 474 (2009). 522–526.
17. K. Ikeda, Y. Nakamori, S. Orimo, Acta Mater, 53 (2005). 3453–3457.
18. K. Ikeda, Y. Kogure, Y. Nakamori, S. Orimo, Solid State Chem. 35 (2007). 329–337.
19. G. Kresse, J. Furthmüller, Phys. Rev. B 54 (1996) 11169.
20. P. E. Blochl, Phys. Rev. B, Condensed matter 50 (1994) 17953.
21. J. P. Perdew, K. Burke, M. Ernzerhof, Phys. Rev. Lett. 77 (1996) 3865.
22. T. Fischer, J. Almlof, J. Phys. Chem. 96 (1992) 9768
23. Bowen, K. H.; Castleman, A. W. J. Phys. Chem. 1996, 100, 12911.
24. Andres, R. P.; Averback, R. S.; Brown, W. L.; Brus, L. E.; Goddard,W. A., III; Kaldor, A.; Louie, S. G.; Moscovitz, M.; Peercy, P. S.; Riley,S. J.; Siegel, R. W.; Saepen, F.; Wang, Y. J. Mater. Res. 1989, 4, 704.

Chapter 5

Metal and Alloy Materials

Morphological evolution of γ' precipitates in wrought nickel-based superalloy GH4742 during long-term thermal exposure

Bao-Ning Zhang, Chao Yuan* and Jian-Ting Guo

Institute of Metal Research, Chinese Academy of Sciences, Shenyang City, Liaoning Province, 110016/Shenhe District, China

**E-mail: ychao@imr.ac.cn*

He-Yong Qin and Guang-Pu Zhao

High Temperature Materials Research Institute, Central Iron & Steel Research Institute, Beijing, 100081/ Haidian District, China

Nickel-based wrought superalloy is applied widely in aero-engine turbine disk and gas turbine disk for its superior mechanical property which is mainly derived from solid solution strengthening and the precipitation hardening caused by its γ' phase. In the process of long-term thermal exposure, morphology change of γ' is essential to the performance. Therefore, in this paper, we study the long-term aging γ' morphology. Microstructure stability of nickel-based superalloy GH4742 exposed at 700 °C and 750 °C for 1000 h, 3000 h and 5000 h were investigated. It was found that during long-term thermal exposure at 700°C, γ' precipitates grow in an instable manner due to enough supersaturation in the γ matrix and the number of protrusion on the flower-like precipitate increases. While, during long-term thermal exposure at 750 °C, with the increase of exposure time, the splitting of γ' precipitates in the alloy occurs.

Keywords: Long-term thermal exposure; Wrought Ni-base superalloy; Microstructure.

1. Introduction

Nickel-based wrought superalloy have applied widely in aero-engine turbine disk and gas turbine disk for their superior mechanical property which are mainly derived from solid solution strengthening and the precipitation hardening caused by its γ' phase [1-3]. With the development of industrial gas turbines, turbine inlet temperature continues to increase and our country started to introduce the wrought superalloy and powder superalloy as the main material of the turbine. Compared with traditional alloy Inconel718 (GH4169), Ni-Fe based Inconel718 superalloy is strengthened by γ' and γ phase, which has been widely

---

* Corresponding author.

used in high temperature environment under 650 °C, due to possessing excellent mechanical and process performance [4]. While the excellent mechanical properties of wrought superalloy and powder superalloy depend on a high volume fraction of γ' phase strengthening. GH4742 alloy is imitated firstly by Russia, which is applied to a surface ships gas turbine, long life up to 30000 hours. In the process of long-term thermal exposure, morphology change of γ' is essential to the performance. Therefore, in this paper, we study the long-term aging γ' morphology. The microstructural characteristics of the γ' phase, including its morphology, size distribution and composition, play an important role in the mechanical properties [5-7]. The morphology of γ' particles in wrought nickel-based superalloy is complex, including sphere, cubic, flower-like, irregular-shape and fan-type etc. [1,5,6,8-12]. According to these literatures[1,2,10-12], the formation of complex γ' precipitates is due to the different cooling rate and various heat treatment system. The morphological instability of γ' in nickel-based superalloy occurs whenever there is enough supersaturation in the γ matrix [2,11,12].

The morphological instability of γ' was reported previously [1,2,12] and recently [9,11] to be correspond to the different cooling rates and various heat treatment system. This work shows the growth and splitting of γ' precipitates of wrought nickel-based superalloy during long-term aging time at different temperature.

2. Experimental procedure

The materials used in this study were the standard heat-treated (SHT) turbine disk GH4742. The chemical composition of the alloy is shown in Table 1. The Integral turbine disc was subjected to a homogenization for 8 hours at 1080 °C followed by air cooling to room temperature; subsequently, annealing treatments 16 hours at 780 °C, producing the secondary and tertiary γ' precipitates to optimize mechanical properties. Test samples were taken from the chord wise direction of (SHT) turbine disk. Then the specimens were exposed at temperatures of 700 °C and 750 °C for time of 1000 h, 3000 h and 5000 h, respectively.

Table 1. The chemical composition of the alloy (wt. %).

C	Cr	Ti	Al	Nb	Co	Mo	Ni
0.06	14.00	2.67	2.63	2.48	10.03	5.02	Bal.

For microstructure observations the samples were grounded, mechanically polished using standard metallographic techniques. Subsequently, they were chemical etched in a solution of 50 ml aquadistillate, 50 ml C_2H_5OH, 50 ml HCl and 10 g $CuSO_4$ for morphology of grain boundary observation. To clearly observe the precipitates, they were electrolytic etched in a solution of 12% H_3PO_4, 40% HNO_3 and 48% H_2SO_4 to strip away γ matrix at 15 V for 12-15 s. Furthermore, γ' size and morphology characterization was carried out using a scanning electron microscope.

3. Results and discussion

3.1. *Microstructure of the standard heat treatment alloy*

Figure 1 shows the microstructure of the standard heat-treated specimen. The main phase composition of the alloy is carbide, carbonitride, γ' precipitates and γ matrix phase.

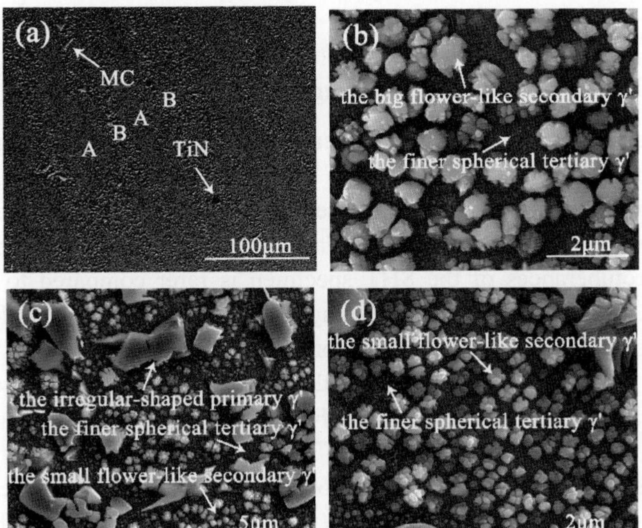

Fig. 1. SEM image of the alloy: (a) MC, MC (N) and γ' distribution of the alloy after STH; (b) magnification of part A: the big flower-like secondary γ' precipitates; (c) magnification of part B: the irregular-shaped primary γ' precipitates; (d) magnification of part B: the small flower-like secondary γ' precipitates.

Carbide, carbonitride is shown in Figure 1 (a). Morphological of γ' precipitates in the specimen (SHT) is mainly consisted of the big flower-like secondary γ' precipitates (as indicated part A in Figure 1 (a)), the irregular-shaped primary γ' precipitates and the small flower-like secondary γ' precipitate(as indicated part B in Figure 1. (a)). In addition to three kinds of γ' strengthening phase, it also includes the finer spherical tertiary γ' precipitates in the γ matrix channel. The part A, namely the big flower-like secondary γ' precipitates with an average diameter of about 600-800 nm and the finer spherical tertiary γ' precipitates in the matrix channel as shown in Figure 1. (b). The part B concludes the irregular-shaped primary γ' precipitates with an average diameter of about 800-1300 nm and the small flower-like secondary γ' precipitate with an average diameter of about 200-450 nm as illustrated in Figure 1. (c) and Figure 1. (d) respectively. The complete solution temperature of the γ' phase for the GH4742 superalloy is about 1120 °C [5]. During the solid solution heat treatment at 1080 °C which lower than solution temperature, some irregular-shaped primary γ' precipitates dissolved partially and others were residual. In this paper, the morphological evolution of big flower-like secondary γ' precipitates were observed only during long-term thermal exposure at the temperature of 700 °C and 750 °C.

3.2. *Morphology evolution of γ' during long-term aging at 700 °C*

Figure 2 shows that the morphological development of γ' during long-term thermal exposure at the temperature of 700 °C as ageing time increasing. It can be seen that the internal sphere and external protrusion of flower-like secondary γ' precipitate are one body. Figure 2. (a) and (b) shows that the number of the external protuberances of the SHT specimen is rarer compared to the ageing specimens. With the increase of ageing time, the γ' precipitates clearly show morphological instability with many protuberances developing on the sphere. Long-term aging at 700 °C for 1000 h, the protuberances is obviously increasing based on γ' matrix phase, as shown in Figure 2. (d). With aging time extending to 3000 h, the protuberances grow up more apparently on the any parts of γ' precipitates matrix and the number of the external protuberances is more. Similar to aging to 3000 h (Figure 2. (f)), the number of the protuberances of specimen aging to 5000 h also increases and are more than specimen aging to 3000 h, the size of the protuberances is more length, as shown in Figure 2. (h). Compared to aging to 3000 h, the second γ' precipitates volume fraction of specimen aged to 5000 h increases and the tertiary γ' precipitates in the γ matrix channel coarsen, the number decreases, size increases, as shown in Figure 2. (h). Y.S. Yoo [12] found that the morphological instability was driven by the point effect of diffusion and spherical γ' precipitates showed morphological instability with abrupt temperature drop. According to Qiu's study [13], the combined effect of both elastic strain energy and interface energy determines γ' precipitates morphology.

Fig. 2. Morphological of γ' precipitates and magnification of the flower-like secondary γ' precipitates during long-term thermal exposure at the temperature of 700 °C for: (a), (b) 0 h-SHT; (c), (d)1000 h; (e), (f) 3000 h; (g), (h) 5000 h.

S. Behrouzghaemia [2] maintained that at high cooling rates, γ' showed morphological instability since the low nucleation of new γ' precipitates lead to high supersaturation in the matrix. When the cooling rate decreased gradually, the growth of γ' precipitates occurs in a stable manner. The morphology changed from a cubic shape to a flower-like shape due to the increase of cooling rate. However, in this paper, our study found that the growth of the protuberances of γ' precipitates is sensitive to the long-term aging time at 700 °C. Studies have found that the protrusions of flower-like γ' precipitates is without any indication of particular crystallographic orientation and can grow on the any parts of γ' precipitates matrix [2,11,14]. The findings are consistent with the experimental results. It indicated that the growth pattern is not controlled by the elastic strain energy and is caused by the local concentration of solute elements in the matrix γ. As long as there is sufficient saturation in the matrix γ, γ' precipitates could grow unstably, which has nothing to do with lattice mismatch of γ/γ'.

According to Mullins and Sekerka's analysis [15], it is found that the stability of the shape of a spherical particle depended on the normal concentration gradient G and capillary constant Γ_D. The rate of growth $\dot{\delta}=d\delta/dt$ of the amplitude of the spherical harmonic the expression is [15]:

$$\dot{\delta}_l = \frac{D\,(l-1)}{(C-C_R)R^2}\left[G - \frac{c_0\Gamma_D}{R^2}(l+1)(l+2)\right]\delta_l. \tag{1}$$

The critical γ' precipitates radius Rc for the growth of a spherical harmonic and the critical nucleation radius R^* are [15]:

$$R_c = \left[\tfrac{1}{2}(l+1)(l+2)\right]R^*. \tag{2}$$

$$R^* = 2\Gamma_D/S. \tag{3}$$

Where $\Gamma_D = \sigma\Omega/RT$ is a capillarity constant in which σ is the γ-γ' interfacial energy and Ω the increment of precipitate volume per mole of added solute, R the gas constant, and T the absolute temperature; typically $\Gamma_D \approx 10^{-7} cm$. The meaning of l is the number of protuberances. The relative supersaturation S is equal to $(c_\infty - c_0)/c_0$, where c_∞ is the initial solute concentration in the supersaturated matrix and c_0 is the solute concentration in the matrix in equilibrium with the precipitate.

Equation (1) shows that $\dot{\delta}_l$ is proportional to the normal concentration gradient G, and is negative proportional to capillary constant Γ_D. In general both normal concentration gradient and capillary effects determined the morphology of precipitate phase. So it is necessary to study who is the dominant factor between concentration gradient G and capillary constant Γ_D in affecting the morphology of γ' phase [15]. When the G plays a dominant role in the growth of γ' precipitates, point effect of diffusion can explain the morphological instability of γ' precipitates. While the capillary effect takes a leading in the growth of γ' precipitates, the growth of γ' precipitates occurs in a stable manner.

Figure 3 shows the critical radius of morphological instability of γ' precipitates concerned supersaturation. Under the R_c line, the growth of γ' phase is stable manner. Above the R_c line, the appearance of γ' precipitates is not stable growth. Y.S. Yoo [12] explained the morphological evolution of γ' precipitates followed the path 1, 2 and 3, respectively. When the samples were exposed at 700 °C during a long-term aging time, according to the statistical computing, experimental points should be located in the dotted line area (as shown in Figure 3), γ' precipitates in unstable growing region. So we can find that substrate supersaturation of solute elements prompts γ' precipitates to grow with an instable manner. With the increase of aging time, the number of protrusions of γ' precipitates increases and the size of protrusions also increases.

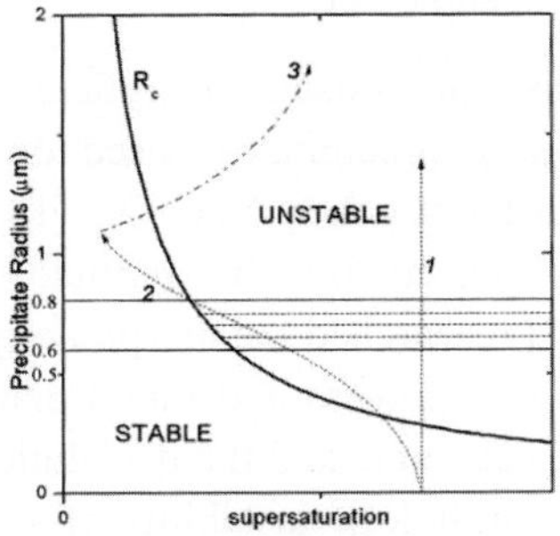

Fig. 3. Schematic stability diagram indicating stable and unstable regions: Possible growth trajectories of spherical precipitates are shown as well.

3.3. *Morphology evolution of γ' during long-term aging at 750 °C*

The morphological development of γ' during long-term thermal exposure at the temperature of 750 °C is shown in Figure 4. It can be found that the size of external protrusions of flower-like secondary γ' precipitate (Figure 4 (c)) are not large than that of the specimen after SHT (Figure 4 (a)). On the contrary, long-term aging at 750 °C for 1000 h, the edges of γ' precipitates aging to 3000 h began to dissolve, leading to the size of flower-like secondary γ' precipitate a slightly smaller and the number of the protrusions is more, as shown in Figure 4. (c) and (d). Figure 4 (e) and (f) show that γ' protrusions of specimen aging time for 3000 h start to divide with the matrix, tertiary γ' precipitates in the γ matrix channel ripen significantly (Figure 4 (b) and (f)). Until aging to 5000 h, the protrusions almost entirely split from γ' precipitates matrix and γ' precipitates is nearly spherical shape. Along with aging time extending, γ' precipitates dissolves more and more obviously and tertiary γ' precipitates in the substrate channel grow significantly, while the number of its declines. P.R. Cha [14] studied the splitting mechanism of coherent mismatch precipitates in anisotropic elastic medium by phase-field simulation. Cubic γ' precipitates have priority to grow along the crystal orientation <110>, which has high chemical driving force. Grooves is formed along the four <100> directions of cubic γ' precipitates and the stress concentration distributes at the edge of groove. Grooves can trap the solute atoms around the solution of γ phase, and interact with elastic stress field enriched solute atoms, so γ' precipitates dissolve partially and result in γ-phase invasion, γ' precipitates phase segmentation. It becomes the starting point of instable phase morphology splitting of γ' precipitates. With curve radius of groove increasing, γ phase along the <100> direction gradually invade γ' precipitates phase, causing γ' precipitates phase splitting, gradually forming doublets, quadruple of cubes and octet.

Khachaturyan [16] also analysised theoretically that the morphological evolution process of cubic γ' precipitates based on strain energy is: cubic-butterfly-double of plates-octet of cubes. Y.Y. Qiu [13] maintained that splitting mechanisms may be influenced by the particle shape and the particle arrangements. J.S. Hou [17] researched the morphology change of γ' precipitates of K44 in the long-term aging process and found that long-term aging process large cubic γ' precipitates degenerate and the dissolution and splitting of smaller γ' precipitates occur. The morphology instability of γ' precipitates is believed to be controlled by the total minimum energy principle, which includes interfacial energy and elastic strain energy. The splitting of γ' precipitates shown in Figure 4 (e-h) is determined by elastic strain energy elastic, similar to the results of Cha's study cubic γ' splitting. Figure 4. (g) and (h) shows that the splitting of γ' precipitates is almost complete and most of the protrusions of γ' precipitates have disappeared. At the moment, interfacial energy increases and strain energy decreases, while the decrease of total free energy prompt γ' precipitates to split, since the decrease of strain energy is more than the increase of interfacial free energy.

During long-term aging at 700 °C, γ' precipitates grow with an unstable manner, due to the supersaturation of the substrate γ. While the long-term thermal exposure at 750 °C, the stress concentration between the protuberances of γ' precipitates results in solute elements invade γ' precipitates and γ' precipitates split. Generally, study considered the morphology change of γ' precipitates depends mainly on the interfacial energy and elastic strain energy caused by lattice mismatch.

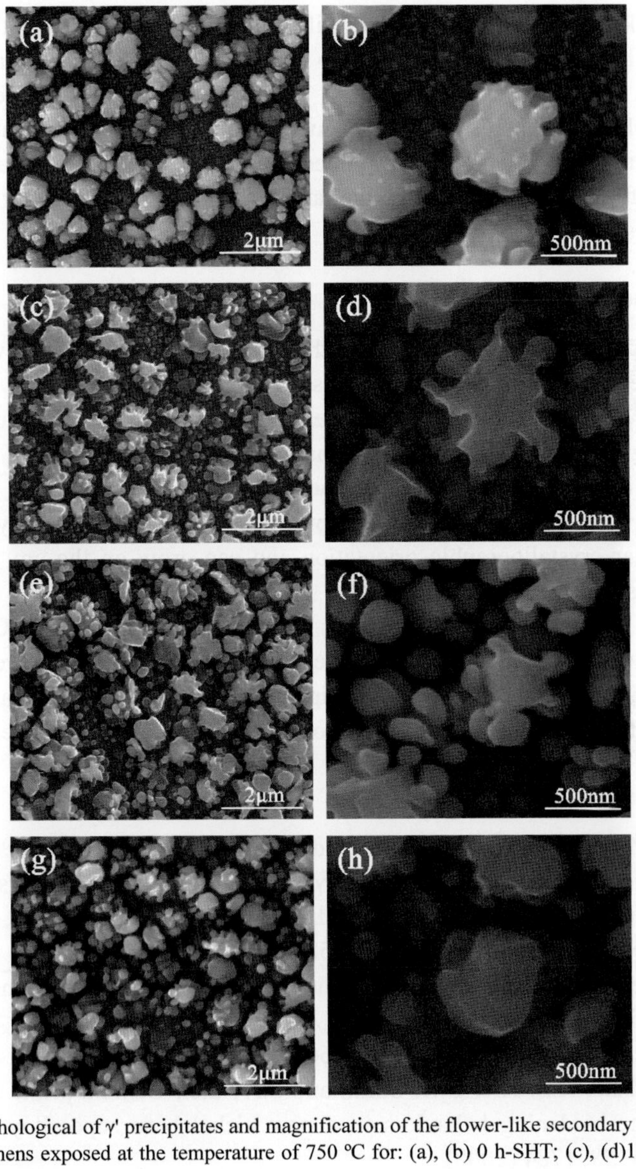

Fig. 4. Morphological of γ' precipitates and magnification of the flower-like secondary γ' precipitates in the specimens exposed at the temperature of 750 °C for: (a), (b) 0 h-SHT; (c), (d)1000 h; (e), (f) 3000 h; (g), (h) 5000 h.

The study found that instable growth of γ' precipitates develop into dendritic morphology, at the moment, the lattice misfit between the γ'/γ is nearly zero [12]. This paper found that in the case of long-term aging at 700 °C and 750 °C, the morphology changes of γ' precipitates arise from two different mechanisms and are likely to be caused by the diffusion of elements, while the element diffusion has a direct relationship to temperature and stress concentration. Therefore, long-term aging at 700 °C the substrate sufficient supersaturation causing solution elements uphill diffusion prompt γ' precipitates instable growth, long-term aging at 750 °C, γ' precipitates with stress concentration causes solute element of γ matrix invade γ' precipitates and the protrusions split from of γ' precipitates.

4. Conclusion

During long-term thermal exposure, the morphology instability evolution of flower-like secondary γ' precipitates was researched.

(1) The γ' precipitates after long-term aging at 700 °C clearly shown morphological instability developing many protuberances based on the sphere. With the increase of ageing time, the size of protuberances of flower-like secondary γ' precipitate increases. The protrusions of γ' is without any indication of particular crystallographic orientation and can grow on the any parts of γ' precipitates matrix.

(2) The γ' precipitates after long-term thermal exposure at 750 °C began to dissolve and split. With aging time extending, γ' precipitates dissolve more and more obviously and tertiary γ' precipitates in the substrate channel grow significantly, with volume fraction of flower-like secondary γ' declines.

References

1. R.J. Mitchell, M. Preuss, S.Tin and M.C. Hardy, The influence of cooling rate from temperatures above the γ′ solvus on morphology, mismatch and hardness in advanced polycrystalline nickel-base superalloy, *Materials Science and Engineering A*. **473**, 158 (2008).
2. S. Behrouzghaemi and R.J. Mitchell, Morphological changes of γ′ precipitates in superalloy IN738LC at various cooling rates, *Materials Science and Engineering A*. **498**, 266 (2008).
3. S.Q. Zhao, X.S. Xie, G.D. Smith and S.J. Patel, Gamma prime coarsening and age-hardening behaviors in a new nickel base superalloy, *Materials Letters*. **58**, 1784 (2008).
4. X.S. Xie, J.X. Dong, S.H. Fu and M.C. Zhang, Reaserch and development of γ″ and γ′ strengthened Ni-Fe base superalloy GH4169, *Acta Metall Sin.* **46**, 1289 (2010).

5. X.D. Lu, Q. Deng, J.H. Du, J.L. Qu, J.Y. Zhuang, and Z.Y. Zhong, Effect of slow cooling treatment on microstructure of difficult deformation GH4742 superalloy, *Journal of Alloys and Compounds*. **477**, 100 (2009).
6. X.D. Lu, J.H. Du, Q. Deng and Z.Y. Zhong, Effect of slow cooling treatment on hot deformation behavior of GH4742 superalloy, *Journal of Alloys and Compounds*. **486**, 195 (2009).
7. H.Y. Li, L.L. Zuo, X.P. Song, Y.L. Wang and G.L. Chen, γ' particles coarsening kinetics in nickel-base superalloy GH742, *Materials Engineering*. **8**, 8 (2007).
8. R.A. Ricks, A.J. Porter and R.C. Ecob, The growth of γ' precipitates superalloy, *Acta Metall.* **31**, 43 (1983).
9. C.L. Qiu and P. Andrews, On the formation of irregular-shaped gamma prime and serrated grain boundaries in a nickel-based superalloy during continuous cooling, *Materials Characterization.* 76, 28 (2013).
10. D.U. Furrer, γ' formation in superalloy U720Li, *Scripta Materialia*. **40**, 1215 (1999).
11. B.F. Hu, G.Q. Liu, K. Wu and G.F. Tian, Morphological instability of γ' phase in nickel–based powder metallurgy superalloy, *Acta Metallurgica Sinica*. **48**, 257 (2012).
12. Y.S. Yoo, Morphological instability of spherical γ' precipitates in a nickel base superalloy, *Scripta Materialia*. **53**, 81 (2005).
13. Y.Y. Qiu, The splitting behavior of γ' particles in Ni-based alloys, *Journal of Alloys and Compounds*. **270**, 145 (1998).
14. P.R. Cha, D.H. Yeon and S.H. Chung, Phase-field study for the splitting mechanism of coherent misfitting precipitates in anisotropic elastic media, *Scripta Materialia*. **52**, 1241 (2005).
15. W.W. Mullins and R. F. Sekerka, Morphological stability of a particle growing by diffusion or heat flow, *Journal of Applied Physics.* **34**, 323 (1963).
16. A.G. Khachaturyan, S.V. Semenovskaya and J.W. Morris Jr, Theoretical analysis of strain-induced shape changes in cubic precipitates during coarsening, *Acta metal.* **36**, 1563 (1988).
17. J.S. Hou, J.T. Guo, L.Z. Zhou, H.Q. Ye, Effects of γ' precipitate coarsening on tensile properties of Ni-base superalloy K44 during long-term aging, *Acta Metall Sin*. **42**, 481 (2006).

On the characterization of mechanical properties of metal materials corroded by microorganisms and weak acids

Hong-qian Chen[a*] and Chun-fang Xue[b]
Department of Mechanical Engineering,Academy of Armored Forces Engineering, Beijing,100072,China
E-mail: [a]chenhongqian@126.com, [b]xchunfang@sina.com

Xin-li Tian
National Key Laboratory for Equipment Remanufacturing, Academy of Armored Forces Engineering,
Beijing,100072,China
E-mail: tianxli719251@126.com

The purpose of this work is to investigate the effect of microorganism corrosion on the mechanical property of metal material. The wide application of metal materials results in complicated failure mechanisms, among which the microorganism corrosion is the most serious one due to the abrupt perforation in the local area of the structure. In this study, the mild steel was corroded 20 days in the condition of weak acid and vitriol bacillus cultivated in the weak acid, and tensile test was performed on the mild steel specimens. Experimental results demonstrate that the strength of specimen has little effect after the corrosion in the dilute sulphuric acid, but significant drop after the microorganism corrosion, revealing that the microorganism corrosion plays a key role in structure stability.

Keywords: Microorganism corrosion; Stress-strain behavior; Tensile strength

1. Introduction

Metal materials have been widely used in many industries such as mechanical engineering, military engineering, transportation engineering and civil engineering. The widely application of metal materials results in complicated failure mechanisms, among which the microorganism corrosion is the most serious one due to the abrupt perforation in the local area of the structure. This may lead to catastrophic consequences, and subsequently tremendous economic losses. To protect people's lives and properties, researchers throughout the world start to explore the microbial flora phagocytosis to increase the service life of metal materials. In this study, the mild steel was corroded 20 days in the condition of weak acid and vitriol bacillus cultivated in the weak acid, and tensile test was performed on the mild steel specimens, Comparison on the mechanical behaviors of specimens under different corrosion conditions has been made.

---

[*] Corresponding author.

2. Methodology

2.1. *Preparation of test specimen and etchant solution*

A total number of 60 standard tests were performed in this work. Before testing, Static culture method was employed for the standard tensile specimens in the air atmosphere.

The etchant solution used in this study are the dilute sulphuric acid with a PH value of 2.0, and the sulphuric acid bacillus solution cultivated in such solution. The initial volume of the dilute sulphuric acid was 150 ml. Both the volume (50 ml) and the PH value (1.4) decreased after 20 days evaporation. For the sulphuric acid bacillus solution, the PH value dropped to 1.8 after 20 days stationary at room temperature.

2.2 *Experimental design*

The specimens were divided into 3 subgroups. The PH values for both dilute sulphuric acid and the sulphuric acid bacillus solution are 2.0.

The specimen configurations can be found in the figure 1. The aspect ratio for tensile specimen is 5.0.

Fig. 1 The mild steel tensile specimen

It can be seen from the figure 2 and figure 3 that there are insignificant changes in shape during the 30 corrosion days of the mild steel corroded by dilute sulphuric acid. however, the volume is maximum on the first microbial corrosion day. The volume decreases with the corrosion elapse gradually.

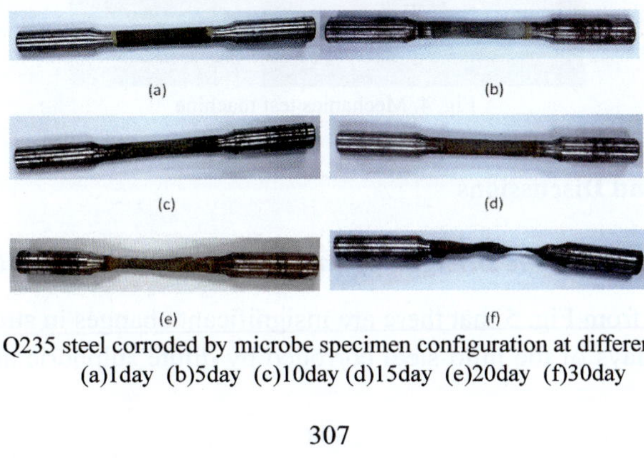

Fig. 2. Q235 steel corroded by microbe specimen configuration at different stages (a)1day (b)5day (c)10day (d)15day (e)20day (f)30day

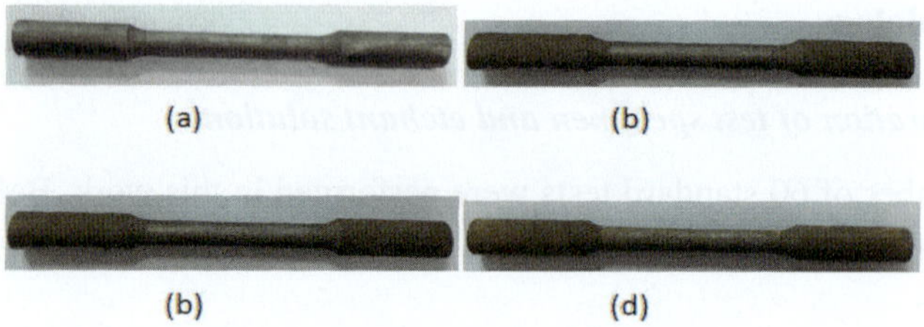

Fig. 3. Q235 steel corroded by dilute sulphuric acid specimen configuration at different stages (a)1day (b)5day (d)15day (f)30day

The quasi-static experiment was carried out in the mechanics test machine referred to Figure 4, whose loading capacity ranging from -50kN to 50kN (negative represents compressions) with a system resolution of 1%.

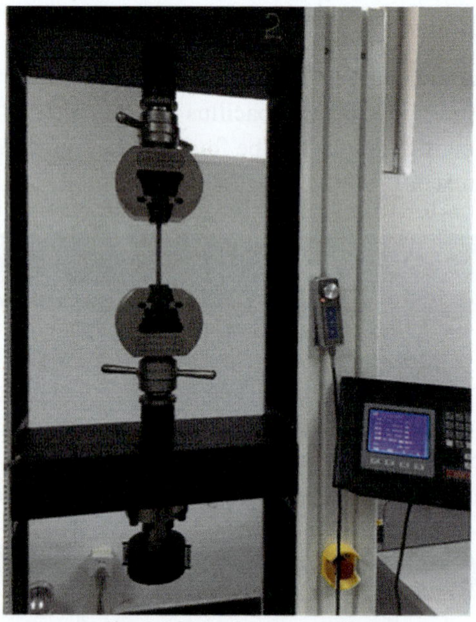

Fig. 4. Mechanics test machine

3. Results and Discussions

3.1. *Quasi static tensile strength of corroded mild steel materials*

It can be seen from Fig. 5 that there are insignificant changes in stress during the 30 corrosion days of the mild steel corroded by dilute sulphuric acid. however, the tensile strength is maximum on the first microbial corrosion day. The tensile strength decreases with the corrosion elapse gradually. It can be concluded that the sulphuric acid bacillus plays an important role in the corrosion for the mild steel, and the effect of the dilute sulphuric acid on the material property can be neglected. Therefore, in a special field, there are two possible ways to increase the service life of metal materials. One is to suppress the microorganism reproduction on the engineering component, the other is to remove the microorganism in the fabrication.

3.2. *Quasi static tensile response of corroded mild steel materials*

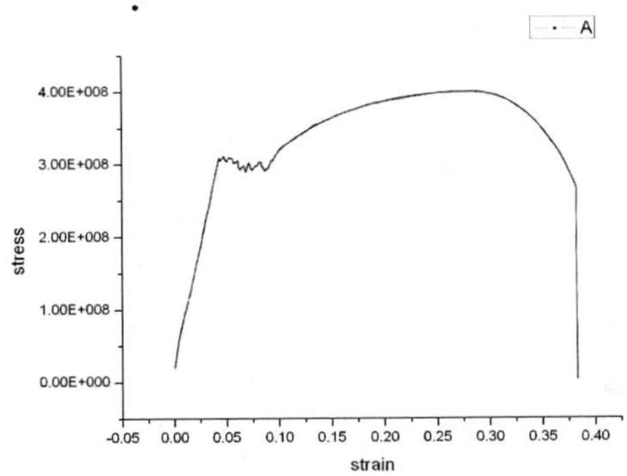

(a)

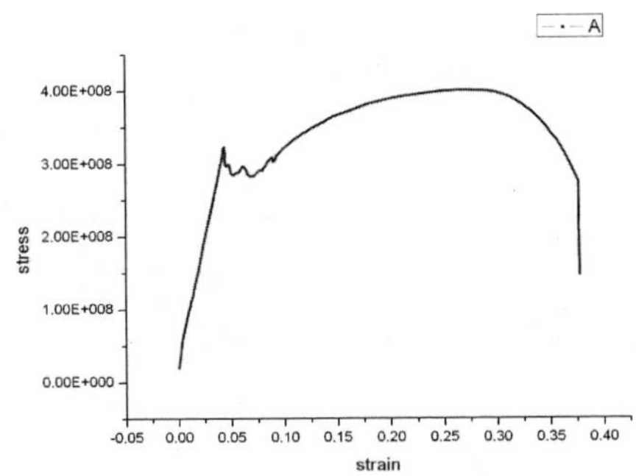

(b)

Fig. 5. Stress-strain diagram of the mild steel corroded by dilute sulphuric acid

(a)1 day (b)5 day (c)15 day (d)30 day

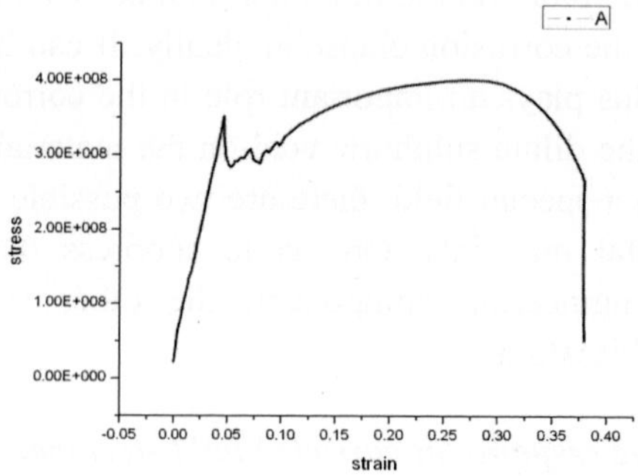

(c)

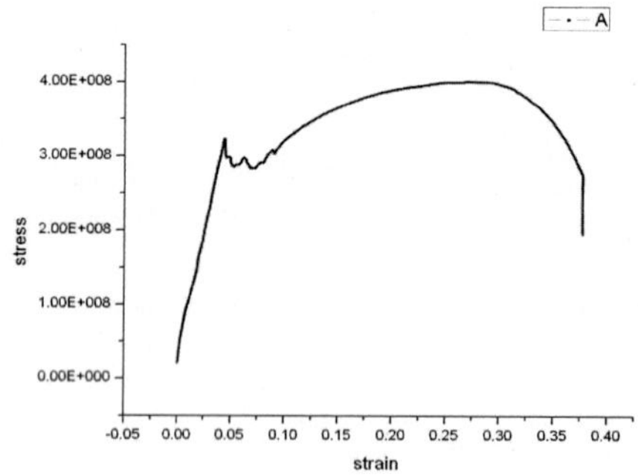

(d)

Fig. 5. (Cont'd)

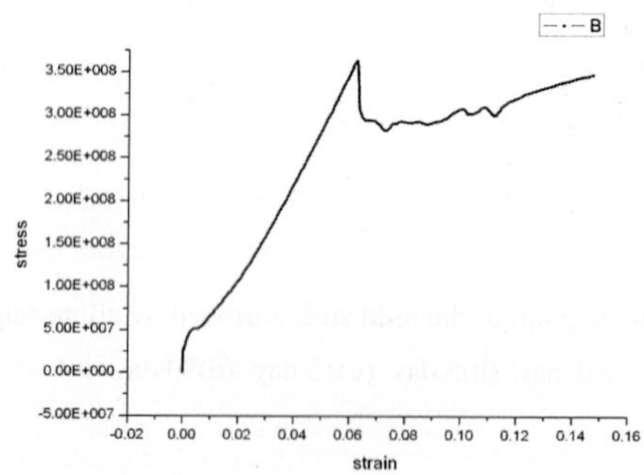

(a)

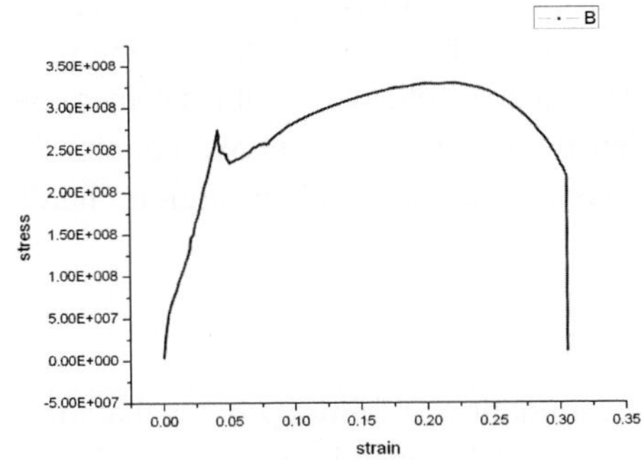

(b)

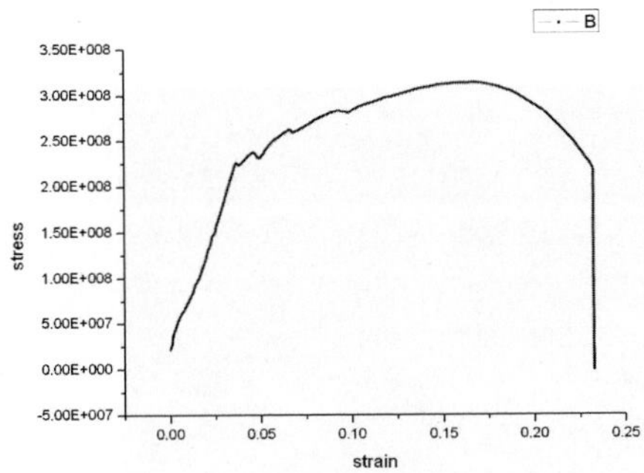

(c)

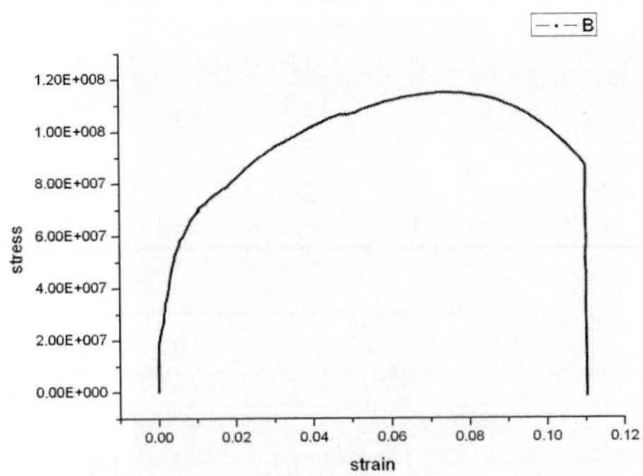

(d)

Fig. 6 Stress-strain diagram of the mild steel corroded by microbe
(a)1 day (b)5 day (c)15 day (d)30 day

It can be seen from the stress-strain behaviour that: The tensile strength of specimen remains the same whether the material corroded by the dilute sulphuric acid or not. With the corrosion time increases, the theoretical tensile strength of specimen corroded by microorganism decreases rallentando. The tensile strength on the 8th day is about 40% of that on the first day. Then tensile strength becomes to 25% on the 30th day. With the increase of the corrosion time, the profile of tensile stress-strain curves remain the same for both conditions, including elastic region, yielding region, hardening region and necking region.

3.3. *Failure analysis*

It can be seen from the examination of failure process that the failure strength of microbial corroded metal materials depends on the corrosion depth.

Post-examination of the fracture surface was conducted under the SEM image.

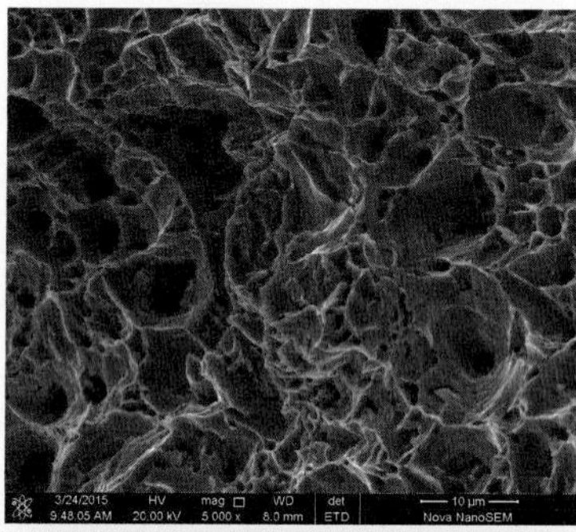

(a)

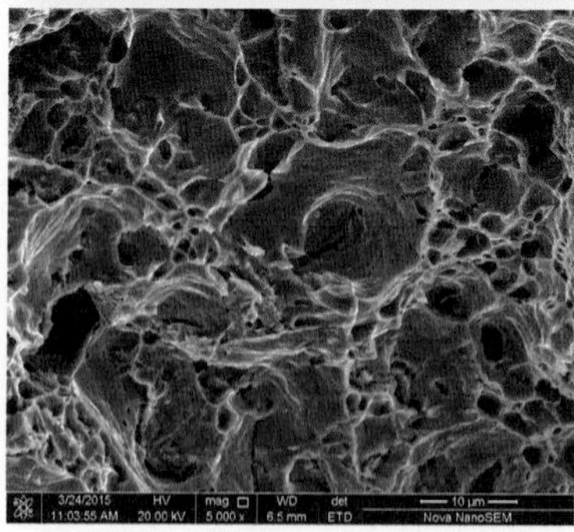

(b)

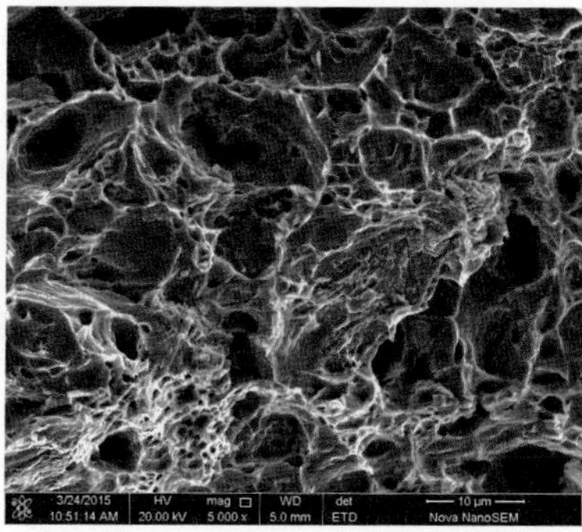
(c)

Figure 7 demonstrates the fracture surfaces of mild steel corroded by JQA and JQB at different corrosion stages. Macroscopically, the cup cone shape fracture of the tensile failure specimen demonstrates threadiness, which is the apparent characteristic of the ductile deformation. Microscopically, the fracture surfaces are full of big and deep dimples, which is the typical characteristic of ductile failure, implying that the ductility of mild steel has been insignificant affected by the microbial flora, and no hydrogen brittleness can be observed.

4. Conclusion

The tensile strength of specimen corroded by microorganism depends on the corrosion time. The longer the corrosion elapse is, the lower the specimen strength is.

The tensile strength of specimen corroded by dilute sulphuric acid and the corrosion time are independent. The generation of sulphate film on the surface of mild steel at the initial stage, protect the material away from further corrosion.

The sulphuric acid in the sulphuric bacillus solution can decrease the strength of mild steel material. However, this affect is insignificant and can be overlooked. The drop of the strength is mainly due to the sulphuric bacillus corrosion.

The sulphuric bacillus can seriously corroded the mild steel material, but can't affected the micro structure of materials. Only ductile failure can be observed after the specimen corrosion.

Acknowledgments

The research is supported by the 12th 5 years research Foundation (NO.513260402)

References

1. Starosvetsky J, Starosvetsky D, Armon R. Identification of icrobiologically influenced corrosion(MIC) in industrial equipment failures[J]. Engineering Failure Analysis,2007,14(8):1500-1504.
2. Lucian Brito V R, Ricardo Coutinho, Eduardo Cavalcanti H S, et al. The influence of macrofouling on the corrosion behaviour of API 5L X65 carbon steel[J]. Biofouling, 2007,23(3):193-201.

Study on biology properties of the bioceramic coating on titanium alloy by cold spraying

Sheng-Qiang Feng[1*†], Lang Cui[1], Zhen-Lin Wang[2], Yong-Ming Xing[2] and Guang Liu[1]

[1] *Department of Welding Engineering, Ningbo Branch of China Academy of Ordnance Science, Ningbo 315103, China*

[2] *Department of orthopedics, 113rd Hospital, CPLA, Ningbo 300500, China*

E-mail: †fsq7980@hotmail.com

It is crucial to find a kind of method with environment protection, economics, high efficiency and feasibility to prepare the HA coating on the surface of the titanium and its alloy. In this study, a novel method, the high speed high pressure cold spraying was employed to prepare the HA coating on the surface of TC4 (a kind of titanium alloy). The cytotoxicity behaviors of coating and titanium alloy substrate were investigated. Hydroxyapatite-zirconia (HA-ZrO2) coating was deposited on TC4 titanium alloy substrate by cold spray (CS). According to the performances of the activity detection of the alkaline phosphatase (ALP) and the activities of aminopherase on the surface of the coating, the biology properties of coatings and substrate were investigated. The results show that the contents of neutyophil alkaline phosphatase on the surface of coating are much more than that on the surface of titanium alloy substrate. In addition, the absorption capacity data of albumin and fibronectin of the coating are much better obviously than that of the titanium substrate. The results indicate that the biology properties of the coating are much better than that of the substrate.

Keywords: Cold Spraying; Biology Property; Bioceramic Coating.

1. Introduction

With the development of the medical research, the reparation of the defective organ and the transplantation of the new organ are much more emphasized than ever before. For instance, such the bioceramic materials as the titanium and its alloys, Hydroxyapatite (HA), zirconia (ZrO_2), titanium matrix composites have been more emphasized than ever before [1-3]. However, because of the inherent brittleness of the titanium, its alloy and HA, both of them are difficult to be used as the transplants.

---

* Corresponding author.

In order to solve the problems above, researchers had carried out many studies in which they prepared the HA bioactive ceramic coatings on the surface of the titanium and its alloys[4].However, there are such defects as poor process repeatability, obvious micro-capacities and cracks and so on. In addition, because of the mismatch heat expansion, low bonding strength, low bioactivity and biodegradability, it is difficult for HA coating prepared by methods above to meet the clinical application requirements of lifetime, safety and reliability.

Therefore, it is crucial to find a kind of method with the environment protection, economics, high efficiency and feasibility to prepare the HA coating on the surface of the titanium and its alloy. In this study, a novel method, the high speed high pressure cold spraying was employed to prepare the HA coating on the surface of TC4 (a kind of titanium alloy). The cytotoxicity behaviors of coating and titanium alloy substrate were investigated.

2. Experiments

In this study, TC4 titanium alloy was used as the substrate material. Commercially HA and ZrO_2 powders were used as feedstock materials for coatings. Two powders were mixed by the mechanical mixing. As shown in Figure 1, the two powders are in a uniform size, which is in accordance with the requirements of the cold spray process.

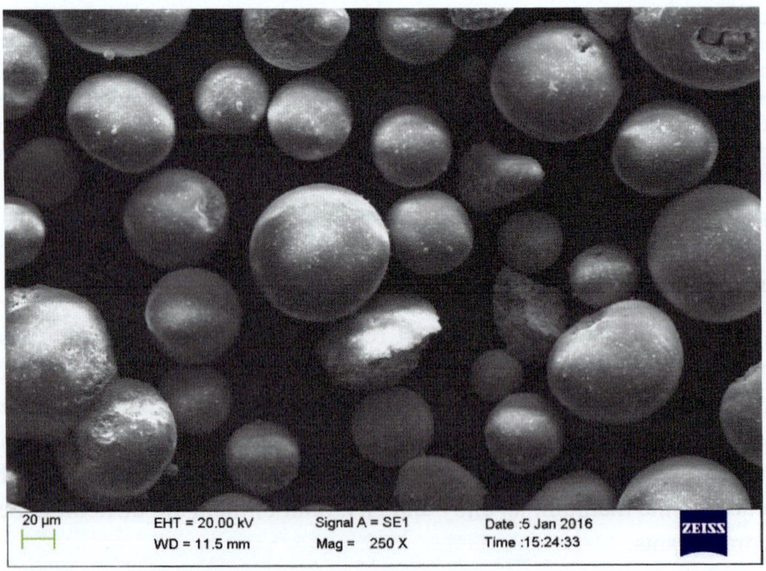

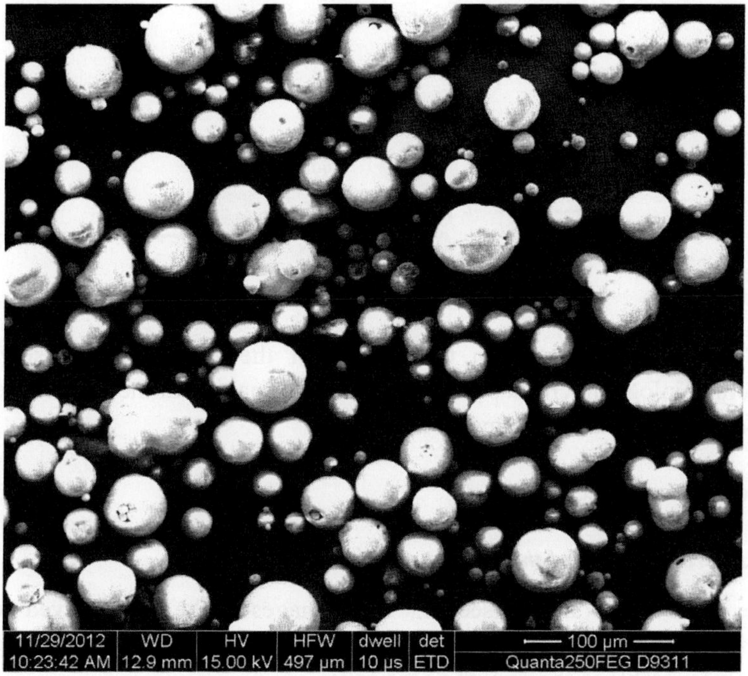

Fig. 1. Morphologies of the HA and ZrO_2

The coating was prepared by Kinetics3400 cold gas dynamic spray system developed by CGT in Germany. The surface of the block samples was prepared by grinding with 1000 SiC paper, followed by ultrasonic cleaning with acetone. Nitrogen was used as acceleration gas and carrier gas for powder feeding with a stagnation pressure of 2.2MPa. During the entire process, the acceleration gas was pre-heated to 350°C. The spraying standoff distance was maintained at 15 mm away from the nozzle exit.

The bonding strength of the coatings and the titanium alloy substrate were tested by a micro-scratch tester (MST). The process of the test is shown in Figure 2. The hemispherical head with 0.2mm dimension slides on the surface of the coating. At the same time, the force is added gradually. When the force reaches to a definite value, the coating is broken and flakes off. In this case, the force value is regard as the bonding strength. After calculation, the bonding strength is 55MPa.

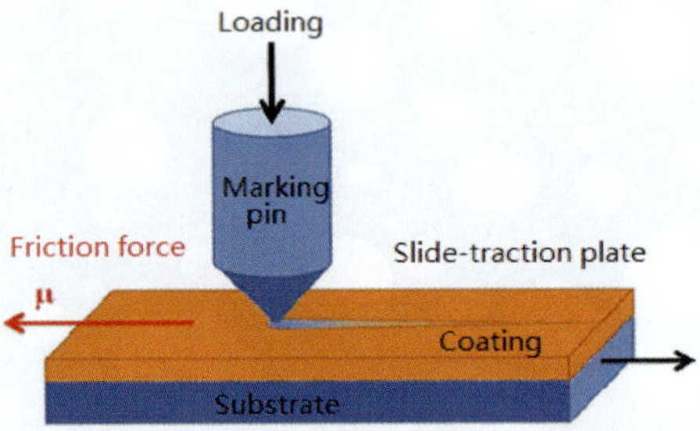

Fig. 2. Schematic diagram of the scratch test

The cell morphology on the surface of the coating was observed by electronic biologic microscope. The quantitative analysis of the cell proliferation was performed by the MTT method.

3. Results and Discussions

3.1. *Microstructure of the cold-sprayed coating*

Figure 3 shows the surface sections SEM images of the cold-sprayed coating. Seen from the Fig.3, the coating surface was rugged and had various porosities, which is favorable to the growth of the cells.

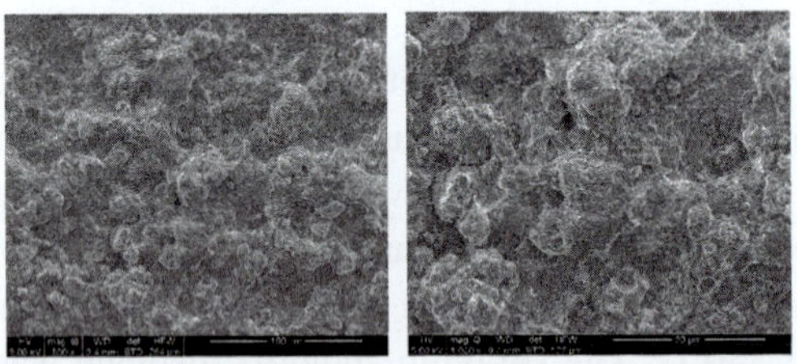
Fig. 3. Surface sections SEM images of the cold-sprayed coating

Meanwhile, from the Figure 3, there were amount of particles incomplete deformation due to lack of impact of follow-up particles. The pits on the surface of coating were generated by the impact of high speed particles. Because particles did not reach its critical speed, some particles at the edge of the nozzle bounced off. The coating performed dense and uniform microstructure. The porosity of the whole coating is 0.25%. Coating thickness was determined to be 50μm.

3.2. *Analysis of cytotoxicity behavior of the cold-sprayed coating*

3.2.1. *Sub-subheadings*

The osteoblasts were extracted from the medical mices. After the mices were anaesthetized and died in the alcohol liquids, their skulls were extracted, and then the tissues such as vascular adventitia were removed. The skulls were sheared to pieces and putted into a bottle where the 10% calf serum MEM (SH30265.01B, HyClone, USA) was poured. All of them were cultivated in a cell culture box where the temperature was 37 degree and the volume fraction of the carbon oxidation was 5%. The liquids were changed every other day. Figure 4. shows the process of the cultivation.

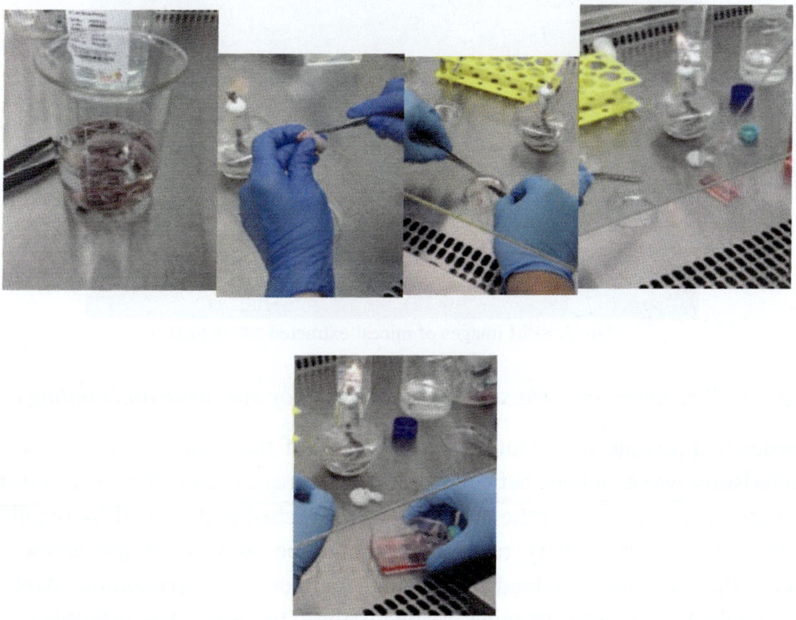

Fig. 4. Cultivation process of the mices' osteoblasts

About one week, the osteoblasts stripped from the mices' skull and grew on the pining of the cell culture box. When the value of the cellular fusion reached to at least 80%, the cell passage was carried out. Figure 5 shows the extracted mices' osteoblasts. Seen from this figure, the osteoblasts are wholesome. They have the abundant rough endoplasmic reticulum, which illustrates the characteristics of the abundant protein synthesis. The mitochondria are long and thin and their amounts have increased. The dissociated ribosomes in the intracytoplasmic are abundant. The nucleuses located the end of the cell are big and roundish. And the nucleuses have the single obvious nucleolus. The chromatin took on reticular shape. The microfilaments in the cytoplasm of a cell distributed near to the cell membrane.

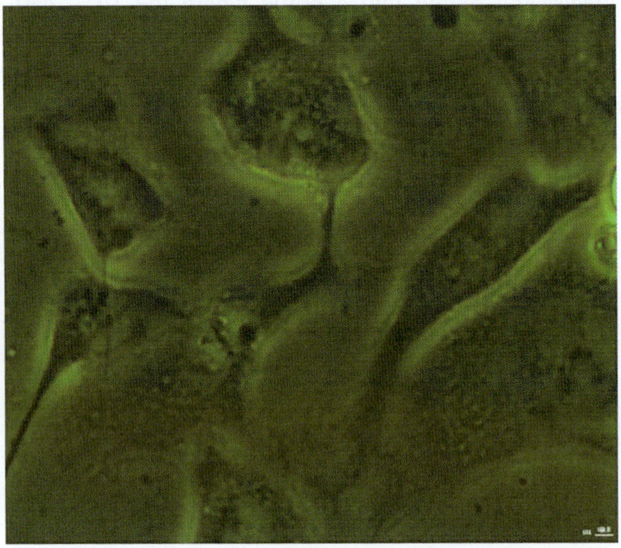

Fig. 5. SEM images of mices' extracted osteoblasts

3.2.2. *Cell seeding on surface of the titanium alloy substrate and coatings*

In order to illustrate the cytotoxicity behavior of the cold-sprayed coating, the comparisons were taken between the cold-sprayed and the titanium alloy substrate. The coating and the substrate were washed and soaked by the alcohol for three times. And every time, they should be soaked for ten hours. After soaked, they would be solved by the high temperature sterilization. And then, they would be put into 24 cups, respectively. The cultivated osteoblasts were planted onto the surface of the titanium alloy substrates and coatings, respectively. The process is shown in Figure 6.

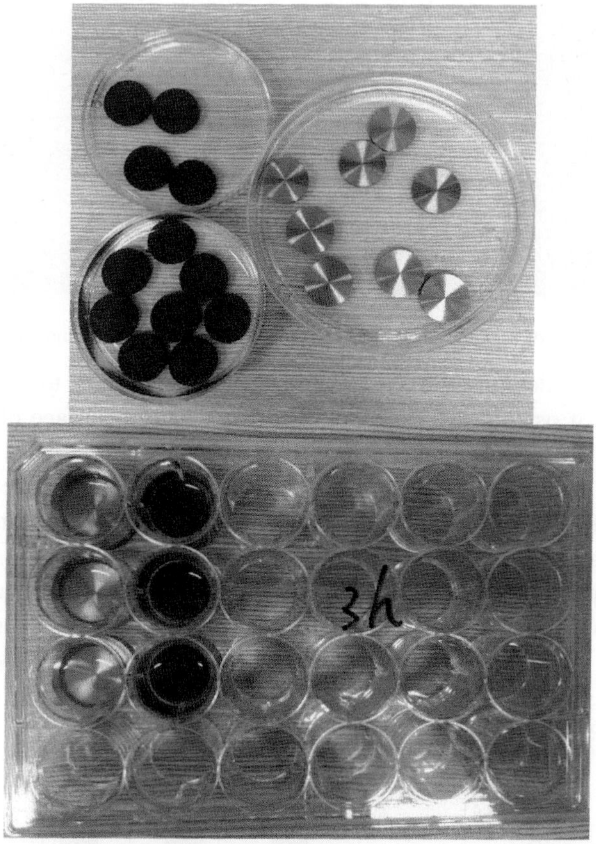

Fig. 6. Plant process of cells on the surface of titanium alloy substrates and coatings

Figure 7 shows the morphology of the cells attachment on the surface of the titanium alloy substrate and coating after 3 hours. Seen from this figure, compared with the substrate, the surface of the coating improved the cells attachment and outgrowth.

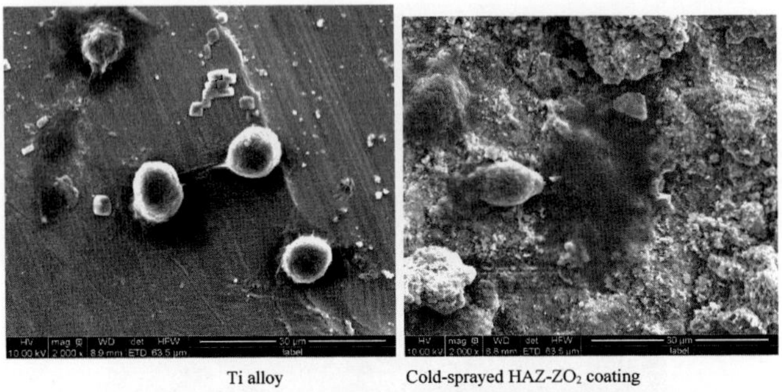

Fig. 7. Morphologies of the cells attachment on the surface of the substrate and coating after 3 hours

In order to demonstrate the advantages of the coating to the growth and proliferation of osteoblasts, they were cultivated on the surface of the titanium alloy substrate and coating for 1 day, 3 days, 5 days and 7 days, respectively. Fig. 8 shows the growth of the cells on the surface of the substrate and coatings by method of MTT. Through observation and analysis on the results, the coating also obviously improved the outgrowth. After cultivated for 3 days, the osteoblasts quickly overgrow the surface of the coating. The results illustrated that the cold-sprayed HA-ZrO_2 coating had no cytotoxicity and had much bioactivities such as improving the cells attachment, outgrowth and cell proliferation.

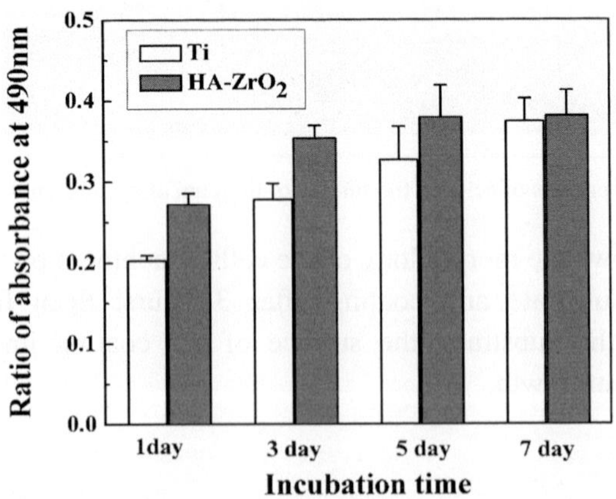

Fig. 8. Growth curves of cells on the surface of the substrate and cold-sprayed coating after 1day, 3days, 5days and 7days

Figure 9 shows the morphologies of cells on the surface of the titanium alloy substrate and cold-sprayed coating after 1 day, 3days, 5days and 7days. Seen from this figure, it was the same as the results by the method of MTT shown in the Figure 8. The cells grew more quickly on the coating than that on the titanium alloy substrate after 1day. And from the third day, the surface of the coating was covered by the cells rapidly. Especially on the fifth day, the whole surface of the coating was covered by cells, which sufficiently illustrated that the cold-sprayed coating has particularly better conditions for the cells' growth than the titanium alloy substrate.

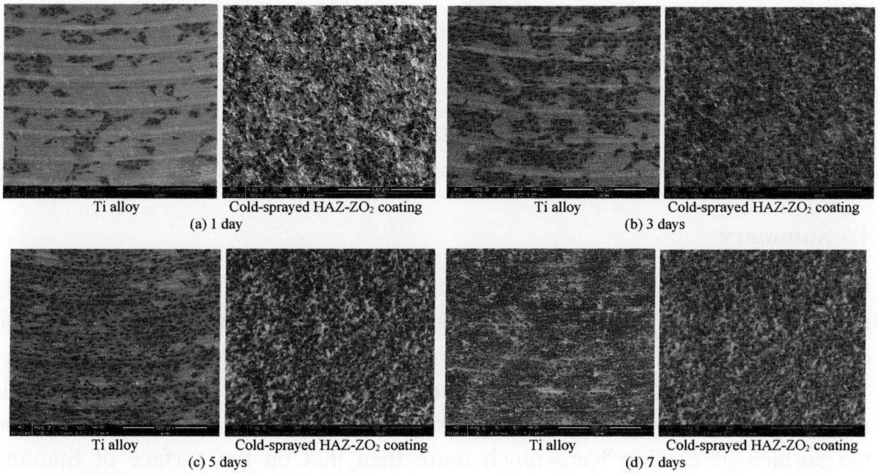

Fig. 9. Morphologies of cells on the surface of the substrate and cold-sprayed coating after 1day, 3days, 5days and 7days

3.2.3. *Activity detection of the alkaline phosphatases*

The osteoblast grows and differentiates with the increase of the alkaline phosphatase (ALP). The alkaline phosphatase, the most important demonstrations, can be reflective of the actual differentiation situations of the osteoblast to a certain extent. The detection of the alkaline phosphatase was carried out to the osteoblast cultivated for 7 days. The results showed that, compared with that of the titanium alloy substrate, the more contents of the alkaline phosphatase were secreted by the osteoblast on the surface of the cold-sprayed coating.

Figure 10 shows the comparisons of the alkaline phosphatase secreted by the osteoblast on the surface of the cold-sprayed coating and the titanium alloy substrate. Seen from this figure, the contents of the alkaline phosphatase secreted by the osteoblast on the surface of the cold-sprayed coating are about 1.5 time than that on the surface of the titanium alloy substrate. This situation fully explains that the cell differentiations are more complete on the surface of the cold-sprayed coating that that on the surface of the titanium alloy substrate.

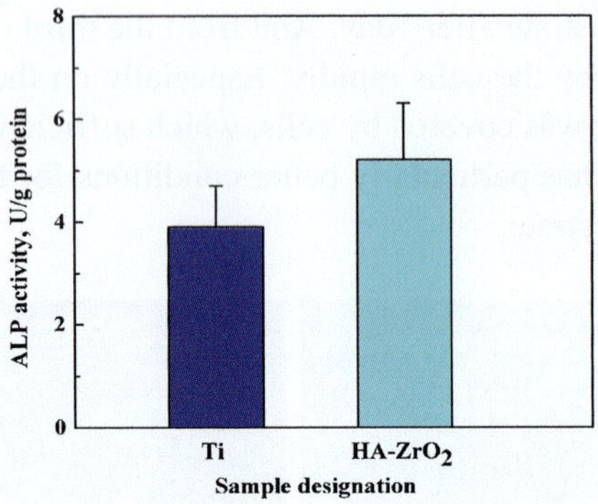

Fig. 10. Effects of cold-sprayed coating on the ALP activities of the osteoblast

4. Summary

It is feasible for the cold spray to prepare the HA-ZrO_2 coating. The coating surface was rugged and had various porosities, which is favorable to the growth of the cells.

It is obvious for the surface of the HA-ZrO_2 coating to improve the cells attachment and outgrowth. The contents of neutyophil alkaline phosphatase on the surface of coating were much more than that on the surface of titanium alloys substrate. The absorption capacity data of albumin and fibronectin of the coating were much better obviously than that of the titanium substrate.

The biology properties of the coating were much better than that of the substrate.

Acknowledgements

This work is supported by Ningbo key scientific and technological project (2012C5018).

References

1. Z.H. Wu, L.Z Liu, The study on the cellular responses around artificial ear ossicles of nano-bioactive ceramics in rats. *Journal of Xinjiang Medical University,* 32,1(2009).
2. C. Li, S.Q. Tao, C.L. Zhou, et al. Histocompatibility evaluation of nano hydroxyapatite-40% zirconia composite bioceramic. *Orthopedic Journal of China,* 17,23(2009).

3. R.D. Bloebaum, D. Beeks, L.D. Dorr, et al. Complications with hydroxyapatiye particulate separation in total hip arthroplasty. *Clin orthop,* 298(1994).
4. Y.M. Wang, Y. Zhou, D.C. Jia, et al. Porous bioactive ceramic coatings of TiO_2 in combination with CaP compounds fabricated on Ti6Al4V surface by microarc oxidation process. *Rare Metal Materials and Engineering,* 11,34(2005).

The oxidation of Fe-35Al alloys in 1 atm of pure O_2 at 700, 800 and 900°C

Long Li [a], Xian-Chao Xu[a], Hong-Hua Zhang[a,b], Yao Chen[a] and Jun-Huai Xiang[a,b,1]

[a]*School of Materials and Mechanical & Electrical Engineering, Jiangxi Science and Technology Normal University,*
Nanchang 330013, China

[b]*Jiangxi Key Laboratory of Surface Engineering, Jiangxi Science and Technology Normal University,*
Nanchang 330013, China
[1]*xiangjunhuai@163.com*

For the application of inner wall layer of carbonization furnace, the components are subjected to repeated thermal cycles and the operation temperature of carbonization furnace is so high that the oxidation resistance of its inner wall layer is becoming an important issue. In this paper, the high temperature oxidation behavior of binary Fe-35Al (at. %) alloys in 1 atm of pure O_2 at 700, 800 and 900 °C has been investigated. It is of great significance to enable understanding of the oxidation behavior of more complex systems, and will provide fundamental information for understanding the basic high temperature oxidation mechanism of Fe-Al intermetallics. The oxidation resistance of Fe-35Al alloys decreases when the temperature increases from 700 to 900 °C. The mass gain of the alloy at 700, 800 and 900 °C after 24 h oxidation is 0.098, 0.170, 0.360 mg/cm^2, respectively. Although all the kinetic curves are not very regular, they can be approximately considered as composed of a rapid transient-oxidation stage and a slow steady-oxidation stage. All the scales are mainly composed of Al_2O_3, coupled with a little Fe_2O_3.

Keywords: Fe-35Al; Isothermal Oxidation; Microstructure; High Temperature.

---

[1] Corresponding author.

1. Introduction

Iron-based superalloys have attracted considerable attention in the past decades for the potential applications in the inner wall layer of carbonization furnace and other high temperature components, because of their excellent high temputure oxidation [1] and sulfldation resistance [2, 3]. Moreover, iron-based superalloys possess numerous attractive and unique properties and distinctive advantages, such as low cost, low density, good oxidation resistances, excellent wear resistance performance and hot corrosion resistance to degradation in corrosive or oxidizing environments [4]. However, limited room temperature ductility (<5%) and a sharp drop in strength above 600°C have limited their consideration for use as structural materials. Thus, in order to solve that problems, it was reported by Sang-Heon Kim1 et al. in Nature in 2015 that FeAl-type brittle but hard intermetallic compound can be effectively used as a strengthening second phase in high-aluminum low-density alloys, while alleviating its harmful effect by controlling its morphology and dispersion [5]. In recent years, a series of Fe-based amorphous or nanocrystalline alloys such as Fe_3Al [6-8] and Fe-40Al [9, 10] have been widely studied as better advanced materials for hydrogen storage, since the 1930's. Given the purpose on this new energy source, it is significant to improve our knowledge regarding the properties of Fe-Al alloys.

For the applications in inner wall layer of carbonization furnace, the components are subjected to repeated thermal cycles, and the operation temperature of carbonization furnace is so high that the oxidation resistance of its inner wall layer is becoming an important issue.

Generally, the protective performance of the oxide scale depends highly on certain features of the scales, such as density, uniformity, and defect structure as well as scale adherence to the base metal [11]. The formation of a continuous inner alumina layer tends to improve isothermal oxidation resistance by reducing the amount of cation transport [12].

In this paper, the high temperature oxidation behavior of binary Fe-35Al (at. %) alloys in 1 atm of pure O_2 at 700, 800 and 900 °C has been investigated. This research is of great significance for understanding the oxidation behavior of more complex systems, and will provide fundamental information for understanding the basic high temperature oxidation mechanism of Fe-Al intermetallics.

2. Experimental Details

The alloys were prepared by repeated melting 99.9% Fe and 99.99% Al in a vacuum induction furnace and casting into a water-cooled mould. The ingots were subsequently annealed for 24 h at 900 °C in vacuum annealing furnace

(~1.3 Pa), and cut into 10×10×1.2 mm^3 pieces by spark-erosion machining, polished on SiC waterproof abrasive papers, then washed in distilled water, alcohol and acetone and dried in warm air immediately before use.

Continuous mass change measurements of the specimens were carried out for 24 h by a Setaram Setsys Evo in 1 atm of pure O_2 at 700, 800 and 900 °C, respectively. The oxidized specimens were characterized using field emission scanning electron microscopy (FESEM) with energy-dispersive X-ray analysis (EDS).

3. Results and Discussion

3.1. *Oxidation kinetics*

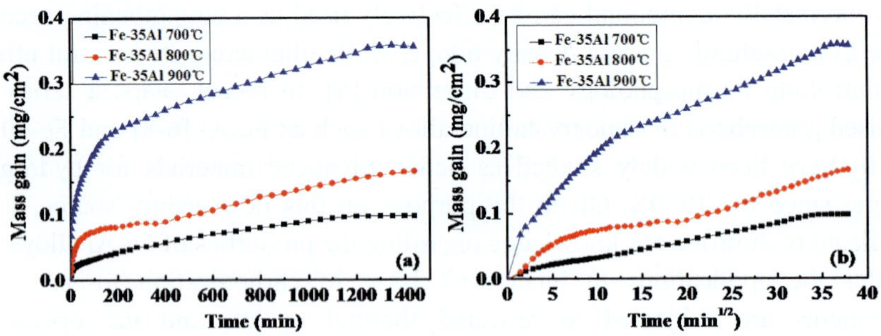

Fig.1. Oxidation kinetics of as-cast Fe-35Al alloys at 700, 800 and 900 ℃ (a) and corresponding parabolic polts (b)

The isothermal oxidation kinetics curves and the corresponding parabolic plots for the oxidation of Fe-35Al alloys at 700, 800 and 900 °C for 24 h are shown in Figure 1(a) and 1(b), respectively. The oxidation resistance of Fe-35Al alloys decreases when the temperature increases from 700 to 900 °C. The mass gain of the alloy at 700, 800 and 900 °C after 24 h oxidation is 0.098, 0.170, 0.360 mg/cm^2, respectively. Although all the kinetic curves are not very regular, they can be approximately considered as composed of a rapid transient-oxidation stage and a slow steady-oxidation stage.

3.2. *Cross-sectional microstructure and composition of the scales*

Cross-sectional morphology and corresponding element mapping of oxide scales formed on Fe-35Al for 24 h oxidation at 700, 800 and 900 °C are shown in Figure 2, Figure 3. and Figure 4, respectively. Although the thickness of the scales formed at 700-900 ˚C is different, the microstructure of the scales is similar. All the scales are mainly composed of Al_2O_3, coupled with a little Fe_2O_3. It has been proposed that for a protective behavior at least 8%Al (mass fraction) is necessary for Fe-Al alloys at 1073 K [13]. The present content of 35 at.% Al is far beyond that required to form a protective outer Al_2O_3 layer on the surface of the alloy. No internal oxidation region is found at all temperatures. Thus, at the given temperatures from 700-900 ˚C, Fe-35Al is of good oxidation resistance by forming a continuous thin protective Al_2O_3 layer on the surface, just like the case of Fe-30Al and Fe-40Al, whose oxidation behavior has also been investigated at the same time. The scales formed at 700 and 800 ˚C are more homogenous and integrated than that formed at 900 ˚C, resulting in their much better oxidation resistance.

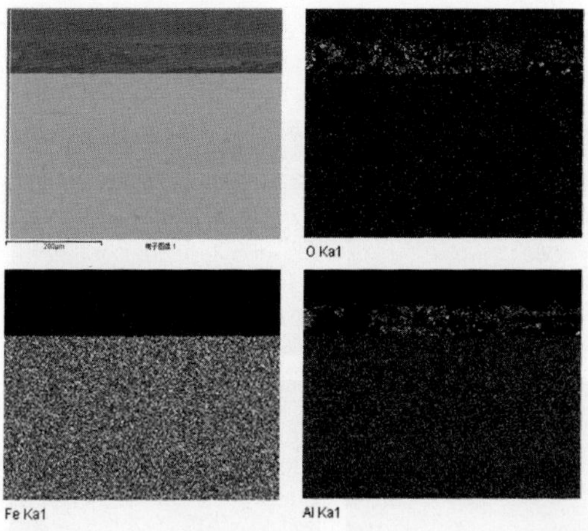

Fig.2. Cross-sectional morphology and corresponding element distribution maps of oxide scales formed on as-cast Fe-35Al alloys at 700 ˚C for 24 h

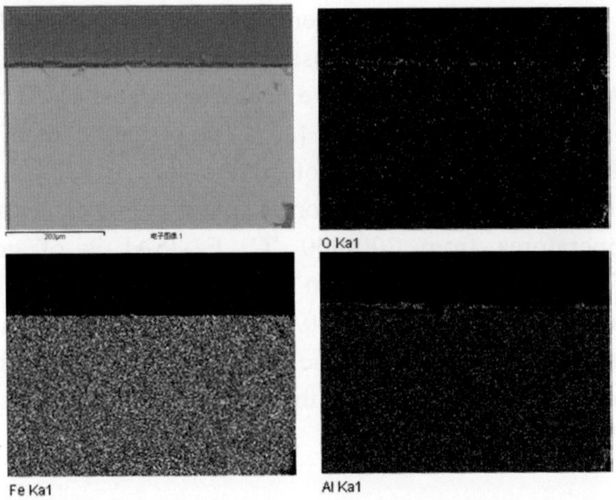

Fig.3. Cross-sectional morphology and corresponding element distribution maps of oxide scales formed on as-cast Fe-35Al alloys at 800 °C for 24 h

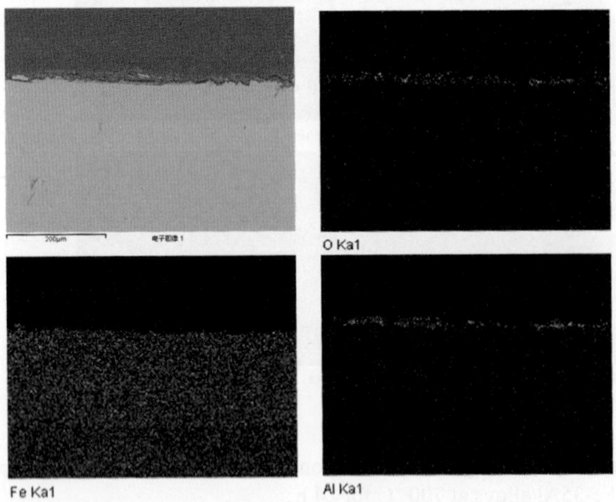

Fig.4. Cross-sectional morphology and corresponding element distribution maps of oxide scales formed on as-cast Fe-35Al alloys at 900 °C for 24 h

4. Conclusion

The present content of 35 at.% Al is far beyond that required to form a protective outer Al_2O_3 layer on the surface of the alloy at 700-900 °C. No internal oxidation region is found at all temperatures. The scales formed at 700 and 800 °C show much better oxidation resistance than that formed at 900 °C.

Acknowledgements

This work was financially supported by the Key Program of Science and Technology of Ministry of Education (Grant No. 209069), Natural Science Foundation of Jiangxi Province (20132BAB206015, 20122BAB216018), and the Masters' Innovative Foundation of Jiangxi Province (YC2014-S428).

References

1. K.Gschneidner, A.Russell, A.Pecharsky, J.Morris, Z.Zhang, T.Lograsso, et al, A family of ductile intermetallic compounds, *Nature Materials,* 2 (2003) 587-590.
2. E.Airiskallio, E.Nurmi, M.H. Heinonen, I.J. Väyrynen, K.Kokko, M.Ropo, et al, High temperature oxidation of Fe-Al and Fe-Cr-Al alloys: The role of Cr as a chemically active element, *Corrosion Science,* 52 (2010) 3394-3404.
3. Z.G. Yang and P.Y. Hou, Wrinkling behavior of alumina scales formed during isothermal oxidation of Fe-Al binary alloys, *Materials Science and Engineering: A,* 391 (2005) 1-9.
4. C.G. McKamey, J.H. DeVan, P.F. Tortorelli and V.K. Sikka, A review of recent developments in Fe_3Al-based alloys, *Journal of Materials Research,* 6 (1991) 1779-1805.
5. S.H. Kim, H.Kim, and N.J. Kim, Brittle intermetallic compound makes ultrastrong low-density steel with large ductility, *Nature,* 518 (2015) 77-79.
6. Z.G. Zhang, F.Gesmundo, P.Y. Hou, and Y. Niu, Criteria for the formation of protective Al_2O_3 scales on Fe-Al and Fe-Cr-Al alloys, *Corrosion Science,* 48 (2006) 741-765.
7. M. Palm and G.Sauthoff, Deformation behaviour and oxidation resistance of single-phase and two-phase L2(1)-ordered Fe-Al-Ti alloys, *Intermetallics,* 12 (2004) 1345-1359.
8. N.Babu, R.Balasubramaniam, and A.Ghosh, High-temperature oxidation of Fe_3Al-based iron aluminides in oxygen, *Corrosion Science,* 43 (2001) 2239-2254.
9. Q.Zeng, and I.Baker, Magnetic properties and thermal ordering of mechanically alloyed Fe-40at% Al, *Intermetallics,* 14 (2006) 396-405.
10. F.Lang, Z.Yu, S.Gedevanishvili, S.C. Deevi, S.Hayashi, and T.Narita, Sulfidation behavior of Fe-40Al sheet in H_2–H_2S mixtures at high temperatures, *Intermetallics,* 12 (2004) 469-475.
11. J.H. Xiang, Y.Niu, and F.Gesmundo, The oxidation of two ternary Fe-Cu-10 at.% Al alloys in 1atm of pure O_2 at 800-900°C, *Corrosion Science,* 47 (2005) 1493-1505.

12. B.Liand, and B.Gleeson, Effects of Silicon on the Oxidation Behavior of Ni-Base Chromia-Forming Alloys, *Oxidation of Metals,* 65 (2006) 101-122.
13. P.Tomaszewicz, and G.R. Willwork, Observation of nodule growth during the oxidation of pure binary iron-aluminum alloys, *Oxidation of Metals*, 19 (1983) 165-185.

Texture and planar anisotropy of AZ31 magnesium alloy sheets processed by different final rolling routes

Tao Zhou*, Deng-Hui Song, Dong Hu, Wei Jiang, Zhi-Cong Liu, Zhao Yang, Ming-Bo Yang, Jian Tu, Tao Feng and Xiao-Bing Zhai

College of Material Science and Engineering, Chongqing University of Technology, Chongqing, 400054, China

*E-mail address: *zt19811118@cqut.edu.cn, tzhou1118@163.com*

In the present work, the microstructures, textures and tensile mechanical anisotropy of AZ31 magnesium alloy sheets processed by different final rolling routes (Route A, B and C) at 550° were investigated, in which the final rolling direction is parallel, perpendicular and at angle of 45° to initial rolling direction (RD) of the sheet, respectively. The results showed that the AZ31 alloy sheets processed by route A, B and C exhibited double-type textures with basal poles tilting at approximately from 9.5° to 12° in the RD, TD and 45° to RD, respectively. By changing the final rolling route from Route A to C, the grain sizes of the sheets were not refined and the basal texture intensity was slightly increased, but the planar anisotropy in strength gradually decreased. The main reason was that the difference of the Schmid factors for basal plane slips among tensile tests conducted along RD, 45° and TD was decreased, which was determined by texture characteristics, including the final tilted directions and angles of basal pole, and the texture intensity.

Keywords: AZ31 Magnesium Alloy; Microstructure; Texture; Planar Anisotropy; Final Rolling Route.

1. Introduction

As the lightest structural metal material, magnesium alloys, which are characteristics of low density, high specific strength and outstanding damping

---

* Corresponding author.

capacity, have attracted increasing interests in many years [1]. However, the ductility of wrought Mg alloys at room temperature is insufficient to manufacture complex components, since only two unique basal slip systems and twinning can be active at room temperature [2]. So many researchers have focused on the improvement of the tensile ductility of the Mg alloy sheets at ambient temperatures by means of texture control through ECAR [3], ECAE [4], differential speed rolling (DSR) [5], High-Speed Rolling (HSR) [6], high strain rate rolling (HSRR) [7], high temperature rolling [8,9], etc. High temperature rolling is an effective method to improve the stretch formability of the Mg alloy sheets, in which the rolling temperature is commonly higher than 500 °C, beneficial to the remarkable basal texture weakening [8-10], which may be also beneficial to decrease in mechanical anisotropy in three dimensions.

It has been reported that cross rolling (CR) is an effective way to weaken the basal texture and anisotropy, and to improve stretched formability of the Mg alloy sheets [11-15]. Tang et al [12] suggested that the unidirectional rolling (UR) sheet had obvious planar anisotropy from strong basal-type texture, but the CR sheet exhibited the weakened basal texture and nearly isotropic properties. Zhang et al [16, 17] found that compared with UR, CR and three-directional rolling (TDR) produced finer grain microstructures and weaker basal textures, leaded to the enhanced mechanical properties and stretched formability of AZ31 alloy sheet. However, the studies related to changing final rolling routes are very few, in which the rolling direction changes only at the finished pass, which is different from CR and TDR, where the rolling direction changes alternately. The effect of the changing rolling routes at the finished pass on the planar anisotropy is unclear, especially at high rolling temperatures. Therefore, in the present work, the microstructures, textures and planar anisotropy of AZ31 magnesium alloy sheets processed by different final rolling routes at 550°C were investigated.

2. Materials and methods

The initial material used in the study was hot rolled AZ31 alloy (Mg-3Al-1Zn-0.3Mn, in wt.%) billets with a thickness of 4.3mm. The billets were preheated to 400°C for 20min prior to rolling, and then were unidirectional rolled from 4.3mm to 1.5mm in thickness by 7 passes at 400 °C. Finally, the billets were preheated to 550 °C for 1min and then rolled down to 1.2mm by a single pass with three different final rolling routes, i.e. route A, B and C, in which the final rolling direction is parallel to, perpendicular to and at angle of 45° to the initial rolling direction (RD) of the sheet, respectively, as seen in Figure 1. After rolling, the sheets were annealed at 350 °C for 1 h.

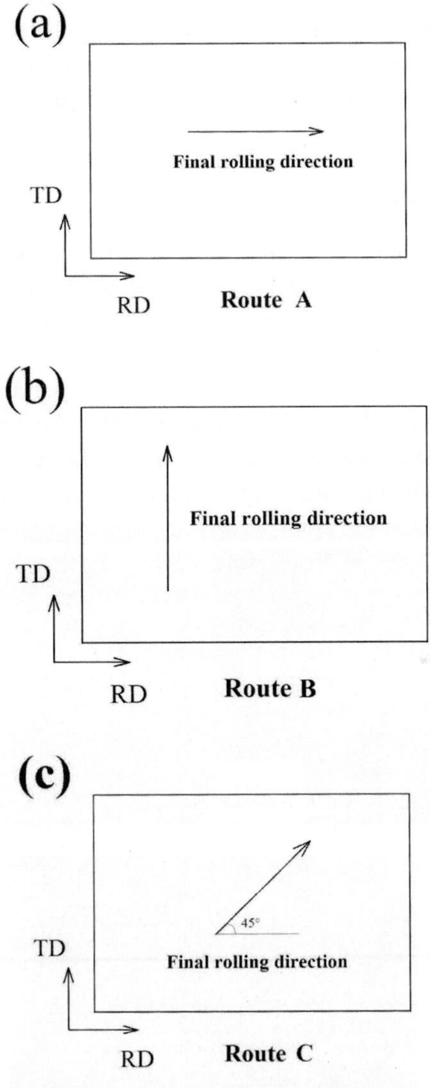

Fig. 1. The schematic diagram of different final rolling routes: (a) Route A, (b) Route B and Route C.

Microstructure evolution was examined on RD-transverse direction (TD) plane by optical microscopy (OM) after mechanical polishing and then etched with a solution of 1 g oxalic acid, 1 ml nitric acid and 98 ml water. The textures of the as-rolled and as-annealed sheets were tested at the mid-layers of the sheets (RD-TD plane) and by X-ray diffraction on a Rigaku D/Max 2500PC.

In order to examine the planar anisotropy of the sheets, uniaxial tensile tests were carried out at the angle of 0°, 45° and 90° to the RD on a SANS testing machine with an initial strain rate of 5.6×10^{-4}/s at room temperature. The specimens for tensile tests had a gauge length of 15 mm, a width of 4 mm and a thickness of 1.2mm.

3. Results and discussions

The microstructures of the as-annealed AZ31 alloy sheets with different final rolling routes are shown in Figure 2. All the sheets exhibited the full static recrysallized grain structures. Zhang et al [16] found that the CR was much more effective in grain refinement than UR. However, in the present work, after annealing the average grain size was about 12.2μm, 13.4μm and 13μm for Route A, B and C, respectively. So it can be concluded that the grain sizes of the sheets both in as-rolled and annealed conditions are hardly affected by changing the final rolling route, which is consistent with the results of ref. [11, 12].

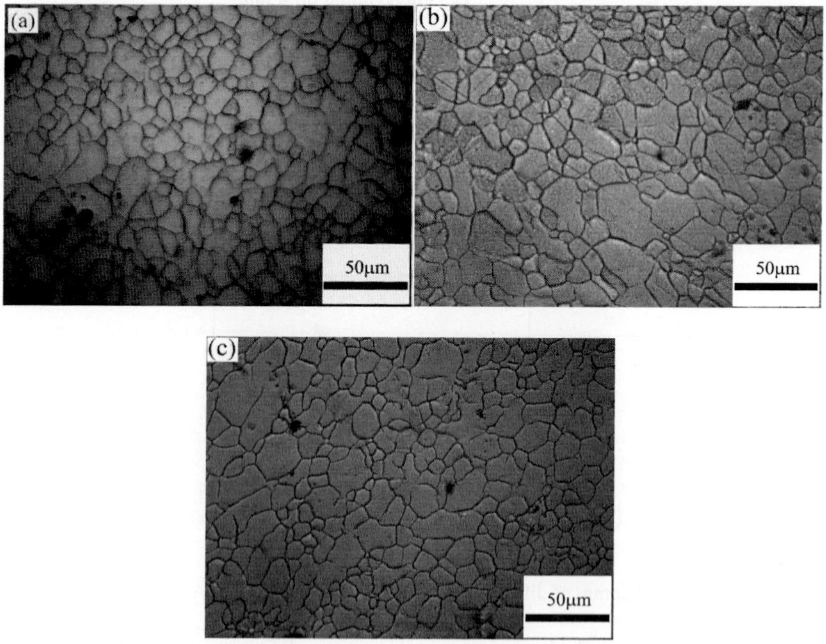

Fig. 2. The Microstructures of the as-annealed AZ31 Mg sheets processed by different final rolling routes: (a) Route A, (b) Route B and (c) Route C.

Figure 3 shows the (0002) and (10-10) pole figures of AZ31 Mg sheets processed by different final rolling routes in annealed condition. The annealed sheets processed by Route A, B and C exhibited double-type textures with basal poles tilting in the RD, TD and 45° to RD, respectively, in (0002) pole figure, which indicates that the c-axis of the some grains is titled away from ND to the final rolling direction, which is consistent with the results of ref.[12, 16-17]. Zhang et al [16] suggested that the main reason for the rotation of the basal pole was that the direction of the tensile stress and compressive stress changed as rolled by CR. Thus, the different texture characteristics formed in the sheets processed by Route A and B is mainly due to the difference between the direction of the tensile stress and compressive stress for the sheets during the final rolling.

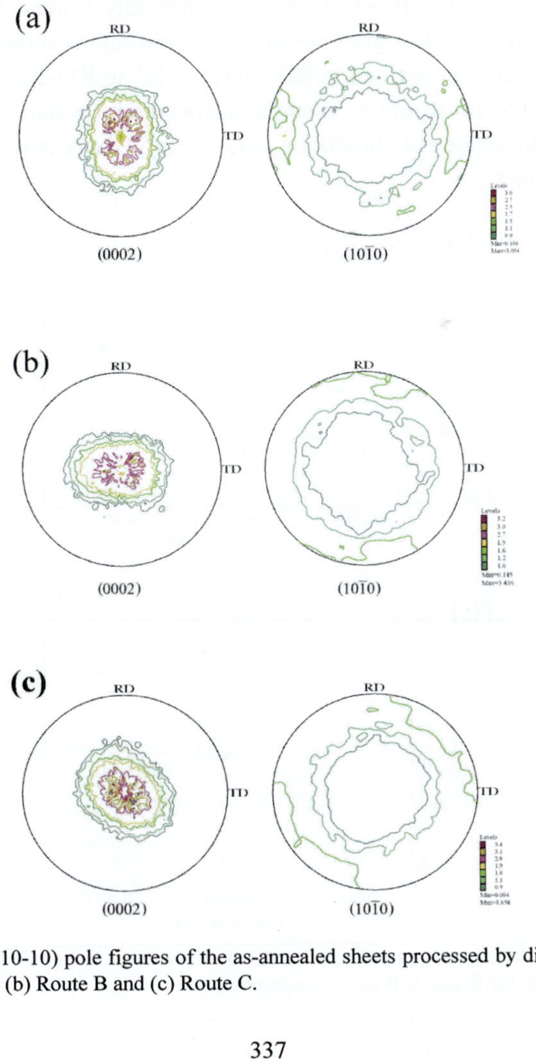

Fig. 3. (0002) and (10-10) pole figures of the as-annealed sheets processed by different final rolling routes: (a) Route A, (b) Route B and (c) Route C.

As shown in Figure 4(a), the texture intensity of the as-annealed sheets was 3.1, 3.4 and 3.7 for Route A, B and C, respectively, which seems that the basal texture intensity is not weakened by only changing the final rolling route form Route A to B and C. In opposition, the texture intensity for Route B and C is slightly higher than that for Route A. It was reported that CR could effectively weaken the basal texture intensity [1, 16]. The main reason was that the alternating changing the direction of tensile stress and compressive stress during CR leaded to the direction of grain rotation was continuously changing [16]. But in the present work, the rolling direction changed only once during Route B and C, which means that the changing the direction of tensile stress and compressive stress only occurs at the finished pass, that possibly limited the continuously changing of the direction of grain rotation. In addition, as shown in Figure 4(b), the angle of the basal pole tilted in RD, TD and 45° to RD directions, was about 12°, 10.4° and 9.5° for Route A, B and C, respectively, which means that the tilt angle in the corresponded directions was slightly decreased as changing the Route A to Route C.

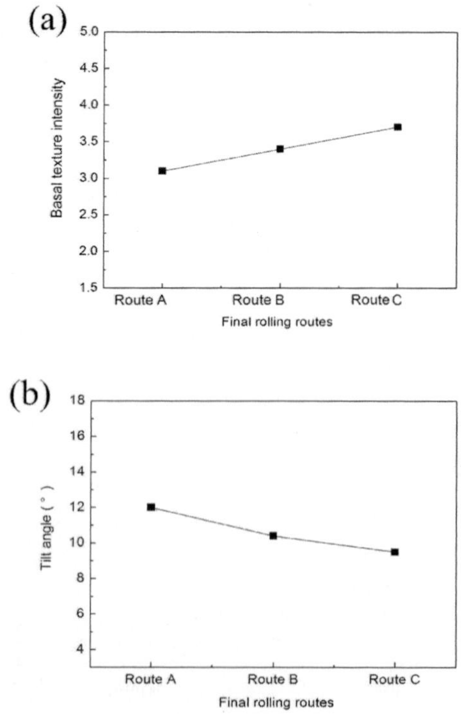

Fig. 4 Changes in (a) the basal texture intensity; (b) the angle of the basal pole tilted in RD, TD and 45° to RD directions, for Route A, B and C, respectively.

The details of mechanical properties in the tensile directions of RD, 45° and TD of the annealed sheets processed by Route A, B and C are shown in table 1. As mentioned above, the difference among the average grain sizes of the sheets processed by the different final rolling routes was few, so the discussions on the mechanical properties and its anisotropy are focused on texture characteristics. The corresponding mechanical properties, i.e. the 0.2% proof stress (YS), the ultimate tensile strength (UTS) and the fracture elongation (FE) are presented in Figure 5. It was obvious that both the sheets exhibited mechanical anisotropy. For the sheet processed by Route A, the lowest YS and UTS, and the highest FE were measured along the RD. For the sheet processed by Route B and C, the lowest YS and UTS were measured along the TD and 45°, respectively. But for route B and C, the difference of FE among the related tensile directions was small, respectively. Moreover, the lowest YS for the sheets processed by Route A, B and C was 122, 137 and 147Mpa, respectively, which means that as changing the Route A to Route C, the corresponded lowest YS is increased. The main reasons will be discussed in the following sections.

Table 1. The results of the tensile tests carried out in the tensile directions of RD, 45° and TD of the as-annealed AZ31 alloy sheets processed by different final rolling routes

Rolling routes	YS(MPa)			UTS(MPa)			FE(%)		
	RD	45	TD	RD	45	TD	RD	45	TD
Route A	122	158	161	255	268	268	23.7	19.1	19.4
Route B	157	155	137	268	273	265	22.2	22.7	23.6
Route C	160	147	152	277	270	270	21.1	20.3	20.6

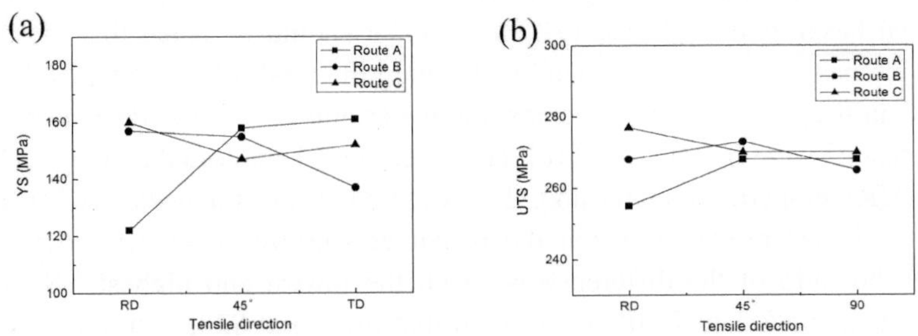

Fig. 5. The changes in (a) tensile yield stress, (b) ultimate tensile stress and (c) elongation of the Route A, B and C samples in the tensile directions of RD, 45° and TD.

(c)

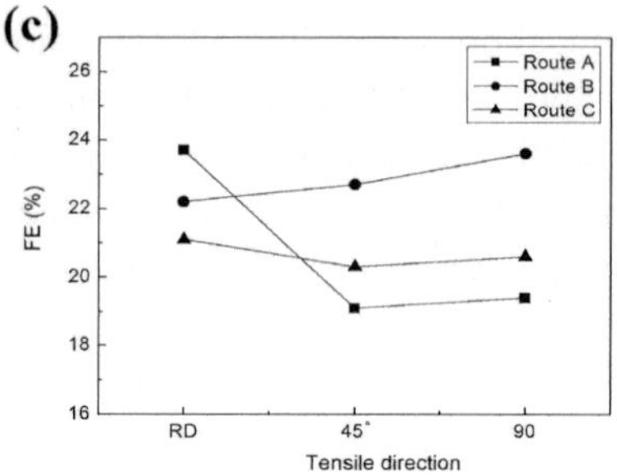

Fig. 5. (Cont'd).

Moreover, the values of planar anisotropy in mechanical properties are calculated by $(X_{max}-X_{min})\times100\%/X_{min}$, which are shown in Figure 6. For the sheet rolled by Route A, the difference between the lowest (RD) and highest YS (TD) was 39MPa, which was 32% of the lowest YS. But for the sheet rolled by Route B, the difference between the lowest (TD) and highest YS (RD) was decreased to 20MPa, which was 14.6% of the lowest YS. For the sheet rolled by Route C, the difference between the lowest (45°) and highest YS (RD) was further decreased to 13MPa, which was 8.8% of the lowest YS. The ratio of the difference between the lowest and highest UTS to the lowest UTS was 3% and 2.6% for the sheet rolled by Route B and C, respectively, lower than that (5.1%) of the sheet rolled by Route A. So it can be seen that by changing the final rolling routes the planar anisotropy of the sheets can be decreased, which is similar to the results in CR [12], and the sequence for the degree of the planar anisotropy of the sheets was Route A> Route B> Route C, similar to that in ref. [16, 17]. They suggested that the CR and TDR sheets exhibited lower mechanical properties anisotropy, which was likely to be responsible for the weakened basal texture [1, 16, 17]. However, it should be noted that the basal texture intensity was not weakened by changing the final rolling route from A to B and C in the present work, whereas the planar anisotropy was decreased. The similar results were found in ref.[18]. Huang et al [19] suggested that as the final rolling (UR) temperature increasing, the basal texture of as-annealed AZ31 alloy sheets weakened remarkably, but the planar anisotropy in YS was increased, because the ratio of the difference between the lowest and highest YS to the lowest YS was 4.8% at 723K, lower than that (6.9%) at 828K. In our previous study [20], by increasing the final rolling (UR) speeds from 0.1m/s to 0.4m/s, the texture intensity of the as-annealed sheets was decreased from 4.4 to 3.1, but the ratio of the difference between the lowest and highest YS to the lowest YS was increased from 13% up to 40%. The similar phenomenon was also found in DSR [21, 22]. So there may be some other reasons for the decrease in planar anisotropy. Kim et al [23] found that AZ31 alloy sheets processed by ECAR exhibited the largely weakened basal fiber texture with the basal pole rotated from the ND in the RD, which resulted in an increase in planar anisotropy in strength, compared with that of the sheet processed by normal rolling. The main reason was that as basal planes rotate around the transverse axis during ECAR, the Schmid factors (SF) for basal and prismatic plane slips along the 0° and 45° directions were higher than those along TD, whereas for the normal rolling sheet, all the tensile directions were on the basal plane so that the related SF were approximated as nearly zero [23]. It was reported that at room temperature, the {0001} <11$\bar{2}$0> basal slip dominates the deformation mechanism of magnesium alloys since its CRSS is much lower than that of the non-basal slips, prismatic slips and <a + c> pyramidal slips [24]. The non-basal slips may be activated for the Mg alloy sheet at room temperature as the grain size is much lower than 10μm [25]. But as seen in Figure 2, the average grain sizes of all the as-annealed sheets processed by route A, B and C was larger than 10μm, which means that the influence of non-basal slips on the planar anisotropy may be small. In addition, the at stress near UTS, the slip plans with low SF remained unmotioned and leaded to the high level of dislocation piles-up in the grains [26], beneficial to increase in the UTS. So the planar anisotropy in strength for the AZ31 alloy sheets is determined by the difficulty of the motion of the basal slips during tensile along different directions, which is related to the SF for the basal slip. Wang et al [24] investigated the effect of initial orientation on the microstructure and mechanical properties of textured AZ31 Mg alloy during torsion and annealing. For the extruded, annealed PAR (torsion axis parallel to ED) and annealed PER samples (torsion axis perpendicular to ED), the difference of the SF for basal slip between tensile tests conducted along ED and perpendicular to ED were 0.215, 0.028 and 0.026, respectively. It is obvious that the difference for the latter two is much lower than that for the former one, leaded to the noticeable decrease in the anisotropy after torsion and annealing. Moreover, Ma et al [27] suggested that mechanical anisotropy of the AZ31 alloy was more dependent on the type of texture than the texture intensity, especially after high-temperature compressed deformation. Dong et al [26] found that the style of the texture was the critical factor to determine the anisotropy of extruded Li-containing AZ31 alloys, neither grain size nor peak texture value. Therefore, it can be concluded that the planar anisotropy in strength may have a positive correlation with the difference degree of the SF for basal plane slips among tensile tests conducted along RD, 45° and TD, in other words, the planar anisotropy in strength was possibly increased with an increase in the difference of the SF between the tensile directions, which is determined by the texture characteristics, including the tilted directions and angles of basal pole, and the texture intensity.

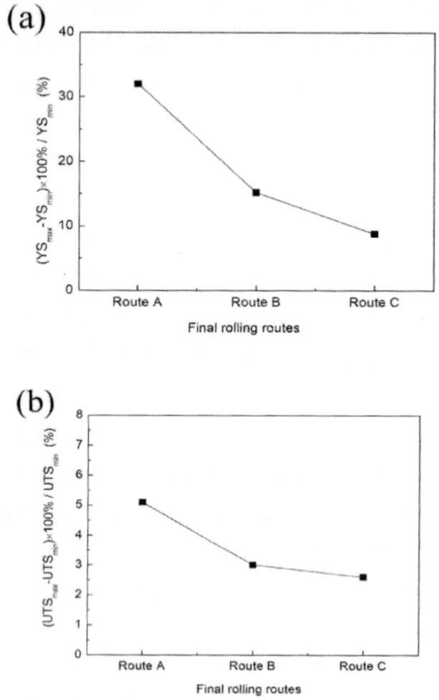

Fig. 6 The (X_{max}-X_{min})×100%/Xmin of the sheets processed by different final rolling routes: (a) tensile yield stress, (b) ultimate tensile stress.

As mentioned above, the final tilted directions and angles of basal pole, and the texture intensity are different for the sheets processed by Route A, B and C, leads to the different degrees of the planar anisotropy. The detailed analyses are as follows: Firstly, taking into account the final tilted directions of basal pole, for the Route A (Figure 4(a)), the basal poles tilting in the RD, which indicates that the c-axis of the some grains is titled away from ND to RD, leads to the highest SF along the RD, and the lowest SF along the TD. So the lowest YS and UTS were measured along the RD. For the Route B (Figure 4(b)), the c-axis of the some grains is titled away from ND to TD, which leads to the highest SF along the TD, and the lowest SF along the RD, results in the lowest YS and UTS were measured along the TD. For the Route C (Figure 4(c)), the c-axis of the some grains is titled away from ND to the direction at angle of 45° to RD, leads to the highest SF along the 45°, and the lowest SF along the direction which is perpendicular to the 45°. So the lowest YS and UTS were measured along the

45°. Due to the angle between the RD, TD and 45° is much smaller than 90°, respectively, the values of SF along the RD and TD are between the highest and lowest SF for the Route C, so the maximum difference of the SF among the tensile tests is smaller than that for Route A and B, which means that the degree of the planar anisotropy in strength for the former one is smaller than that for the latter two. But it is hard to distinguish that the maximum difference of SF among tensile tests for the Route A is whether larger than that for Route B nor not, only according to the final tilted directions of basal pole. Secondly, taking into account the tilted angles of basal pole and the texture intensity, as seen in Figure 5, as changing the final rolling routes from Route A to C, the tilted angles of basal pole in the corresponded directions are slightly decreased, but the texture intensity is slightly increased, which lead to the corresponded lowest YS is increased. Meanwhile, the decreased tilted angle and increased texture intensity is beneficial to a decrease in the difference of SF among tensile tests, which leads to the decrease in planar anisotropy in strength from Route A to C.

In a word, compared with the Route A, the planar anisotropy in strength for the Route B is decreased, results from the decreased tilted angle of basal pole in the corresponded directions and increased texture intensity, which leads to the difference of SF among tensile tests decreasing. For the Route C, the c-axis of the some grains is titled away from ND to the direction at angle of 45° to RD, which leads to the values of SF along the RD and TD are between the highest and lowest SF, respectively. So the maximum difference of the SF among the tensile tests is smaller than that for Route A and B. In addition, compared with Route B, the tilted angle of basal pole in the corresponded directions was decreased and the texture intensity was increased slightly, also beneficial to the decrease in the difference of SF among tensile tests. So the planar anisotropy in strength for the Route C is further decreased. Therefore, it can be concluded that the texture characteristics of the sheets, including the final tilted directions and angles of basal pole, and the texture intensity, is the critical factor to determine the planar anisotropy in strength of the AZ31 alloy sheets processed by the different final rolling routes.

4. Conclusions

The microstructures, textures and tensile mechanical anisotropy of AZ31 magnesium alloy sheets processed by different final rolling routes (Route A, B and C) at 550 °C were investigated. The main results are as follows:

(1) The as-annealed sheets processed by Route A, B and C exhibited double-type textures with basal poles tilting at approximately from 9.5° to 12° in the RD, TD and 45° to RD, respectively, mainly due to the difference between

the direction of the tensile stress and compressive stress for the sheets during the final rolling.

(2) Compared with the Route A, the planar anisotropy in strength for the Route B is a decreased, result from the decreased tilted angle of basal pole in the corresponded directions and increased texture intensity, which leads to the difference of Schmid factors for basal slips among tensile tests along RD, TD and 45° decreasing.

(3) For the Route C, the c-axis of the some grains is titled away from ND to the direction at angle of 45° to RD, which leads to the maximum difference of the Schmid factors among the tensile tests is smaller than that for Route A and B. In addition, compared with Route B, the tilted angle of basal pole in the corresponded directions was decreased and the texture intensity was increased slightly, also beneficial to the decrease in the difference of Schmid factors among tensile tests. Thus, the planar anisotropy in strength for the Route C is further decreased.

(4) The texture characteristics of the sheets, including the final tilted directions and angles of basal pole, and the texture intensity, is the critical factor to determine the planar anisotropy in strength of the AZ31 alloy sheets processed by the different final rolling routes.

Acknowledgements

This work was funded by the National Natural Science Foundation of China (Grant No. 51301213, 51501026) and Chongqing Research Program of Basic Research and Frontier Technology (No. cstc2014jcyjA50008, cstc2014jcyj A50017).

References

1. T. Chen, Z.Y. Chen, L. Yi, J.Y. Xiong and C.M. Liu, Effects of texture on anisotropy of mechanical properties in annealed Mg–0.6%Zr–1.0%Cd sheets by unidirectional and cross rolling, *Mater. Sci. Eng. A*.615, 324 (2014).
2. A.J. Carpenter, A.R. Antoniswamy, J.T. Carter, L.G. Hector Jr and E.M. Taleff, A mechanism-dependent material model for the effects of grain growth and anisotropy on plastic deformation of magnesium alloy AZ31 sheet at 450 °C, *Acta Mater*.68, 254 (2014).
3. Y.Q. Cheng, Z.H. Chen and W.J. Xia, Effect of crystal orientation on the ductility in AZ31 Mg alloy sheets produced by equal channel angular rolling, *J. Mater. Sci*.42, 3552 (2007).

4. J.Suh, J.Victoria-Hernandez, D.Letzig, R.Golle, S.Yi, J.Bohlen and W.Volk, Improvement in cold formability of AZ31 magnesium alloy sheets processed by equal channel angular pressing, *J. Mater. Process Tech.*217, 286 (2015).
5. W.Y. Kim, W.J. Kim, Fabrication of ultrafine-grained Mg–3Al–1Zn magnesium alloy sheets using a continuous high-ratio differential speed rolling technique, *Mater. Sci. Eng. A.*594, 189 (2014).
6. Y. Onuki, K. Hara, H. Utsunomiya and JA. Szpunar, High-Speed Rolling of AZ31 Magnesium Alloy Having Different Initial Textures, *J. Mater. Eng. Perform.*24, 972 (2015).
7. S.Q. Zhu, H.G. Yan, J.H. Chen, Y.Z. Wu, J.Z. Liu and J. Tian, Effect of twinning and dynamic recrystallization on the high strain rate rolling process, *Scripta Mater.*63, 985(2010).
8. X.S. Huang, K. Suzuki, Y. Chino and M. Mabuchi, Influence of rolling temperature on static recrystallization behavior of AZ31 magnesium alloy, *J. Mater. Sci.*47, 4561 (2012).
9. X.S. Huang, K. Suzuki and Y. Chino, Influences of initial texture on microstructure and stretch formability of Mg–3Al–1Zn alloy sheet obtained by a combination of high temperature and subsequent warm rolling, *Scripta Mater.*63, 395 (2010).
10. X.S. Huang, K. Suzuki, Y. Chino and M. Mabuchi, Influence of aluminum content on the texture and sheet formability of AM series magnesium alloys, *Mater. Sci. Eng. A.* 633, 144 (2015).
11. Y. Chino, J.S. Lee, K. Sassa, A. Kamiya and M. Mabuchi, Press formability of a rolled AZ31 Mg alloy sheet with controlled texture, *Mater. Lett.* 60, 173 (2006).
12. W.Q. Tang, S.Y. Huang, D.Y. Li and Y.G. Peng, Mechanical anisotropy and deep drawing behaviors of AZ31 magnesium alloy sheets produced by unidirectional and cross rolling, *J. Mater. Process Tech.* 215, 320 (2015).
13. T. Al-Samman and G. Gottstein, Influence of strain path change on the rolling behavior of twin roll cast magnesium alloy, *Scripta Mater.* 59, 760 (2008).
14. Y. Chino, K. Sassa, A. Kamiya and M. Mabuchi, Stretch formability at elevated temperature of a cross-rolled AZ31 Mg alloy sheet with different rolling routes, *Mater. Sci. Eng. A.* 473, 195 (2008).
15. X. Li, T. Al-Samman and G. Gottstein, Mechanical properties and anisotropy of ME20 magnesium sheet produced by unidirectional and cross rolling, *Mater. Design* 32, 4385 (2011).

16. H. Zhang, G..S. Huang, H.J. Roven, L.F. Wang and F.S. Pan, Influence of different rolling routes on the microstructure evolution and properties of AZ31 magnesium alloy sheets, *Mater. Design* 50, 667 (2013).
17. H. Zhang, G.S. Huang, L.F. Wang, H.J. Roven and F.S. Pan, Enhanced mechanical properties of AZ31 magnesium alloy sheets processed by three-directional rolling, *J. Alloys Compd.* 575, 408 (2013).
18. X.P. Chen, D. Shang, R. Xiao, G.J. Huang and Q. Liu, Influence of rolling ways on microstructure and anisotropy of AZ31 alloy sheet, *Nonferrous Met. Soc. China* 20, s589 (2010).
19. X.S. Huang, K. Suzuki, Y. Chino and M. Mabuchi, Improvement of stretch formability of Mg–3Al–1Zn alloy sheet by high temperature rolling at finishing pass, *J. Alloys Compd.* 509, 7579 (2011).
20. T. Zhou, Z. Yang, D. Hu, T. Feng, M.B. Yang and X.B. Zhai, Effect of the final rolling speeds on the stretch formability of AZ31 alloy sheet rolled at a high temperature, *J. Alloys Compd.* 650, 436 (2015).
21. X.S. Huang, K. Suzuki, A. Watazu, I. Shigematsu and N. Saito, Improvement of formability of Mg–Al–Zn alloy sheet at low temperatures using differential speed rolling, *J. Alloys Compd.* 470, 263 (2009).
22. X.S. Huang, K. Suzuki, A. Watazu, I. Shigematsu and N. Saito, Mechanical properties of Mg–Al–Zn alloy with a tilted basal texture obtained by differential speed rolling, *Mater. Sci. Eng. A.* 488, 214 (2008).
23. W.J. Kim, S.J. Yoo, Z.H. Chen and H.T. Jeong, Grain size and texture control of Mg–3Al–1Zn alloy sheet using a combination of equal-channel angular rolling and high-speed-ratio differential speed-rolling processes, *Scripta Mater.* 60, 897 (2009).
24. J. Wang, D.X. Zhang, Y. Li, Z.Y. Xiao, J.P. Fouse and X.Y. Yang, Effect of initial orientation on the microstructure and mechanical properties of textured AZ31 Mg alloy during torsion and annealing, *Mater. Design* 86, 526 (2015).
25. J. Koike, T. Kobayashi, T. Mukai, H. Watanabe, M. Suzuki, K. Maruyama and K. Higashi, The activity of non-basal slip systems and dynamic recovery at room temperature in fine-grained AZ31B magnesium alloys, *Acta Mater.*51, 2055 (2003).
26. H.W. Dong, F.S. Pan, B. Jiang, J.H. Dai and Q.S. Yang, Anisotropy of the extruded and heat-treated Li containing AZ31 magnesium alloys, *J. Alloys Compd.* 590, 233 (2014).
27. J.J. Ma, X.Y. Yang, H. Sun, Q.H. Huo, J.Wang and J. Qin, Anisotropy in the mechanical properties of AZ31 magnesium alloy after being compressed at high temperatures (up to 823 K), *Mater. Sci. Eng. A.* 584, 156 (2013).

Effects of high pressure heat treatment on micro-mechanical properties of TC6 alloy

Jing Li[a,*] and Guo-Li Liang
Department of mechanical and electrical engineering, Tangshan University
Tangshan, 063000, China
E-mail: [a]lijing318@126.com

Yan Chen
Department of Environmental and Chemical Engineering, Yanshan University
Qinhuangdao, 066004, China
E-mail: 108318044@qq.com

Nano indentation technique is an effective means to measure the mechanical properties of materials, which calculates the mechanical properties through the load - indentation depth curve. This measurement can eliminate some human error and ensure a more accurate value of the test. The micromechanical properties of TC6 titanium alloy before and after high pressure treatment were measured by the nanoindentation technique and the microstructures were analyzed by optical microscopy and transmission electron microscopy. The effects of high pressure heat treatment on micromechanical properties were also discussed in the paper. The results show that high pressure treatment can increase the hardness, elastic modulus and hard shells ratio of TC6 titanium alloy, and decrease its creep rate. This is because of high pressure treatment that can refine the microstructure and increase internal dislocations density of TC6 titanium alloy.

Keywords: TC6 Titanium Alloy; High Pressure Heat Treatment; Nano Indentation; Micromechanical Properties.

1. Introduction

High pressure can promote nucleation and inhibit grain growth[1-2], Using of high pressure to improve microstructure of materials has attracted people's attention. According to the literature [3-4], high pressure heat treatment can refine the microstructure of steel, copper and aluminum alloy, thereby improve its mechanical properties. But there are seldom studies on application of aluminum alloy by high pressure treatment. Because of TC6 alloy has the

* Corresponding author.

advantages of low density, high strength, good resistance of corrosion and high temperature, it is the most widely used [5].With the development of modern industry, the mechanical properties of TC6 alloy has higher requirements. Therefore, using high pressure treatment to improve the mechanical properties of TC6 titanium alloy is a worthy subject. In addition, nano indentation technique is an effective means to measure the mechanical properties of materials, which calculates the mechanical properties through load - indentation depth curve. This measurement can eliminate some human error and ensure a more accurate value of the test [6]. In this paper, TC6 titanium alloy was treated by high pressure and its effects on mechanical properties were investigated by nano indentation technique.

2. Experimental

TC6 titanium alloy was annealed and its chemical composition in accordence with GB/T 3620.2-2007 standards. The samples of TC6 titanium alloy with the size of Φ8mm×10mm were sealed into graphite heaters, and then were pressurized to 4GPa and heated to 1000℃ for 20 min on the CS-IB type six-anvil high-pressure equipment which uses pyrophyllite as the pressure-transmitting medium. The pressure was released after the samples were cooled down to room temperature. The micro-mechanical experiments were performed on a triboindenter nano-mechanical test instrument. Berkovich-type indenter with curvature radius of 150nm was chosen to determine the hardness and elastic modulus, in which the maximum applied load was 3000μN and kept this load for 10s, using a loading and unloading rates of 100μN/s. In addition, the changing values of the specimen's hardness and elastic modulus with indenting depth can be obtained by continuous loading on a certain area of the sample's surface. In this experiment, the data was the mean value of five measured results. The microstructures of the TC6 titanium alloy before and after high pressure heat treatment were observed by means of Axiovert200MAT optical microscope and JEM-2010 transmission electron microscopy.

3. Results and discussion

3.1. *Microstructures*

Figure 1 shows that the microstructures of TC6 titanium alloy before and after high pressure heat treatment. It can be seen that the microstructure of TC6 titanium alloy is combined α phase with β phase. The grains before high pressure treatment are rather coarse, however, the grains become refined and more homogeneous after high pressure heat treatment. TEM observations (Fig. 2) find that pre-pressure processing TC6 titanium alloy has a thick α phase, which is arranged in parallel. However, after high pressure treatment, α-phase have a varied shapes that are randomly arranged with the dislocation density increasing. This is due to the high pressure can make TC6 titanium alloy turn out plastic deformation violently, which results in dislocation slipped and proliferated and finally the dislocation density increases. At the same time these dislocations and grain boundaries will provide more site for the new nuclei, bringing about an increase in the number of newly formed nuclei, for the reason that the microstructure of TC6 titanium alloy is refinement.

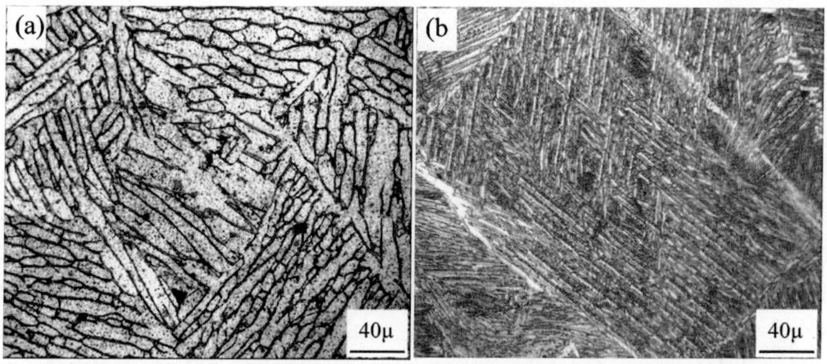

Fig.1. Microstructure of TC6 titanium alloy: (a) original; (b) 4GPa treatment

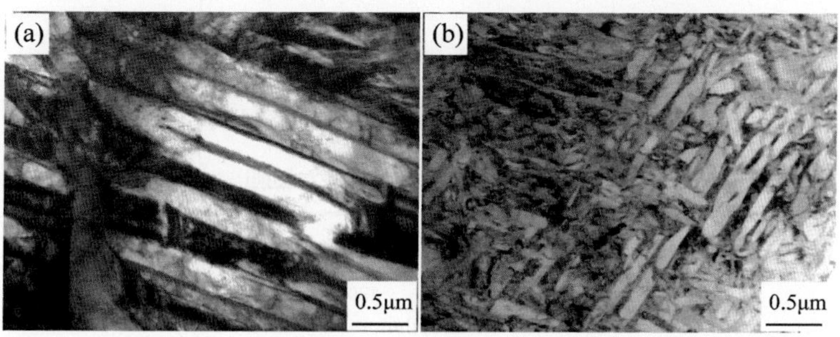

Fig.2. TEM images of TC6 titanium alloy: (a) original; (b) 4GPa treatment

3.2. *Micro-mechanical property*

Figure 3 shows the load-displacement curves of the TC6 titanium alloy before and after 4GPa high pressure heat treatment. Both the loading and unloading curves show no superposition, indicating that the elastic deformation takes place inside the alloy. Testing results are tabulated in Table 1. The nanoindentation hardness (H), elastic modulus(Er) and the ratio of hardness to modulus(H/E) are increased after high pressure treatment , but indentation depth is decreased. After 4GPa high pressure treatment, the H ,Er and H/E of TC6 titanium alloy are 5.29GPa, 124.36GPa and 0.043, respectively, that increase 12.31%, 4.72% and 7.50% compared with pre-high pressure treatment. While the maximum indentation depth and residual indentation depth is 147.81nm and 97.24 nm, respectively, which is decreased 18.53% and 4.78% than before high pressure treatment. It can also be seen from the Figure 3 that there is a tiny platform in the two curves, indicating that TC6 titanium alloy take place creep that high pressure treatment can reduce its rate. The creep amount before and after high pressure treatment of TC6 titanium alloy is 4.99nm and 3.57nm, respectively. The creep rates are 0.499nm/s and 0.357nm/s, wherefore the creep rate of after high pressure treatment is decreased 28.46%. The dependences of nanohardness and elastic modulus on the indent depth are shown in Figure 4. The nanohardness decreases gradually with the indent depth, which indicates that the size effect phenomenon exists in the case of the nanohardness measurement[7]. However, the elastic modulus exhibits fluctuant trend with increasing depth, which indicates that the size effect phenomenon does not exist in the elastic modulus measurement. Although the range of nano indentation is smaller, the results of this paper are averaged by multi-point, it can be said that the experimental data can reflect the average micro-mechanical property of samples.

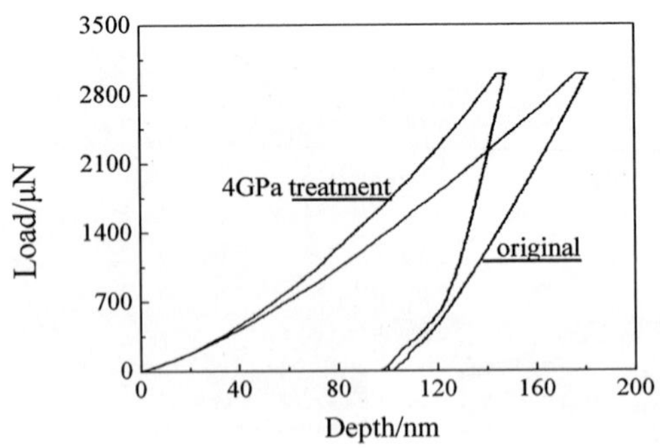

Fig.3. Load-displacement curves for nano-indentation tests of TC6 titanium alloy

Table 1. Nanoindentation results of TC6 titanium alloy

Samples	H/GPa	Er/GPa	H/Er	h_{max}/nm	h_f/nm
Original	4.71	118.76	0.040	181.43	102.12
4GPa treatment	5.29	124.36	0.043	147.81	97.24

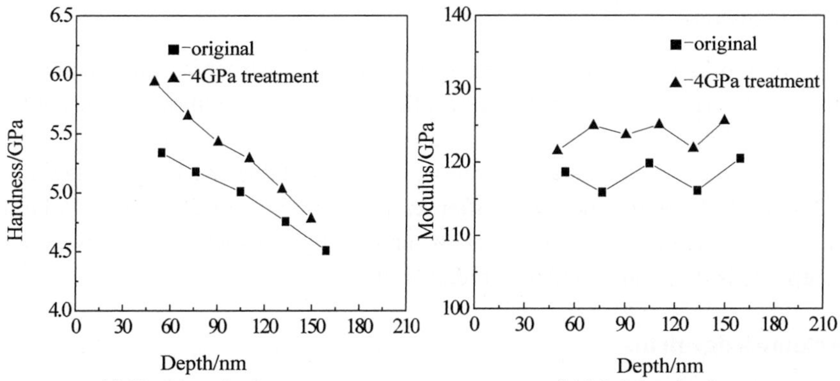

(a) Hardness-depth curves (b) Modulus-depth curves

Fig.4. Hardness, elastic modulus profile curves of TC6 titanium alloy along depth

High pressure treatment can refine the microstructures of TC6 titanium alloy, increase the dislocation density and compactness, strengthen grain refinement and dislocation strengthening effect, while improve the ability to resist deformation. Therefore, the hardness of TC6 titanium alloy is increased after the high pressure treatment. As the indentation depth and width increased, the nucleation rate of dislocation under the indenter will be decreased and its deferred. It is resulted that the dislocation density declined and dislocation strengthening effect weakened, consequently the nanohardness indentation size effect phenomenon was appeared. For the elastic modulus, the relationship between elastic modulus and material porosity is[8]: E =E0(1-1.9P+0.9P2). Where E is the elastic modulus of material, P is the porosity (0<P<1), E0 is elastic modulus without pore. These facts imply that the more porosity in the material, the lower compactness and smaller elastic modulus. High pressure treatment can increase the density of TC6 titanium alloy and reduce porosity. Therefore, the elastic modulus after high pressure treatment is moved up.

It has been confirmed that high pressure treatment can improve resistance to plastic deformation and creep in TC6 titanium alloy, which is due to high pressure processing can refine the microstructures, making the grain boundary that impeded the dislocation motion increased per unit volume. Furthermore, there are high density of dislocations after high pressure treatment, causing dislocation delivery and tangles during the dislocation motion. Consequently, the movement of dislocation is blocked strongly and the creep rate is decreased, the resistance to plastic deformation and creep properties of TC6 titanium alloy are increased.

4. Conclusions

(1) High pressure treatment can refine the microstructure and increase the dislocation density of TC6 titanium alloy, therefore it improves the hardness, elastic modulus and the resistance of creep.

(2) The nano indentation hardness, elastic modulus and hard shells ratio are 5.29Gpa, 124.36GPa and 0.043 after keeping 1000℃ for 20 min at 4GPa, that have been increased 12.31%, 4.72% and 7.50%, respectively. However, the creep rate is 0.357nm/s which is reduced 28.46%.

Acknowledgements

This work is supported by the natural science foundation of Hebei Province, China (E2014203135) and Independent research project of YanShan University Young teachers, China (14LGA015).

References

1. H.Y. Wang, Y. Chen, J.H. Liu and W.K. Wang. *Effects of 3GPa high pressure treatment on the $\beta 1 \rightarrow \alpha + \beta 2$ phase transition dynamics of TC4 titanium alloy. Chinese Physics Letters*, 30(3):036202-1(2013).
2. Wang Zhi-xin, Lu Jin-bin, Xi Yan-jun. Effects of quenching under high pressure on crystallization of Cu60Zr20Hf10Ti10 bulk metallic glass. *Materials Science and Engineering A*, 499(1-2):192(2009)
3. L.L. Wu, L. Liu, J.H. Liu and R.J. Zhang. Effects of High Pressure Heat Treatment on Microstructure and Micro-mechanical Properties of Cu77.96Al22.04 Alloy. *Materials Transactions*, 53(3):504(2012).
4. Y.Q. Ma, J.M. Li and H.J. Lin. Effects of 4 GPa pressure heat treatment on micro-mechanical properties of brass by nanoindentation. *Materials Transactions*, 55(7): 1065(2014).
5. Q.L. Niu, M. Chen, W.W. Ming and Q.L. An. Evaluation of the performance of coated carbide tools in face milling TC6 alloy under dry condition, International. *Journal of Advanced Manufacturing Technology*, 64(5-8): 623(2013).

6. Y. Li, R.K. Kang, H. Gao, J.H. Wang and Y.J. Lang. Nanomechanical behaviors of (110) and (111) CdZnTe crystals investigated by nanoindentation. *Rare Metals*, 28(6): 570(2009).
7. C.S. Zhai, L. Yang and J. Wang. Microstructure and nanoindentation mechanical properties of detonation sprayed coatings derived from $A1_2O_3/TiO_2$ nanocomposite powders. *Journal of Inorganic Materials*, 20 (6):1500(2005).
8. J.P. Chu, J.M. Rigsbee, G. Banas., et al. Laser-shock processing effects on surface microstructure and mechanical properties of low carbon steel. *Materials Science and Engineering A*, 260 (1-2): 260(1999).

Synthesis carbon doped titanium dioxide by sol-gel method

Zhi Wang[a,*], Xing-hui Li[b] and Chun-hong Yu[c]
School of Safety Engineering, Shenyang Aerospace University,
Liaoning, Shenyang, 110136, China
[a]*ychwz@sohu.com,* [b]*lixinghui77@sohu.com,* [c]*yuxiangci@sohu.com*

Titanium dioxide (TiO_2) is attractive because of their excellent photocatalytic activity, low cost, non-toxicity and good stability. Carbon doped titanium dioxide (TiO_2) was prepared using tetrabutyl titanate (TBT) as the source of titanium and tetrabutyl ammonium hydroxide (TABH) as the source of carbon by sel-gol method. The effect of carbon doping on the photocatalytic properties of titanium dioxide was investigated by the degradation of methyl orange under the light of mercury lamp and xenon lamp. The results show that carbon doping can improve the photocatalytic capacity of titanium dioxide under visible light and the optimum TABH additive amount was achieved.

Keywords: Titanium Dioxide, Carbon Doped, Sol-gel Method; Photocatalysis;

1. Introduction

Titanium dioxide (TiO_2) is attracting great interest for their excellent photocatalytic activity, low cost, non-toxicity and good stability. It has been considered the most potential application value of air purification materials [1-3].However, the photocatalysis of titanium dioxide can only be activated by ultraviolet light. The ultraviolet light constitutes only about 5 percent of the total energy emitted from the sun which restricting the applications of titanium dioxide.

The people have found some methods to improve the TiO_2 photocatalytic activity under visible light, some result show that doped impurity particles can reduce band gap and improve photocatalytic activity of TiO_2 under visible light. At present, the doped particles mainly include metallic and nonmetallic particles, because metal particles will affect the TiO_2 photocatalytic activity under ultraviolet range [4-6], so the non-metallic particles doping become a research hotspot in recent years. The non-metallic elements generally include B, N, S, and C [7-13].

---

* Corresponding author.

In the paper, we have synthesized of carbon doped titanium dioxide by sol-gel method and TABH as the source of carbon. The effects of the additive amount on photocatalytic characteristics of TiO_2 under visible and ultraviolet light were studied.

2. Experimental

Carbon doped TiO_2 were prepared using tetrabutyl titanate as titanium source and TABH as the source of carbon by sel-gol method. Firstly, 0.06 mol of tetrabutyl titanate was dissolved in 18.4 ml of ethanol, fully stirring for 10 min to make solution A. 9.2 ml of anhydrous ethanol was mixed with 10 ml of glacial acetic acid and 1.1 ml of deionized water, fully stirring for 10 min to make solution B. The solution A was added drop-wise to solution B under vigorous stirring for 10 min at room temperature. Then, the suitable amount of TABH was added into the resulting solution respectively (0.001 mol, 0.002 mol, 0.003 mol, 0.004 mol and 0.005 mol), and nitric acid was used to adjust the pH value of the solution (pH = 3), the samples were marked as S1, S2, S3, S4 and S5. The doped solution was stirred continuously for 30 min, then the solution was transferred into a thermostat water bath at 40 °C for gel. The wet gel was dried at 80°C for 8 h. Calcinations were carried out in a electric box furnace by heating the system up to 450°C at a rate of 5°C /min and maintaining at the final temperature for 120 min, then the reactor was annealed naturally to room temperature. After calcinations, the carbon doped TiO_2 and un-doped pure TiO_2 powders were obtained.

The photocatalytic properties of samples were characterized by photodegradation of methyl orange (MO) under the light of mercury lamp and xenon lamp. In this experiment, mercury lamp was used to provide ultraviolet light and xenon lamp was used to provide visible light. Methyl orange was dissolved in deionized water to obtain a methyl orange solution and the solution concentration is $20mg.L^{-1}$. Then 5 mg of each samples were add into test tube that contained 30 ml of methyl orange solution and mixed by magnetic stirrer. Solutions containing different samples were left under dark for 40 min in order to establish the adsorption-desorption equilibrium between dye and catalyst surface. Subsequently each solution was irradiated with 500 W mercury lamp or 500 W xenon lamp under vigorous stirring. In every 40 or 80 min, the remaining methyl orange concentration was determined from its absorbance at 465 nm. The absorbance of methyl orange solution was tested using spectrophotometer (UV-2601).

3. Results and discussion

Table1 and Figure1 show the changes of absorbance values of different samples under the light of mercury lamp. Experimental results show that the methyl orange solution is very stable and not easy to degrade under mercury lamp irradiation. In contrast, the absorbance of methyl orange solution mixed with samples decrease quickly. Almost all of methyl orange was degraded according to the absorbance from table 1 when the reaction time over 120 min. The result means all sample have higher photocatalytic activity under ultraviolet light, whether carbon doped or pure TiO_2.

Table 1. The changes of absorbance values of samples under mercury lamp

Specimens	0 min	40 min	80min	120min	160min
MO	0.226	0.226	0.223	0.223	0.223
MO +TiO_2	0.227	0.059	0.032	0.018	0.005
MO +S1	0.225	0.054	0.024	0.012	0.007
MO +S2	0.226	0.042	0.019	0.008	0.004
MO +S3	0.222	0.055	0.028	0.014	0.007
MO +S4	0.226	0.060	0.036	0.019	0.009
MO +S5	0.226	0.064	0.040	0.022	0.006

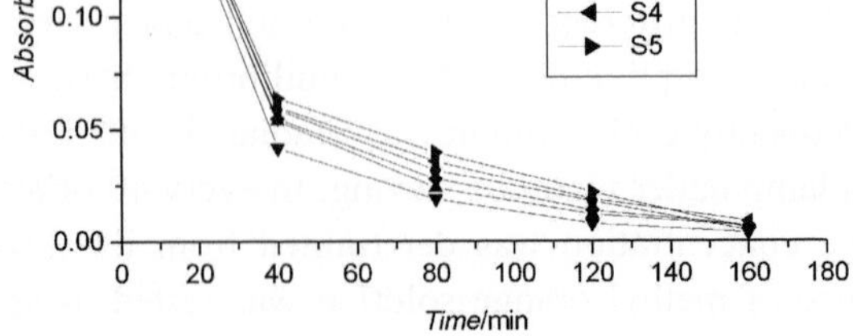

Fig. 1 Curve of absorbance of methyl orange of different specimens under mercury lamp

Table 2 shows the changes of absorbance values of different samples under the light of xenon lamp. The photocatalytic degradation curves of methyl orange obtained for all of the synthesized samples are given in Fig.2. The results show that the absorbance of methyl orange solution decreases slowly and the degradation rates vary with the additive amount of TABH.

From the figure 2 and table 2, we can see that when the additive amount of TABH is 0.002mol (sample S2), the ability of degradation of methyl orange is the strongest and the degradation rate achieves 16% after six hours. That is to say the optimum mole ratio of TBT and TABH is 30:1in this experiment. Under the same conditions, the degradation rate of pure TiO_2 is only 9%. So the carbon doping modifies and improves the TiO_2 photocatalytic activity under visible light, the light response range area may be expanded to sunlight area by ultraviolet light.

Table 2. The changes of absorbance values of samples under xenon lamp

Specimens	0 min	80 min	160min	240min	320min
MO	0.226	0.223	0.222	0.222	0.221
MO +TiO_2	0.227	0.224	0.224	0.220	0.211
MO +S1	0.223	0.220	0.211	0.209	0.204
MO +S2	0.222	0.213	0.206	0.201	0.188
MO +S3	0.227	0.216	0.209	0.206	0.195
MO +S4	0.226	0.219	0.215	0.209	0.202
MO +S5	0.225	0.221	0.219	0.208	0.202

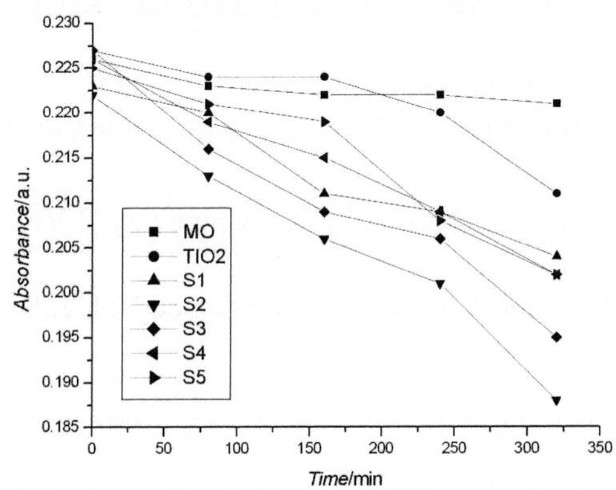

Fig. 2. Curve of absorbance of methyl orange of different specimens under xenon lamp

Conclusions

We have studied the effects of carbon doping on the photocatalytic properties of TiO_2. The result shows the optimum mole ratio of TBT and TABH is 30:1 in this experiment. Carbon doping can improve the TiO_2 photocatalytic activity under visible light. At the same time, carbon doping not effect the photocatalytic activity of TiO_2 under ultraviolet light, which will better than the metal doping did.

Acknowledgements

This work was financially supported by the Liaoning science public research funds (GY2014-C-005).

References

1. A.Fujishima, K. Honda: Nature Vol. 238 (1972), p. 37.
2. A.T.Hodgson, H.Destaillats, D.P. Sullivan: Indoor Air Vol. 17 (2007), p. 305.
3. K.Nakata, A.Fujishima: Journal of Photochemistry and Photobiology C: Photochemistry Reviews Vol. 13 (2012), p. 169.
4. J.Q.Li, D. F.Wang, Z.Y.Guo: Applied Surface Science Vol. 263 (2012), p. 382.
5. H.Y.Liu, L.Gao : Chemistry Letters Vol. 33 (2004), p. 730.
6. Y.H.Peng, G.F.Huang, W.Q.Huang: Advanced Powder Technology Vol. 23 (2012), p. 8.
7. V.D. Binas, K.Sambani, T.Maggos, A.Katsanaki, G.Kiriakidis: Applied Catalysis B: Environmental Vol. 113 (2012), p. 79.
8. R.Asahi, T.Marikawa, T.Ohwaki: Science Vol. 293 (2001), p. 269.
9. H.Irie, Y.Watanabe, K.Hashimoto: Physical Chemistry B Vol. 107 (2003), p. 5483.
10. S.Sakthivel, H.Kisch: Science Vol. 4 (2003), p. 487.
11. J.L.Cole, J.D.Stout, C.Burda: Physical Chemistry B Vol. 108 (2004), p. 1230.
12. D.Oliver, L.Tracy, Z.Tyrkhon: Physical Chemistry B Vol. 108 (2004), p. 6004.
13. M.Behpour, V. Atouf: Applied Surface Science Vol. 258 (2012), p. 6595.

Predicting the strain-hardening behavior in metals

R. K. Nutor*† and Y. Z. Fang

College of Mathematics, Physics and Information Engineering, Zhejiang Normal University,
321004 Jinhua, China.
E-mail: †ray19gh@yahoo.com

When a metal is subjected to increasing applied loads below its yield point, the strains are initially often proportional to the stress, and the material returns to its original shape or dimensions when the load is removed, which is an elastic behavior of the metal. Stress – strain values obtained from tensile tests of aluminium and steel is used to evaluate the true stress – true strain values. The Hollomon's model is then used to predict the strain-hardening behavior in the two specimens. It is clearly seen that the strain-hardening behavior in metals can be described using the Hollomon's model. However, we have assumed that the onset of strain-hardening is at the yield point up until the ultimate tensile strength. The correlation between the experimental true stress – true strain values of aluminium and the calculated values using the Hollomon equation is much better than that of steel..

Keywords: Strain-hardening; Hollomon's model; Tensile strength;

1. Introduction

When a metal is subjected to increasing applied loads below its yield point, the strains are initially often proportional to the stress, and the material returns to its original shape or dimensions when the load is removed, this is an elastic behavior of the metal [1]. As the load is increased, the loads that exceed the yield point cause disproportionality between the strains and the loads. Upon unloading, the strains are not recovered completely. This irreversible deformation characterizes plastic deformation and where the strain-hardening phenomena sets in [2].

The strain-hardening phenomenon is explained on the basis of dislocation–dislocation strain field interactions. The dislocation density in a metal increases with deformation or cold work, due to dislocation multiplication or the

* Corresponding author.

formation of new dislocations. Consequently, the average distance of separation between dislocations decreases. The dislocations are positioned closer together. On the average, dislocation–dislocation strain interactions are repulsive. The net result is that the motion of a dislocation is hindered by the presence of other dislocations. As the dislocation density increases, this resistance to dislocation motion by other dislocations becomes more evident. Thus, the imposed stress necessary to deform a metal increases with increasing cold work [3].

Over the years, several flow relationships have been proposed [5-11] to describe the tensile flow and strain-hardening behavior in metals.

The Hollomon's equation which is a power law relating the true strain to the true stress is given by

$$\sigma_T = K\varepsilon_T^n \tag{1}$$

Where σ_T is the true stress, ε_T is the true strain, n is the strain-hardening exponent and K is the strength coefficient. In the Hollomon's expression, the strain-hardening exponent measures the ability of a metal to strain-harden, larger magnitudes indicate larger degrees of strain hardening.

The Kocks-Mecking [11,12] model assumes a constant strain rate for the evolution of dislocation structure with strain. The strain-hardening is controlled by packing and reordering of dislocations which are assumed to superimpose in an additive manner. The total elongations in a material during deformation has been known to originate from elastic and plastic contributions. Therefore, the plastic strain can be evaluated according to the expressions in equations 2-4.

$$\varepsilon_T = \varepsilon_p + \varepsilon_e \tag{2}$$

Where ε_T is the total strain in the metal, ε_p is the plastic strain and ε_e is the elastic strain.

However, the elastic strain is related to the elastic stress and elastic modulus according to Hooke's law given by:

$$\varepsilon_e = \frac{\sigma}{E} \tag{3}$$

Therefore, the plastic strain is given as

$$\varepsilon_p = \varepsilon_T - \frac{\sigma}{E} \tag{4}$$

The stain-hardening rate (θ) is therefore given as;

$$\theta = \frac{d\sigma}{d\varepsilon_p} = \theta_o \left[1 - \frac{\sigma}{\sigma_s}\right] \tag{5}$$

where $\frac{d\sigma}{d\varepsilon_p}$ is the strain-hardening rate, θ_o is the strain hardening coefficient which corresponds to the initial strain-hardening rate at σ=0 and σ_s is the saturation stress at high strains corresponding to conditions when the strain-hardening rate is zero.

However, in this paper we use the Hollomon's model to predict the strain-hardening behavior in aluminium and steel.

2. Results and discussion

In predicting the strain-hardening behavior using the Hollomon model, we used stress-strain data obtained from tensile tests of steel and aluminium at room temperature. We have assumed that strain-hardening primarily begins after the yield point of the metal and ends before ductile fracture. Therefore, we considered stress-strain values from the yield point up to the ultimate tensile stress. It is worth mentioning that since the power law is expressed in terms of true stress and true strain we calculated the true stress and true strain before plotting the values.

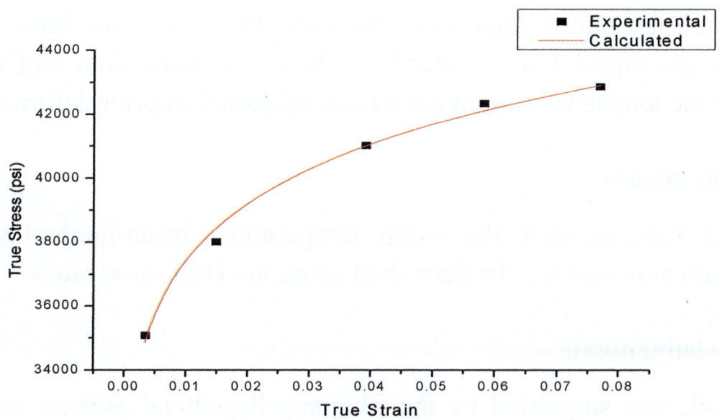

Fig. 1. The Strain-hardening behavior of Aluminium

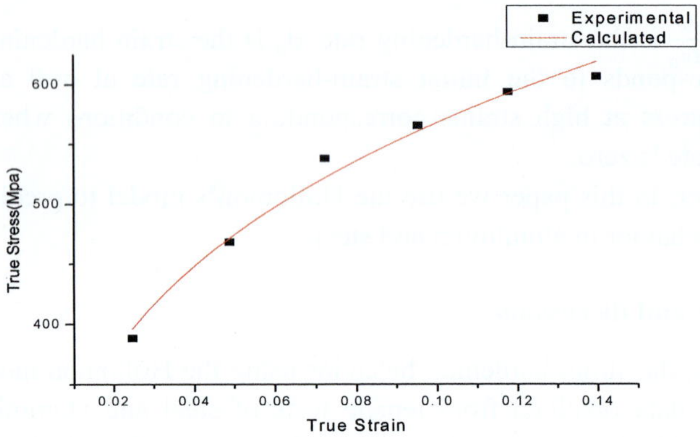

Fig. 2. The Strain-hardening behavior of Steel

Figure 1 and Figure 2 show a comparison of the experimental strain-hardening behavior and the calculated strain-hardening behavior using the Hollomon's model. It was seen that the correlation between the experimental and calculated values was better for aluminium as compared to steel. This can be attributed to steel being an alloy and therefore its deformation behavior is determined by more than one element. Therefore, we have assumed that although the correlation is somehow close, the aluminium and steel specimen used for the tensile test might have been subjected to prior deformation.

3. Conclusions

Analysis indicate that the room temperature strain-hardening behavior in aluminium and steel can be described using the Hollomon model.

Acknowledgements

This work was supported by the Zhejiang Provincial Science and Technology Key Innovation Team (No. 2011R50012), Zhejiang Provincial Key Laboratory (No. 2013E1022) and Project 973 corpus (2012CB825705). The author would also like to thank Professor Samuel Kwofie and Professor Yun Zhang Fang for reading through the manuscript and providing some editorial propositions.

References

1. Mikell, P. Groover, Fundamentals of Modern Manufacturing Materials, Processes and Systems (2010), pp. 383-442.

2. Donald K. Askeland, Pradeep P. Fulay, and Wendelin J. Wright, The Science and Engineering of Materials (2010), pp. 198-307.
3. William D. Callister Jr., Materials Science and Engineering – An Introduction (2007), pp. 131-195.
4. Dowling N.E, Mechanical Behavior of Materials: Engineering Methods for Deformation, Fracture and Fatigue, Prentice-Hall International, New Jersey (1993), pp. 555-690.
5. P. Ludwik, Elements der Technologischen Mechanik 32, Verlag Von Julius Springer, 1909 (Leipzig).
6. J.H. Hollomon, Tensile deformation, Trans. AIME 162 (1945), pp. 268-290.
7. E. Voce, The relationship between stress and strain for homogeneous deformation, J. Inst. Met. 74 (1948), pp. 537-562.
8. H.W. Swift, Plastic instability under plane stress, J. Mech. Phys. Solids 1 (1952) pp. 1-18.
9. E. Voce, A practical strain hardening function, Metallurgia 51 (1955), pp. 219-226.
10. D.C. Ludwigson, Modified stress-strain relation for FCC metals and alloys, Metall. Trans. 2 (1971), pp. 2825-2828.
11. U.F. Kocks, Laws for work-hardening and low-temperature creep, J. Eng. Mater. Technol. 98 (1976), pp. 76-85.
12. H. Mecking, U.F. Kocks, Kinetics of flow and strain-hardening, Acta Metall. Mater. 29 (1981), pp. 1865-1875.

Effect of HCPEB treatment on thermal fatigue resistance of 3Cr2W8V mold steel

Yuan-Fang Chen[a,b,1], Meng Tang[a] ,Tao Zhang[a] and Ding Xie[a]

[a]*College of Materials Science and Engineering, Chongqing University of Technology,*

[b]*Chongqing Key Laboratory of Mould Technology,*

Chongqing, 400054

E-mail:[1]cqitcyf@qq.com;

Tel: +86 15826091695

Thermal fatigue index could be the thermal cycle time that thermal fatigue cracks come out, and could also be the number and average length of fatigue cracks appeared after a certain amount of thermal fatigue cycle. The effect of HCPEB Treatment on the thermal fatigue resistance of 3Cr2W8V mold steel, conventional heat treatment, HCPEB treatment and HCPEB surface alloying were investigated. Thermal fatigue test was conducted by salt bath heat circularly test method. The results indicate that all sample after electron beam irradiation exhibited higher surface residual stress compared to the original sample. The original sample surface residual stress is 42MPa, direct electron beam irradiation sample surface residual stress is 626MPa and electron beam surface alloying Al sample surface residual stress is 597MPa. Moreover, it is found that this seems to increase micro-cracks length and biggest width by experiment. It could cause lower the thermal fatigue resistance. Furthermore, the morphology shows a smooth surface with scattered particles distributed in the amorphous matrix.

Keywords: 3Cr2W8V Mold Steel; HCPEB Treatment; Thermal Fatigue Resistance.

1. Introduction

In recent years, scholars had done a number of study on thermal fatigue behavior of mold steel in order to explore the way to improve its performance, and has

[1] Corresponding author.

achieved good results[2-4]. But the research on intense pulsed electron beam (HCPEB) on thermal fatigue performance were not much[5-6]. In this study, a heat treatment sample, HCPEB treatment sample and HCPEB surface alloying sample were chosen for investigating the effect of surface modification of electron beam thermal fatigue properties of 3Cr2W8V mold steel by thermal fatigue cycle test.

Thermal fatigue index could be the thermal cycle time that thermal fatigue cracks came out, it also could be the number and average length of fatigue cracks appeared after a certain amount of thermal fatigue cycle[7-8]. The test methods to determine the thermal fatigue properties of die steel were different at home and abroad such as Salt bath heat circularly testing method, flame heating testing method, Coffin thermal fatigue testing machine testing method and Uddeholm thermal fatigue testing method. In this study, the salt bath heat circularly test method was used to determine the change of the thermal fatigue properties of 3Cr2W8V mold steel by comparison. At the same time, The maximum width and the total length of the crack would be the standard of judging the thermal fatigue properties combining with the Uddeholm evaluation criteria[8].

2. Experimental Methods

The sample of 3Cr2W8V mold steel on conventional heat treatment, HCPEB treatment and HCPEB surface alloying were taken for the experiment. All samples need Conventional heat treatment and Polishing (Ra0.082μm) before Coating (Magnetron sputtering) and HCPEB treatment. The conventional heat treatment parameters are shown in Table 1 and the HCPEB treatment parameters are shown in Table 2.

Salt bath heat circularly test method: the sample was dipped in the salt bath furnace which temperature was kept at 600 (or ~ 750) °C for 1min and then cooled in water for 20s for one testing cycle. Each sample was investigated by 20, 40, 60 thermal fatigue cycle until the cracks caused. The thermal fatigue cycles when the surface cracks caused were used as the judging standard for thermal fatigue properties of 3Cr2W8V steel. The parameters of thermal fatigue test are shown in Table 3.The morphology of sample surface crack was observed by metallographic microscope after every 20 thermal fatigue cycles.

Table 1. Conventional heat treatment parameters of 3Cr2W8V mold steel

Chemical composition	Heat treatment parameters	Surface pretreatment method
C0.30~0.40,Cr2.20~2.70,W7 .50~9.0,V0.20~0.50,Si Mn≤0.40-wt%	Quench at 900,1050,1150,1250°C; stay 25min for austenitizing, oil cool, then temper 2h at 580 °C	Polishing

Table 2. Parameters of HCPEB process for 3Cr2W8V Mold Steel

Sample	Voltage	Γarget Source Distance	Bombard-ment Times	Thickness of Ceramic Coating
3Cr2W8V(heat treatment)	None	None	None	None
Cr2W8V(HCPEB treatment)	27KV	80mm	25	0μm
2W8V(HCPEB surface alloying t=2μm)	27KV	80mm	25	2μm

Table 3. Parameters of Thermal fatigue test

temperature in furnace	heating time	cooling time	Total Thermal fatigue cycle times
600 °C	1min	20s	60

3. Experiment Results

3.1. *Microstructure*

The morphologies of original sample and HCPEB treated sample are shown in Figure 1. The surface of original sample was smooth as shown in Figure 1(a). and the surface of HCPEB treatment sample had obvious crater phenomenon as shown in Figure 1(b). A lot of craters were appeared on the surface of HCPEB surface alloyed sample and no micro-cracks as shown in Figure 1(c).This could be that surface stress was reduced by the yield form-craters producing by HCPEB surface alloying.

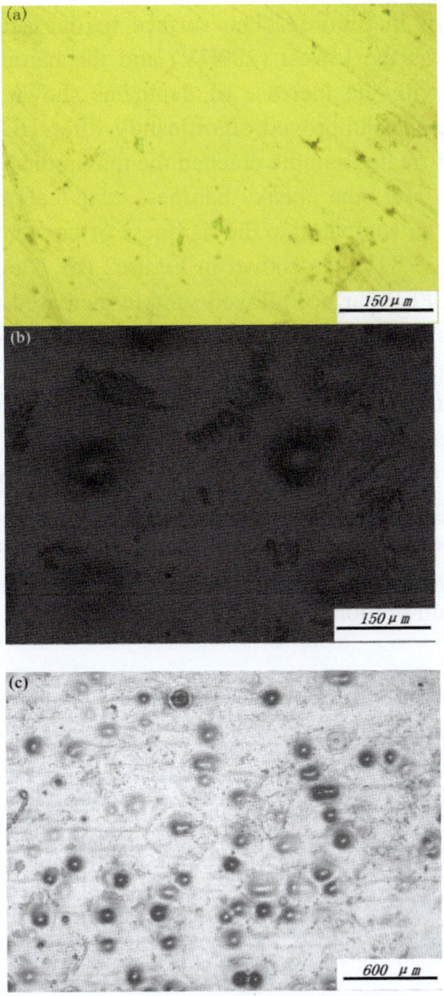

Fig.1. Surface morphology of 3Cr2W8V simple:
(a)Original Sample (b) HCPEB Treatment Sample (c) HCPEB Surface Alloying Sample

3.2. *Section micro-hardness*

The section micro-hardness was tested on the conventional heat treatment sample, HCPEB treatment sample and HCPEB surface alloyed sample, and the thickness of alloy layer pre-coating on HCPEB surface alloying (t) was 2μm.

The result is shown in Figure2. The surface hardness of conventional heat treatment sample was the lowest (289HV) and the hardness fluctuated in the range 330-350HV with the increase of depth, as shown in Figure2 (a). The hardness of sample was improved significantly after HCPEB treatment. The hardness of HCPEB treated sample reached the maximum at the depths from the surface of 45μm and the peak hardness was about 400HV, it was increased by 18.7% as compared to the hardness of conventional heat treatment at the same depth (337HV), as shown in Figure2 (b). The Figure2 (c) showed that hardness of HCPEB surface alloyed sample increased relatively low which reached the maximum value at the depths from the surface of 85μm and the peak hardness was about 383HV, it was increased by 8.4% as compared to the hardness of conventional heat treatment at the same depth (353HV). The result shown that reinforcement on hardness of HCPEB surface alloyed sample was poorer because the electron beam range at Al layer was largest (6μm), and the electron beam energy absorption rate of Al was lower than other metals, so the energy density of electron beam was diluted when it penetrated through Al layer. The result also showed the hardness had not been significantly improved in the depth of 105μm or more.

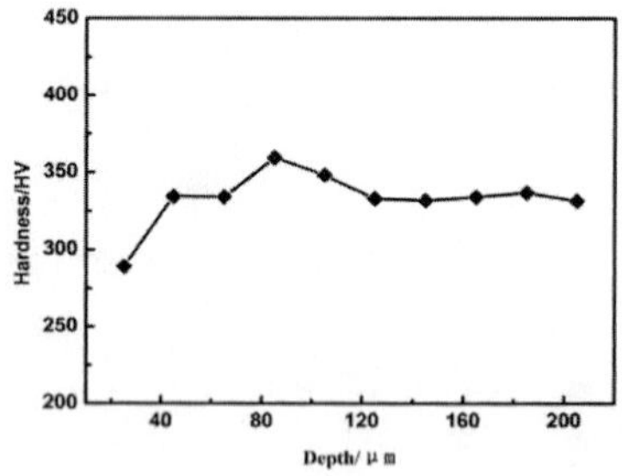

(a) Conventional heat treatment sample

Fig. 2 The changeable hardness of 3Cr2W8V sample with depth

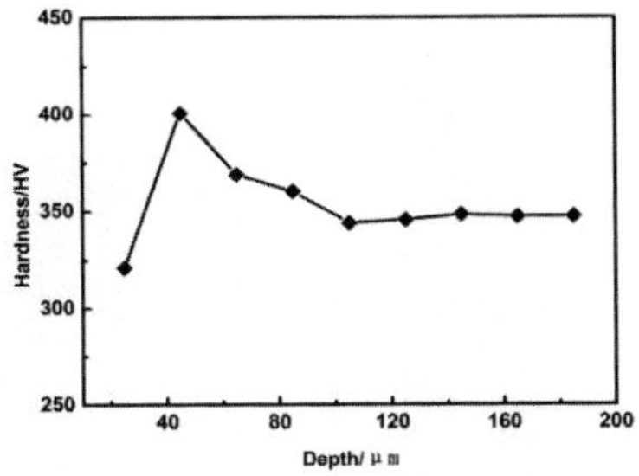

(b) HCPEB treatment sample

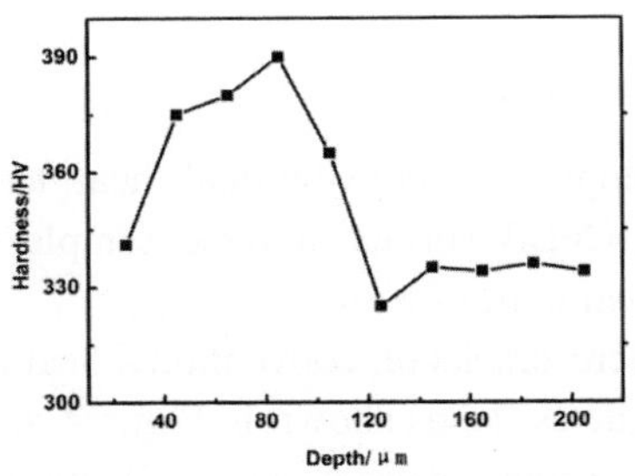

(c) HCPEB surface alloying sample(t=2μm)

Fig.2 The changeable hardness of 3Cr2W8V sample with depth

3.3. *Surface roughness*

The roughness change in the sample of different original roughness after HCPEB is shown in Figure3. The surface roughness of original sample was changed from 0.486μm to 0.455μm after HCPEB as shown in Figure3 (a), the reason may be that the HCPEB remelting the mechanical scratch of the sample surface, so that the surface roughness of sample was reduced. Another original sample roughness was changed from 0.076μm to 0.301μm after HCPEB as shown in Figure3 (b), the reason may be that a lot of uneven craters appeared on the sample surface after HCPEB, and they made the surface roughness significantly increased. So it can be seen that original roughness of sample had great effect on surface roughness of sample after HCPEB. HCPEB could increase the surface roughness for the smooth surface of steel and reduced the surface roughness for the rough steel.

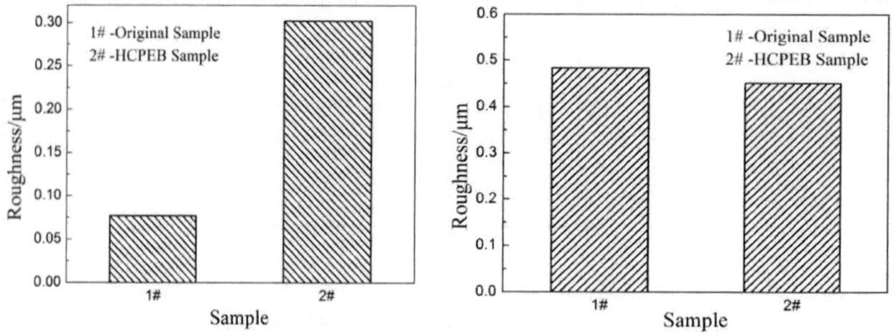

(a) Original sample, Ra=0.486μm (b) Original sample, Ra=0.076μm

Fig.3. The roughness comparison before and after HCPEB

3.4. *Thermal fatigue resistance*

The surface morphologies of conventional heat treatment sample, HCPEB treated sample and HCPEB surface alloyed sample after 20, 40, 60 thermal fatigue cycles are shown in Figure 4-6.

There were no micro-cracks on conventional heat treatment sample after 20, 40 times thermal fatigue cycle, as shown in Figure4. Only in the thermal fatigue cycle to reach 60 times sample there was a small amount of sparse fatigue crack, the maximum width of crack was 6.2μm and the total length of the cracks was 1333.5μm.

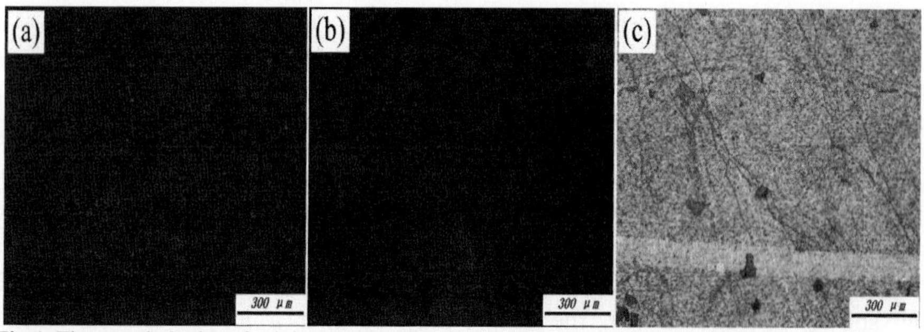

Fig.4. The morphologies of original 3Cr2W8V sample after thermal fatigue cycle test (Ra0.076μm): (a) 20, (b) 40, and (c) 60 cycles

There had been obvious cracks on HCPEB treatment sample after 20 times thermal fatigue cycle, as shown in Figure5. With the increase of the thermal fatigue cycle times, the total length and maximum width of cracks were increased. When the number of thermal fatigue cycle times was up to 60 times, the maximum width of cracks reached 68.9μm and the total length of the cracks was 10141.9μm.

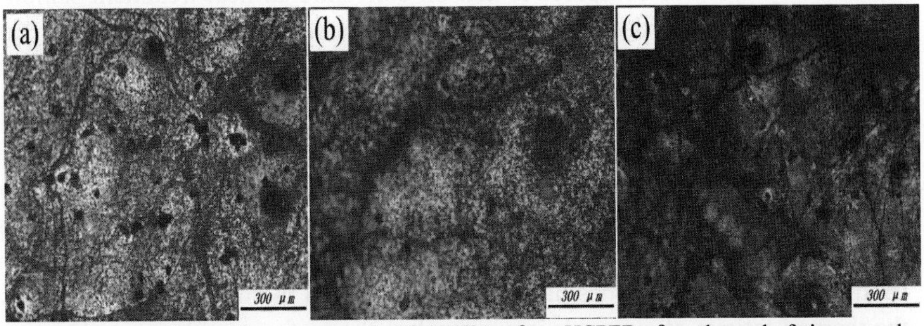

Fig.5. The metallographic picture of 3Cr2W8V after HCPEB for thermal fatigue cycle test:(a)Thermal fatigue cycle 20 times (b) Thermal fatigue cycle 40 times(c) Thermal fatigue cycle 60 times

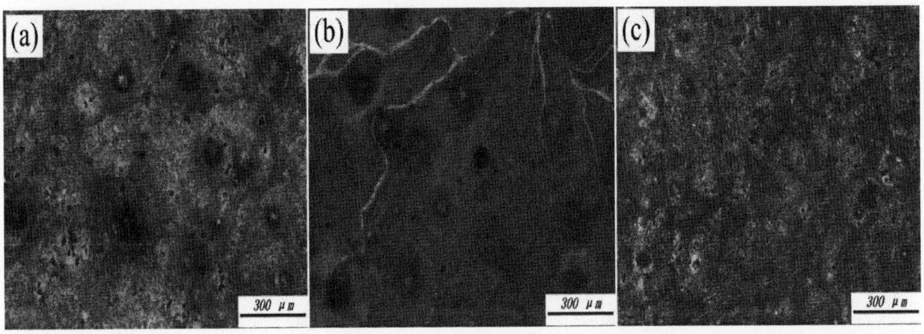

Fig.6. The metallographic picture of 3Cr2W8V processed with HCPEB and Al for thermal fatigue cycle test:(a) Thermal fatigue cycle 20 times (b) Thermal fatigue cycle 40 times (c) Thermal fatigue cycle 60 times

There also had been few cracks on HCPEB surface alloying sample after 20 times thermal fatigue cycle, as shown in Figure6. With the increase of the thermal fatigue cycle times, the total length and maximum width of cracks were increased. When the number of thermal fatigue cycle times was up to 60 times. The maximum width of cracks reached 24.5μm and the total length of the cracks was 5250μm.

The results showed that the maximum width of cracks on the HCPEB surface alloying sample was increased by 295% and the total length of the cracks was increased by 294% as compared to that of the conventional heat treated sample. Also the maximum width of cracks on the HCPEB treated sample was increased by 1011% and the total length of the cracks was increased by 661% as compared to that on conventional heat treatment sample. So thermal fatigue resistance of conventional heat treatment sample was the best, HCPEB surface alloying sample followed, and the HCPEB treatment sample was the worst.

4. Results and Discussion

The surface morphology, sectional hardness, surface roughness, residual stress and other properties of 3Cr2W8V sample is changed after electron beam irradiation and the properties have greater impact on thermal fatigue. Generally the thermal fatigue of materials is worse when the surface roughness and greater residual stress is larger. The craters on the material surface cause stress concentration; therefore, the thermal fatigue resistance of materials is decreased. Table 4 shows the surface roughness, residual stress, surface morphology of 3Cr2W8V sample after electron beam treatment.

Table 4. Thermal fatigue performance factors of 3Cr2W8V sample processed by HCPEB

Sample types	roughness	surface residual stress	surface topography
Conventional heat treatment sample	0.076 μm	42MPa	no craters
HCPEB treatment sample	0.301 μm	626MPa	craters
HCPEB surface alloying sample(Al, t=2μm)	0.274 μm	597MPa	craters

From the point of residual stress, the Table 4 shows the surface residual stress of 3Cr2W8V sample reaches 42MPa after conventional heat treatment, each sample residual stress increases to 626MPa after HCPEB treatment, the residual stress reaches 597MPa when use HCPEB surface alloying Al. These data close to the dynamic yield limit of 3Cr2W8V which is 750MPa. Because of the residual stress increase, the material tend to produce more micro-cracks under thermal cycling shock, so that the sample surface micro-cracks by electron beam irradiation after thermal fatigue cycle test both length and maximum width increases.

From the point of roughness, Figure7. shows the metallographic picture of 3Cr2W8V original sample (Ra0.486μm) HCPEB treatment after 20,40 times thermal fatigue cycle test. Compare with 3Cr2W8V original sample (Ra0.076μm), crack phenomenon has been significantly improved but the experimental results contrast with the expected results. Theoretically, the rougher surface of sample, the more micro-cracks produce, thermal fatigue properties of material will be worse too. But the experimental results show that with the original sample surface roughness increase, the number of cracks significantly reduce and the maximum width also reduce, this phenomenon indicate that after HCPEB treatment the sample with large roughness, the thermal fatigue resistance will be significantly improved. Sample surface residual stress (Ra0.486μm) is tested after treatment , the surface residual stress up to 615MPa, residual stress does not decrease with the increase of roughness, thus the residual stress is not the reason why the bigger roughness, the less micro-cracks.

Fig.7. Themorphologies of 3Cr2W8V directly processed by HCPEB without polishing pretreatment afeter the thermal fatigue cycle test: (a) 20, (b) 40, and (c) 60 cycles.

5. Conclusions

●After polishing, the roughness of 3Cr2W8V was 0.082μm, the surface residual stress was 42MPa; after HCPEB treatment and HCPEB surface alloying, the surface residual stress was enhanced into the values of 626MPa and 597MPa respectively. The micro-cracks after the thermal cycling tests were more likely caused when the residual stress was greater.

●Unpolished 3Cr2W8V sample with HCPEB treatment (roughness was 0.455μm) appeared micro-cracks at 40 times by thermal fatigue cycle, compared with polished sample the thermal fatigue properties have significantly improved and the surface residual stress valued as high as 615MPa, indicating that the reason of improved thermal fatigue properties was not due to residual stress reducing.

●3Cr2W8V had a large number of craters on surface after HCPEB treatment, because of the stress concentration which was produced by thermal Stress during the Thermal fatigue cycle, they had negative effects for thermal fatigue properties.

●Because the residual stress of sample surface by HCPEB treatment would increase dramatically, it had a great interference effect to judge and analysis the thermal fatigue properties, so there was a suggestion for future research, the residual stress of all the samples should down to the same level then follow-up testing and analysis.

Acknowledgements

This work is supported by Chongqing Science and Technology Natural Science Fund Project: cstc2013-yykfc6004.

References

1. Ao Yi, GAo Bo,TU Gan-feng, et al, effect of HCPEB treatment on microstructure and microhardness of hypereutectic Al-20Si alloy. *Transactions of Materials and Heat Treatment*, 2010.31:142-145.
2. KANG Jinwu, YOU Rui, NIE Gang, et al. Study on the Thermal Fatigue Life of H13 Steel for Aluminum Alloy Casting. *Journal of Mechanical Engineering*, 2012, 48(12): 63-68.
3. XIA Peng-cheng, CHEN Yun-bo, GE Xue-yuan, et al. Research status and development trends of thermal fatigue property of hot die steels. *Heat Treatment of Metals*, 2008, 33(12): 1-6
4. HE Rongcui, HU Xinbin, ZHOU Xiao ping, LI Junwan, Influence of Pulse Current Treatment on Thermal Fatigue Crack Propagation of New Type Hot Work Tool Steel.*Hot Working Technology*, 2015, 44(6):28-31
5. C.Oproiu, D.Martin, M.Toma, etc. Instruments and Methods in Physics Research B, 2000, (166-167): 669-675.
6. CHEN Yuan-fang, XU Hong-bin, HU Jian-jun, YANG Chang-hui, HAN Chao, Effect of HCPEB on Frictional Wear Resistance of 3Cr2W8V Mold Steel.*Journal of Wuhan University of Technology*, 2011, 33(11):19-23

7. WANG Hong. Quantitative Assessment of Fatigue and Fracture Properties of Materials. Xi.an: Northwest Industrial University, 2002.
8. HAN Zeng-xiang,Exploration of Testing Method for Thermal Fatigue of Metal. *Physical Testing and Chemical Analysis(Part A:Physical Testing)*, 2008, 44(5):250-254+257.

Corrosion behavior of high nitrogen austenitic stainless steel in the solutions containing different F^- concentration

Feng Shi*†, Yu-Cong Li, Zhi-Wei Liu and Yong-Lin Chen

Institute of Materials Physics and Chemistry, School of Materials Science and Engineering Northeastern University,
Shenyang, 110819, China
E-mail: †shifeng@imp.neu.edu.cn

Stainless steels are attractive due to its high strength, high chemical stability, low gas permeability and wide selection of alloys. However, stainless steels are prone to corrosion where $SO4^{2-}$ and F^- are released from the membrane in the PEFCs, which, as a result, will shorten the age of the bipolar plates and exert a negative influence on the function of the cells. Therefore, great attention is given on high-nitrogen austenitic stainless steels (HNASS) which have lower costs and good localized corrosion resistance. The effects of the different F- concentration on the corrosion behavior in the high nitrogen austenitic stain steel were investigated through polarization curve measurement and electrochemical impedance spectroscopy. The results show that the corrosion resistance of the experimental steel is the best in the 0.05 mol/L SO42- (pH 5.5) + 2 ppm F- solution. As the F^- concentration increases, the passivation film of the experimental steel becomes thicker and denser, leading to stronger corrosion resistance.

Keywords: High Nitrogen; Austenitic Stainless Steel; PEMFC; Corrosion Resistance.

1. Introduction

The fuel cell is known as the fourth generator after hydroelectric generation, thermal generation and nuclear power, while the proton exchange membrane fuel cell (PEMFC) is most likely to be commercialized in the market. PEMFCs are systems that use hydrogen and oxygen or air and they transform chemical energy into electrical energy with high transformation efficiency and low environmental contamination [1-2]. PEMFCs consist of membrane electrode assembly and bipolar plates. The bipolar plates are the key elements of the device which possess the function of separating cells and gas, conductivity, flow concentration and supporting the membrane electrode. Ideal materials for bipolar plates should have high corrosion resistance, high electrical conductivity, chemical stability and heat stability. Consequently, choosing ideal bipolar

* Corresponding author.

material and reducing its cost are critical to propel the advancement in PEMFC [3-4].

Carbonaceous bipolar would be the ideal candidates for PEMFCs but shortcomings such as brittleness and high gas permeability need to be further improved [5]. From this point of view, stainless steels are attractive due to its high strength, high chemical stability, low gas permeability and wide selection of alloys. However, stainless steels are prone to corrosion where SO_4^{2-} and F^- are released from the membrane in the PEFCs, as a result, it will shorten the age of the bipolar plates and exert a negative influence on the function of the cells. Therefore, great attention is now focused on high-nitrogen austenitic stainless steels (HNASS) owning lower cost and good localized corrosion resistance.

In this paper, the corrosion behaviors are discussed in the simulation PEMFC environment to analyze the possibility of direct application in the bipolar plates.

2. Experimental

The experimental steel used in present study was melted by induction furnace and electroslag remelting furnace both filled with N_2 and the chemical compositions were shown in table 1. The ingot was hot forged into 4 mm plate. After the plates were solid solution treated at 1150 °C for 1h and then water quenched, they were machined into specimens with the size of 10 mm×10 mm×2 mm. One side of the specimens connected with copper conducting wire was sealed in plastic pipe with epoxy resin, and the other side was completely exposed in the air. The experiment adopted Reference 600 electrochemical workstation to test polarization curve and electrochemical impedance spectroscopy (EIS) in the 0.05mol/L SO_4^{2-} (pH 5.5) + 2 ppm, 20 ppm and 200 ppm F^- solutions.

Table 1. Chemical composition of experimental steel (mass %)

Cr	Mn	N	C	Mo	Fe
20.17	19.10	0.82	0.065	2.29	Balanced

Figure 1 shows the experimental system. Auxiliary electrode is platinum electrode, reference electrode is saturated calomel electrode (SCE) and working electrode is specimens. The sample is put in the electrolyte solution. If the change of open circuit potential is less than 1 mv/min, the experiment reaches stable state.

In order to eliminate the surface oxidation film, linear scan was started from 250 mV less than self-corrode voltage with a rate of 0.33 mV/s. The scanning terminate electrode voltage, which is relative to saturated calomel electrode potential, is 1500 mV. Alternating-current impedance ranges from 10mHz to 100 KHz and 7 points were tested within each magnitude. Impedance test signal is sine wave with 10 mv amplitude. Also its electrode voltage is relative to open circuit voltage.

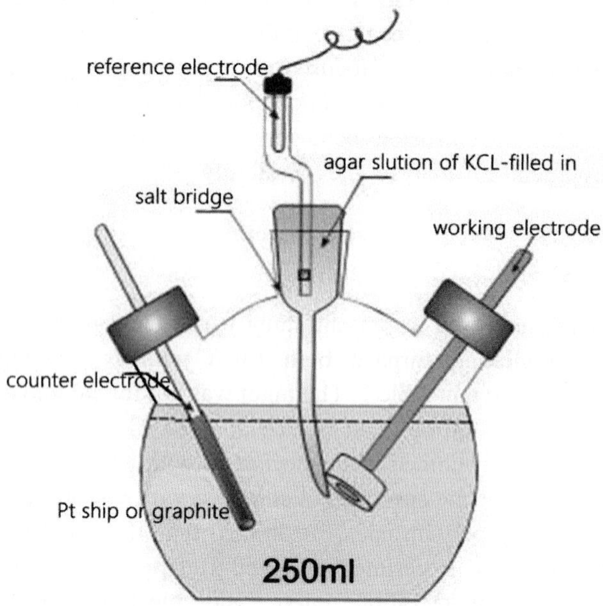

Fig. 1. The diagram of three electrodes system

3. Results and Discussion

Figure 2 shows the test results of polarization curves in the 0.05mol/L SO_4^{2-} (pH 5.5) + 2 ppm, 20 ppm and 200 ppm F^- solutions. Table 2 shows the polarization curve analytical results of the experimental steel in three kinds of solution with different F^- concentration. From Figure 2 and Table 2, it can be seen that the cathodic polarization current density decreases under the cathodic working voltage (-0.1V) with the increase of the F^- concentration, presenting suppression to the cathodic corrosion, which indicates that the increase of the NaF concentration leads to the slowing down of the cathodic reaction process. In addition, as for the properties of the bipolar plates, the United States Department of Energy requires that the corrosion current density of the samples should be less than 16 μA/cm^2 [5], and the passivation current density and cathodic/anodic corrosion current density of the experimental steel in these three solutions all satisfy the requirement.

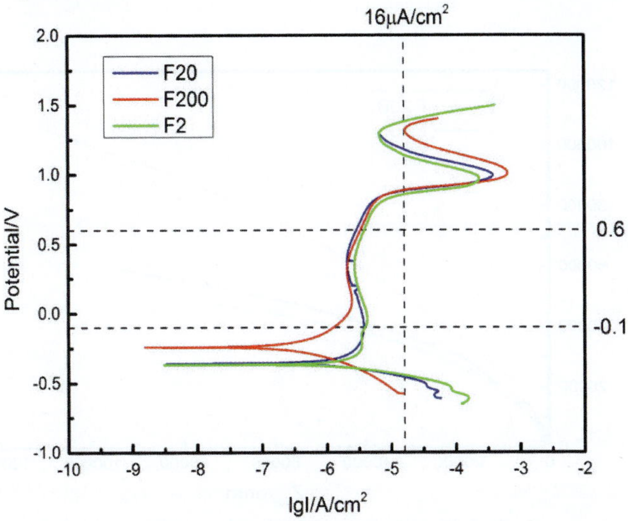

Fig. 2. The polarization curves in the 0.05mol/L SO_4^{2-} (pH 5.5) + 2ppm, 20ppm and 200ppm F^- solutions

Table 2. The polarization curve analytical results of experimental steel in three kinds of solution with different F^- concentration

F^- concentration	2ppm	20ppm	200ppm
E_C/(mV)	366.6	357.3	240.5
I_p/(μA/cm²)	2.824	2.268	2.113
$I_{E-0.1V}$/(μA/cm²)	3.928	3.748	1.275
$I_{E+0.6V}$/(μA/cm²)	3.838	2.896	3.362

Figures 3 shows the test results of electrochemical impedance spectroscopy (EIS) of the experimental steel in the 0.05mol/L SO_4^{2-} (pH 5.5) + 2ppm, 20ppm and 200 ppm F^- solutions. Nyquist diagram emphasizes the relation between the real part and the imaginary part and is used to demonstrate the anticorrosion mechanism. The curve seems like a semi-circle. The radius of the semi-circle can reflect the corrosion resistance of the samples. The bigger the radius of the diagram is, the better the corrosion resistance is. It can be clearly seen that the sample in the solution containing 200 ppm F^- has the largest radius while the least radius in the solution containing 2 ppm F^- (Figure 3). Therefore the corrosion resistance of experimental steel is the best in the 0.05mol/L SO_4^{2-} (pH

5.5) + 200 ppm F^- solution. The analytical result of EIS is in accordance with that of polarization curves in the experimental steel.

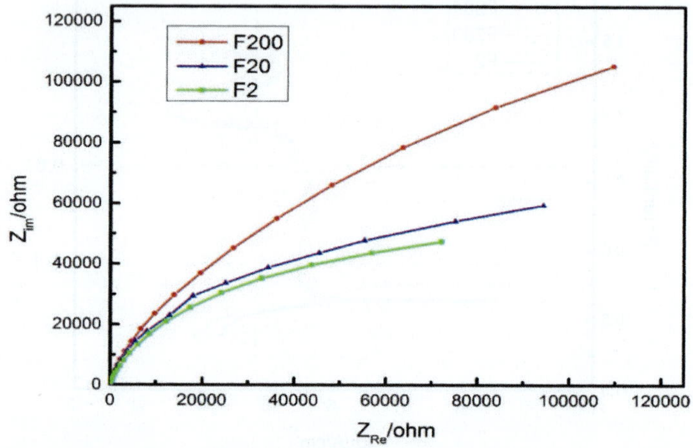

Fig. 3. Nyquist curves in the 0.05mol/L SO_4^{2-} (pH 5.5) + 2ppm, 20ppm and 200ppm F^- solutions

EIS data with a time constant can be interpreted by using parallel circuit of simple Interface capacitance and polarization resistance, seen in Figure 4. Because of the dispersion effect existing on the surface of electrode, capacitance C is often replaced by constant phase angle element (CPE) when interpreted.

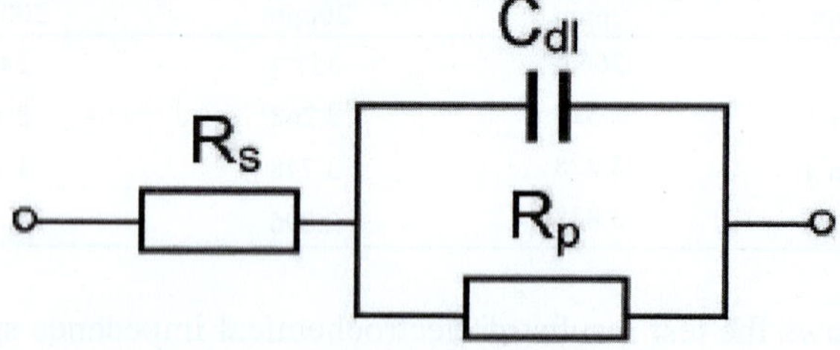

Fig. 4. The equivalent circuit of corrosion system

Thus, analytic expression of impedance spectroscopy analysis can be expressed as [6]:

$$Z = R_S + 1/(Y_F Y_{CPE}) = R_S + R_P/(1 + R_P Y_0 (j\omega)^{\alpha}) \quad (1)$$

Where R_s is the medium resistance, Y_0 and α are both the constants of CPE, relevant to interface capacitance. R_p is the charge transfer resistance. If its value increases, it means the passivation film is harder to penetrate. α value (dispersion index) represents the roughness of passivation film. Table 3 shows the fitting results. From table 3, we can conclude that, with the increase of F^- concentration, the values of α, Y_0 and R_p all increase, which indicates that the passivation film becomes thicker and denser with the F^- concentration increasing. Therefore the experimental steel displays the best corrosion resistance in the solution containing 200 ppm F^-.

Table 3. EIS fitting results of different samples

F^- Concentration	R_S/ohm	CPE,Y_0/ s^α/ohm	α	R_P/ohm
200ppm	27.36	4.29E-05	0.8756	2.06E+05
20ppm	12.47	5.21E-05	0.8672	1.19E+05
2ppm	25.42	7.01E-05	0.8624	1.01E+05

4. Conclusions

The corrosion resistance of the experimental steel is the best in the 0.05 mol/L SO_4^{2-} (pH 5.5) + 200 ppm F^- solution. As the F^- concentration increases, the passivation film of the experimental steel becomes thicker and denser, which leads to stronger corrosion resistance.

Acknowledgements

This work is supported by the National Natural Science Foundation of China (Grant No. 51201027) and the Fundamental Research Funds for the Central Universities (L1502039 and N150504001)

References

1. Z. X. Liu, W. Qian, J. W. Guo, J. Zhang, C. Wang, and Z.Q. Mao, The material of proton exchange membrane fuel cell, *Prog. Chem.* **23**, 487 (2011).
2. M. C. Song, P. C. Pei, H. S. Zha, and H. Xu, Water management of proton exchange membrane fuel cell based on control of hydrogen pressure drop, *J. Power Sources* **267**, 655 (2014).
3. R. Taherian. A review of composite and metallic bipolar plates in proton exchange membrane fuel cell: Materials, fabrication, and material selection, *J. Power Sources* **265**, 370 (2014).
4. R. F. Silva, D. Franchi, A. Leone, L. Pilloni, A. Masci, and A. Pozio. Surface conductivity and stability of metallic bipolar plate materials for polymer electrolyte fuel cells, *Electrochim. Acta* **51**, 3592 (2006).

5. M. Kumagai, S. T. Myung, S. Kuwata, R. Asaishi, Y. Katada, and H. Yashiro, Application of Ni-free high nitrogen stainless steel for bipolar plates of proton exchange membrane fuel cells, *Electrochim. Acta*, **54**, 1127 (2009).
6. M. C. Li, H. C. Lin and C. A.Cao. Electrochemical impedance spectrum characteristics of carbon steel corroding in the soil, *J. Chin. Soc. Corros. Prot.* **20**, 111 (2000).

Calculation and analysis of bolt fatigue reliability based on limited life

Jun-Min Du[1], Jun Liu and Zhong-Gui Yang
The 713 Research Institute of CSIC, Zhengzhou, Henan, 450015, China
E-mail: [1]dujunmin713@163.com

According to the analysis of the load character of the anchor bolt, the simplified method of the actual load that anchor bolt bearing is raised by series load spectrum blocks. Aiming to its character of including two random variables of average value and amplitude, the load spectrum blocks was transformed into a one-dimensional random variable by use of the equal life curve. At the same time, the calculation method for the structure fatigue strength of the anchor bolt under a certain life was deduced, and the fatigue reliability of the anchor bolt within a certain life can be calculated by the series method of stress-strength interference model and reliability unit. Results show that controlling the structural characteristics of the anchor bolt is also an important way to improve its fatigue performance.

Keywords: Load Spectrum Bolt; Fatigue; Reliability; Random Variable Parameter.

1. Introduction

As a significant equipment of wind power field, the wind turbine foundation (hereinafter referred to as WTF) is mainly used to fix the wind turbine generators tightly and bear the multi-directional and complicated alternating load during the wind turbine working. At the beginning of 21st century, the pre-stressed WTF was proposed by some countries such as the USA and Canada and applied in actual projects gradually, which is to pour the metallic components such as pre-stressed anchor bolt and flange and the concrete into a whole to bear the load of wind turbine. The anchor bolt is the core component of the pre-stressed WTF and has the following characteristics:

i) Bearing the complicated high-cycle alternating load. During its whole service life, the anchor bolt must bears more than 10^6 times periodic loads, and, every periodic load is a load spectrum having some randomness instead of definite single-cycle load.

---

[1] Corresponding author.

ii) Excellent fatigue performance. Once the WTF is completed, the anchor bolt cannot be replaced. Therefore, high fatigue reliability is required for the anchor bolt.

iii) A WTF includes 120-180 anchor bolts. Generally, the cost of a single anchor bolt has great influence on the construction cost of the WTF or even the wind farm. As a result, using the unlimited life method to design the anchor bolt is obviously not economical.

Basing on the above characteristics exactly, it is very necessary to study the fatigue reliability of the anchor bolt and identify the normally operates (without fatigue damages) reliability under certain life conditions and complicated loading environment.

As an important direction of the study on product fatigue technology, fatigue reliability is an inter-discipline combined with mechanics, fatigue theory and reliability theory. Some studies have been conducted at home and abroad [1]. For example, literature[2] raised a probability calculation method for structure unlimited life which use the stress-strength interference method basing on one-dimensional fatigue limit theory to evaluate the reliability of structure unlimited life; literature [3]and [4] considered the common effects of both alternating load average value and amplitude, and evaluated the fatigue reliability by defining the composite fatigue strength; literature[5] researched the fatigue reliability of the structure under the effect of random stress spectrum, and brought up method to determine the distribution function of fatigue life when the stress ratio was -1.

In fact, the anchor bolt always bears complicated load spectrum in actual operation, while, the alternating load stress ratio is usually greater than 0.5 after simplification. Thus, a simple and applicable method must be raised to calculate the fatigue reliability of the anchor bolt basing on the fully analyzing the characteristics of load spectrum. In addition, the comprehensive influence of two random variables of the average value and amplitude of fatigue stress and fatigue strength must be considered in this method design.

2. Analysis on Characteristics of Load Spectrum Based on Two-dimensional Random Distribution

The anchor bolt bears periodic load effect during its service life.
Figure 1 represents several periodic of the load. It can be seen that the load characteristics within these period are similar. And, all of the load include various alternating cyclic loads and follow statistics rule. At the same time, for the average value and amplitude of corresponding alternating cyclic loads, there is a definite randomness between different periods.

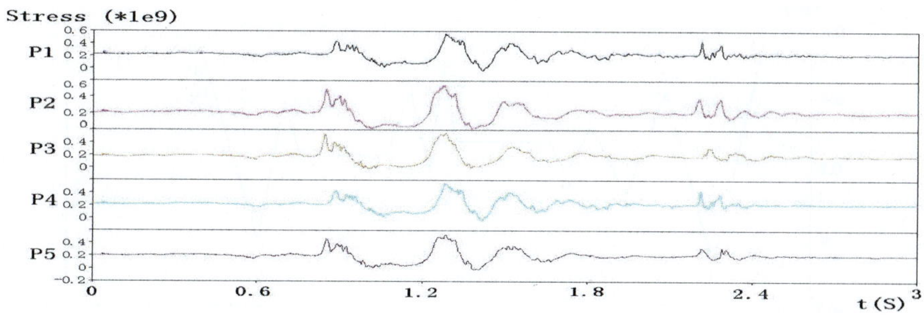

Fig. 1. Comparisons of Anchor Bolt Stresses between Different Periods

Based on the load characteristic of the anchor bolt, the common form in periodic loads is extracted and the normalized form of individual periodic loads is created, as shown in Figure 2. Supposing that the action sequence and mutual influence of loads [6] are not considered, the normalized pattern load is counted using the rain-flow counting method. So, a periodic load shall be equaled to a combination of a series of alternating cyclic loads, as the six alternating cyclic loads of 1-2-1', 3-4-3', 5-6-5', 8-9-8', 10-11-10' and 12-13-12' shown in Figure 3. Within the whole service life, the subsamples of a certain alternating load constitute a load spectrum block. Inside a load spectrum block, the average value Y_{mi} and amplitude Y_{ai} of alternating cyclic loads are not definite but random having certain distribution; therefore, a load spectrum block is actually a random two-dimensional function. In order to simplify the calculation, both Y_{mi} and Y_{ai} can be assumed to subjecting the normal distribution, while their average value $(\bar{Y}_{mi}, \bar{Y}_{ai})$ and mean square deviation $(\sigma_{Ymi}, \sigma_{Yai})$ can be acquired through statistics of several periodic data.

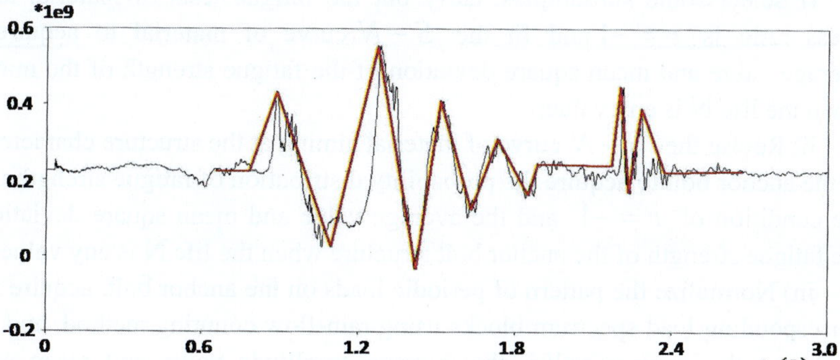

Fig. 2 Normalization of Periodic Loads

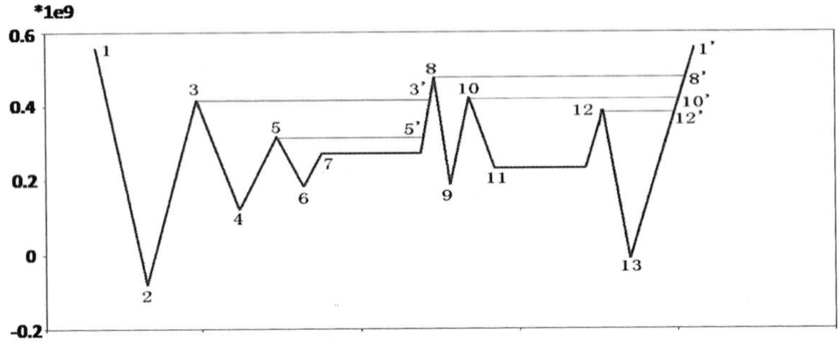

Fig.3 Rain-flow Counting of Loads

It can be deemed that the impact of a periodic load to the anchor bolt is equal to that of six alternating cyclic loads. Thus, the load effect within the whole service life is equal to the effect of six load spectrum blocks, and the fatigue reliability of the anchor bolt within a certain service life is the series effects of the reliability of six load spectrum blocks. Assuming the reliability within the anchor bolt service life is $\mathrm{R_i}$ when the i load spectrum block works independently, the reliability shall be $\prod \mathrm{R_i}(i=1\cdots6)$ when six load spectrum blocks working together. Consequently, the calculation process is greatly simplified by transforming the actual loads into series load spectrum blocks, which change the research of fatigue reliability of the anchor bolt bearing periodic loads to that of the anchor bolt bearing six load spectrum blocks.

3. Calculation Method of Fatigue Reliability of the Anchor Bolt

3.1 *Basic steps of calculation*

The basic steps for calculating the anchor bolt fatigue reliability are as follow:

i) Select some subsamples, carry out the fatigue tests of material as the stress ratio is $r=-1$ and fit the $S-N$ curve of material to acquire the average value and mean square deviation of the fatigue strength of the material when the life N is any value;

ii) Revise the $S-N$ curve of material aiming at the structure characteristic of the anchor bolt to acquire the probability distribution of fatigue strength under the condition of $r=-1$ and the average value and mean square deviation of the fatigue strength of the anchor bolt structure when the life N is any value;

iii) Normalize the pattern of periodic loads on the anchor bolt, acquire some corresponding load spectrum blocks using rain-flow counting method, and then, work out the average value, the average amplitude value and mean square deviation of each spectrum block according to the measure or simulate calculation data;

iv) Carry out equal life transforming to the average value and amplitude of the 1st load spectrum block to acquire the probability distribution of fatigue stress under the condition of $r = -1$ and the same life;

v) Calculate the reliability R_1 by using the stress-strength interference method under the effect of the 1st load spectrum block when the life of the anchor bolt structure is N;

vi) Repeat d) and e), acquire the reliability R_i in turn under the effect of the i st load spectrum block when the life of the anchor bolt structure is N;

vii) Calculate the fatigue reliability $\prod R_i (i = 1 \cdots 6)$ under the combined effect of several load spectrum blocks.

3.2 *Determinate the S-N curve of material*

Using three-parameter power function expression to fit the $S - N$ curve of the anchor bolt material [3]:

$$S = S_{-1} \times (1 + A/N^{\alpha}) \tag{1}$$

Here: S, N indicate the fatigue strength S corresponding the material life is N; S_{-1} is the fatigue limit of material; A, α are material constants.

Suppose that A and α are known and definite values, make k samples and carry out the fatigue life test when the fatigue strength is $S(r = -1)$ to acquire the corresponding life observed value N_k. Put S and N_k into the Eq. (1) to calculate the average value $\overline{S}_{-1}$ and mean square deviation $\sigma_{s\text{-}1}$ of the fatigue limit of material. And, the average value and mean square deviation of the fatigue strength when the life N is any value can be acquired according to the Eq. (2) and Eq. (3):

$$\overline{S} = \overline{S}_{-1} \times (1 + A/N^{\alpha}) \tag{2}$$

$$\sigma_S = \sigma_{S\text{-}1} \cdot (1 + A/N^{\alpha}) \tag{3}$$

3.3 *Probability distribution of structure fatigue strength*

Because the structure fatigue life of the anchor bolt is related to both the material characteristics and the structure characteristics, such as stress concentration, size and surface, and so on. The $S - N$ curve of material is needed to revise to acquire the fatigue strength of the structure by using the empirical Eq. (3):

$$S' = \frac{\varepsilon \beta_1 \beta_2}{k_f} S \tag{4}$$

Here: S' , S are the fatigue strength of the structure and the material respectively; ε is the structure size factor, β_1 is the surface machining factor, β_2 is the surface strengthening factor, k_f is the effective stress concentration factor.

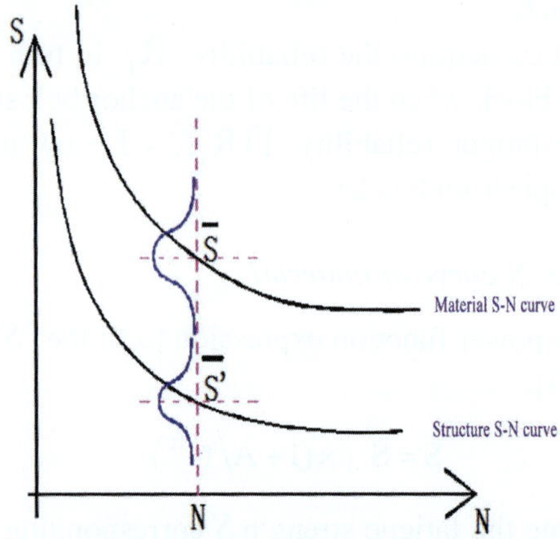

Fig. 4 Material and Structure S-N Curves

In actual working conditions, the material fatigue strength and its various influence elements are random variables that conform to certain distribution instead of definite values. Therefore, the structure fatigue strength is also a random function determined by a number of random variables. It can be known from the central-limit theorem of probability theory: when the variation factor of any random variable satisfy the $C_X = \sigma_X / \mu_X < 0.1$, and all these variables are independent of each other and not the main control element, the structure fatigue strength shall obey to the normal distribution rule [7] In addition, the approximation solution of its average value and mean square deviation can be acquired by using Taylor Series method:

$$\overline{S'} = \frac{\overline{\varepsilon} \cdot \overline{\beta_1} \cdot \overline{\beta_2}}{\overline{k_f}} \overline{S} \tag{5}$$

$$\sigma_{S'}{}^2 = \left(\frac{\overline{S} \cdot \overline{\beta_1} \cdot \overline{\beta_2}}{\overline{k_f}} \sigma_\varepsilon\right)^2 + \left(\frac{\overline{S} \cdot \overline{\varepsilon} \cdot \overline{\beta_2}}{\overline{k_f}} \sigma_{\beta 1}\right)^2 + \left(\frac{\overline{S} \cdot \overline{\varepsilon} \cdot \overline{\beta_1}}{\overline{k_f}} \sigma_{\beta 2}\right)^2 - \left(\frac{\overline{S} \cdot \overline{\varepsilon} \cdot \overline{\beta_1} \cdot \overline{\beta_2}}{\overline{k_f^2}} \sigma_{kf}\right)^2 \tag{6}$$

Synthesize Eq.(2), (3), (4), (5), and (6), the average value $\overline{S'}$ and mean square deviation $\sigma_{S'}$ of the structure fatigue strength of the anchor bolt under some certain life N condition ($r=-1$) can be acquired, just as shown in Figure 4.

3.4 *Characteristic parameter transformation of load spectrum blocks based on constant life*

§2.2 and §2.3 shows the probability distribution calculation of structure fatigue strength of the anchor bolt when the stress ratio is $r=-1$. However, the stress ratio of load spectrum block that the anchor bolt bears is not -1 in any actual working condition. So, the structure fatigue strength and the load spectrum block calculated in §2.3 cannot be used to calculate the reliability of the anchor bolt directly, while the stress ratio of the load spectrum block and the fatigue strength needs to be unified first. Literature 3 and 4 suggested acquire the average value and amplitude of fatigue strength under the same fatigue stress ratio as the actual load spectrum block by transforming the fatigue strength, and then, carry out the reliability calculation by constructing composite fatigue strength and composite fatigue stress. But there is a very obvious defect of this method that the distribution of two random variables from the distribution of one random variable does not conform to the assumption that two random variables are mutually independent. As a result, this paper raise a new reliability calculation method by contrasting the probability distribution of fatigue load and the fatigue strength, which needs transform the load spectrum of the anchor bolt first, then acquire the probability distribution of fatigue load under the condition of $r=-1$ (i.e., the average value is 0) through the probability distribution of the average value and amplitude of the load spectrum.

Under a certain life N condition, no matter what is r, the average value Y_{mi} and amplitude Y_{ai} of any load spectrum block will conform to the following relations [8], as shown in Figure 5:

$$(Y_{ai}/Y_{i(r=-1)})+(Y_{mi}/S_u)=1 \tag{7}$$

The above Eq. is called Goodman curve, it can also be expressed as:

$$Y_{i(r=-1)}=(S_u\cdot Y_{ai})/(S_u-Y_{mi}) \tag{8}$$

Here: $Y_{i(r=-1)}$ is the fatigue load with the same life N as the fatigue effect of (Y_{mi}, Y_{ai}) when the anchor bolt is bearing the fatigue load of $r=-1$. It means that the $Y_{i(r=-1)}$ has the same fatigue effect on the anchor bolt as (Y_{mi}, Y_{ai}). S_u is the damage strength (yield or tensile strength) of material under static load condition, supposing its probability distribution is known.

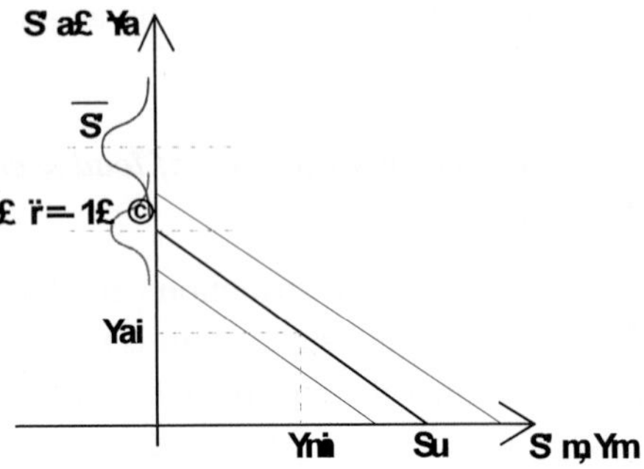

Fig. 5 Constant Life Curve of Load Spectrum Block

Using the above Eq.5, the effect of fatigue stress under random stress ratio can be transformed into that of $r=-1$. Because the distributions of Y_{mi}, Y_{ai} and S_u are already known, the average value $\overline{Y}_{i(r=-1)}$ and mean square deviation $\sigma_{Yi(r=-1)}$ of $Y_{i(r=-1)}$ can be acquired expediently by using Taylor Series method.

3.5 *Calculation of fatigue reliability of the anchor bolt*

Under a certain life N condition, the average value and mean square deviation of the structure fatigue strength of the anchor bolt can be calculated by Eq. (5) and Eq. (6), farther more, the average value and mean square deviation of the load spectrum block of each anchor bolt under the condition of $r=-1$ can be calculated by Eq.(8), thus the fatigue reliability of anchor bolt is acquired through the stress-strength interference method.

Assume that both the structure stress of the anchor bolt Y_i and the material strength S' conform to the normal distribution rule, and the limit state function is defined to $Z_i = S' - Y_i$, so the structure reliability of the anchor bolt under the effect of the ith load spectrum block is[7]:

$$R_i = P(Z>0) = \int_0^{\infty} f(Z)dZ = \int_0^{\infty} \frac{1}{\sigma_Z\sqrt{2\pi}} e^{-\frac{1}{2}\left(\frac{Z-\mu_Z}{\sigma_Z}\right)^2} dZ \qquad (9)$$

Here: $\overline{Z}$, σ_Z are respectively the average value and square deviation of the limit state function Z_i.

Transform Eq.(9) into standard normal distribution and consider the

$$R_i = \int_{-\infty}^{\beta} \frac{1}{\sqrt{2\pi}} e^{-\frac{1}{2}u^2} du = \Phi(\beta) \qquad (10)$$

symmetry of the normal distribution, Eq. (10) is reached:

Here: $\beta = \overline{Z} / \sigma_Z$ is also known as reliability factor.

According to the reliability factor β_i, the reliability R_i of the anchor bolt under the effect of the ith load spectrum block can be acquired by searching the normal distribution function table. And the working reliability during the whole life of the anchor bolt can be calculated according to $\prod R_i (i = 1 \cdots 6)$.

4. Calculation Example

4.1 *Calculation requirement*

To acquire the working reliability of the anchor bolt under the 10^6 periodic loads, each periodic load shall be simplified into six load spectrum blocks. The average value and square deviation of load spectrum block are in Table 2. And the simplification process of load using rain-flow counting method can reference the literature[6].

4.2 *Detailed calculation*

I) Anchor bolt material test and data processing

Select m samples for the static load strength test to acquire the damage strength S_u (900, 80)MPa of anchor bolt material under the static loading condition.

Select 6 samples for fatigue life ($r = -1$) test with the amplitude being 180MPa and acquire the life observation value N_k. Then supposing $A = 8.128 \times 10^3$ and $\alpha = 0.8046^2$, the material fatigue limit S_{-1} can be calculated according to Eq. (1), as shown in Table 1.

Table 1. Fatigue life observation value and fatigue limit of materials.

Times (k = 6)	Fatigue life observation value of materials N_k ($\times 10^6$)	Fatigue limit of materials S_{-1} (MPa)
1	1.087	104.61
2	1.062	103.79
3	1.440	114.30
4	1.445	114.42
5	0.776	92.53
6	1.328	111.53

It can be calculated that the average value and of the mean square deviation of material fatigue limit is $\overline{S}_{-1}$=106.86 and $\sigma_{s\text{-}1}$=8.42 respectively. In addition, the average value and variance of the material fatigue strength of anchor bolt when the life is 10^6 can be calculated to $\overline{S}$=119.78 and σ_s=9.44 according to Eq. (2) and Eq. (3).

II) Transform the date of structure fatigue strength of the anchor bolt

Supposing the effective stress concentration factor is $k_f - N(1.6, 0.0384^2)$, the structure size factor is $\varepsilon - N(0.79, 0.069^2)$, the surface machining factor is $\beta_1 - N(1,0)$, and the surface strengthening factor is $\beta_2 - N(1,0)$, the average value and variance of the structure fatigue strength of anchor bolt when the life is 10^6 can be calculated to $\bar{S}'=59.14$ and $\sigma_{s'}=6.48$ according to Eq. (5) and Eq. (6).

III) Transform the data of load spectrum blocks with equal life

Transform the data of load spectrum blocks in table 2 with equal life according to §2.4 and acquire its equivalent fatigue stress under the condition of $r=-1$.

IV) Reliability calculation

Basing on the equivalent fatigue stres Y_i and fatigue strength $S(59.14, 6.48)$ in Table2, calculate the reliability factor β_i of each load spectrum block by Eq. (10). Then refer the normal distribution function table to acquire corresponding factor. At last, the reliability of the anchor bolt structure under the condition of $N=10^6$ can be calculated to be 0.9932 according to $\prod R_i (i=1 \cdots 6)$.

Table 2. Average value and variance of load spectrum block. Unit: MPa

Load pectrum block (i = 6)	**Average value**	**Amplitude**	**Equivalent fatigue stress ($r=-1$)**	**Reliability R**
1	(400, 38)	(17, 3)	(29.36, 8.74)	0.9979
2	(480, 45)	(14, 2)	(28.30, 9.95)	0.9953
3	(520, 50)	(5, 1)	(11.04, 4.54)	1
4	(320, 29)	(12, 2)	(18.10, 5.32)	1
5	(290, 27)	(9, 1)	(12.95, 3.33)	1
6	(300, 18)	(11, 2)	(16.08, 4.76)	1

4.3 *Calculation analysis*

The structure static strength calculated using stress-strength interference model is a definite value during the whole life. It does not change along with the working time of the structure. But when the anchor bolt is bearing the effect of fatigue stress, the structure must be affect at the accumulated damage effects. Therefore, the reliability of anchor bolt would gradually reduce following the increase of the number of use.

Supposing the design life of the anchor bolt is N_1, the corresponding fatigue strength of the structure shall be $S'_{(N1)}$, then the fatigue reliability of the structure under actual working stress Y_1 can be calculated, which express the reliability of the anchor bolt at N_1 time cyclic loads effect. If the actual life of the anchor bolt is N_2 ($N_2 < N_1$) then its reliability should be calculated with the fatigue strength $S'_{(N2)}$ and the working stress Y_1 corresponding to N_2. In addition, because of $S'_{(N2)} > S'_{(N1)}$, the less the anchor bolt is used, the calculation is more reliable, as shown in Figure 6.

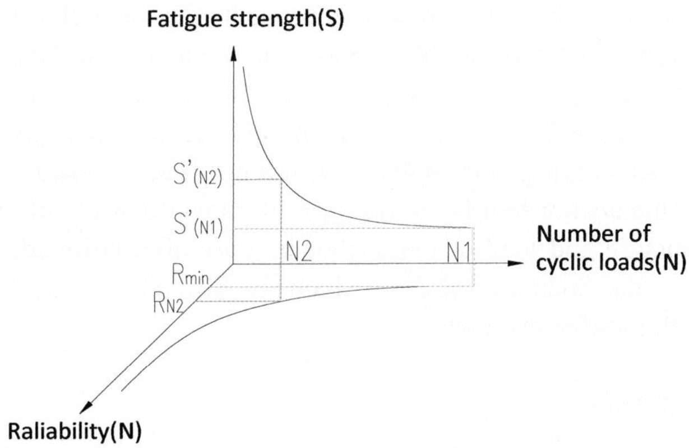

Fig. 6 Relationship between Fatigue Reliability and Number of Cyclic Loads

For traditional analyzing method of fatigue life, the damage caused by the fatigue stress is usually described by $D = n/N$ that is the ratio of load cyclic times to fatigue strength life, which treat the structure life as a process of linear change. In this paper, the damage caused by the fatigue stress is described by the reliability changes with the number of cyclic loads, and the $S-N$ curve shape of the material and the influence of each random variable can be considered fully.

5. Conclusion

i) The complicated periodic load spectrum that the anchor bolt bears can be simplified into the series form of many load spectrum blocks by rain-flow counting method, and the effects of series load spectrum blocks are considered equal to the effect of complicated periodic load that the anchor bolt bears during its whole life. Because of the average value of the load spectrum block is normally not 0, it actually is a random function that includes the two random variables of average value and amplitude. Therefore, in order to compare with the structure fatigue strength, the load spectrum block usually needs to be transformed into a one-dimensional random variable that the average value of which is 0 (i.e. $r = -1$).

ii) The structure fatigue performance of the anchor bolt is not only determined by the fatigue characteristics of its material, but also determined by the size, structure and the surface quality of anchor bolt, while the difference between the structure fatigue strength and the material fatigue strength can reach to many times usually. Therefore, controlling the structure characteristics of the anchor bolt is also an important way to improve its fatigue performance.

iii) Supposing not consider the action sequence and the mutual influences of the load spectrum blocks, the fatigue reliability of the anchor bolt within a certain life period can be described by the product of the reliability of some load spectrum blocks working independently. But it needs to be noted that the fatigue reliability of the anchor bolt is the function of its working life N. So, when the material, structure and load of the anchor bolt are all confirmed, its reliability reduces with the increase of N, which is different from the traditional stress-strength interference method.

Acknowledgments:

This work is supported by State Administration of science, Technology and Industry for Nation Defence, PRC

Reference

1. Xiao-Li ZHANG, Xue-Feng CHEN, Bing LI, etc. Review of Life Predication for Mechanical Major Equipments *J.* Journal of Mechanical Engineering, 2011, 47(11).
2. Jun-Jiang XIONG. Fatigue Life Design For Aircraft Structure. Beijing: Beihang University Press, 2004(In Chinese).
3. Da-Tong QIN, Li-Yang XIE. Fatigue Strength and Reliability Design. Beijing: Chemical Industry Press, 2013(In Chinese).
4. Zong-Miao DAI, Jun-Min DU. Calculation and Analysis for Screw Bolt Fatigue Reliability *J.* Mechanical Engineer, 2015, (3) .
5. Yan-Jiang SU, Guang-Qin WANG, Qing GAO. Miner's Rule-Life Distribution Function Method for Fatigue Reliability Analysis of Component Under Random Stress Spectrum *J.* Journal of Mechanical Strength, 2002, 24(3).

6. Le-Yi Dong, Jun LUO, Li CHENG. Rain Flow Count Method and Its Realization in Programming *J*. Journal of Computer Application, 2004, 24(3).
7. Liang-Qiao LI, Wei-Ming GU. Mechanical Reliability Design and Analysis. Beijing: National Defend Industry Press, 1998: 63(In Chinese).
8. Chuan-Yao CHEN. Fatigue and Fracture. Wuhan: Huazhong University of Science and Technology Press, 2001 (In Chinese).

Chapter 6

Electrical Materials

Tuning electronic and magnetic properties of fluorinated ZnO nanosheets under strain field

Mei-Xia Xiao*, Ting-Zhen Yao and Lei Wang
College of Materials Science and Engineering, Xi'an Shiyou University, Xi'an, 710065, People's Republic of China
E-mail: [a]mxxiao@xsyu.edu.cn

[a]Mei-Xia Xiao, [b]Cheng He
State Key Laboratory for Mechanical Behavior of Materials, Xi'an Jiaotong University, Xi'an 710049, People's Republic of China
E-mail: [b]hecheng@mail.xjtu.edu.cn

We systematically investigated the electronic and magnetic properties of fluorinated ZnO nanosheets (NSs) through first-principles calculations. Unlike bare ZnO NSs terminating with polar {0001} surfaces, fluorination on the surfaces preserve the initial wurtzite configuration. The two-bilayer ZnO NS with full fluorination (F–ZnO–F) is a small direct-gap semiconductor without nonmagnetism, while the ZnO NS with fluorinated Zn atoms (F–ZnO) can be made metal with the ferromagnetic (FM) state. More surprisingly, the F–ZnO NS exhibits intriguing magnetic transitions between FM and antiferromagnetic coupling by applying strain, even giving rise to AFM metal under the compression of 4%. Our work offers a new avenue to facilitate the design of controllable and tunable nanoelectronic and optoelectronic devices.

Keywords: Electronic and Magnetic Properties; Strain Field; Surface Modification.

1. Introduction

Low-dimensional zinc oxide (ZnO) nanostructures, such as nanowires[1], and nanosheets (NSs) [2-3], have attracted considerable interest due to their potential applications in nanoelectronic and optoelectronic devices. Recently, the progress in graphene with surface modification was predicated to be an effective way to tune the electronic and magnetic properties of graphene in theory[4-5]. These studies on graphene activate significant interest in other honeycomb lattices. The current reports show that the graphitic-like structure can be more stable than the

* Corresponding author.

wurtzite structure for thinner ZnO NSs[6]. The monolayer ZnO NS with full hydrogenation is an indirect band gap semiconductor with nonmagnetic (NM) state[2], whereas those thicker-bilayer ones become metallic independent of thickness[3]. However, the semihydrogenated ZnO NSs on Zn atoms transform into ferromagnetic (FM) states. Recently, semifluorinated 2D NSs of group-IV and III-V binary compounds have been found that these NSs can exhibit AFM or FM states[7]. It is naturally expected to investigate what will occur in the case of decorated ZnO NSs with fluorination. Is it a semiconductor for fully fluorinated ZnO NS or still a FM semiconductor for semifulorinated one? Moreover, recent studies verify that, the semifluorinated BN NSs exhibit intriguing magnetic transitions between FM and antiferromagnetic (AFM) couplings by applying strain[7-8]. Thus, How can the relative stability of FM and AFM states be further modulated by applying external strain field.

2. Computational methods

In this paper, our calculations employ spin–polarized DFT with the generalized gradient approximation (GGA) of Perdew–Burke–Ernzerhof (PBE) as the exchange–correlation functional, as implemented in the DMol^3 code[9-10]. The 2×2 unit cells of ZnO NSs are adopted with a vacuum space of 15 Å to avoid interactions between neighboring cells. The atoms are all relaxed without any constraints. The DFT semicore pseudopots is implemented for relativistic effects, which replaces core electrons by effective potentials for Zn elements; all–electron is used for the core treatment for O and F elements. In addition, the double numerical plus polarization (DNP) is as basis sets in all calculations. The Brillouin zone integration is performed with 17×17×1 *k*–points. The convergence tolerances of energy of 1.0×10^{-5} Ha, the maximum force of 0.002 Ha/Å, and the maximum displacement of 0.005 Å are taken in the geometry optimization.

3. Results and discussion

The geometrical configurations of ZnO NSs terminated with polar (0001) surfaces are constructed from the bulk wurtzite structure. We investigate the electronic and magnetic properties of two–bilayer ZnO NSs tuned by surface modification of F atoms. Because of relatively weak binding between F and O atoms, we only consider two types of models are considered, that is, ZnO NSs with full fluorination on both Zn and O atoms (F–ZnO–F), and semifluorination on only Zn atoms (F–ZnO). In order to take into account the contributions of the van der Waals (vdW) interactions between different layers, the DFT–D (D stands for dispersion) approach within the Grimme scheme is adopted for the vdW correction[11].This method has been successfully applied in previous studies of the interaction, such as silicene/silicon NSs[12], and silicene/Ag NSs[13], and benzene/metal[14]. In this paper, we use the DFT–D method to study the electronic and magnetic properties of the two-bilayer ZnO NSs with fluorination.

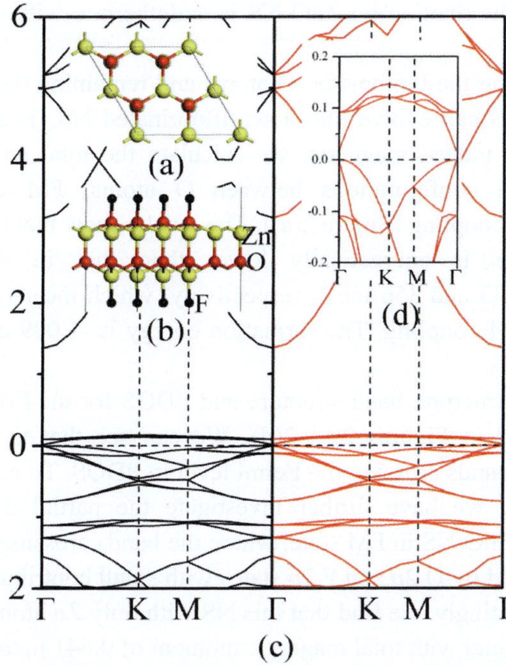

Fig. 1. The relaxed atomic configurations in top (a) and side (b) views, and band structures (c) for F–ZnO NS, where (d) is the insert of the Dotted frame. The Fermi energy is set to zero.

The relaxed atomic structures of the two–bilayer F–ZnO–F NS with full fluorination from both top and side views are shown in Figure 1(a)–1(b). As mentioned above, the F–ZnO–F NS prefers the wurtzite structure rather than graphite one, whose stable configuration is buckled, similar to monolayer ZnO NSs with full hydrogenation[2]. The buckled height is 0.871 Å, 1.883 Å, and 0.722 Å between Zn_1 and O_1 layers, between O_1 and Zn_2 layers, between Zn_2 and O_2 layers, respectively. Especially, the bond lengths of Zn_1–O_1, O_1–Zn_2, and Zn_2–O_2 bonds are 2.096 Å, 1.883 Å, and 2.309 Å. The band structure of F–ZnO–F NS exhibits a direct band gap at Γ point with energy gap of 0.02 eV, as shown in Figure 1(c) and 1(d). The spin-polarized calculations show that the F–ZnO–F NS has no magnetism, which can also be seen from paired spin-up and spin-down channels of the band structure. The results are in good agreement with previous studies of 2D fully hydrogenated ZnO NS[2-3], indicating the accuracy of our computational approaches. To investigate the relative stability of F–ZnO–F NS, the calculated formation energy is defined as the energy difference between the F–ZnO–F sheet, pristine ZnO sheet, and F2 molecules. We find that the fluorination on ZnO NS is endothermic with energy of 0.745 eV/F.

When removing the F atoms on O atoms and remaining the rest ones on B atoms, the ZnO NS can evolve into a semifluorinated NS (F–ZnO). To verify that the system is indeed magnetic, we calculate the total energies for three magnetic coupling configurations between O atoms: FM coupling, AFM coupling, and NM coupling [Figure 2(a)]. The results show that the FM coupling between O atoms is energetically lower than that in AFM and NM configurations by 33 and 356 meV, respectively, which means that the F–ZnO NS exhibits the FM coupling. The formation energy is -1.009 eV/F for F–ZnO NS in FM state.

The atomic structure, band structure and PDOS for the F�ZnO NS in FM state are presented in Figures 2(b)–2(d). We can see that the F–ZnO NS is metallic with two bands crossing the Fermi level in PDOS. To explore the origin of the metallicity, we have further investigate the partial density of states (PDOS) of the F–ZnO NS in FM state, where the bands crossing the Fermi level are mainly occupied by O 2*p* and F 2*p* states with small contribution from Zn 3d states. Most interestingly, we find that this NS with only Zn atoms passivated by F atoms is a FM metal with total magnetic moment of 0.641 μ_B/unit cell.

Because strain effect induced by radial size is always important in nanostructures, we investigate the strain dependence of spin coupling in F–ZnO NS by varying the isotropic strain, where all crystal symmetries and overall honeycomb-like structure are maintained. The tensile or compressive strain is uniformly applied along both zig-zag and armchair directions on the periodic boundary. The isotropic strain is defined as $\varepsilon = \Delta a/a_0$, where a_0 and Δa_0 are the lattice constants of the unstrained and strained supercell, respectively. The variation in energy difference ΔE ($\Delta E = E_{FM} - E_{AFM}$) of F–ZnO NS under strain field show that the energy difference ΔE between FM and AFM states increases with higher tensile strain and FM is always more stable than AFM state. However, we find that the energy EAFM of AFM state increases slower than the energy E_{FM} of FM state and AFM state is energetically more favorable

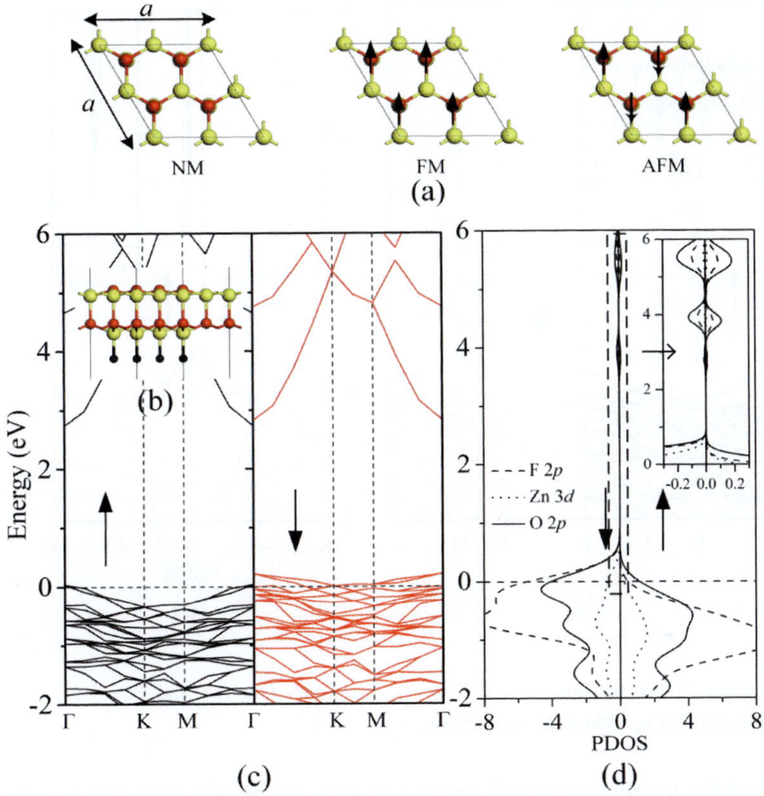

Fig. 2. Three magnetic coupling configurations (a) between O atoms with FM coupling, AFM coupling, and NM coupling, optimized atomic structure in side view (b), band structure (c), and PDOS (d) of two-bilayer F–ZnO NS in FM state, respectively.

than FM state by 6 meV at 4% in-plane compression. Thus, the magnetic transition between FM and AFM orders in F–ZnO NS can be induced by applying strain.

Figure 3 illustrates the atomic structure, band structure, and PDOS for F–ZnO NS with AFM state under compressive strain of 4%, where bands in spin-up and spin-down states cross the Fermi level. So, electric transport through F–ZnO NS with full-metallicity under compressive stress of 4% is mainly contributed by the F 2*p* and O 2*p* states.

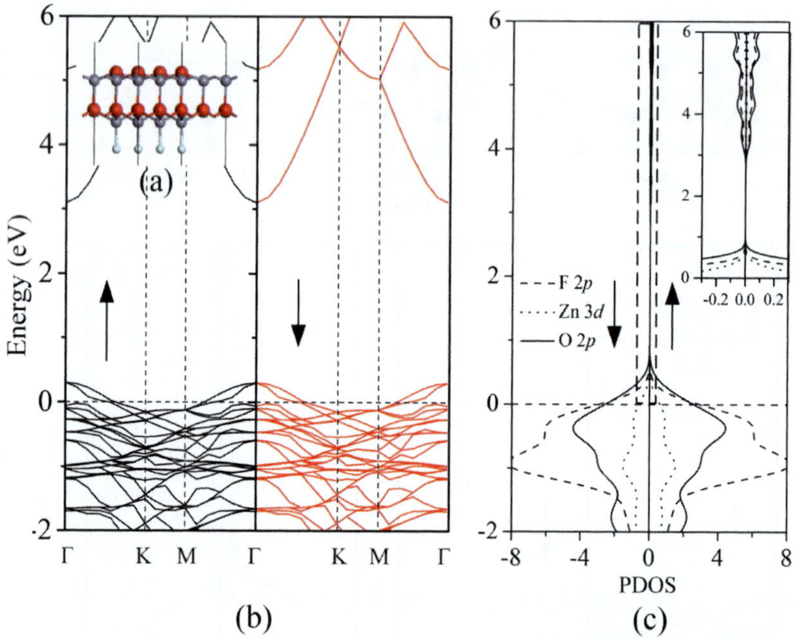

Fig. 3. Optimized atomic structure in side view (a), band structure (b), and PDOS (c) of two-bilayer F—ZnO NSs in AFM state, respectively.

We find the magnetism transition mainly induced by *p–p* direct interaction, where a N atom with spin-up (spin-down) density induces a spin-down (spin-up) density on the nearest-neighboring N atom directly. Thus, the *p–p* direct spin polarization couples antiferromagnetically the spins of the dangling bonds of N atoms. The mechanism is also suitable for semifluorinated BN NSs and semifluorinated graphene[7].

4. Conclusion

In summary, we have systematically investigated the electronic and magnetic properties of fluorinated ZnO NSs with the strain field. Fluorination on the surface Zn and O atoms can transform the two–bilayer F–ZnO–F NS into wurtzite configurations, and the F–ZnO–F NS is a small direct-gap semiconductor with NM state. However, the F–ZnO NS with fluoination on Zn atoms shows FM state, and exhibits metallic behavior. Under the compressive strain of 4%, the F–ZnO NS has a transition from FM into AFM, and remains the metallic behavior. Furthermore, the fluorinated ZnO NSs are favorable energetically. Thus, the strain field coupling to surface modification by fluorine plays a key role in tuning the electronic and magnetic properties of low-dimensional ZnO NSs, which facilitate the design of nanoelectronic and optoelectronic devices.

Acknowledgments

This work is supported by Supported by State Key Laboratory for Mechanical Behavior of Materials (20151708), Scientific Research Program Funded by Shaanxi Provincial Education Department (2013JK0894), Natural Science Foundation of Shaanxi Province (2015JM6327), Youth Science and Technology Innovation Fund Project at the Xi'an Shiyou University (2012BS004), and the Provincial Superior Discipline for Materials Science and Engineering of Xi'an Shiyou University, China (Grant No. 312010005).

References

1. S. Li, Q. Jiang and G.W. Yang, Uniaxial strain modulated band gap of ZnO nanostructures, *Appl. Phys. Lett.* **96**, 213101 (2010).
2. Y. Zhang, S.Q. Wu, Y.H. Wen and Z.Z. Zhu, Surface-passivation-induced metallic and magnetic properties of ZnO graphitic sheet, *Appl. Phys. Lett.* **96**, 223113 (2010).
3. Q. Tang, Y.F. Li, Z.Zhou, Y.S. Chen and Z.F. Chen, Tuning electronic and magnetic properties of wurtzite ZnO nanosheets by surface hydrogenation, *ACS Appl. Mater. Inter.* **2**, 2442 (2010).
4. J. Zhou, Q. Wang, Q. Sun, X.S. Chen, Y.Kawazoe and P. Jena, Ferromagnetism in semihydrogenated graphene sheet, *Nano Lett.* **9**, 3867 (2009).
5. Y.F. Zhu, Q.Q. Dai, M. Zhao and Q. Jiang, Physicochemical insight into gap openings in graphene, *Sci. Rep.* **3**, 1524 (2013).
6. C.L. Freeman, F.Claeyssens, N.L. Allan and J.H. Harding, Graphitic nanofilms as precursors to wurtzite films: theory, *Phys. Rev. Lett.* **96**, 066102 (2006).
7. Y.D. Ma, Y. Dai, M. Guo, C.W. Niu, L. Yu and B.B. Huang, Strain-induced magnetic transitions in half-fluorinated single layers of BN, GaN and graphene, *Nanoscale* **3**, 2301 (2011).
8. J. Zhou, Q. Wang, Q. Sun and P. Jena, Electronic and magnetic properties of a BN sheet decorated with hydrogen and fluorine, *Phys. Rev. B* **81**, 085442 (2010).

9. B. Delley, From molecules to solids with the DMol3 approach, *J. Chem. Phys.* **113**, 7756 (2000).
10. B. Delley, Hardness conserving semilocal pseudopotentials, *Phys. Rev. B* **66**, 155125 (2002).
11. S. Grimme, Semiempirical GGA-type density functional constructed
12. S. Li, Y. Wu, W. Liu and Y. Zhao, Control of band structure of van der Waals heterostructures: Silicene on ultrathin silicon nanosheets, *Chem. Phys. Lett.* **609**, 161 (2014).
13. Z.X. Guo and A. Oshiyama, Structural tristability and deep Dirac states in bilayer silicene on Ag (111) surfaces, *Phys. Rev. B* **89**, 155418 (2014).
14. W. Liu, S.N. Filimonov, J. Carrasco and A. Tkatchenko, Molecular switches from benzene derivatives adsorbed on metal surfaces, *Nat. Commun.* **4**, 2569 (2013).

The evaluation of static characteristics of pressure sensor based on conductive rubber

Jian-Rui Duan[a], Bin Li*, Shuai-Zhen Li and Qin Li
Faculty of Chemical Engineering, Kunming University of Science and Technology, Kunming, 650500, P.R.China
*E-mail: [a]djryan@live.cn, *kmlb@vip.sina.com*

The static characteristics of the conductive rubber pressure sensor are very important in design, manufacturing and use process, and usually index of sensor static characteristics calculation requires manual calculation. However, in the application of study of carbon black filled conductive rubber pressure sensor, the workload is larger, and the efficiency is very low with its reliability being questionable. In this paper, carbon black filled conductive rubber samples and sensor arrays were prepared. It introduces the definition and calculation method of the static characteristics of the conductive rubber pressure sensor, including measurement range, linearity, sensitivity and hysteresis, and realizes that the automatic calculation of the static characteristic and the display of the calibration curve with the data processed by MATLAB. The test results show the algorithm and the feasibility of the program, which can proceed the static characteristics evaluation of pressure sensor simply and intuitively and provide the theory and research foundation for the application of the composite conductive polymer material in the flexible sensor.

Keywords: Conductive Rubber; Static Characteristics; MATLAB; Pressure Sensor; piezoresistive properties

1. Introduction

In 1970s, Hideki Shirakawa, Heeger, MacDiarmid et al first synthesized the polyacetylene films [1], then they found conductive polymer by doping, also known as conductive polymer materials, which the conductive property between the insulator and the metal, and the electrical conductivity range is 10^{-7}~10^{3}s/m, with good force-sensitive and heat-sensitive properties. Conductive polymer composite is a kind of conductive polymer materials, which commonly synthetise used polymer as the matrix with conductive fillers by physical and chemical methods with electrical conductivity and good mechanical properties.

* Corresponding author.

Conductive polymer composite material has the characteristics that raw material easily obtained, simple production process, low cost, large adjustment range of resistivity and so on. When it is influenced by the external force or thermal load, the resistance of the material value will change, which has a good piezoresistive effect [2-4] and thermo-resistive effect [5].

Sensor is a detection device, which can feel the information and output the law of a certain number of electrical signals or other forms of information accordingly. As the development of conductive polymer materials, the research of flexible sensor has made great progress. Flexible sensor not only has the advantages of ordinary sensor array, but also has the flexibility, extension, free bending and folding, convenient carrying, wear, etc., to detect a variety of rules or irregular surface, which is widely used in nondestructive testing, contact measurement, robot skin and biomechanical field [6-8]. Composite conductive polymer materials with silicone rubber as a matrix by filled carbon black have good flexibility and have good piezoresistive properties, which the resistance value is changed with the change of the stress within the scope of a certain stress. Conductive rubber pressure sensor uses the piezoresistive properties of conductive material to the design of a new type of sensor which is a very valuable research content in the field of flexible sensing technology.

Conductive rubber pressure sensor static characteristic indexes includes measurement range, linearity, sensitivity, hysteresis and so on, which is the relationship between input and output and can reflect the various functions of the sensor when each number of input of sensor is steady state. So the static characteristics of the sensor are very important in design, manufacturing and use process, and usually index of sensor static characteristics calculation requires manual calculation. But in the application of study of carbon black filled conductive rubber pressure sensor, the workload is larger, and the efficiency is very low and the reliability is not high[9]. In view of this, in this paper, we use MATLAB to realize the calculation and evaluation about pressure sensor static characteristic based on conductive rubber, which realize the rapid and accurate assessment of the number of static characteristics and also verify the validity of the algorithm and the program.

2. Experimental

2.1. *Materials*

Base material is two component RTV silicone rubber GMX-331D,modified filler is nano SiO_2 (particle diameter of 40nm): Bluestar New Material Co., Ltd.;

Conductive filler is acetylene carbon black (particle diameter of 10-30 nm) : Beijing Boyu Gaoke new materials technology Co., Ltd.; The diluent is petroleum ether (boiling range of category III 90-120°C): Tianjin FengChuan chemical reagent technology Co., Ltd.; Silane coupling agent Si69, Dibutyltin dilaurate (DBTD) as catalysts,tetraethyl orthosilicate (TEOS), Tianjin Guangfu Institute of Fine Cataly.

2.2. *Experimental equipment*

Digital display type electronic Pressure meter:SLJ-B, Wenzhou Shangdu Instrument Co.,Ltd.;Intelligent digital mixing pot heating:ZNCL-G-C, Zhengzhou Huate instrument and Equipment Co.,Ltd.;Electrothermal blowing dry box:GZX-9023,Shanghai Boxun Industrial Co.,Ltd.; Circulating water vacuum pump: SHZ-D (III), Gongyi Yuhua Instrument Co.,Ltd.;Digital multimeter: VC9801A, Shenzhen Shengli Electronic Technology Co.,Ltd.; Precision force electric blender: JJ-1, Changzhou Guohua electrical Co.,Ltd.

2.3. *Sample preparation*

Experimental materials ratio(mass ratio): filled silicone rubber with 6% acetylene carbon black (CB) and 2% nano SiO_2, 3% silane coupling agent Si69, 3% DBTD, 7% TEOS. Sample preparation process is:mixing the silicone rubber and diluent, followed by adding acetylene CB , SiO_2, Si69, DBTD, stirring 5 min to fully mix. Then adding curing agent(TEOS), at same time continue stirring and vacuum pump evacuated the mixture into then evacuated, the final pour it in the model, it were vulcanized at room temperature 22h.

2.4. *Sensor array*

The sensitive unit of conductive rubber pressure sensor array from the previous text, the pressure sensitive element is made into a 9mm in diameter and thickness is 4mm.

In order to measurements over a large area of the contact surface of the force, a plurality of discrete conductive rubber unit design into combination array structure, it can effectively improve sensor to obtain the pressure information of the contact area. Designed a flexible array stratified structure about 10×10 array of printed circuit board and corresponding hierarchical structure of the flexible sensor array, as shown in figure 1 and 2.

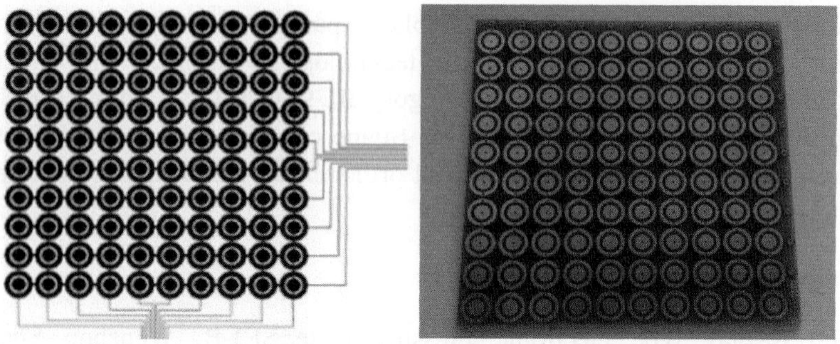

Fig. 1. Schematic diagram and physical diagram of printed circuit array

Figure 2 shows the hierarchical structure of the flexible sensor array. 1and 5 are protective covers, 2 is the array of conductive silicon rubber, 3 is flexible mesh liner layer, 4 is a printed circuit.protective covers are mainly used to protect the conductive silicone rubber and the upper surface of the printed circuit board, conductive silicone rubber layer array elements according to the printed circuit board unit structure, flexible mesh liner layer is mainly used for fixing a wafer shaped conductive silicone rubber sensing unit,in order to increase the contact stability of silicon rubber power unit and the printed circuit, the size and quantity of the hole mainly according to the size of the printing unit circuit to determine.

2.5. *Performance and characterization of samples*

The array of conductive silicon rubber and flexible mesh liner layer level will be placed in the test rig and up and down surface placement electrode,and wire is connected to high precision multimeter, then pressure meter stable loading or unloading, through high precision multimeter measuring resistance. Flowchart of test for the piezoresistive properties is shown in figure 3.

Loading and unloading order: loading→full load→waiting for 10min→unloading→no load, 3 times of cycle. Which maximum test load is 20N, the minimum test load is 0.2N, due to the conductive rubber has the viscoelasticity, When a pressure load to the conductive rubber, there is a period of time when electrical signal from began to electrical signal is stable,So the data must keep 10min when each record. The Piezoresistive properties curves of sample under loading and unloading is shown in figure 4.

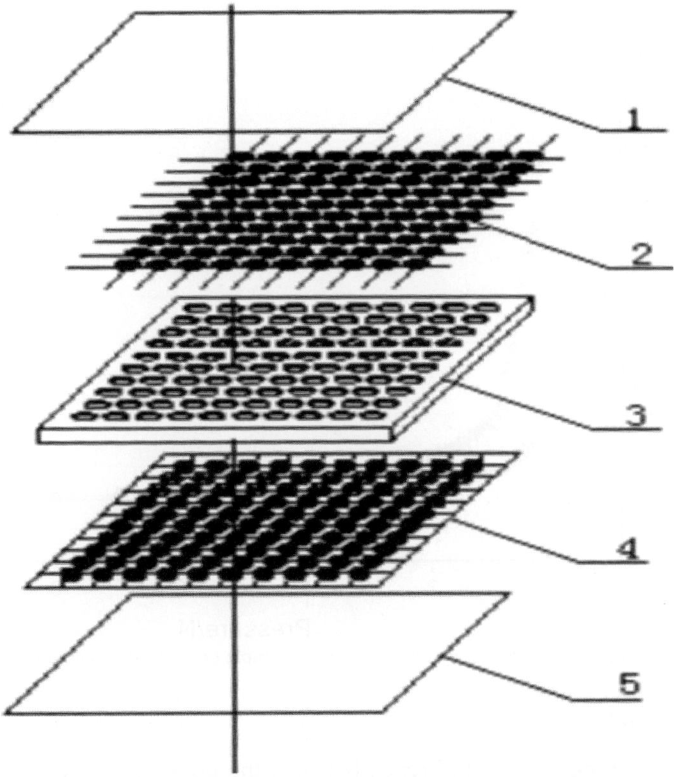

Fig. 2. Layered structure of conductive rubber pressure sensor array

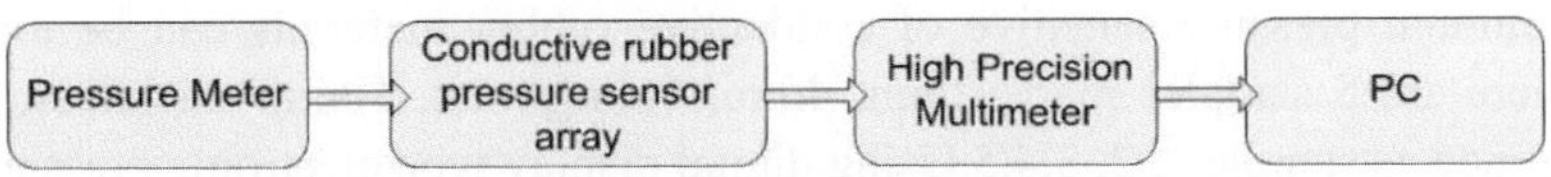

Fig. 3. Flowchart of test for the piezoresistive properties

3. Results and discussion

3.1. *Static characteristics and data processing*

Measurement system of static characteristics is to be measured in steady state measurement the relationship between the input and output, typically used indexes to characterized which include measurement range, linearity, sensitivity and hysteresis[10].

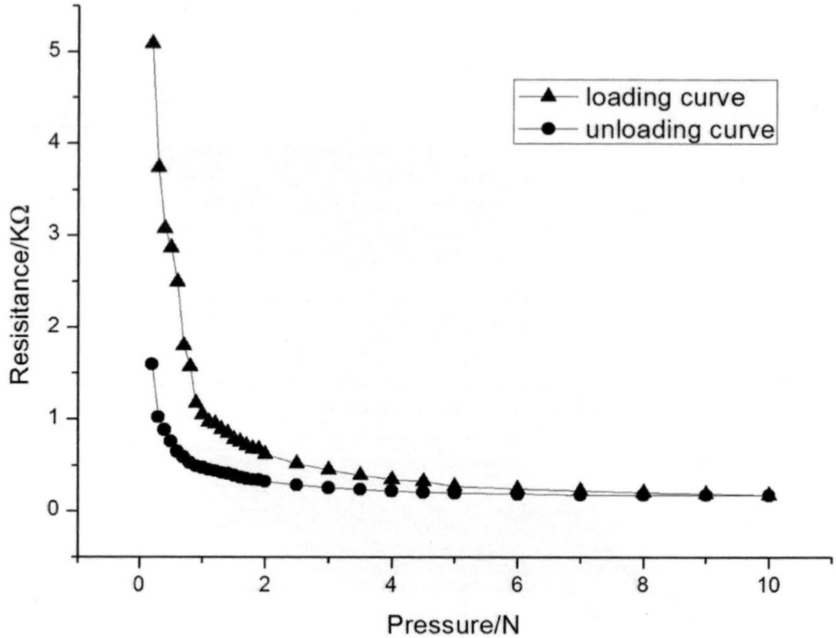

Fig . 4. The Piezoresistive properties curves of sample under loading and unloading

3.1.1. *Measurement range*

Measurement range is the difference between the upper and lower limits of the sensor's measurement range. For conductive rubber pressure sensor measurement range refers to the difference between the maximum and the minimum pressure sensitive of conductive rubber materials can be measured. There are 5 samples made 2mm, 4mm, 6mm, 8mm, 10mm thickness, and are defined as sample 1, 2, 3,4,5.Using digital display type electronic pressure meter to exert pressure. When increased 0.5N measuring the resistance value of relative variation of less than 10Ωor increase 1N the measured resistance value relative variation of less than 20Ωthen defined resistance value change tends to be stable, the end of the measurement. The pressure range of the measurement is the measurement range of the sample, the measurement value is shown in figure 5.

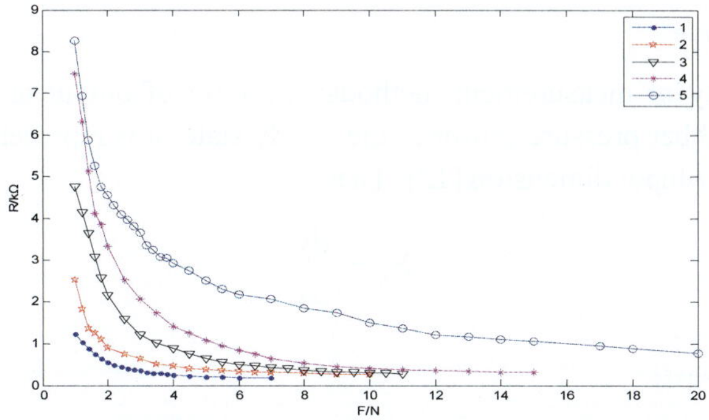

Fig. 5. The Piezoresistive properties curves of samples with different thickness

According to our defined,when resistance value change tends to be stable,can get the resistance range of each sample is shown in table 1.

Table 1. The measurement range of samples with different thickness

Sample thickness	2mm	4mm	6mm	8mm	10mm
measurement range	1-7N	1-10N	1-11N	1-15N	1-20N

According to the table 1 shows that with the increase of the thickness of the sample, the measurement range of conductive rubber is also increased accordingly. This is due to the fact that with the increase of the thickness of the conductive rubber, conductive rubber resistance value increases, the ability of pressure measurement is also a corresponding increase. The thickness of the conductive rubber can be changed, and the range of the measuring pressure of the sensor can be changed.

3.1.2. *Linearity*

Linearity is the nonlinear error, when got sensor input and output characteristic curve, linear fitting of the characteristic curve of the sensor, uses the least squares method to obtain fitting straight line. In the full range($y_{F.S.}$), the difference characteristic curve fitting curve and the actual are compared, $|\Delta L|_{max}$ is the maximum difference[11]. The linearity e_L of the sensorcan be expressed as:

$$e_L = \frac{|\Delta L|_{max}}{y_{F.S.}} \times 100\% \quad (1)$$

3.1.3. *Sensitive*

The sensitivity of measurement methods: the ratio of output and input of the conductive rubber pressure sensor in the steady state, it is represented by Sn and has an output / input dimension [12]. That

$$S_n = \frac{dy}{dx} \tag{2}$$

3.1.4. *Hysteresis*

Hysteresis is defined as when the sensor under the same working condition full measurement range calibration, the loading and unloading curve subtraction, recorded the maximum difference ΔY_{max}, the maximum output is $Y_{F.S}$, hysteresis error of flexible array pressure sensor can be expressed as δ_H:

$$\delta_H = \frac{\Delta Y_{max}}{Y_{F.S.}} \times 100\% \tag{3}$$

3.2. *Program realization of Matlab*

This paper calculated by MATLAB language under Windows operating system implementation, the key sub module of the program design includes: Method of finding fitting straight line by least squares fitting, the matrix was constructed with deducing coefficients of least-square method, calculate the regression value, linearity, hysteresis, and drawing the calibration curve and the loading and unloading curves etc.. Evaluation curve of static parameter of conductive rubber is shown in figure 6.

The fitting curve equation is:

$$\begin{aligned} y = &-0.0002\chi^7 + 0.0072\chi^6 - 0.1121\chi^5 + 0.9016\chi^4 - 4.0269\chi^3 + \\ &9.8967\chi^2 - 12.4654\chi + 6.9441 \end{aligned} \tag{4}$$

Using the above formula the conductive rubber sensor of 4mm thickness which the evaluation result of the static characteristic parameters are: measurement range 1-10N and linearity error is 0.0538, hysteresis for is 0.6866, through fitting curve equation can be known that between the output and the input of the conductive rubber pressure sensor is nonlinear relationship, so sensitivity Sn (KΩ/ N) is not a constant, it will change with input changes, that equation independent front coefficient.

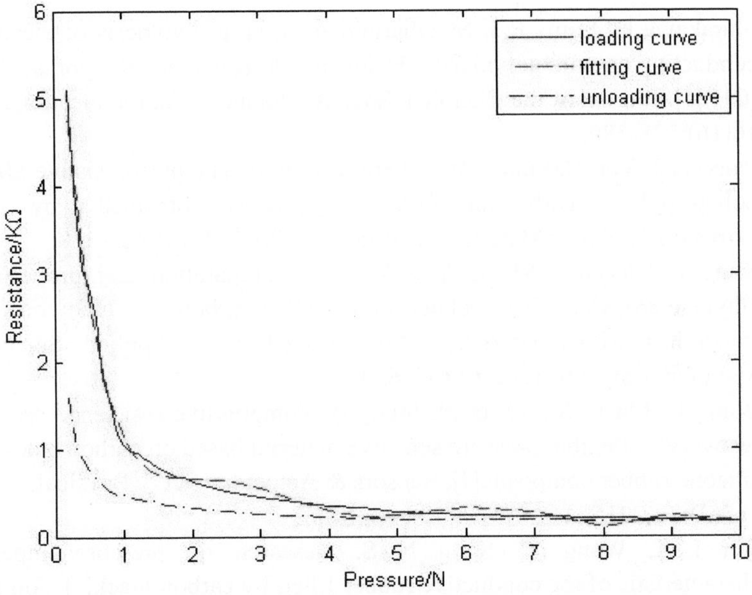

Fig. 6. Evaluation curve of static parameter of conductive rubber

4. Conclusion

This paper introduces the static characteristics evaluation method of conductive rubber pressure sensor, which obtains a number of indicators of sensor static characteristics by using MATLAB to calculate the data, and draws the static calibration curve, which has the advantages of simple evaluation, efficient, rapid, and accurate when compared with the conventional manual calculation. The results show that the hysteresis performance of the material is bad and should be improved. In addition, this article is to evaluate the conductive rubber static characteristic of pressure sensor. The process part is suitable to other formulations of the conductive rubber pressure sensor and any sensor static evaluation, which has the certain promotion application value.

Acknowledgements

This work is supported by the National Natural Science Foundation of china(project No.51167008).

References

1. Shirakawa H, Louis E J, Macdiarmid A G, et al. Synthesis of electrically conducting organic polymers: Halogen derivatives of polyacetylene, (CH)[J]. Journal of the Chemical Society Chemical Communications, 1977, 16(16):578-580.
2. Paleo A J, Van Hattum F W J, Pereira J, et al. The piezoresistive effect in polypropylene—carbon nanofibre composites obtained by shear extrusion[J]. Smart Materials & Structures, 2010, 19(6):322-328.
3. Kang I, Khaleque M A, Yoo Y, et al. Preparation and properties of ethylene propylene diene rubber/multi walled carbon nanotube composites for strain sensitive materials[J]. Composites Part A Applied Science & Manufacturing, 2011, 42(6):623-630.
4. Wang L, Ma F, Shi Q, et al. Study on compressive resistance creep and recovery of flexible pressure sensitive material based on carbon black filled silicone rubber composite[J]. Sensors & Actuators A Physical, 2011, 165(2):207-215.
5. Bin L I, Wang Y, Song N S. Research on pressure/temperature characteristic of the conductive rubber filled by carbon black[J]. Journal of Functional Materials, 2013, 44(13):1871-1875.
6. Hardy L, Billat A, Villermain-Lecolier G. Flat eddy-current matrix sensor for detecting metallic objects[J]. Sensors & Actuators A Physical, 1991, 29(1):13-19.
7. Uesaka M, Hakuta K, Miya K, et al. Eddy-current testing by flexible microloop magnetic sensor array[J]. IEEE Transactions on Magnetics, 1998, 34(4):2287-2297.
8. Ding X, Kuribayashi K, Hashida T. Development of a new type tactile sensor using micro electromagnetic coil for human robot[C]// Micromechatronics and Human Science, 2000. MHS 2000. Proceedings of 2000 International Symposium on. IEEE, 2000:181-187.
9. Sui W T, Zhang D. Evaluation of static characteristics of sensor[J]. Transducer & Microsystem Technologies, 2007,26(3):80-81.
10. GB/T 18459-200 Calculation method of main static performance index of sensor[S].
11. Jiegang P, Jing Ni, Gang D. Sensor Principles and Applications[M]. Beijing:Publishing House of Electronics Industry,2012:4-14.
12. Qingquan L,Xiang G,Chuanyin C,Zhonglei X.Water vapor measurement system based on a MEMSsensor[J].Chinese Journal of Sensors and Actuators,2014,27(6):853-856.

Aluminum electrolysis multi-objective control system based on quantum optimized

Jie-Jia Li*, Zhao-Jun Wang and Jing-Lin Zhu

Information and Control Engineering Faculty, Shenyang JianZhu University
Shenyang,Liaoning,China
E-mail: ljj_0123@sjzu.edu.cn

Aiming at get the objective of saving energy reducing consumption. We proposed Aluminum electrolytic multi-objective optimization control strategy. In this control system, we adopt BP neural network controller to improve the control accuracy and dynamic tracking accuracy and BP neural network controller is optimized by the approach of quantum optimization. Simulation result shows that the DC power consumption of the aluminum electrolytic multi-objective control system which based on quantum improvement has significantly decreased.

Keywords: Aluminum Electrolysis; Multi-objective Optimization; BP Neural Network; Quantum Optimization Algorithm.

1. Introduction

Aluminum electrolytic industry as energy-intensive polluting industries, and how to solve the problem of saving energy and reducing consumption is the important research topics within the industry. Select the appropriate control strategy of electrolytic aluminium production, it is a key to solve the problem of aluminium electrolysis energy saving way. Multi-objective optimization was carried out on the aluminum electrolysis production control is an important way to complete the overall goal.

2. The control scheme of aluminum electrolysis multi-objective control system and quantum optimization algorithm

For comprehensive optimization of all parameters of the aluminum electrolytic cell, we need to make sure the overall scheme of the control system of aluminum electrolytic multi-objective optimization, which can be based on specific circumstances to improve the controller behind.

* Corresponding author.

According to the analysis of electrolytic cell control related influence to the process, especially combining with the predictive control model of established Elman neural network. We raise the following aluminum electrolytic multi-objective optimization control scheme.

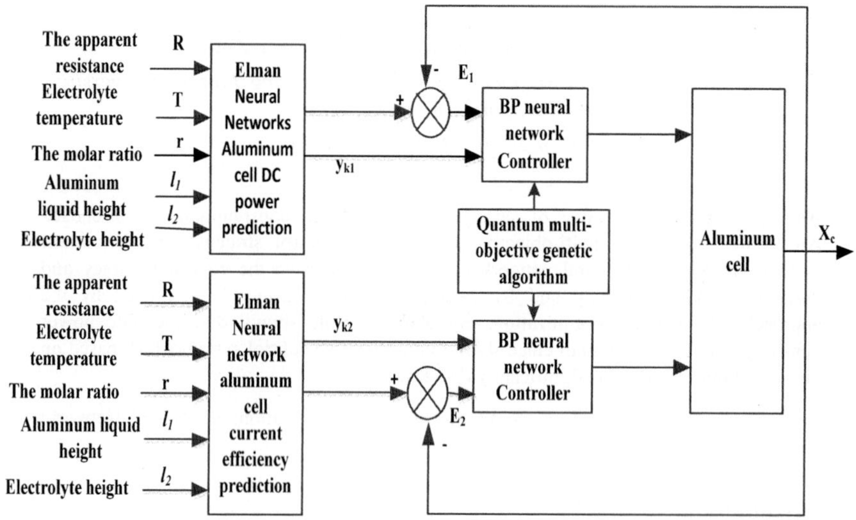

Fig.2.1 Aluminum electrolysis multi-objective control system based on quantum optimized

As shown in figure 2.1, the above is the aluminum electrolysis overall model of the control of multi-objective optimization system. The model of the overall construction well shows that multi-objective optimization control ideas, combining with the organic. It generally consists of these three parts:

The first layer is based on the forecasting model of single objective optimization. we selects the Elman neural network as prediction model. Then analyses the aluminum electrolysis process, selecting five variables as the input value (T apparent resistance R, electrolyte temperature, mole ratio R, height of liquid aluminum l1, electrolyte highly l2). About these five variables set, we have discussed in chapter 2, the neural network has mainly affected the control system preliminary prediction model of the dc power consumption and the current efficiency of modeling and optimization of the output.

The second floor is the optimization model of system, which through the quantum optimization algorithm to control optimize. Considering optimize, the input indexes of multi-objective and then produce a set of variables of inferior quality later.

The third layer is output layer; the optimized parameters control the forecast model of controlled object, create a set of final optimized dc power consumption, which can make the dc power consumption and current efficiency being able to meet the requirements of control system meanwhile. After that, the feedback of controlled object conduct to the front of the current model and the DC power model, respectively, and then compare the two signals.

In order to build the aluminum electrolytic multi-objective optimization control model, choosing suitable controller is very important. Neural network has become nonlinear system controller for its unique advantages .

BP neural network of strong robustness, and can be close to real value, vector-valued and discrete values, and so good at dealing with nonlinear systems, and has strong learning ability and the ability of organization, can be distributed to store information, and can tolerate a long period of training. A large number of scientific practice shows that the BP neural network still has some problems .

3. Fusion quantum multi - objective genetic algorithms and BP neural network controller

Quantum optimization algorithm belongs to a kind of heuristic search algorithm, and its theoretical ideas come from quantum mechanics, which based on quantum evolutionary algorithm to improve multi-objective genetic algorithm. The algorithm adopts the quantum selection operator, crossover operator; mutation operator and population update operators. Now we discuss the specific design process of the algorithm.

BP neural network to determine the initial value and the threshold of complete quantum multi-objective optimization algorithm optimization operation use the global search capability to determine the specific binding process.

Therefore, the combination of quantum multi-objective Genetic Algorithm BP Neural Network should include the following major steps:

Step1: Initialization BP neural network to determine the number of input and output, network layers, network training times, learning initialization parameter precision, regulatory factors, population size, maximum number of iterations, crossover and mutation probability, the cumulative number of iterations t is set to 0 .

Step2: the weights and thresholds by BP network input layer and the hidden layer weights and thresholds, right hidden layer and output layer with a threshold value concatenated sequence, randomly generated n quantum chromosome, forming initial population.

Step3: calculate fitness of each chromosome. Population produce offspring with the parent population, and the union of the two populations together to form a new population, carried the gravitational class classifying operation, the next generation population.

Step4: The qualified solution, into the next generation, to continue its implementation of selection, crossover and mutation genetic manipulation.

Step5: to determine whether the maximum number of iterations proceed if it reaches down, if not return to step 3.

Step6: The best individual coding sequence by decoding, as the initial weights and thresholds of BP network.

Step7: ago BP network to spread, the global error calculation, if they meet the requirements to end.

Step8: is less than the number of cycles, if less than the proceeds back propagation BP neural network. Otherwise, the algorithm ends.

4. The experiment simulation and results analysis

To verify the performance of quantum many-objective evolutionary algorithm, compare this algorithm with NSGAII algorithm, program based on NSGA II algorithm for BP neural network optimization has been described indetail in relevant literature.

4.1. *The experiment simulation parameter settings*

Based on a pre-baked anode electrolytic aluminum plant when to 350kA, take the relevant production data in a month as the reference value simulation inputs, when the plant during normal operation. Use the Matlab software for simulation experiments under ideal conditions.

This paper discusses the BP neural network optimization program, Learning and training methods, as for the Elman neural network which most role of prediction model is less involved, this is because the role of front-end mainly from the modeling of Elman neural network control does not demand much, input five variables represent resistance R, electrolyte temperature T, the molar ratio r, aluminum liquid height l1, Electrolyte height l2, about NSGA II algorithm we can set scale population to Np = 200, set Mutation rate as 0.7,ser crossover probability to CR = 0.5, set the maximum number of iterations of evolution to 500.

As for quantum multi-objective optimization algorithm, crossover probability PC=0.9, mutation probability Pm=1/21, decided by the weights of BP neural network controller to optimize, population size is set to 100, evolution algebra is still set to 500.

4.2. *Results analysis*

Based on the input data before, input respectively, we can get two control effect comparison chart of the algorithm, as shown in figure 4.1. and figure 4..2, respectively:

The average of current efficiency is calculated using equation 4.1,

$$\overline{\eta} = \frac{\sum_{\eta=T_1}^{T_{31}} \eta_n}{T} \qquad (4.1) \qquad\qquad \overline{W} = \frac{\sum_{W=T_1}^{T_{31}} W_n}{T} \qquad (4.2)$$

In this sheet, $\overline{\eta}$ is the average current efficiency, η_n is the current efficiency of every single day, T is time.

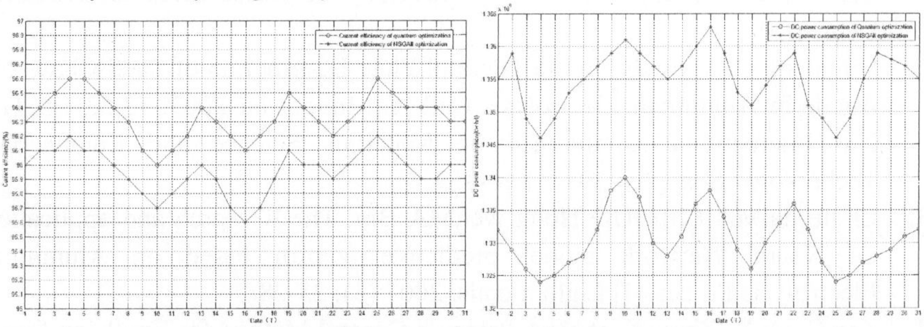

(a) (b)

Fig.4.1 (a)The area chart of two algorithms of current efficiency optimization effect (b)The area chart of two algorithms of DC power dissipation optimization effect

Picture 4.1(a) is the current efficiency of the two algorithms optimization results comparison chart, based on actual operating history data in a month aluminum plant as a reference inputs, simulation diagram clearly see the quantum optimization algorithm (in circle) the current efficiency is improved, after calculation increased by 0.39%. And the corresponding dc power consumption, as shown in figure 4.1 (b), calculated using quantum optimization (in circle) of the dc power consumption compared to the average of the average of the dc power consumption was optimized by using the NSGA II reduces 244.55 KWH.

5. Conclusion

The quantum optimization method put forward in this paper is to improve controller, multi-objective control of DC power consumption of front-end Elman neural network and current efficiency prediction model, this method is good for Aluminum electrolytic energy saving. Computer simulations show that the current efficiency of control system Based on Quantum electrolytic multi-objective optimization is improved as to the consumption of NSGAII electrolytic multi-objective optimization control system, and the DC power consumption is reduced. It has a certain practical significance for the production of electrolytic.

Reference

1. Cheng Chunyan. Study on the Development Strategies of China's Aluminum Industry from the Background of Economic Transition[D]. A Dissertation Submitted to China University of Geosciences for Doctoral Degree.2013,6.
2. Dingdong QIN. Study on the Thermal Field Distribution of Steel Claws for Anode in Aluminum Electrolysis Cell [C]. Indoor air 2003, 13:53-64. The 2nd International Conference on Materials Science and Engineering Technology: Shanghai 2015: 848-853.
3. LI Jie jia,MA Chi,GUO Hong wei,MA Bin. The Research and application Of Neural Network model predictive control based on Aluminum Electrolysis Process[J] Light metals, 2007 (3): 25-28.
4. Tan C, Qi N, Zhou X, et al. A Pressure Control Method for Emulsion Pump Station Based on Elman Neural Network[J].Computational intelligence and neuroscience, 2015, 12(2) 153-160.
5. Deb K, Agrawal S, Pratap A, et al. A fast elitist non-dominated sorting genetic algorithm for multi-objective optimization: NSGA-II [J]. Lecture notes in computer science, 2000, 1917: 849-858
6. Jiejia Li, ZhiChao Fang, Lulu Sun. Rough set - neural network application in aluminum electrolysis fault diagnosis [J]. Journal of shenyang architectural university (natural science edition), 2009, 792:792-795.

7. Li Li, Shi Zhongping, Pan Feng. Modeling of penicillin ferment process based on the combination of non-dominated sorting genetic algorithm II and error back propagation network [J] Micro computer information, 2009, 25(4): 287-289.

Dual-responses electrochemical detection of Cu^{2+} using a novel terpyridine functionalized GC electrode

Yong-Ping Luo*†, Shun-Jian Xu, Zong-Hu Xiao, Wei Zhong, Huan-Wen Wu and Hui Ou

Xinyu Institute of New Energy, Xinyu University,
Xinyu, 338004, P. R. China
E-mail: † luoyongpingxy@163.com

A novel and facile method for the electrochemical detection of Cu^{2+} was first developed by using a modified glass carbon (GC) electrode with 4'-(4-hydroxyphenyl)-2,2':6',2"-terpyridine (p-HO-ptpy). P-HO-ptpy was first synthesized by Kröhnke protocol and then modified onto GC electrode steadily. The modified GC electrode with unexpected high affinity for Cu^{2+} was used for the electrochemical analysis of trace Cu^{2+} by differential pulse voltammetry (DPV). The modified GC electrode was found to be highly selective to Cu^{2+} in the range of 0.2 – 30 μM with a detection limit of 60 nM. The modified electrode offered an excellent approach for the selective detection of Cu^{2+} in practical samples based on the dual responses of anode and cathodic currents.

Keywords: Dual-responses; Electrochemical Analysis; Copper ion; Terpyridine.

1. Introduction

Cu^{2+} is the third in abundance among the essential heavy metal ions in the human body and plays a significant role in many fundamental physiological processes ranging from bacteria to mammals [1]. If the level of Cu^{2+} exceeds the cellular needs, it will be high toxic to organisms and biological systems [2]. Excessive Cu^{2+} can cause oxidative stress and disorders associated with neurodegenerative diseases, such as Alzheimer's disease, Wilson's diseases, and Parkinson's disease, and so on [3,4]. Due to the continuing concern over Cu^{2+} contamination in the environment and food, the sensitive and on-site monitoring of Cu^{2+} in biological and environmental samples is highly desirable.

Various methods have been used for the determination of $Cu^{2+,}$ including atomic absorption spectroscopy [5], inductively coupled plasma mass spectrometry [6], inductively coupled plasma optical emission spectroscopy [7],

* Corresponding author.

chemosensor [8], and electrochemical approaches [9,10]. Because of the good analytical properties, electrochemical approaches have been gained increasing attentions for affording high selectivity, sensitivity, and sufficient accuracy without the need for expensive and specialized equipment [11].

In this work, 4'-(4-hydroxyphenyl)-2,2':6',2"- terpyridine (p-HO-ptpy), which could exhibit Cu^{2+} selectivity by electrochemical signal, was designed, synthesized and anchored onto the surface of glass carbon (GC) electrode. The characteristics and the electrocatalytic activity of the p-HO-ptpy modified GC electrode for the detection of Cu^{2+} were studied. Then the determination of Cu^{2+} with the p-HO-ptpy modified GC electrode by differential pulse voltammetry (DPV) was performed as a novel approach to the development of electrochemical sensor for the determination of Cu^{2+}. Finally the proposed method was applied to analysis of practical samples.

2. Materials and methods

2.1. *Preparation of p-HO-ptpy/GC electrode*

P-HO-ptpy was synthesized in an aldol reaction from 2-acetylpyridine and 4-Hydroxy benzaldehyde by Kröhnke protocol [12]. The GC electrode modified with p-HO-ptpy was labeled as p-HO-ptpy/GC. The p-HO-ptpy/GC electrode was used for the determination of Cu^{2+} in various water samples.

2.2. *Electrochemical measurements*

The anodic and cathodic currents were monitored and recorded by dipping the modified electrodes into ~10 mL of a stirred sample solution containing different concentration Cu^{2+}. The calibration curves were plotted according to the relationship between peak current of Cu^{2+} and its concentration. The relationship between peak current of Cu^{2+} and its concentration was plotted. Various raw water samples were determined with the same method, and their concentrations extrapolated from the calibration plot.

3. Results and discussion

3.1. *Electrochemical activity of Cu^{2+} on p-HO-ptpy/GC electrode*

As shown in Figure 1(A), curves a and b displayed the CV curves of the bare GC electrode and p-HO-ptpy/GC electrode in 10 mM PBS (pH 7.0), respectively. No voltammetric peaks, but only charge currents were observed for them. When 10 μM Cu^{2+} was added into 10 mM PBS (pH 7.0), an oxidation peak was observed at GC electrode (curve c). For the p-HO-ptpy/GC electrode (curve d) in 10 mM PBS (pH 7.0) containing 10 μM Cu^{2+}, a pair of well-defined and redox peaks were observed with the corresponding oxidative and reductive peak potential located at -0.02 mV and -0.17 mV, respectively. The results indicated that as expected that the p-HO-ptpy/GC electrode showed good response toward Cu^{2+} due to the effective combination of designed functional molecule p-HO-ptpy with Cu^{2+} through coordination interaction. Thus, the novel p-HO-ptpy/GC provided a strong basis to construct a sensor for the analysis of Cu^{2+} in natural water.

Figure 1(B) showed CVs of the p-HO-ptpy/GC electrode in 10 mM PBS (pH 7.0) containing 10 μM Cu^{2+} at various scan rates. The peak current was found to increase proportional to the scan rate with the increase of scan rate in the range of 10 – 500 mV s^{-1} as shown in the inset of Figure 1B, which suggested that a typical adsorption-controlled process on the surface of the p-HO-ptpy/GC electrode.

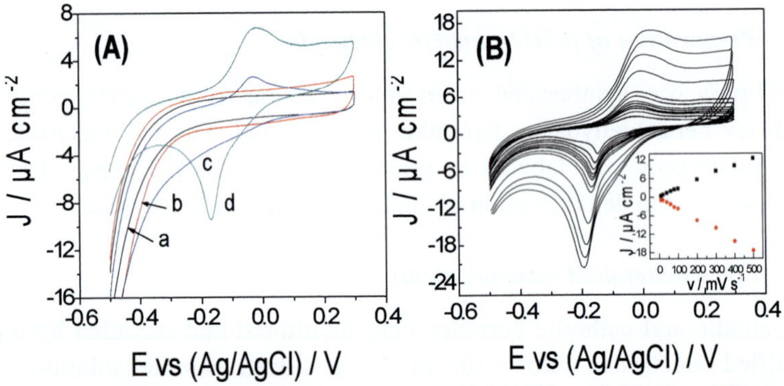

Fig. 1. (A) CVs of (a) GC, (b) p-HO-ptpy/GC in 10 mM PBS (pH 7.0), and GC (c), p-HO-ptpy/GC (d) in 10 mM PBS (pH 7.0) containing 10 μM Cu^{2+} at 100 mV s^{-1}. (B) CVs of p-HO-ptpy/GC electrode in 10 mM PBS (pH 7.0) containing 10 μM Cu^{2+} at different scan rates: 10 – 500 mV s^{-1}. Inset: plot of peak current versus scan rates.

3.2. *DPV of Cu^{2+} on p-HO-ptpy/GC electrode*

In order to prove the sensitivity for the quantitative analysis of Cu^{2+}, DPV had been adopted to record the redox peak current under the optimized experimental conditions. Figure 2(A) and Figure 2(C) depicted the DPV responses at the p-HO-ptpy/GC electrode in 10 mM PBS (pH 7.0) containing different concentrations of Cu^{2+} for anodic and cathodic peak current, respectively. The insets of Figure 2(A) and (C) showed the DPV responses of (A) anodic and (C) cathodic currents at the p-HO-ptpy/GC electrode in 10 mM PBS (pH 7.0) containing 0.2 μM Cu^{2+}. Discernible peaks could be seen. The anodic and cathodic peak currents increased linearly with the increasing concentration of Cu^{2+}. The relationship between peak current of Cu^{2+} and its concentration was plotted in Figure 2(C) and (D). A good linearity was obtained in the range of $2 \times 10^{-7} - 3.0 \times 10^{-5}$ M with a correlation coefficient of 0.994. The detection limit and sensitivity were found to be 60 nM (S/N = 3) and 1.79×10^3 μA M^{-1} cm^{-2} under optimal experimental conditions, respectively. The p-HO-ptpy/GC electrode showed excellent analytical performance, for instance, wider and perfect linear detection range, relatively low detect limit and higher selectivity.

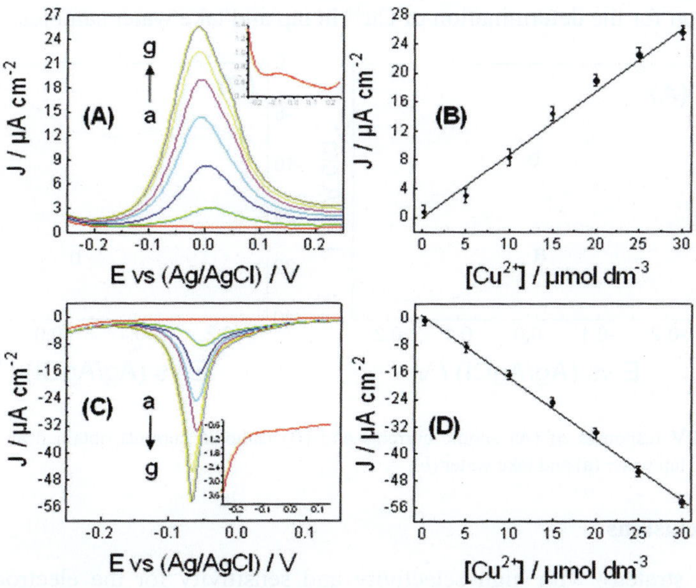

Fig. 2. DPV responses of (A) anodic currents and (C) cathodic currents obtained at the p-HO-ptpy/GC in 10 mM PBS (pH 7.0) containing different concentrations of Cu^{2+}. The concentration of Cu^{2+} is (a) 0.2, (b) 5, (c) 10, (d) 15, (e) 20, (f) 25, (g) 30 μM. The insets in (A) and (C) show DPV responses of (A) anodic currents and (C) cathodic currents in 10 mM PBS (pH 7.0) containing 0.2 μM Cu^{2+}. Linear plots of anodic peak currents (C) and cathodic currents (D) vs. concentration of Cu^{2+}.

3.3. *Analysis of Cu^{2+}in natural water*

As demonstrated above, the developed electrochemical method based on p-HO-ptpy/GC electrode for Cu^{2+} detection with high selectivity and sensitivity provided an excellent platform for assaying Cu^{2+} in tap water and lake water without pretreatment. As shown in Figure 3, the obvious anodic and cathodic peak currents were obtained. According to the calibration curves shown in Figure 2(B) and (D), the concentrations of Cu^{2+} in tap water and lake water were estimated to be 2.65 ± 0.06 μM and 16.51 ± 0.09 μM, respectively. The concentration of Cu^{2+} in tap water was about one magnitude lower than the World Health Organization (WHO) recommended level (2.0 mg L^{-1} or 31.25 μM) for Cu^{2+} in drinking water [13], while that in lake water was the same magnitude. The detected results of Cu^{2+} in tap water and lake wake by the new method were compared with those obtained by using the traditional AAS method. The concentrations of Cu^{2+} determined by the present method were in good agreement with those obtained by the AAS method. The results demonstrated the developed p-HO-ptpy/GC electrode was potentially application for the determination of Cu^{2+} in tap and lake water samples.

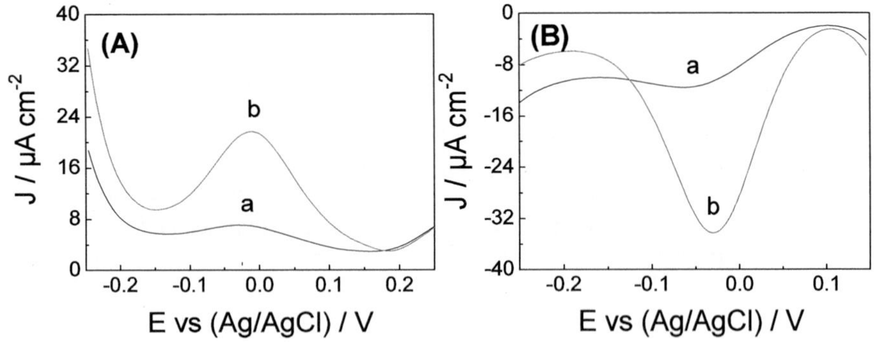

Fig. 3. DPV responses of (A) anodic currents and (B) cathodic currents obtained at the p-HO-ptpy/GC in tap water (a) and lake water (b).

4. Conclusions

A novel strategy with high selectivity and sensitivity for the electrochemical detection of Cu^{2+} was developed based on the specific molecular recognition of p-HO-ptpy toward Cu^{2+} due to p-HO-ptpy on GC that led to the formation of an organic film with unexpected high affinity for Cu^{2+}. Using the p-HO-ptpy/GC electrode for determination of Cu^{2+}, the linear range was $2 \times 10^{-7} – 3.0 \times 10^{-5}$ M and the detection limit was 60 nM. This novel method with theoretical simplicity, and less instrumental demands, combined with the characteristics of electrochemistry such as easy to real-time and on-line detection, led us to achieve a sensitive method for the detection of Cu^{2+} in natural water with dual responses of anodic current and cathodic current, with good selectivity towards some common interfering ions.

Acknowledgements

This work is supported by National Natural Science Foundations of China (51263021, 51462035), Natural Science Foundation of Jiangxi (2014BAB216014), Educational Commission of Jiangxi (GJJ151214) and China Postdoctoral Science Foundation (2014M561436)

References

1. M.C. Linder and M. Hazegh-Azam, Copper biochemistry and molecular biology, *Am. J. Clin. Nutr.* **63**, 797S (1996).
2. Z.L. Harris and J.D. Gitlin, Genetic and molecular basis for copper toxicity, *Am. J. Clin. Nutr.* **63**, 836S (1996).
3. G. Multhaup, A. Schlicksupp, L. Hesse et al., The amyloid precursor protein of Alzheimer's disease in the reduction of copper(II) to copper(I). *Science* **271**, 1406 (1996).
4. R.A. Løvstad, A kinetic study on the distribution of Cu(II)-ions between albumin and transferring, *Biometals* **17**, 111(2004).
5. X. Wen, Y. Zhao, Q. Deng et al., The coupling of rapidly synergistic cloud point extraction with thermospray flame furnace atomic absorption spectrometry, *Microchim. Acta* **178**, 139 (2012).
6. J.S. Becker, A. Matusch, C Depboylu et al., Quantitative imaging of selenium, copper, and zinc in thin sections of biological tissues (Slugs-genus arion) measured by laser ablation inductively coupled of plasma mass spectrometry, *Anal. Chem.* **79**, 6074 (2007).
7. Y. Liu, P. Liang, L Guo, Nanometer titanium dioxide immobilized on silica gel as sorbent for preconcentration of metal ions prior to their determination by inductively coupled plasma atomic emission spectrometry, *Talanta* **68**, 25 (2005).
8. S.J. Lee, S.S. Lee, J.Y. Lee et al., A functionalized inorganic nanotube for the selective detection of copper(II) ion, *Chem. Mater.* **18**, 4713 (2006).
9. H. Zhao, Y. Jiang, Y. Ma et al., Poly(2-amino-4-thiazoleacetic acid)/multiwalled carbon nanotubes modified glassy carbon electrodes for the electrochemical detection of copper(II), *Electrochim. Acta* **55**, 2518 (2010).
10. M. Lin, M. Cho, W.S. Choe et al., Electrochemical analysis of copper ion using a Gly-Gly-His tripeptide modified poly(3-thiopheneacetic acid) biosensor, *Biosens. Bioelectron.* **25**, 28 (2009).

11. X. Shao, H. Gu, Z. Wang et al., Highly selective electrochemical strategy for monitoring of cerebral Cu^{2+} based on a carbon dot-TPEA hybridized surface, *Anal. Chem.* **85**, 418 (2013).
12. F. Krhnke, The specific synthesis of pyridines and oligopyridines, *Synthesis* **1**, 1 (1976).
13. X. Ma, Z. Tan, G. Wei et al., Solvent controlled sugar–rhodamine fluorescence sensor for Cu^{2+} detection, *Analyst* **137**, 1436 (2012).

Application of leaves derived biochars into counter electrode in dye-sensitized solar cells

Shun-Jian Xu

Xinyu Institute of New Energy, Xinyu University,
Xinyu, 338004, China
E-mail: xushunjian@126.com

Four biochars, which were obtained by one-step pyrolysis of leaves, were employed as catalytic materials for counter electrodes in dye-sensitized solar cells (DSCs). The leaves that were involved were pine needles, camphor leaves, firry leaves and red after-wood leaves. The results show that the remaining weight of the pine needles, camphor leaves, firry leaves and red after-wood leaves are respectively 9.49, 12.45, 10.39 and 10.96%. The pore size in the biochar derived from camphor leaves has an obvious bimodal distribution, with both macrospores (up to 10 μm) and mesopores (2~6nm), which is main reason for higher catalytic activities of biochar than graphite through providing more active sites. The S_{BET} of the biochar is 186.3 cm^2/g. The conversion efficiency (η) of each biochar based DSC is higher than that of graphite based one. Moreover, the camphor leaves derived biochar used in DSCs shows the maximum efficiency values in the four biochars.

Keywords: Leaves; Pyrolysis; Biochar; Dye-sensitized Solar Cells; Counter Electrode.

1. Introduction

Global climate change and the inevitable depletion of fossil fuel reserves are major challenges facing the 21st century which has led to a boom in research relating to alternative energy sources. As a kind of devices to harness the limitless power of the Sun, dye-sensitized solar cells (DSCs) have attracted increasing fundamental and commercial interest for their low cost, easy fabrication and moderate efficiency [1]. A typical DSC consists of a dye-sensitized TiO_2 photoanode, an electrolyte and a Pt counter electrode. The role of Pt counter electrode is to collect electrons arriving from the external circuit and to catalyze the reduction of triiodide ion. However, the resource scarcity, poor stability and high costs of Pt could inhibit its large-scale application in the future [2].

Recently, several carbon materials such as carbon nanotubes, carbon black, graphite, activated carbon (AC), porous carbon and graphene have been utilized as catalyst to replace Pt for counter electrode, and their devices show favorable photovoltaic performances [3-5]. In such application, the disadvantages in the catalytic activity of carbon compared with Pt can be compensated by increasing the active surface area of catalytic film via a porous structure. Therefore, carbon materials with well-developed pores (i.e. porous carbon) should be more favorable than dense or undeveloped-pore carbon materials.

Biochar is the carbon-rich product with porous structure when biomass (such as wood, leaf or crop residues) is heated in a closed container with little or no available air. The biochar have attracted increasing fundamental and commercial interest for their advantage of cheap, abundant, renewable and environmental friendly [6]. In this research, four biochars obtained by one-step pyrolysis of leaves were employed as catalytic materials for counter electrodes in DSCs to investigate theirs potential application. The leaves were involved in pine needles, camphor Leaves, firry leaves and red after-wood leaves.

2. Experimental

2.1. *Biochar synthesis and counter electrode fabrication*

Four leaves were collected from Xinyu University and were individually selected as precursors to prepare the biochars. The leaves were first thoroughly washed with deionized water in order to remove the impurity, and dried at 120 °C to constant weight. The leaves were then pyrolyzed by slowly heating to 700 °C for 1h in a N_2 atmosphere, and mechanically milled to obtain biochar powders.

The Biochar counter electrode was prepared by coating a blended biochar paste on FTO substrate and then heated at 120 °C for 2h. The blended biochar paste was obtained through grinding 300mg biochar powders, 30mg carboxymethyl cellulose (CMC) and 5ml ethyl alcohol aqueous solution. For comparison, graphite counter electrode was fabricated by the same procedures.

2.2. *Devices configuration*

The DSC was configured by assembling the N719 sensitized TiO_2 photoanode and the biochar or graphite counter electrode, between which was separated by using a 50 μm thick spacer. The surface area of both TiO_2 film and biochar or graphite film was ca. 0.7 cm × 0.7 cm. The preparation of the photoanode can be seen in the previous work [7]. The electrolyte was 0.5 M LiI; 0.05 M I_2; 0.5 M TBP in a solvent of 3-methoxypropionitrile.

2.3. *Measurements and characterizations*

The morphology of biochar was observed by scanning electron microscopy (SEM; SIRION200, FEI, USA). Pore characteristics of the biochar were examined on an automatic adsorption instrument (ASAP 2020M, Micromeritics, USA), using nitrogen at 77 K. Pore size distributions and surface areas were calculated by BJH and BET methods respectively from N_2 desorption isotherms.

Electrochemical impedance spectroscopy (EIS) measurements were conducted with an electrochemical workstation at 10 mV of amplitude. A symmetric cell consisting of two identical counter electrodes and an electrolyte was used in the EIS measurements. The electrolyte was same as the one used in the fully functional DSCs. Photocurrent-voltage performances of DSCs were measured under the illumination of a Xenon light source. The light intensity was 100 mW/cm^2.

3. Results and discussion

3.1. *Photovoltaic Performances of Biochars based DSCs*

The photovoltaic parameters of DSCs consisting of biochar counter electrodes derived from different leaves are listed in Table 1. Table 1 also shows the photovoltaic parameters of DSCs with graphite counter electrode for comparison. The conversion efficiency (η) of each biochar based DSC is higher than that of graphite based one. The η value of different biochar based cells varies from 1.07 to 1.37% while that of graphite based cell is 0.77%. The high efficiency for the biochar based DSCs is attributed to the enhancement in Fill Factor (*FF*) and open circuit voltage (V_{oc}). The values of *FF* and V_{oc} of the biochar based DSCs distribute respectively in the range of 0.169~0.227 and 0.529~0.623 V while the values of *FF* and V_{oc} of the graphite based ones are respectively 0.175 and 0.447 V. Moreover, the camphor leaves derived biochar used as catalytic material in DSCs shows the maximum η value in the four biochars.

Table 1. Photovoltaic parameters of DSCs consisting of biochar counter electrodes derived from different leaves and graphite counter electrode.

Leaves	J_{sc} (mA/cm^2)	V_{oc} (V)	FF	η (%)
Pine needles	10.14	0.623	0.169	1.07
Camphor Leaves	11.98	0.532	0.215	1.37
Firry leaves	9.07	0.529	0.227	1.09
Red after-wood leaves	10.07	0.615	0.184	1.14
Graphite	9.9	0.447	0.175	0.77

3.2. *Catalytic Properties of Biochar Counter Electrode*

To insight into the reasons for the higher efficiency of biochar than graphite as counter electrode in DSCs, the EIS was employed. The typical Nyquist plots for camphor Leaves derived biochar and graphite counter electrodes are shown in Figure 1. Two impedance parameters, charge transfer resistance (R_{ct}) and the ohmic serial resistance (R_s) can be derived from plot fitting by using the equivalent circuit. The R_{ct} characterizes the catalytic activity for the triiodide reduction, and the R_s is mainly related to the sheet resistance of the two identical counter electrodes. The biochar electrode presents lower R_{ct} than the graphite electrode, implying higher catalytic activity toward the triiodide ion. The R_{ct} values of biochar and graphite electrodes are 24.71 and 52.2 Ω cm^2, respectively. Moreover, from biochar to graphite, the R_s value decreases from 31.33 Ω cm^2 to 45.4 Ω cm^2, indicating lower sheet resistance for the biochar electrode.

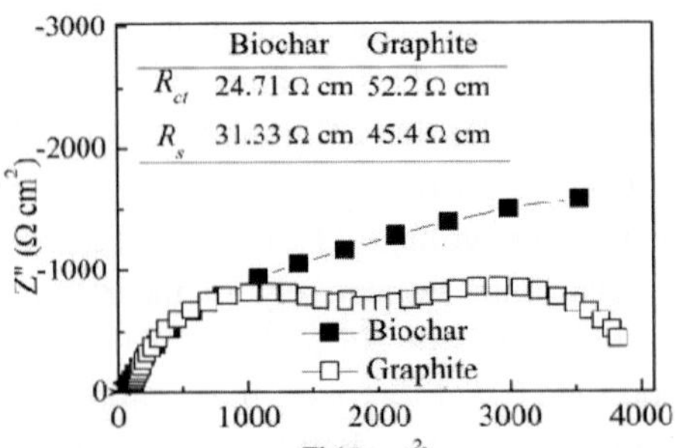

Fig. 1. Nyquist plots of biochar derived from camphor leaves and graphite counter electrodes.

3.3. *Morphologies and Pore Structure of Biochar*

Through the pyrolysis at 700 °C in N_2 atmosphere, the leaves are converted into the biochar consisting of major carbon and minor other substances, such as calcite and kaolinite [8]. The remaining weight of the pine needles, camphor leaves, firry leaves and red after-wood leaves is respectively 9.49, 12.45, 10.39 and 10.96%. The SEM morphologies of biochar derived from camphor leaves are illustrated in Fig. 2. It is obvious that the convex parts with dispersive distribution are observed on the surface of the biochar. The cross-section morphology of the biochar reveals that there exist a lot of pore channels with disorder distribution. A closer observation further shows that the pore channels has a size of 2~10 μm and coarse pore walls, which is beneficial to provide more active sites for the triiodide reduction than dense graphite. Fig. 3 demonstrates the pore size distribution of biochar derived from camphor leaves. The biochar contains a great quantity of mesopores with size from 2 to 50 nm, which is a reason for higher catalytic activities of the biochar electrode than the graphite electrode through providing more active sites. The S_{BET} of the biochar is 186.3 cm^2/g.

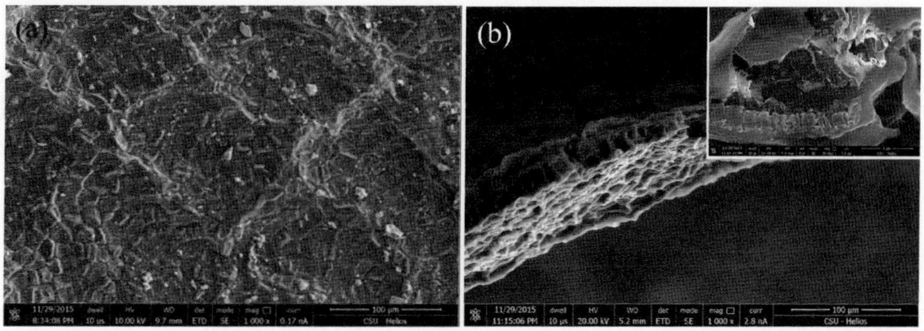

Fig. 2. SEM morphologies of biochar derived from camphor leaves: (a) plane (surface); (b) cross-section.

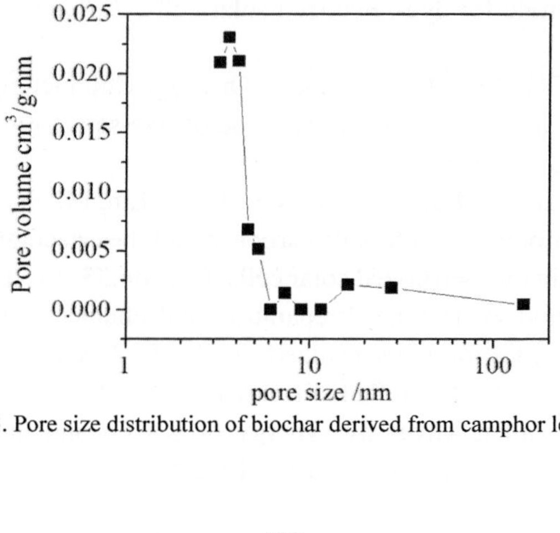

Fig. 3. Pore size distribution of biochar derived from camphor leaves.

4. Conclusion

Four biochars derived from leaves were employed as catalytic materials for counter electrodes in DSCs. The pore size in biochar derived from camphor leaves has an obvious bimodal distribution, which is main reason for higher catalytic activities of biochar than graphite through providing more active sites. The η of each biochar based DSC is higher than that of graphite based one. The high efficiency for the biochars based DSCs is attributed to the enhancement in the *FF* and V_{oc}. Moreover, the camphor leaves derived biochar used as the catalytic material in DSCs shows the maximum efficiency values in the four biochars.

The low temperature (120 °C) treatment process can be employed to fabricate a low-cost and efficient counter electrode for both glass based DSCs and flexible DSCs.

Acknowledgements

This work is supported by National Natural Science Foundation of China (51462035), Training Programme Foundation for Young Scientist of Jiangxi, China (20133BCB23035) and Educational Commission of Jiangxi, China (KJLD13100).

References

1. B. O'Regan and M. Grätzel, A low-cost, high-efficiency solar cell based on dye-sensitized colloidal TiO_2 films, *Nature* **353**, 737 (1991).
2. G.Q. Wang, W. Xing and S.P. Zhuo, Application of mesoporous carbon to counter electrode for dye-sensitized solar cells, *J. Power. Sources* **194**, 568 (2009).
3. B.H. Fan, X.G. Mei, K. Sun and J Ouyang, Conducting polymer/carbon nanotube composite as counter electrode of dye-sensitized solar cells. *Appl. Phys. Lett.* **93**, 143103(2008).
4. G. Yue, J. Wu, J.Y. Lin, Y. Xiao, S.Y. Tai, J. Lin, M. Huang and Z. Lan, A counter electrode of multi-wall carbon nanotubes decorated with tungsten sulfide used in dye-sensitized solar cells, *Carbon* **55**, 1 (2013).
5. A. Tetsuya and H. Hajime, Preparation and characterization of Pt/carbon counter electrodes for dye-sensitized solar cells, *Mater. Lett.* **94**, 15 (2013).
6. Y.T. Li, Y.T. Pi, L.M. Lu, S.H. Xu and T.Z. Ren, Hierarchical porous active carbon from fallen leaves by synergy of K_2CO_3 and their supercapacitor performance, *J. Power Sources* **299**,519 (2015).

7. S.J. Xu, Y.F. Luo, G.W. Liu, G.J. Qiao, W. Zhong, Z.H. Xiao, Y.P. Luo and H. Ou, Bifacial dye-sensitized solar cells using highly transparent PEDOT:PSS films as counter electrodes, *Electrochim. Acta* **156**, 20 (2015).
8. Y.S. Yin, Z.Z. Wang, H. Tian, W. Zhang, J.Q. He, L. Liu and X. Yan, Evolution of mineral matter and carbonaceous structure during lignocellulosic municipal solid waste pyrolysis, *J. Fuel Chem. Tech.* **43**, 160 (2015).

Effect of negative bias on photoelectric properties of NiO thin films prepared by magnetron sputtering

Zong-Hu Xiao*†, Shun-Jian Xu, Yong-Ping Luo, Wei Zhong, Hui Ou,
Huan-Wen Wu and Er-Shuai Jiang

Xinyu Institute of New Energy, Xinyu University
Xinyu 338004, China
E-mail: †xiaozonghu@126.com

Magnetron sputtering has been the most widely useful technique having high deposition rates, uniformity over large areas of the substrates and easy control over the composition of the deposited films. The films properties depend on various deposition process parameters such as substrate temperature, sputtering power, oxygen partial pressure, sputtering pressure and post deposition conditions. Using reactive radio-frequency magnetron sputtering method, NiO thin films were prepared on glass substrates. The influence of negative bias on the crystal structure, surface morphological, optical property was investigated using X-ray diffraction, scanning electron microscopy and ultraviolet-visible spectrophotometer and Hall Effect tester, respectively. It is found that the negative bias can improve the crystallinity, flatness and photoelectric properties of the NiO films in a certain range. An optimal transmittance of 78% and resistivity of 0.6 Ω·cm is obtained when the negative bias is 100 V. This indicates that the negative bias is an important means to control the properties of the films.

Keywords: Nickel Oxide; Magnetron Sputtering; Negative Bias; Photoelectric Properties

1. Introduction

The transparent conductive electrode usually consists of n-type semiconductor material, the lack of p-type transparent conductive electrode has limited the development of fully transparent active devices. Therefore, researchers around the world are stepping up the development of p-type transparent conductive electrode. Nickel oxide (NiO) is a promising material for various applications, such as electrochromic devices, p-type transparent conductive thin films, chemical sensors and solar cells [1-2]. Although stoichiometric NiO is an insulator, it is reported that the resistivity of NiO can be lowered by increasing Ni^{3+} ions, being ascribed to doping of monovalent atom such as Li, Na, K or Ni

* Corresponding author.

vacancies and interstitial oxygen in NiO crystallites[3-4]. However, the properties of NiO thin films for the p-type transparent conductive films is low until now.

Magnetron sputtering has been most widely useful technique having high deposition rates, uniformity over large areas of the substrates and easy control over the composition of the deposited films. The films properties depend on various deposition process parameters such as substrate temperature, sputtering power, oxygen partial pressure, sputtering pressure and post deposition conditions. In addition, applying a negative bias on the substrate can increase the flow and energy of cations [5], leading to the film growth process, structure and properties change.

In this work, we report on the preparation of NiO thin films by reactive radio frequency magnetron sputtering system and applying a negative bias on the substrate, investigating the influence of negative bias on the crystal structure, surface morphology, optical and electrical properties of NiO thin films.

2. Experimental

2.1. *Preparation of NiO thin films*

The NiO thin films were deposited on Corning 1737 glass substrates by reactive radio frequency magnetron sputtering method from a metallic nickel target (size: 60 mm in diameter with a thickness of 3 mm) of 99.99 % purity. The chamber base pressure was approximately 8×10^{-4} Pa. Pure argon was used as sputter gas and oxygen as reactive gas.

All substrates were ultrasonically cleaned with acetone, ethanol and deionised water. In addition, the target was pre-sputtered for 10 min to remove the surface oxide layers if any on the surface of the target. The distance between the target and the substrate was approximately 70 mm. Sputtering deposition was performed under a gas pressure of 2 Pa. The parameters during the deposition are listed in Table 1.

Table 1. Deposition conditions of the reactive radio frequency magnetron sputtered NiO thin films

Sample	Negative bias/V	T/(°C)	Power/W	Gas flow/sccm		Time/h
				O_2	Ar	
NB-0	0	25	80	10	20	2
NB-50	50	25	80	10	20	2

Table 1. (Cont'd)

NB-100	100	25	80	10	20	2
NB-150	150	25	80	10	20	2
NB-200	200	25	80	10	20	2
NB-250	250	25	80	10	20	2

2.2. *Measurements and characterizations*

The structural properties of NiO thin films were characterized by X-ray diffractometer (XRD, Bruker Advance D8), using Cu K_α radiation (λ=0.1546 nm). The surface morphology of the films was analysed by scanning electron microscopy (SEM, JEOL JSM-6700F). The thickness of the films were carried out by thin film analysis system (Aquila NKD 7000V). For the optical properties of the films, the transmittance measurements were measured by using ultraviolet-visible spectrophotometer (Purkinje General TU-1901) with a wavelength resolution better than ±0.1 nm. The electrical properties of the NiO thin films were carried out by Hall effect tester (Ecopia HMS-3000). All the measurements were performed at room temperature.

3. Results and Discussion

3.1. *Crystal structure*

The crystal structure of the films were performed on X-ray diffraction measurements, shown in Figure 1. It is found that the NiO thin films is polycrystalline and retains the cubic structure (JCPDS data No. 78-0643) with a strong (200) plane orientation. From the XRD patterns, it could be seen that the intensity of (200) diffraction peak is enhanced with the increase of negative bias. While the negative bias voltage exceed 200 V, diffraction peak disappears.

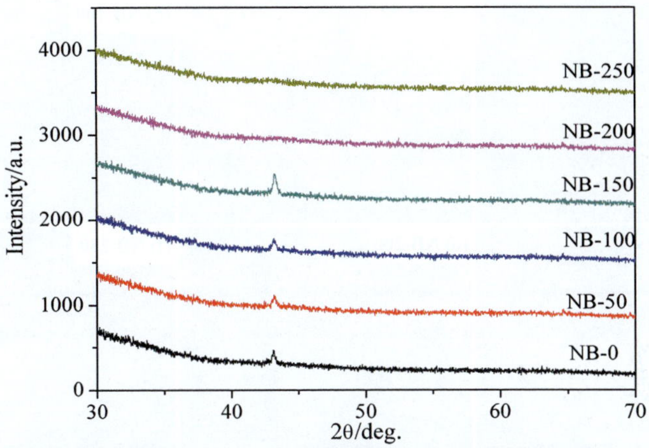

Fig.1. XRD patterns of the NiO thin films at various negative bias

The presence of negative bias can attract a portion of the argon ion moving onto the substrate, to bombard the substrate surface. With the increase of the negative bias, the bombardment effect is gradually enhanced. On the one hand, the films crystallinity is better, showing diffraction peaks getting stronger, since the bombardment energy is transferred to the growth units deposited on the substrate, and the growth units have enough energy to migrate to the appropriate location. On the other hand, when the negative bias excess, argon ion bombardment transforms into a destructive effect, resulting in that the flim exists a lot of crystal defects and the grains are refined gradually until amorphous growth appears.

3.2. *Surface morphology*

The surface morphology variation of NiO thin films with sputtering power was observed by SEM , shown in Figure 2. It can be seen that the most significant impact of negative bias on the films morphology is flatness. With the gradual increase of the negative bias, the flatness is higher and higher, and finally to achieve the atomic level flatness. One possible explanation is that there are many holes on the film growth surface with island growth and poor surface flatness, due to the original crystalline defects and atom self-shadowing effect, when there is no or low negative bias. And when the negative bias is large enough, energetic particle bombardment energy can migrate Ni or O unit deposited on the islands to these holes, showing an improved surface flatness.

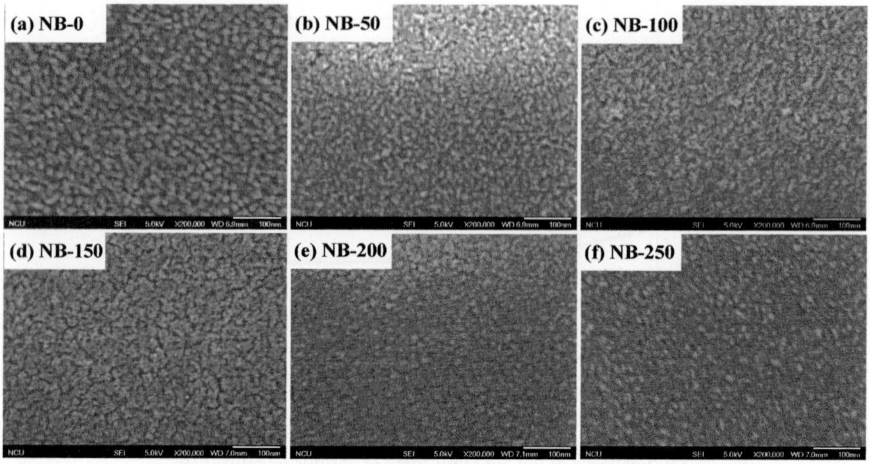

Fig.2. SEM images of the NiO thin films at various negative bias

3.3. *Optical and electrical properties*

The optical transmittance of all as-prepared NiO thin films is more than 60 % at the visible range, shown in Figure 3. Overall, the transmittance of the films decreases with the increase of negative bias, except for the abnormity of the 100V, shown in the inset of Figure 3. There are two ways , light absorption and scattering, for the optical loss in thin film transmission [6]. The light absorption mainly depends on the film thickness, while the light scattering is related to the defects of the film. It is found that the transmittance of the films vary obviously when the bias voltage is less than 150 V, and decreased significantly after 150 V.

SEM and XRD analysis results show that the negative bias does not significantly affect t the films flatness and grain size before 150 V. While the negative bias continues to increase, the film flatness is improved significantly and grains refine rapidly, which well explains the decrease of transmittance. In addition, the increase of transmission is caused from the reduction of light absorption when the negative bias is 100 V, duing to the reduction of crystal defects and a better crystallinity.

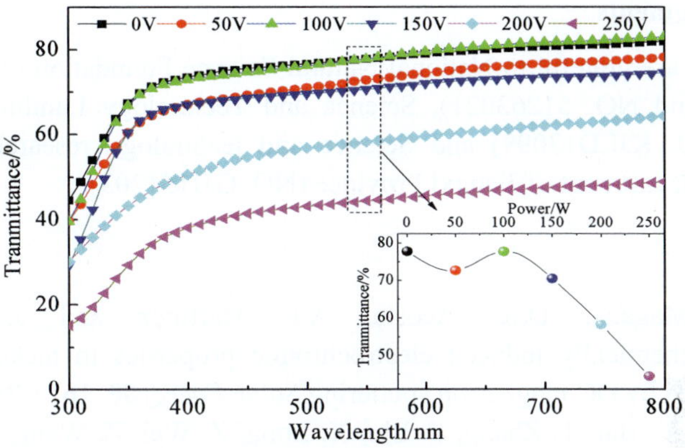

Fig. 3. The transmittance of the NiO thin films at various negative bias

The electrical properties of the NiO thin films is shown in Table. 2. It is clear that the resistivity of the NiO thin films performs decreasing previously and increasing later with increasing the negative bias, having minimum resistivity of 0.6 Ω·cm at negative bias of 100 V. The electrical resistivity of NiO thin films has a strong dependence on the microstructural defects existing in NiO crystallites, such as nickel vacancies and interstitial defects [7]. This result is in agreement with the optical properties.

Table 2. Resistivity of the NiO thin films at various negative bias

Sample	NB-0	NB-50	NB-100	NB-150	NB-200	NB-250
Resistivity/(Ω·cm)	4.1	2.9	0.6	1.8	34	77

4. Conclusions

The NiO thin films have been preapred by reactive radio frequency magnetron sputtering method on glass substrates. The influence of negative bias on the crystal structure, surface morphological, optical and electrical properties was investigated. Conclusions are as follows:

(i)the crystallinity of the NiO films increases with the increase of negative bias voltage in a certain range. While the negative bias excess, the films grows into an amorphous.

(ii)Negative bias can improve the flatness of the NiO films and decrease, especially a larger negative bias excess 200V.

(iii)Negative bias can decrease the transmittance and resistivity of the film in a certain extent, the minimum resistivity of 0.6 Ω·cm is obtained at negative bias of 100 V.

Acknowledgments

The authors are grateful to National Natural Science Foundation of China (NO. 51462035 and NO. 51263021), Science and Technology Landing Project in Jiangxi (NO. KJLD13099) and Science and technology research project of Education Department of Jiangxi Province (NO. GJJ151203).

References

1. C.R. Magana, D.R. Acosta, A.I. Martinez and J.M. Ortega, Electrochemically induced electrochromic properties in nickel thin films deposited by DC magnetron sputtering,*Solar Energy***80**, 161 (2006).
2. Z. Zhu, Y. Bai, T. Zhang, Z. Liu, X. Long, Z. Wei, Z. Wang, L. Zhang, J. Wang and S. Yang, High-Performance Hole-Extraction Layer of Sol–Gel-Processed NiO Nanocrystals for Inverted Planar Perovskite Solar Cells,*Angew. Chem. Int. Ed.***53**, 1 (2014).
3. S. Kerli and U. Alver, Effect of F-doping on structural, electrical, and optical properties of NiO thin films,*Crystallography Reports***59(7)**, 1103 (2014).
4. Y. Reddy, B. Ajitha and P.S. Reddy, Influence of thermal annealing on structural, morphological, optical and electrical properties of NiO–Cu composite thin films,*Mater. Express***4(1)**, 32 (2014).
5. S. Kerli, U. Alver and H. Yaykssli, Investigation of the properties of In doped NiO films,*Applied Surface Science***318**, 164 (2014).
6. Z. Xiao, M. Zhang and M. Guo, Investigation on zinc oxide thin films prepared by sol-gel progress,*Journal of Synthetic Crystals***37(4)**,1003 (2008).
7. Y. Zhao, H. Wang, C. Wu, Z.F. Shi, F.B. Gao, W.C. Li, G.G. Wu,B.L. Zhang and G.T. Du, Structures, electrical and optical properties of nickel oxide films by radio frequency magnetron sputtering,*Vacuum***103**, 14 (2014)

Preparation of transparent carbon counter electrode with PEG800 as carbon precursor and its bifacial photoelectric properties

Wei Zhong*†, Shun-Jian Xu, Zong-Hu Xiao, Yong-Ping Luo, Huan-Wen Wu and Hui Ou

Xinyu Institute of New Energy, Xinyu University,
Xinyu, 338004, China
†E-mail: zhongweiwen@126.com

An integrated transparent carbon catalyst layer was prepared in situ on conductive subtract with polyethylene glycol 800 (PEG800) as carbon precursor. The effect of spin coating rate on transmittances, morphologies, catalytic activity of carbon counter electrodes and bifacial photoelectric properties of assembled Dye-sensitized solar cells (DSCs) were emphatically investigated. Results show that prepared carbon counter electrodes possess high light transmittance property, with transmittances of over 70% for visible range (400nm~700nm) basically. DSCs assembled with carbon counter electrodes show great biface light absorption property, their rear conversion efficiencies reaches 80%~90% of front efficiencies. Spin coating rate shows a reverse relationship with catalytic activity of carbon counter electrodes and photoelectric properties of corresponding device.

Keywords: Carbon Counter Electrode; In-Situ Synthesis; Dye-Sensitized Solar Cells; Bifacial Photoelectric Properties.

1. Introduction

Dye-sensitized solar cells (DSCs) with relatively high conversion efficiency while relatively low cost provide an effective way of exploring solar energy. Basically, a DSC is composed of a photoelectrode, a counter electrode, dye and electrolyte. A counter electrode acts to accept electrons from outer circuit and catalyze electrolyte. It is typically made up with transparent conductive substrate and deposited Pt catalyst layer. Although the assembling devices based on Pt counter electrodes show excellent property, keeping the record of highest conversion efficiency of DSCs[1], Pt counter electrodes could hardly be applied

* Corresponding author.

into mass production due to its high price, low corrosion resistance and harsh preparation condition[2].

Recently, more and more efforts were devoted to develop new counter electrodes to replace Pt ones, related catalyst materials include carbon[3] (including traditional carbon materials such as activated carbon, graphite, carbon black and new carbon materials such as carbon nanotubes, graphene and mesoporous carbon), conducting polymers[4] (PEDOT, polyaniline et al.), transition metal compounds[5] and their composites. Carbon materials become outstanding and attract extensive attention for their wide sources, low cost, good catalytic activity and high stability. However, carbon counter electrode show two obvious deficiencies limited by its preparation technique: carbon particles may easily fall off from conductive substrate after long time soaking in electrolyte resulting in decrease of stability of the device, and the non-transparency of carbon electrode limits its application in bifacial DSCs[6]. If counter electrode could be made transparent or translucent, the assembled device could acquire light from both photoanode and counter electrode, bringing about higher conversion efficiency. Moreover, this kind of DSCs may be prospective to be applied in building-integrated photovoltaics (BIPV).

In this paper, in-situ synthesis technique was applied to form an integrated transparent carbon catalyst layer on transparent conductive substrate through processes of solution preparation, spin coating and pyrolysis, with polyethylene glycol 800 (PEG800) as carbon precursor. The influence of spin coating rate on the transmittances, catalyst activity, morphologies of carbon counter electrodes and photoelectric properties of assembling DSCs were studied, intrinsic causes for diverse conversion efficiencies were analyzed.

2. Experimental

2.1. *Preparation of carbon counter electrode*

Carbon counter electrodes were prepared as following: 2g PEG800 was added into mixed solvent of 2ml acetylacetone and 2ml isopropanol to form a mixed solution using stirring; then the mixed solution was spin coated onto FTO substrate with a spin coating rate of 1000 rpm, 2000rpm, 3000rpm, respectively, and a spin coating time of 20s; the samples were kept at 60°C for 2h, and ultrapure water was used to clean the edge of samples; finally, after pyrolysis in N_2 atmosphere at 500°C for 30min, carbon counter electrodes were obtained.

2.2. *Configuration of device*

A photoanode and a carbon counter electrode were piled up to form a sandwich structure with a spacer of 25um thick inserted between, electrolyte was injected slowly and absorbed spontaneously into the space between these two electrodes due to capillary force. After sealing process, a DSC was obtained. A 3-Methoxypropionitrile solution with 0.5mol/L LiI, 0.05mol/L I_2 and 0.5mol/L 4-tert-butylpyridine was used as electrolyte. The preparation and sensitization of photoanode have been introduced in previous works[7].

2.3. *Measurements and characterization*

UV-visible absorption spectrums of counter electrodes were measured by a UV-Vis spectrophotometer (UV-Vis; TU-1901, Persee, China). The morphologies of carbon film on the surface of carbon counter electrodes were characterized by an atomic force microscopy (AFM; SPM-9600, Shimadzu, Japan). Photoelectric properties and electrochemical impedance spectroscopy (EIS) were tested with an electrochemical workstation (CS310, Corrtest, China). Photocurrent-voltage performances were measured under the illumination of a Xenon light source (CHF-XM500, Trusttech, China) with light intensity of 100 mW/cm^2.

3. Results and Discussion

3.1. *Transmittance of carbon counter electrodes*

UV-Vis absorption spectrums of carbon counter electrodes are showed in Figure 1, compared with absorption spectrum of Pt electrode. It's obvious that the prepared carbon counter electrodes possess excellent light transmittance property. The transmittances for visible range (400nm~700nm) are above 70% basically, and increase slowly as the wavelength rises, reaching a stable level of 90% at near infrared range (700nm~900nm).

Compared with carbon counter electrodes, Pt counter electrode shows relatively poor transmittances, stable at around 45% and reaching 46.78% at 550nm. For bifacial DSCs, when the sunlight incidences from counter electrode side, the light shall transmits through counter electrode and electrolyte, then reaches photoanode and gets absorbed. Thus the transparency of counter electrode is a decisive factor for conversion efficiency under counter electrode illuminations. Therefore, the carbon counter electrodes with high transmittance property could be qualified to assemble bifacial DSCs, and it can be predicted that the efficiency acquired from rear illuminations (with light illuminates from counter electrode side) shall be comparable with the efficiency obtained from front illuminations (light illuminates from photoanode side).

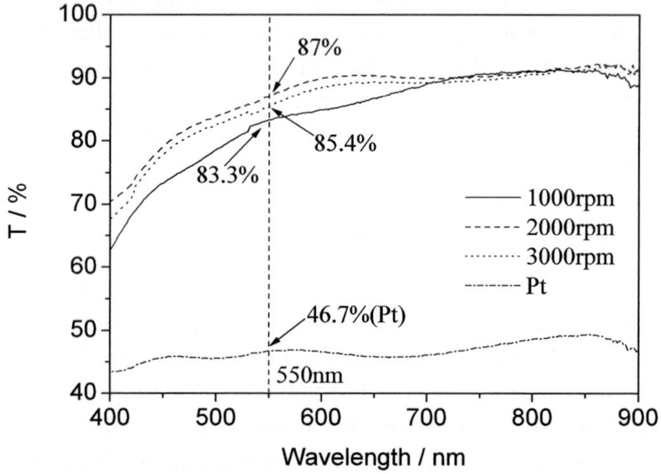

Fig.1. Effect of spin coating rate on UV-Vis spectrum of carbon counter electrodes

3.2. *Photoelectric properties of DSCs*

Figure 2 illustrates the IV curves of DSCs assembling with carbon counter electrodes, Table 1 summarizes corresponding photoelectric properties. Photoelectric properties of device based on Pt electrode are also listed for comparisons. Under photoanode illuminations, as spin coating rate rises from 1000rpm to 3000rpm, open circuit voltage (V_{oc}) stays around 0.69V~0.71V, short-circuit photocurrent density (J_{sc}) decreases from 9.88mA/cm^2 to 8.83mA/cm^2, fill factor (FF) declines from 0.32 to 0.27, finally resulting in conversion efficiency (η) to drops from 2.20% to 1.65%.

Under counter electrode illuminations, carbon counter electrodes with superior transmittance property guaranteed devices with relatively high rear efficiencies. As the spin coating rates are 1000rpm, 2000rpm, 3000rpm, respectively, the rear efficiencies attain 1.76%, 1.55%, 1.33%, reaching 80.0%, 79.9%, 80.6% of corresponding front efficiencies. The efficiency differences between two illumination conditions are caused mainly by Jsc with a large variation of 30%~39% between two illumination conditions. While for the device based on Pt counter electrode, the efficiency for rear illuminations only reaches 44% of the efficiency for front illuminations, caused by poor

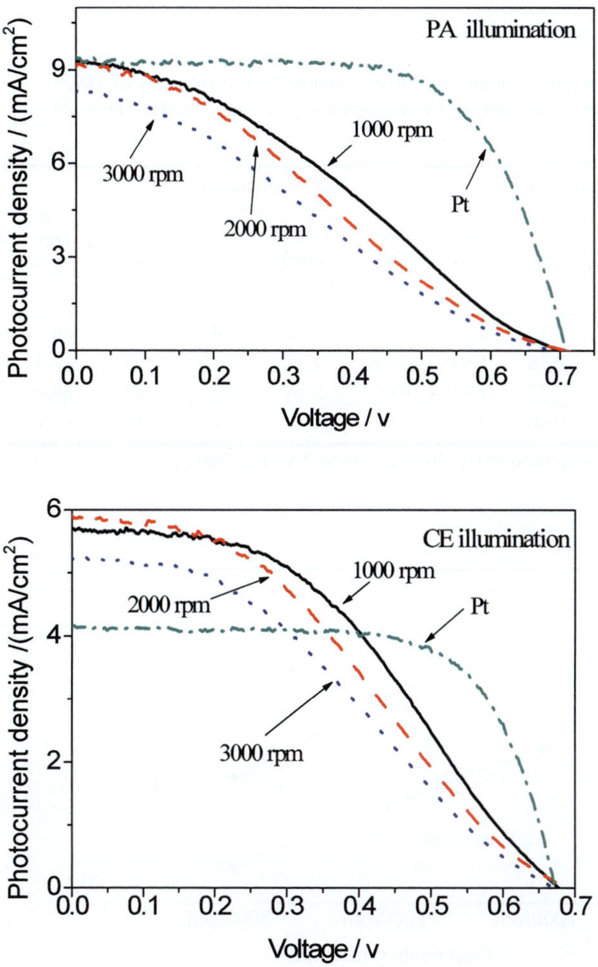

Fig. 2 Influence of spin coating rate on the J-V curves of DSCs based on carbon counter electrodes under photoanode (PA) and counter electrode (CE) illuminations

transmittance property of Pt electrode. Therefore, the DSCs assembled with prepared carbon counter electrodes show relatively high bifacial light absorption. Regardless of either illumination condition, the conversion efficiency declines approximately linearly as the spin coating rate increases, as shown in Figure 3.

Table 1 Influence of the spin coating rate on the photovoltaic parameters of DSCs based on carbon counter electrodes under photoanode (PA) and counter electrode (CE) illuminations

Spin coating rate/rpm	PA illumination				CE illumination				h_{CE}/h_{PA} (%)
	V_{oc} (V)	J_{sc}/(mA /cm^2)	FF	η (%)	V_{oc} (V)	J_{sc}/(mA /cm^2)	FF	η (%)	
1000	0.70	9.88	0.32	2.20	0.68	6.08	0.43	1.76	80.0
2000	0.71	9.76	0.28	1.94	0.69	6.31	0.36	1.55	79.9
3000	0.69	8.83	0.27	1.65	0.67	5.56	0.36	1.33	80.6
Pt	0.71	10.03	0.66	4.70	4.64	0.67	0.67	2.10	44.1

Note: h_{CE}/h_{PA}(%) means efficiency ratio of CE illumination to PA illumination

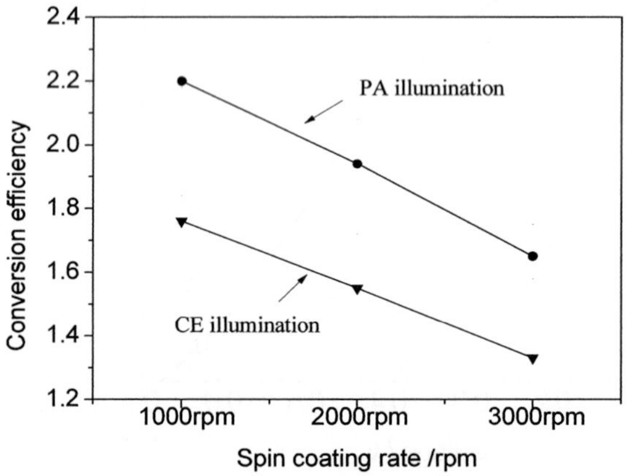

Fig. 3. Effect of spin coating rate on the conversion efficiency of DSCs based on carbon counter electrodes under photoanode (PA) and counter electrode (CE) illuminations

3.3. *Electrochemistry properties of carbon counter electrodes*

EIS was applied to explore the relationship between the front conversion efficiency of the device and spin coating rate. The charge transfer resistance (R_{ct}) representing catalytic activity of carbon counter electrode could be acquired from Figure 4[8]. R_{ct} increases from $9067\Omega\cdot cm^2$ to $17944\Omega\cdot cm^2$ with spin coating rate rises from 1000rpm to 3000rpm, indicating the catalytic activity is decreasing, so that the conversion efficiency of corresponding DSCs shall decrease.

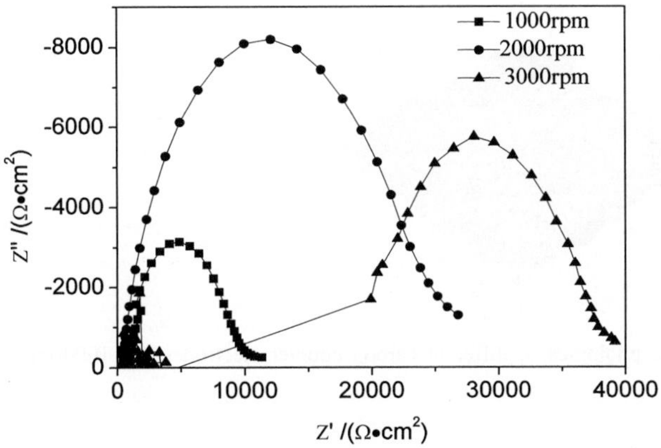

Fig. 4. Nyquist plots of carbon counter electrodes with different spin coating rate

3.4. *Morphologies of carbon counter electrodes*

AFM was conducted to study the cause of diverse catalytic activity of carbon electrodes under various preparation conditions. Figure 5. reveals that Spin coating rate has a great effect on the morphologies of counter electrodes. Counter electrode with 1000rpm shows a round ball-like surface with a relatively high average roughness (R_a) of 8.855nm, while with 1000rpm a silkworm strip-like surface with a R_a of 4.965nm, and with 3000rpm a square tuber-like surface with a R_a of 3.617nm. A smaller spin coating rate brings about a bigger average roughness, and a rougher surface here means more catalytic active sites which will lead to better catalytic activity of the counter electrode and eventually higher photoelectric properties of corresponding device.

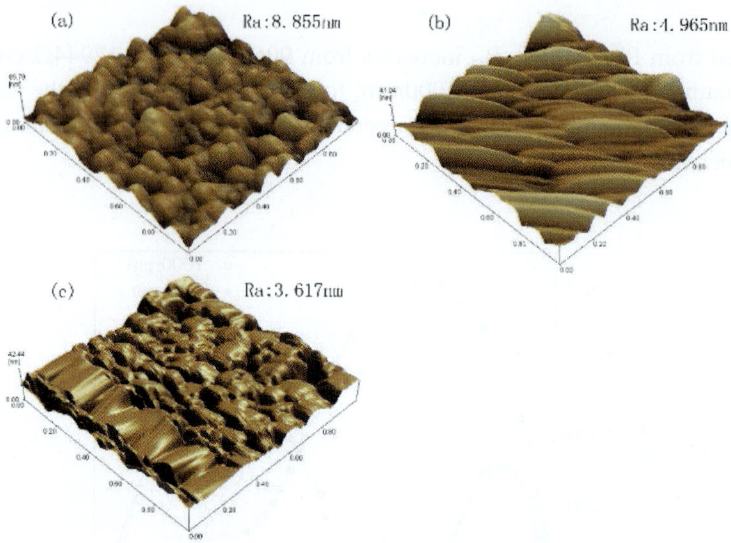

Fig. 5 AFM morphologies of different carbon counter electrodes. (a) 1000rpm; (b) 2000rpm; (c) 3000rpm

4. Conclusion

The prepared carbon counter electrodes show high light transmittance property, with transmittances of over 70% for visible range (400nm~700nm) basically. DSCs assembled with carbon counter electrodes show great biface light absorption property, their rear conversion efficiencies reaches 80%~90% of front efficiencies, compared with merely 44% for Pt electrodes based device. Results also show that surface roughness should be considered as the main reason for diverse catalytic activity of carbon electrodes under various preparation conditions. Therefore, a smaller spin coating rate brings about a bigger average roughness, and a rougher surface here means more catalytic active sites which will lead to better catalytic activity of the counter electrode and eventually higher photoelectric properties of corresponding device.

Acknowledgements

This work is supported by National Natural Science Foundation of China (51462035), Training Programme Foundation for Young Scientist of Jiangxi, China (20133BCB23035) and Educational Commission of Jiangxi, China (KJLD13100, GJJ151203).

References

1. A. Yella, M. Gratzel, Porphyrin-sensitized solar cells with cobalt (II/III)-based redox electrolyte exceed 12 percent efficiency,*Science***334**,629(2011).
2. M. Grätzel, Solar Energy Conversion by Dye-Sensitized Photovoltaic Cells, *Inorganic Chemistry***44**,6841(2005).
3. G. Wang, J. Zhang. Nitrogen-doped porous carbon prepared by a facile soft-templating process as low-cost counter electrode for high-performance dye-sensitized solar cells, *Materials Science in Semiconductor Processing***38**,234(2015).
4. Q. Tai, B. Chen, In-Situ prepared transparent polyaniline electrode and its application in bifacial Dye-Sensitized solar cells, *American Chemical Society***5**,3795(2011).
5. P. Yang, Z. Zhao, Counter electrodes from Mo-Se nanosheet alloys for bifacial dye-sensitized solar cells, *Journal of Alloys and Compounds***648**,930(2015).
6. J. Liu, Q. Tang, Cost-effective bifacial dye-sensitized solar cells with transparent iron selenide counter electrodes, *Journal of Power Sources***275**,288(2015).
7. S. Xu, Y. Luo, Investigation of catalytic activity of glassy carbon with controlled crystallinity for counter electrode in dye-sensitized solar cells, *Solar Energy***85**,2826(2011).
8. M. Grätzel, Dye-sensitized solar cells, *Journal of Photochemistry and Photobiology C:Photochemistry Reviews***4**,145(2003).

Charge storage mechanisms in nanoparticles of hydrogen ruthenium oxide

Jim P. Zheng

Department of Electrical and Computer Engineering
Florida A&M University and Florida State University
Aero-Propulsion, Mechatronics and Energy Center and Center for Advanced Power Systems
Florida State University
Tallahassee, FL 32310, USA
Email: zheng@eng.fsu.edu
Phone: (850)410-6464

A model of proton transfer and storage in amorphous $RuO_2.xH_2O$ was proposed to explain the relationship between the specific capacitance and the structure and water content in $RuO_2.xH_2O$ materials. The morphologies and crystalline structures of $RuO_2.xH_2O$ annealed at different temperatures were investigated using electrochemical (EC) methods, transmission electron microscope (TEM) and selected area electron diffraction (SAED) techniques, respectively. From EC measurements, it was found that the specific capacitance increased with increasing annealing temperature, and reached the maximum value at a critical temperature, at which the crystalline $RuO_2.xH_2O$ started to be formed. It was also found from EC measurement that proton conductivity is low for $RuO_2.xH_2O$ at steady-state. From the SAED patterns, the structure disorder brought about by proton insertion was obtained from a sample annealed at the critical temperature. From ^1H nuclear magnetic resonance (NMR) spectroscopy, it was demonstrated that the activation energy of proton in $RuO_2.xH_2O$ samples is directly associated with the proton transport and storage ability of these materials. The proton intercalation mechanism was proposed.

Keywords: Hydrous RuO_2; Crystalline Structure; Proton Insertion.

1. Introduction

Since the nanoparticles of hydrous ruthenium oxide ($RuO_2.xH_2O$) were first developed using sol-gel method and demonstrated a high specific capacitance [1], much effort has been focused on further improvement of specific capacitance and reduction of the cost for electrode materials by using alloy of ruthenium and other metal oxides or coating ruthenium oxide on high surface materials [2-4]. The nanostructures of composed $RuO_2.xH_2O$ were also developed for enhancing the specific capacitance[5-7]. The charge storage mechanisms for $RuO_2.xH_2O$ were also studied. A model of proton transfer in amorphous $RuO_2.xH_2O$ was proposed to explain the relationship between the specific capacitance and the structural water content in $RuO_2.xH_2O$ materials[8]. Recent discussions on utilizing the unique properties (both electron- and ion-conductive) of $RuO_2.xH_2O$ by introducing $RuO_2.xH_2O$ into the catalytic electrodes to improve fuel cell performance have been purposed[9].

In this paper, we are going to summarize our studies on understanding of $RuO_2.xH_2O$ properties, particularly, on the proton transfer and storage mechanisms.

2. Experimental

Experimental samples of $RuO_2.xH_2O$ were prepared by annealing commercial $RuO_2.2.76H_2O$ powders (Alfa) in an air circulation oven for about 24 hours. Samples with different water contents (x) and crystalline structures were obtained after annealed at different temperatures. Values of x were calculated from the weight change of annealed samples.

The specific capacitance reported in this paper was measured by cyclic voltammogram (CV). To prepare a $RuO_2.xH_2O$ electrode, Teflon emulsion (60% solid) as binder was first mixed with the $RuO_2.xH_2O$ powder. The mixture was then rolled into a film 100-200 μm thick, from which a square of about 0.4×0.4 cm was cut. The platinum grid was then pressed into the square from both sides for mechanical support and current collection. Three-electrode cells were used for CV experiments. The working electrode was formed Teflon bound $RuO_2.xH_2O$ powders. The counter electrode was made of a Pt grid of a size much larger than that of the working electrode. A standard saturated calomel electrode (SCE) (+0.2412 V vs. the standard hydrogen electrode) was used as the reference electrode. Aqueous solutions of 0.5 M H_2SO_4 were used for the electrolyte. The electrolyte was bubbled with nitrogen gas for at least 30 minutes before any electrochemical measurement.

In order to characterize the difference of $RuO_2.xH_2O$ materials before and after proton insertion on a microscopic scale, two groups of TEM samples were studied. The first group of sample was prepared by pressing the thin carbon film coated TEM Cu mesh grid into the dry $RuO_2.xH_2O$ powders, by which the sample powder particles are attached to the carbon film. The second group of sample was prepared by mixing powders in methanol to form a suspension, and then TEM sample was made by dipping the TEM carbon grid into the suspension, and letting it dry in air. Samples were investigated by a JEOL-2010 TEM system, operated at 200 kV, with the resolution of 0.23 nm. In order to prevent electron irradiation damage (i.e., sample heating) and contamination, weak illumination condition and cold trap were used. The selected area electron diffraction (SAED) pattern was employed to analyze the crystallinity.

3. Results and Discussion

3.1. *Microstructural properties*

Figures 1(a) and (b) shows the TEM image and SAED pattern from $RuO_2.2.76H_2O$, respectively. It can be seen from Figure. 1(a) that the $RuO_2.2.76H_2O$ was formed in particles mostly with a size of about 50 nm, which is about 2 times larger than reported previously[7] in which a diameter of sphere particle is used to estimate the size of the particle based on the measurement of the specific surface area. The difference can be explained by the fact that the surface of particles is quite rough as can be seen from TEM image, which resulted that the actual specific surface area was larger than that based on smooth sphere particle model. From Figure 1(b), the characteristic feature of the diffraction pattern that has one broad-ring and diffused background indicated that $RuO_2.2.76H_2O$ is an amorphous phase material. The position of the maximum of the ring corresponded to real space distance of 0.25 nm. However, further quantitative analysis of the diffraction data through Fourier transform is needed to obtain the nearest neighbor correlation distances and coordination numbers.

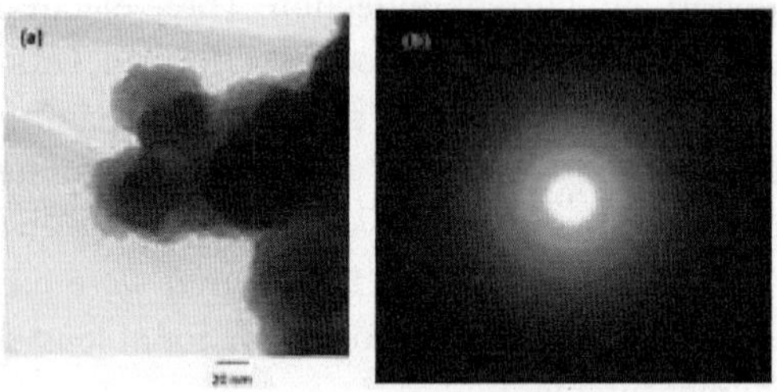

Fig. 1(a) TEM image and (b) SAED pattern from the sample of $RuO_2.2.76H_2O$.

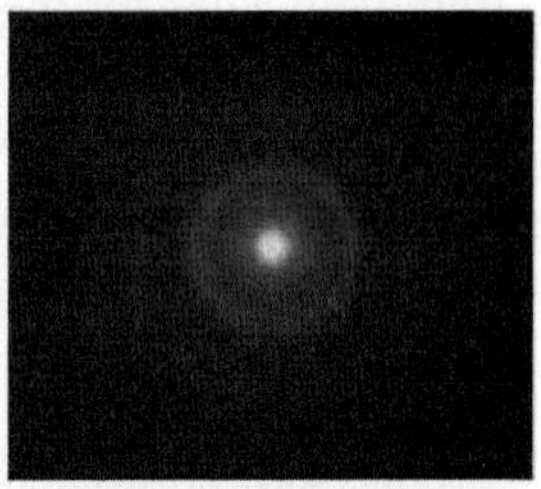

Fig. 2. The SAED pattern measured from the sample of $RuO_2.0.60H_2O$ prepared at 116°C.

Figure 2 shows the SAED pattern from $RuO_2.0.60H_2O$ annealed at 116 °C. The real space distance was 0.22 nm from the position of the maximum of the ring. From Figure 2, it can be concluded that the $RuO_2.0.60H_2O$ was still an amorphous phase. The larger diameter of the first ring compared to the sample without thermal annealing indicate the shorter distance of the nearest neighbors of Ru-Ru atoms in the clusters. It became progressively more ordered in local structure than that without thermal annealing. Figures 3(a) and (b) were obtained from $RuO_2.0.06H_2O$ prepared at 300 °C. From Figure 3(a), it can also be seen that in each cluster a large numbers of crystalline RuO_2 with size less than 5 nm were formed (Figure 3(b). The orientations of nano-sized crystals were random; however, the size of crystals increased with increasing annealing temperature. In Figure 3(b), there are three rings present, and they corresponded to (110), (101) and (121) of the rutile RuO_2 structure. At 500 °C, larger crystalline RuO_2 particle was formed (Figure 4).

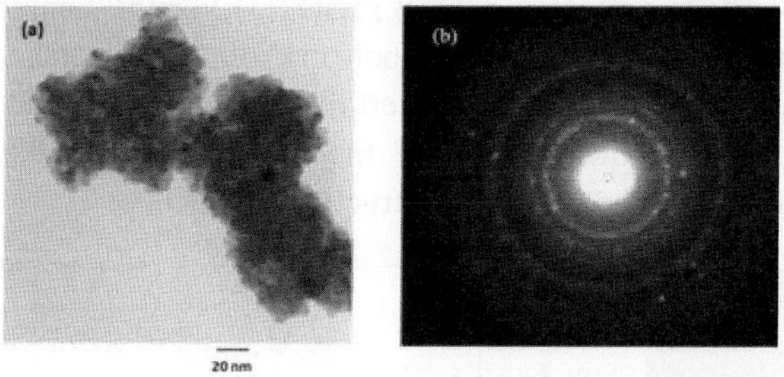

Fig. 3 (a) TEM image and (b) SAED pattern from the crystalline $RuO_2.0.06H_2O$ prepared at 300 °C.

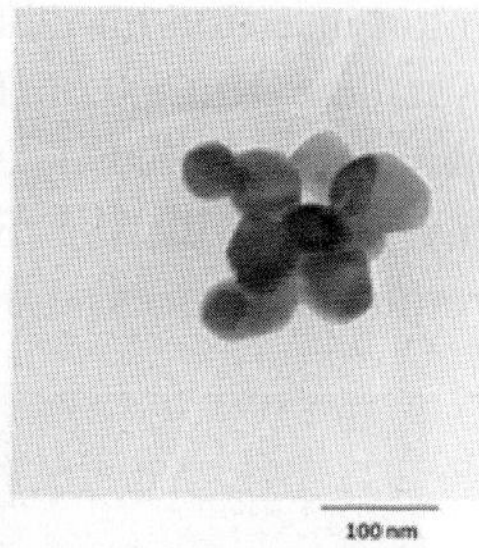

Fig. 4 TEM image of RuO_2 sample annealed at 500 °C.

Similar measurements were also performed on protonated samples after the chemical reaction with methanol. For samples prepared at 25 and 300 °C, no difference was observed for both TEM image and SAED pattern from the samples with and without protonation process. However, a different SAED pattern was obtained for the protonated sample annealed at 116 °C as shown in Figure 5. By comparing SAED patterns in Figures 2 and 5, it can be seen that the diffraction ring from protonated sample was more diffused and broader than that from non-protonated samples. The changes of SAED patterns from samples with and without pontonation processes can be explained as follows: (1) For the sample of $RuO_2.2.76H_2O$, the structure of ruthenium oxide was filled with larger amount of H_2O molecules. The broad-ring in SAED pattern indicated that the sample was highly disordered. The EXAFS study[8] also proved that no three-dimensional networks of linked RuO_6 octahedral chains existed in $RuO_2.xH_2O$ with high structural water content. The same SAED pattern observed on the $RuO_2.2.76H_2O$ sample before and after protonation can be understood that the proton insertion into $RuO_2.2.76H_2O$ didn't cause further disorder in already highly disordered lattices, and further disorder in the lattice could not be detected within the resolution of SAED pattern. (2) For the sample of $RuO_2.0.60H_2O$, a local structure of RuO_2 was formed. The EXAFS study also showed that no long-range order of RuO_2 structure was formed; however, local structure consisting of rutile RuO_6 octahedral interconnected in chains. From OCV_2 potential measurements of protonated samples, it was found about 50% of Ru-O bonds was converted to Ru-OH bonds after protons inserted into $RuO_2.xH_2O$ structure. The proton insertion process caused an inhomogeneous lattice expansion and more degree of disordering, and was reflected by the observation of the broadening of the diffraction ring in SAED pattern. Therefore, the broadening in SAED pattern is a direct evidence of the proton insertion into the bulk of amorphous $RuO_2.xH_2O$. (3) For the sample of crystalline $RuO_2.0.06H_2O$, a three-dimensional network of RuO_6 octahedral chains was formed. The proton was not able to insert into the bulk of crystal but only react with the Ru ion at the surface of the crystal. Therefore, the protonation process did not affect the crystalline structure of RuO_2.

Fig. 5. SAED pattern measured from $RuO_{1.0}(OH)_{1.0}.0.60H_2O$ that was annealed at 116 °C and then reacted with methanol.

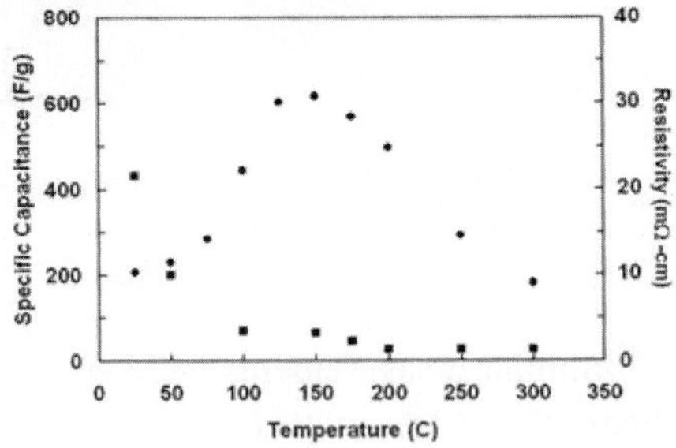

Fig. 6. The specific capacitance (cycle) and electrical resistivity (square) of $RuO_2 \cdot xH_2O$ as a function of sample preparation temperature.

The microstructural changes with increasing annealing temperature directly affect the specific capacitance of $RuO_2.xH_2O$ samples. It is believed that the proton diffusion can only occur in amorphous phase of $RuO_2.xH_2O$ and is dependent on the Ru-Ru distance. For $RuO_2.xH_2O$ samples prepared at low annealing temperatures, the Ru-Ru distance was relatively long due to the high structural water content (e.g. x=2.76 at T_{anneal}=25 °C), which isolated Ru ion from each other. SAED and EXAFS[11] results proved that $RuO_2.xH_2O$ prepared at 25 °C had no short-range order. Therefore, to transport a proton from one Ru site to another was rather difficult. For $RuO_2.xH_2O$ sample annealed at 116 °C, the Ru-Ru distance decreased while the water content reduced to x=0.6. The SAED results indicated that the Ru-Ru distance is indeed shorter in the local structure. The proton could be diffused easily in $RuO_2.0.60H_2O$. Finally for the samples annealed at temperatures above 116 °C, the crystalline structure was formed. The lattice of crystalline RuO_2 was too rigid to allow proton insertion, which prohibited the proton diffusion. Only the Ru ions at the crystal surface involved the protonation process. Therefore, $RuO_2.0.60H_2O$ sample annealed at 116 °C, which contained the lowest structure water in among amorphous phase of $RuO_2.xH_2O$, has the highest specific capacitance. The recent study[12] on ^1H nuclear magnetic resonance spectroscopy at various temperatures also showed a strong correlation between the activation energy of proton and specific capacitance of $RuO_2.xH_2O$. A minimum value of activation energy was obtained from the sample annealed at 116 °C, which indicated that the potential barrier was low and proton could be easily diffused in and out from the bulk of $RuO_2.xH_2O$ sample.

It was found that the BET surface area (a_{BET}) for the powder decreased when the annealing temperature was increased. The surface roughness of nanoparticles decreases with increasing the temperature, which is consistent with the results obtained from BET specific surface area measurement.

3.2. *Electrical properties*

The resistivity of $RuO_2.xH_2O$, calculated based on the resistance measurement of the pellet, is shown in Figure 6. It can be seen that resistivity of the $RuO_2.2H_2O$ pellet is relatively high. But for $RuO_2.xH_2O$ pellets prepared at temperature higher than 100 °C, the resistivity is less dependent on the annealing temperature and is in the order of 10^{-3} Ω-cm. The resistivity measured is believed to be dominated by the contact resistance of powders. The bulk resistivity of $RuO_2.xH_2O$ would be much lower than this value. For example, the resistivity of the RuO_2 single crystal was reported to be about $4x10^{-5}$ Ω-cm[13]. However, the resistivity of $RuO_2.xH_2O$ pellets is sufficiently low for use in EC capacitors.

3.3. *Electrochemical properties*

Figure 7 shows a CV curve of $RuO_2.xH_2O$ electrodes annealed at the critical temperature. The practical potential window is from -0.2 to 1.15 V vs. SCE in diluted H_2SO_4 electrolyte. The specific capacitance increases initially with increasing the annealing temperature and reached maximum value of 720 F/g at the critical temperature, but when the annealing temperature is higher than the critical temperature, the value of specific capacitance dropped significantly. The specific capacitance is found to be insensitive to the specific surface area but sensitive to the crystalline structure, which indicates that proton inserts into the bulk of $RuO_2.xH_2O$. It is also found that the $RuO_2.xH_2O$ with the maximum specific capacitance was always obtained at the temperature near the phase transition from amorphous to crystalline. The maximum value of theoretical specific capacitance for RuO_2 should be about 1000 F/g based on the assumption that the proton exchange rate occurs evenly in the potential range of 1.4 V[1].

3.4. *Proton conductivity*

The proton conductivity properties in $RuO_2.xH_2O$ were measured using a modified direct methanol fuel cell (DMFC) system as shown in Figure 8 [14]. In modified DMFC, a $RuO_2.xH_2O$ layer was added in between anode catalytic electrode and Nafion membrane. The state-steady polarization curve and ac impedance spectra of modified DMFC with various loadings of $RuO_2.xH_2O$ layer were measured. The proton resistance of modified DMFC was obtained by fitting polarization curve using Tafel equation and ac impedance spectra using an equivalent circuit model. Figure 9 shows proton resistance attributed by $RuO_2.xH_2O$ layer as different loadings. $RuO_2.xH_2O$. Based on resistance measurements and mass density estimation of $RuO_2 \cdot xH_2O$ layer (about 1 g/cm^3), the proton conductivity of $RuO_2 \cdot xH_2O$ layer is about 5.3×10^{-3} S/cm. The specific proton resistance of $RuO_2 \cdot xH_2O$ layer was about 0.19 Ω-cm^4/mg, which is close to the value obtained from mixture of Pt-C and Nafion ionomer (0.15 Ω-cm^4/mg) [15]. It must be pointed out that the proton conductivity is much lower than the electrical conductivity for $RuO_2.xH_2O$ at any water content.

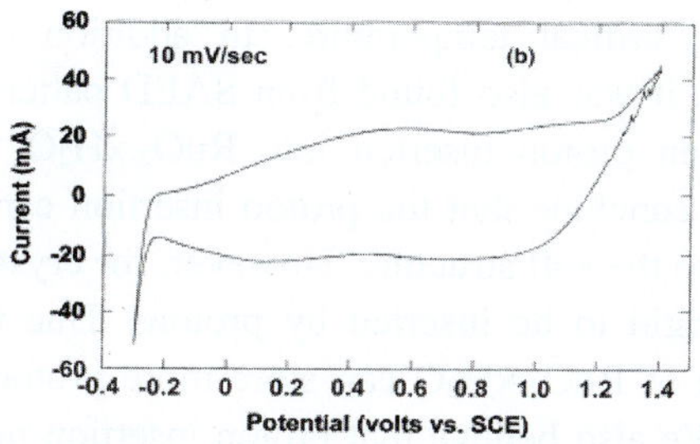

Fig. 7 The CV curves for amorphous $RuO_2.xH_2O$ powder electrodes in 4.6 M H_2SO_4 solution. The voltage scan rate is 10 mV/sec. The weight and size of the amorphous $RuO_2.xH_2O$ electrode is 2.87 mg and 0.3x0.3 cm^2, respectively.

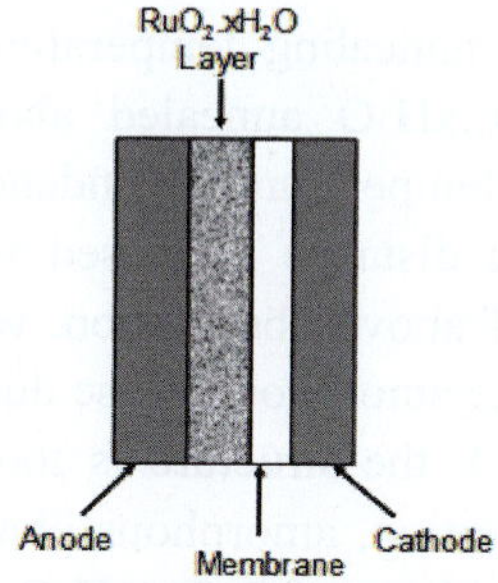

Fig. 8. The sketch of a modified direct methanol fuel cell used for measurement of proton conductivity in $RuO_2.xH_2O$ layer.

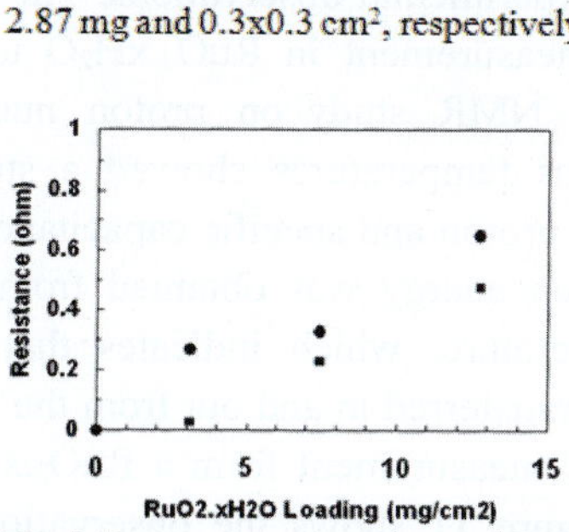

Fig. 9. Proton resistance obtained from dc polarization curve (circle) and ac impedance spectra (square) as a function of $RuO_2.xH_2O$ loading.

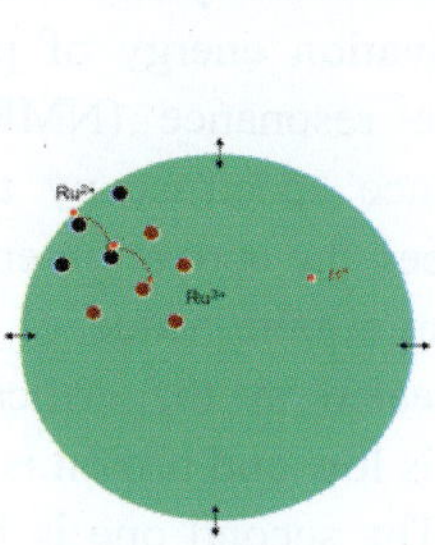

Fig. 10. Illustration of proton intercalation processes. The black arrows indicate that the proton insertion may cause a lattice expansion.

3.5. *Proton transfer and storage*

In order to understand the proton transfer and storage in $RuO_2.xH_2O$, it is necessary to summarize the changes of structure of $RuO_2.xH_2O$ and proton storage capacity as a function of annealing temperature. For $RuO_2.xH_2O$ without thermal treatment, it is amorphous. The structure contains larger amount of water and larger Ru-Ru distance. The SAED pattern indicated that the Ru-Ru distance decreased with increasing annealing temperature. The CV measurement found that the specific capacitance increased with increasing annealing temperature, and reached the maximum value at the critical annealing temperature. However, when annealing temperature is above the critical temperature, the crystalline phase of $RuO_2.xH_2O$ was formed, and it was found that the crystal size and content in $RuO_2.xH_2O$ particle increased with increasing annealing temperature. The specific capacitance dropped rapidly after $RuO_2.xH_2O$ annealed above critical temperature. In addition to the annealing temperature dependence, it was also found from SAED pattern that the Ru-Ru distance increased when proton inserted into $RuO_2.xH_2O$. After analysis of above observation, we conclude that the proton insertion can only occur in the amorphous phase due to the soft structure. However, for crystalline $RuO_2.xH_2O$, the structure is too rigid to be inserted by protons. Due to the proton insertion, amorphous phase of $RuO_2.xH_2O$ can store more proton than crystalline phase of $RuO_2.xH_2O$. We also believe that proton insertion process was realized by proton hopping mechanism from one Ru site to its neighbor Ru site as shown in Figure 10. The rate of hopping should be strongly dependent on the Ru-Ru distance and decreased with increasing Ru-Ru distance. This statement was confirmed by two different experimental observations. The first one is from activation energy of proton measurement in $RuO_2.xH_2O$ using nuclear magnetic resonance (NMR). The NMR study on proton nuclear magnetic resonance spectroscopy at various temperatures showed a strong correlation between the activation energy of proton and specific capacitance of $RuO_2.xH_2O$. A minimum value of activation energy was obtained from the $RuO_2.xH_2O$ annealed at the critical temperature, which indicates that the potential barrier is low and proton is easily transferred in and out from the bulk of $RuO_2.xH_2O$. The second one is from EC measurement from a $RuO_2.xH_2O$ sample with different Ru-Ru distances. Figure 11 shows the observation of current decay after a small potential step from a $RuO_2.xH_2O$ electrode. The $RuO_2.xH_2O$ electrode prepared at T_{annel}=25 °C at different potentials. The current decay after a 10 mV potential step was measured. When the potential decreased, due to the proton insertion, the average Ru-Ru distance increased. Figure 10 clearly indicated that at high potentials (>0.5 V vs. SCE), the sample reached to a steady-state within 10s seconds; however, at low potential such as at 0 V vs. SCE, the sample could reach to a steady-state even more than 100s seconds.

The unexpected low proton conductivity for $RuO_2.xH_2O$ layer at steady-state can be understood, because there is no vacant Ru site available at steady-state. Even there is proton concentration change outside $RuO_2.xH_2O$ particles, proton can hop between Ru sites, since all Ru sites are occupied by a proper numbers of proton which corresponds to the vacant state. However, when the potential of $RuO_2.xH_2O$ electrode is changed, new vacancy is created and proton hopping can occur.

4. Conclusion

The $RuO_2.xH_2O$ samples annealed at different temperature were systematically studied through microcrystalline analysis, electrical and EC properties. For sample without thermal treatment, $RuO_2.xH_2O$ was amorphous. The structure contained larger amount of water and larger Ru-Ru distance. Annealed at the critical temperature, the sample possesses highest specific capacitance. Smaller Ru-Ru distance in the local structure was formed; however, the structure can become more disordered by proton insertion. At high temperatures, crystalline RuO_2 was obtained, and the proton insertion was prohibited. It is believed that proton insertion is realized through proton hopping between Ru sites.

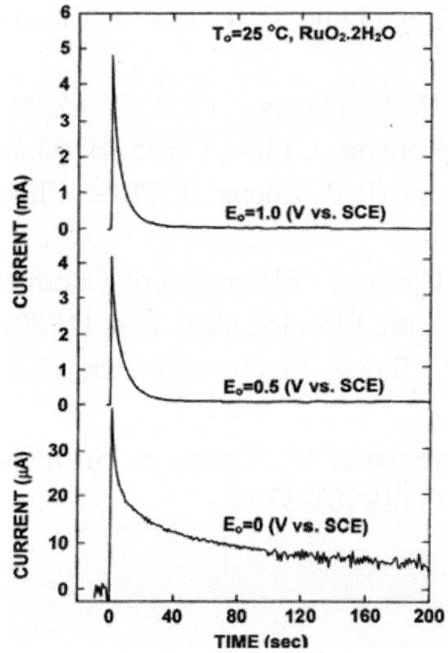

Fig. 11. The current decay profiles for $RuO_2.xH_2O$ electrode at different potentials.

Author thanks Dr. Y. Xin for performing TEM and SAED measurements, Dr. Q. Fu for performing NMR measurement, and Y. Wang for performing the proton conductive measurements.

References

1. J. P. Zheng and T. R. Jow, J. Electrochem. Soc. 142 (1995), L6-L8.
2. Y. U. Jeong and A. Manthiram, Electrochem. and Solid-State Lett. 3 (2000), 205-208.
3. H. Kim and B. N. Popov, J. Power Sources, 104 (2002), 52-61.

4. M. Min, K.Machida, J. H. Jang, and K. Naoi, J. Electrochem. Soc. 153 (2006), A334-A338.
5. N. Lin, J. Tian, Z. Shan, K. Chen, and W. Liao, Electrochimica Acta, 99 (2013), 219-224.
6. S. Mohamad, F. Shaikh, J.Y. Lim, and O.S. Joo, Current Applied Physics, 13 (2013), 758-761.
7. C.-C. Hu, H.-Y Guo, K.-H. Chang, and C.-C. Huang, Electrochemistry Communications, 8 (2009), 1631-1634.
8. J. P. Zheng and Y. Xin, J. Power Sources, 110 (2002), 86-90.
9. Z. Chen, X. Qiu, B. Lu, S. Zhang, W. Zhu, and L. Chen, Electrochem. Comm, 7 (2005), 593-586.
10. J. P. Zheng, P. J. Cygan and T. R. Jow, J. Electrochem. Soc. 142 (1995), 2695-2703.
11. D. A. McKeown, P. L. Hagans, L. P. L. Carette, A. E. Russell, K. E. Swider, and D. R. Rolison, J. Phys. Chem. B103 (1999), 4825-.
12. R. Q. Fu, Z. Ma, and J. P. Zheng, J. Phys. Chem. B. 106 (2002), 3592-3596.
13. A. Cimino and S. Carra in "Electrodes of Conductive Metallic Oxides-Part A", p. 97, ed. S. Trasatti, Elsevier, New York (1980).
14. Y. Wang and J. P. Zheng, Electrochem. and Solid-State Lett. 10 (2007), B26-B30.
15. C. Boyer, S. Gamburzev, O. Velev, S. Srinivasan, and A. J. Appleby, Electrchim. Acta, 43 (1998), 3703-.

Compact TE/TM polarization splitters in $LiNbO_3$ photonic wires at 1.31μm wavelength

Xue-Wei Sun[+] and Ming Chen[*]
School of Electronic Engineering, Xi'an University of posts & telecommunications, Xi'an 710121, China
*Email: *chenming5628@sina.com,* [+]1274220857@qq.com

TE/TM polarization splitters in $LiNbO_3$ photonic wires with compact structure at 1.31μm wavelength have been proposed. Its operating principles are based on the directional coupling mode. Their optimization designs are achieved by using commercial software Optiwave FDTD. Calculated steady-state field distributions indicate that guided lightwaves in both quasi-TM and quasi-TE polarizations can be spatially separated from different output waveguide ports. The transmittances of quasi-TE and quasi-TM polarization at 1.31μm wavelength are 95% and 97.8%, respectively. The total length of the devices is only 46μm. Moreover, the coupling length dependence of the transmittance, the bandwidth dependence of transmittance, and fabrication sensitivity are evaluated, respectively. The simple and compact polarization splitters are expected to be applied to highly dense LN photonic integrated circuits.

Keywords: Polarization Splitters; $LiNbO_3$ Photonic Wires; Directional Coupling Mode.

1. Introduction

$LiNbO_3$ (LN) photonic wires (i.e., LN waveguide) [1–8] are promising candidates for future integrated photonics because of their ultra-small size, excellent electro-optic, acousto-optic, non-linear optical properties [9], easily being doped with rare-earth ions to get a laser active material [10], especially being expected high efficiency devices even at moderate optical power levels, and so on. Therefore, some efforts have been made to investigate passive elements in the form of LN photonic wires, such as optical directional coupler [11], optical wavelength splitters [12], and so on. It is equally important, however, to develop optical polarization splitters in the form of LN photonic wires which play an important part in integrated optical circuits.

[*] Corresponding author

In this paper, optical polarization splitters based on directional coupler for 1.31μm wavelength with compact structures formed by LN photonic wires have been proposed. To fabricate LN optical wires a LNOI (LN on Insulator) sample is used. It consists of a 730 nm thick single crystalline LN layer direct bonded to the surface of a 1.3 μm thick SiO2 layer deposited by plasma enhanced chemical vapor deposition (PECVD) on the –Z-face of a congruent Z-cut LN substrate. The thin bonded LN layer (i.e., LN thin film) has the same crystal orientation as the substrate; its surfaces have a ms-roughness of 0.5 nm obtained by chemical-mechanical polishing (CMP). The height and top width of the LN waveguide are respectively taken as 0.73μm and 0.5μm respectively so that the single mode propagation can be achieved [8]. A commercial software (Optiwave FDTD) is used to optimally design the structural parameters at 1.31μm wavelength and simulate the light propagation characteristics in the proposed structure.

2. Theory and Design

The proposed polarization splitter with straight LN waveguide and bending LN waveguides is shown in Figure1 (a) and (b).

Where Lc expresses the length of the coupling region, L the total length of the device, Ls the length of the transition segment with straight output waveguide, Sc the spacing of constant branch separation, Sp the spacing of output port, W the width of the waveguide. The height and top width of the LN waveguide is 0.73μm and 0.5μm, respectively.

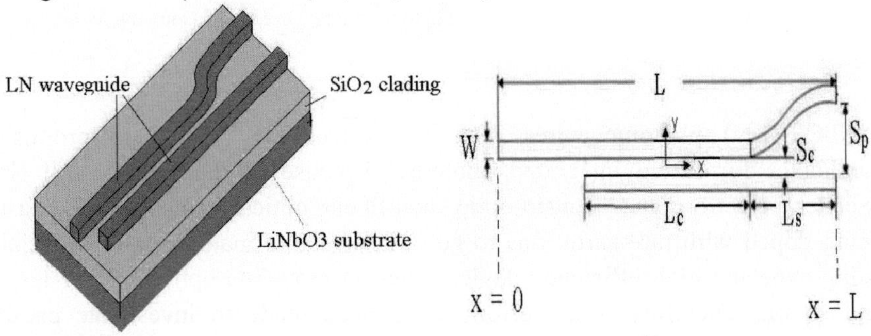

(a) (b)

Fig.1. Polarization splitters with straight LN waveguide and bend LN waveguide: (a) the 3D view; (b) the top view.

A commercial software (Optiwave FDTD) is used to optimally design the structural parameters at 1.31μm wavelength and simulate the light propagation characteristics in the proposed structure. Calculated parameters for the polarization splitter are as follows: L=46μm, Lc = 37μm, Ls = 7μm, Sp =

2.25μm, and Sc = 0.24μm. Figure 2 shows intensity distributions of quasi-TE (upper) and quasi-TM (lower) at 1.31μm wavelength in the structure. Figure 3 shows the steady-state field distribution of quasi-TE mode and quasi-TM mode at 1.31μm wavelength.

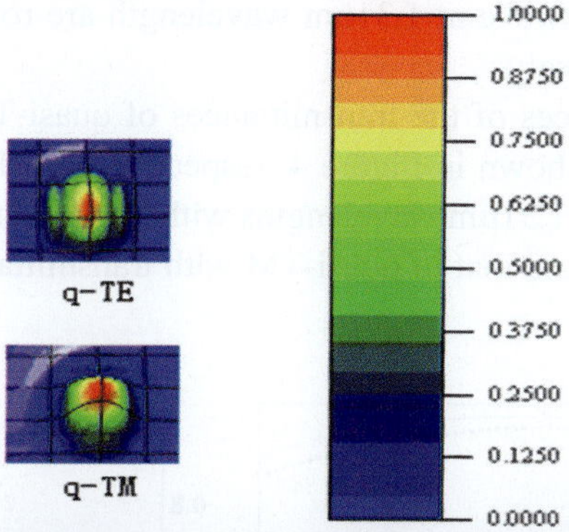

Fig.2. Intensity distribution of quasi-TE (upper) and quasi-TM (lower)

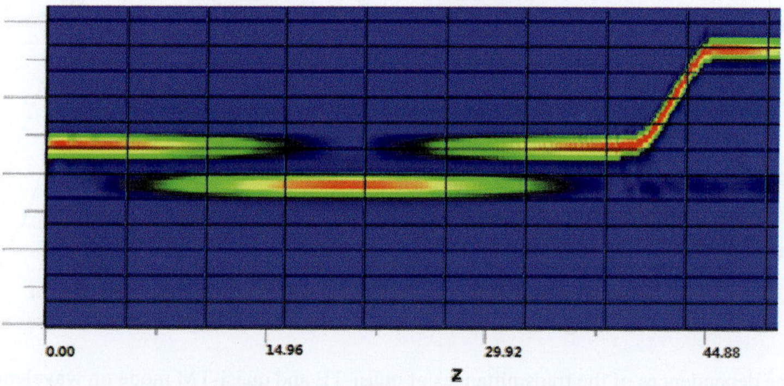

(a)

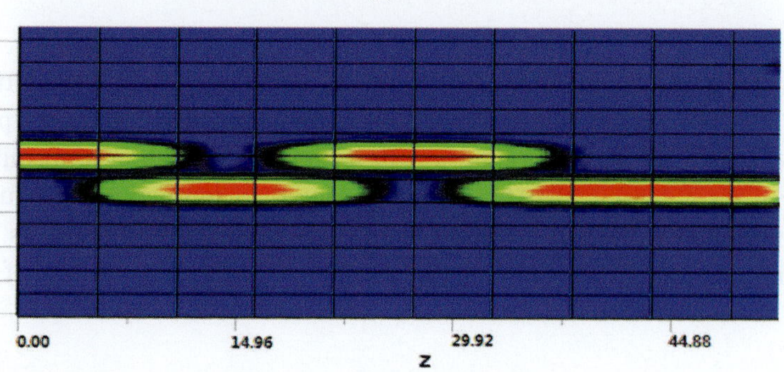

(b)

Fig.3. Steady-state field distribution of quasi-TE (a) and quasi-TM (b)

In Figure 3, the quasi-TE lightwave is guided to the bar output waveguide via a coupling length, while the quasi-TM lightwave is guided to the cross output waveguide via two coupling length. The corresponding transmittances at 1.31μm wavelength are 95% and 97.8%. Thus the quasi-TE mode lightwave and quasi-TM mode lightwave at 1.31μm wavelength are routed to different output waveguides, respectively.

The dependences of the transmittances of quasi-TE and quasi-TM mode on wavelengths are shown in Figure 4, respectively. In Figure 4, the bandwidth of quasi-TE mode at 1.31μm wavelengths with transmittances more than 90 % is greater than 50nm. And that of quasi-TM with transmittances more than 90 % is greater than 60nm.

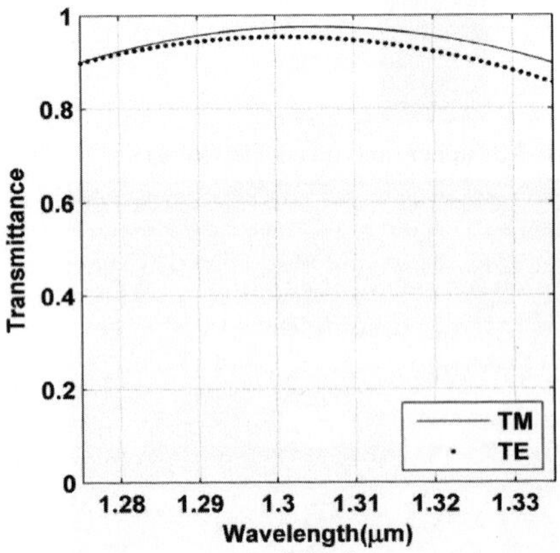

Fig.4

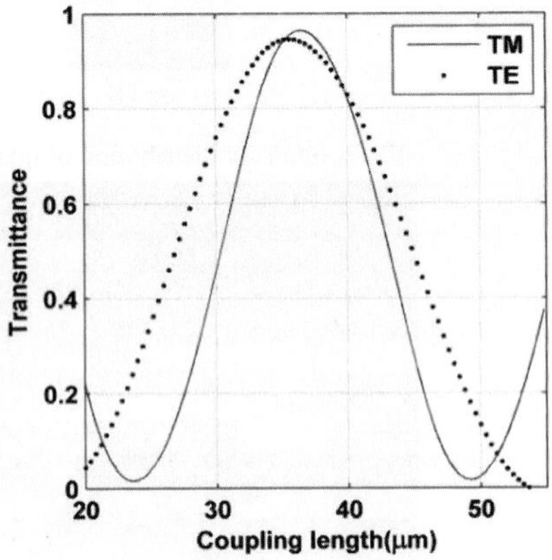

Fig.5

Fig. 4 The dependences of the transmittances of quasi-TE and quasi-TM mode on wavelength
Fig. 5 The dependences of the transmittances of quasi-TE and quasi-TM mode on coupling length

Figure 5 shows the coupling length dependences of the transmittances of quasi-TE and quasi-TM mode. The transmittances change periodical approximately as the coupling length changes. Maximum transmittance is achieved on condition that the coupling length is approximately equal to the even multiple of the half coupling length. Minimum transmittance is achieved on condition that it is approximately equal to the odd multiple of the half coupling length. The approximately maximum transmittances for both quasi-TE and quasi-TM mode are achieved simultaneously under condition of Lc = 37μm.

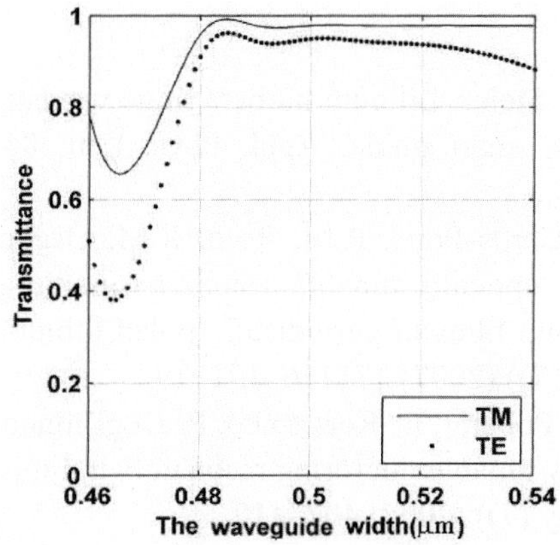

Fig.6 The waveguide width dependence of transmittance for the quasi- TE and quasi- TM mode

Figure 6. shows the waveguide width dependences of transmittances of the quasi-TE and quasi-TM mode. In Figure 6, as the waveguide width changes from 0.5μm to 0.54μm and from 0.5μm to 0.48μm, the changes of transmittances are not significant. However, as the waveguide width decreases from 0.48μm, the transmittances decrease sharply, especially for quasi-TE mode.

3. Conclusions

In conclusion, ultra-compact TE/TM modes polarization splitters with compact structures in LiNbO3 photonic wires are investigated analytically and numerically, respectively. The steady-state field distributions in proposed structure show that the quasi-TE lightwave and quasi-TM lightwave at 1.31μm wavelength from same input port are routed to bar and cross output waveguides, respectively. The polarization splitters have wide bandwidths, high transmittances, and stable optical characteristics with fabrication error in a certain range. The simple and compact polarization splitter is expected to be applied to highly dense LN photonic integrated circuits.

Acknowledgements

This work is supported by Natural Science Foundation of China (Grant No. 61040064)

References

1. P. Rabiei, W.H. Steier, Lithium niobate ridge waveguides and modula-tors fabricated using smart guide, Appl. Phys. Lett. 86 (April (16)) (2005) 161115–161118.
2. D. Djukic, G. Cerda-Pons, R.M. Roth, R.M. Osgood Jr., S. Bakhru, H. Bakhru, Electro-optically tunable second-harmonic-generation gratings in ion-exfoliated thin films of periodically poled lithium niobate, Appl. Phys. Lett. 90(April (17)) (2007) 171116–171119.
3. A. Guarino, G. Poberaj, D. Rezzonico, R. Degl'innocenti, P. Günter, Electro-optically tunable microring resonators in lithium niobate, Nat. Photonics 1(May (7)) (2007) 407–410.
4. F. Schrempel, T. Gischkat, H. Hartung, T. Höche, E.B. Kley, A. Tünnermann, W. Wesch, Ultrathin membranes in x-cut lithium niobate, Opt. Lett. 34 (April (9))(2009) 1426–1428.
5. T. Takaoka, M. Fujimura, T. Suhara, Fabrication of ridge waveguides in LiNbO3thin film crystal by proton-exchange accelerated etching, Electron. Lett. 45 (18)(2009) 940–941.
6. G. Poberaj, M. Koechlin, F. Sulser, A. Guarino, J. Hajfler, P. Günter, Ion-sliced lithium niobate thin films for active photonic devices, Opt. Mater. 31 (7) (2009)1054–1058.
7. G.W. Burr, S. Diziain, M.-P. Bernal, Theoretical study of lithium niobate slab waveguides for integrated optics applications, Opt. Mater. 31 (10) (2009)1492–1497.
8. H. Hu, R. Ricken, W. Sohler, Lithium niobate photonic wires, Opt. Express 17 (December (26)) (2009) 2426–242681.
9. R.S. Weis, T.K. Gaylord, Lithium niobate: summary of physical properties and crystal structure, Appl. Phys. A: Mater. Sci. Process. 37(March (4))(1985)191–203.
10. W. Sohler, B. Das, D. Dey, S. Reza, H. Suche, R. Ricken, Erbium-doped lithium niobate waveguides lasers, in: IEICE Trans. Electron. E88(C), 2005, pp. 990–997.

11. Ming Chen, Zichen Liu, Rong Lu, Yi Yang, Meng Liang, Tian-tong Tang, "An ultracompact optical directional coupler based on lithium niobate photonic wires" Optik , 124 (2013) 1974– 1976.
12. Ming Chen, An ultracompact wavelength splitters in LiNbO3 photonic wires, Submitted to Optical Engineering.

Chapter 7

Biomaterials

Relationship between uric acid stone formation and urinary crystallite components —an investigation of 109 cases of patients with UA stones

Qin Gui, Poonam Bhadja and Jian-Ming Ouyang †
Department of Chemistry, Jinan University,
Guangzhou, 510632, China
E-mail: †toyjm@jnu.edu.cn

The formation of kidney stones was related closely to the properties of urinary crystallites. This study aims to investigate the components of urinary crystallites in 109 cases of patients with uric acid (UA) stones using High-Resolution Transmission Electron Microscopy (HRTEM). Urinary crystallites in patients with UA stones, that contain calcium oxalate monohydrate, calcium phosphate, and ammonium urate crystals, were detected and confirmed by X-ray Diffraction (XRD) and Fast Fourier transformation (FFT) in HRTEM, even in small quantities, showing the presence of high UA concentration and low pH in urine.

Keywords: Urine Crystallite; Uric Acid; XRD; HRTEM; Urinary Calculi.

1. Introduction

Several scholars used urinary crystallites to evaluate the risk index of urinary stone disease [1]. However pure uric acid stone formation is rare and mostly accompanied by urates, phosphates or calcium oxalate. Reports state that uric acid can influence calcium oxalate/phosphate crystallization and deposition. The most striking evidence for this association came from the apparent success of allopurinol, a drug that reduces urinary output of urate, is capable of reducing the calcium oxalate stone recurrence. In an attempt to explain this apparent association of uric acid and calcium oxalate stone formation, several theories have been proposed. However, to date, a plausible rationale for these observations is still without sound scientific foundation.

Therefore, specific components of stones and their urinary crystallites should be detected to elucidate the cause of renal stone formation; relevant information can also be obtained, particularly regarding appropriate remedies and prevention of stone recurrence.

---

† Corresponding author.

2. Materials and methods

Absolute ethanol was of analytical purity. Glass vessels were cleaned with distilled water.

X-ray diffraction (XRD) results were recorded using a D/max 2400 X-ray diffractometer (Rigaku, Tokyo, Japan). TEM was conducted on a HRTEM (JOEL 2100F), and fast Fourier transformation (FFT) analysis in the Digital Micrograph software was also conducted to obtain the patterns.

109 cases of uric acid stones were collected from patients with calculi after surgery and their urine samples were also collected. Approximately 30 mL of urine was centrifuged at 4000 r/min for 15 min, and the residual was dried in a vacuum desiccator for further XRD characterization.

Approximately 5 μL of urine was submerged in a copper mesh by using a microsyringe; the mesh was then stored in a desiccator for 2 d before HRTEM analyses were performed.

3. Results and discussion

3.1. *X-ray diffraction (XRD) analysis of urinary stones*

The XRD patterns of 109 cases of UA stones were analyzed. Several representative results of XRD patterns are shown in Figure 1. The diffraction peaks at d=6.54, 5.63, 4.91, 3.86, 3.19, 3.10, 2.87, 2.80, 2.57 and 2.24 Å were assigned to the (200), $(\bar{1}01)$, (210), $(\bar{2}11)$, (021), $(\bar{1}21)$, $(\bar{2}21)$, (112), (321), and $(\bar{2}22)$ faces of UA crystals, respectively [2], and the peaks at d=5.93, 3.65, 2.97 and 2.35 Å were detected and assigned to the $(\bar{1}01)$, (020), $(\bar{2}02)$ and (130) face of COM crystals.

3.2. *Component analysis of urinary crystallites*

XRD, FT-IR, HRTEM, FFT of HRTEM, SAED, and EDS were performed to investigate the components of urinary crystallites of UA stones. Urinary crystallites occurred in 92 cases of patients among the 109 cases of patients with UA stone as the main component (Figure 2); these crystallites mainly contained UA, COM, ammonium acid urate, and CaP. They are present either separately or in combination.

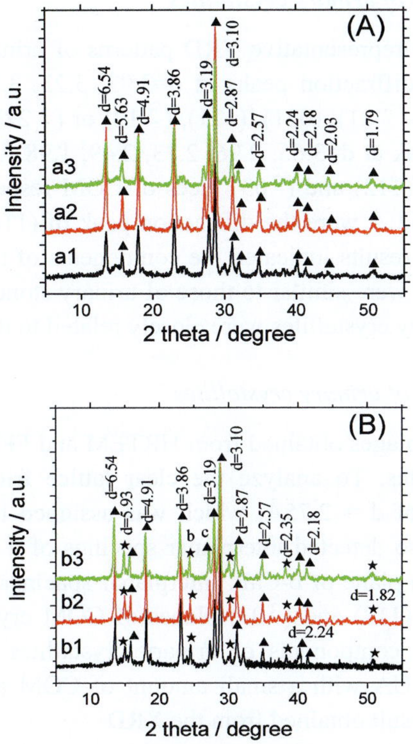

Fig. 1. XRD patterns of three pure UA stones (A) and three UA-COM mixed stones (B).

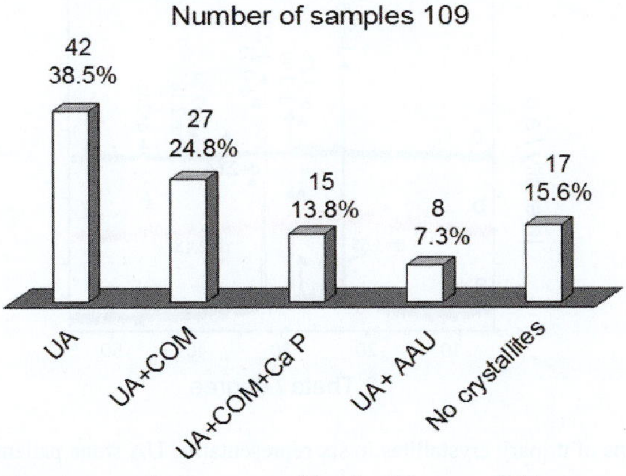

Fig. 2. Components of urinary crystallites in 109 cases of UA stone patients.

[*] UA: uric acid; COM: calcium oxalate monohydrate; AAU: ammonium urate; CaP: calcium phosphate.

3.2.1. *XRD analysis of urinary crystallites*

Figure 3 showed the representative XRD patterns of urinary crystallites of UA stone formers. The diffraction peaks at d=3.93, 3.23, 3.10, 2.66, and 2.25 Å were assigned to the $(\bar{2}11)$, (021), $(\bar{1}21)$, $(\bar{2}12)$ or $(\bar{2}22)$ faces of UA crystals respectively. The peak at d=3.65, 2.96, 2.73, 2.49, 1.98 Å were assigned to the (020), $(\bar{2}02)$, (420), (112) and $(\bar{3}03)$ faces of COM respectively. The peaks at d=5.21, 3.46 and 2.52 Å were the diffraction peak of (110), (1010) and (2110) faces of CaP. These results indicated the components of the urinary crystallites of UA stone formers were similar to those of urinary stones. In other words, the components of urinary crystallites were closely related to the type of stone.

3.2.2. *FFT analysis of urinary crystallites*

Figure 4 shows the images obtained from HRTEM and FFT in different areas of the urinary crystallites. To analyze the clear lattice fringes, we detected an interplanar spacing of d = 2.75Å, which was assigned to the strongest plane (112) of UA. We also detected interplanar spacings of d = 2.60 Å, which was assigned to the (220) plane of ß-CaP. Interplanar spacings of d = 2.49 and 1.95 Å were assigned to (112) and $(\bar{3}03)$ planes of COM crystal [3], respectively. Therefore, the main components of urinary crystallites in patients with UA stones were mainly UA with a small amount of COM and CaP , which was consistent with the result obtained from the XRD.

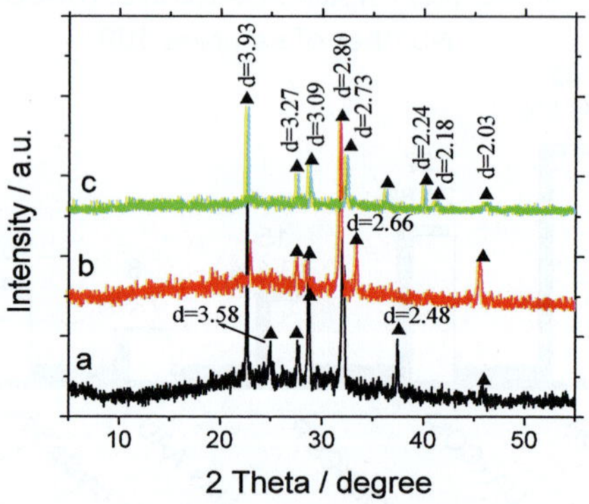

Fig. 3. XRD patterns of urinary crystallites in six representative UA stone patients. Symbols: ▲: uric acid, ★: COM.

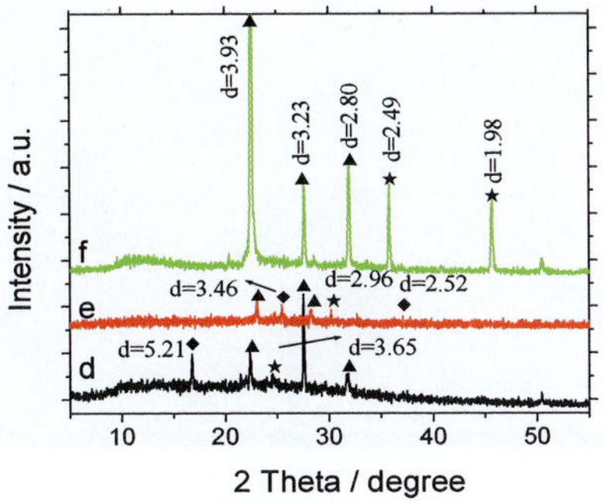

Fig. 3. (Cont'd).

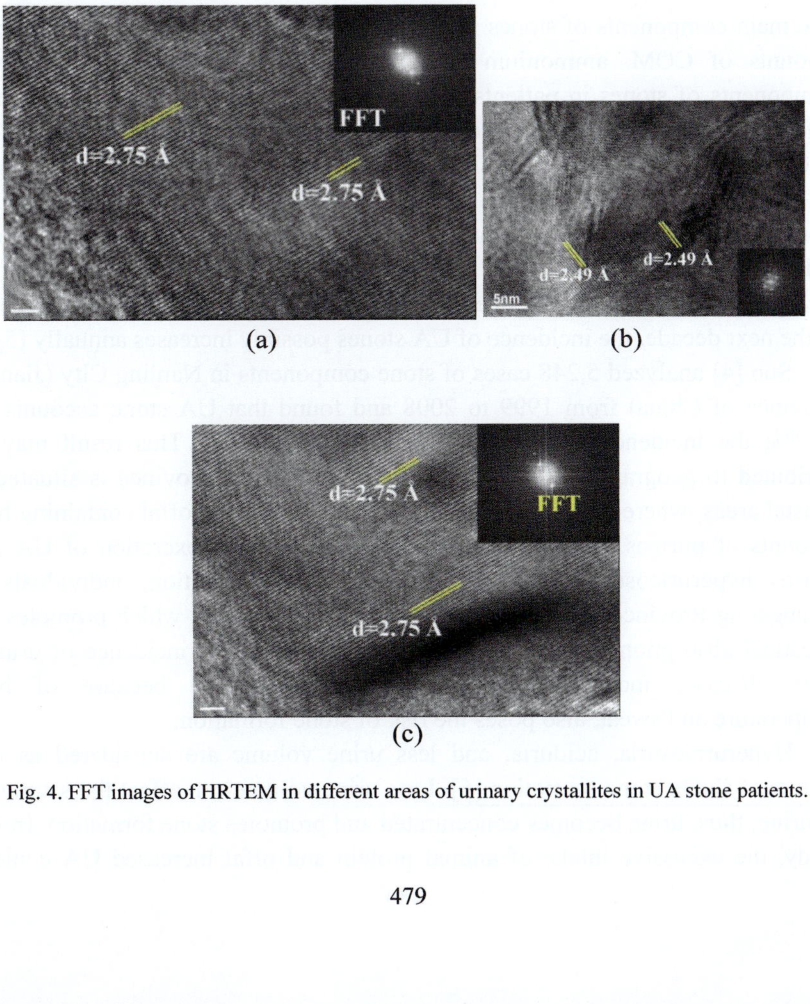

Fig. 4. FFT images of HRTEM in different areas of urinary crystallites in UA stone patients.

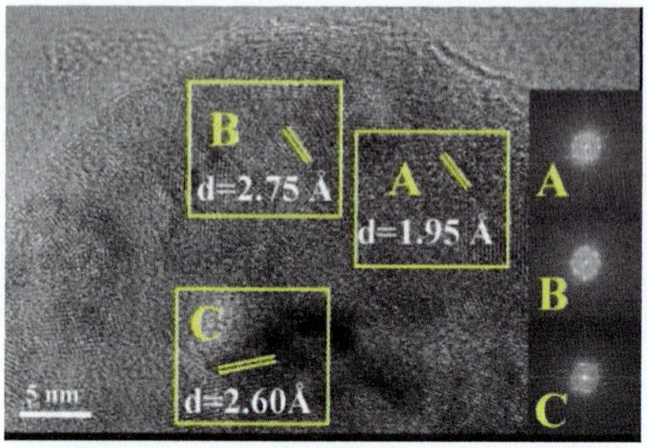

(d)

Fig. 4. FFT images of HRTEM in different areas of urinary crystallites in UA stone patients.

4. Discussion

The main components of stones in patients with UA stone were UA with small amounts of COM, ammonium acid urate, and CaP (Figure 3); the main components of stones in patients with urinary crystallites were UA with small amounts of COM, ammonium acid urate, and CaP crystals (Figsure 4 and 5). The components of stones were similar to those of urinary crystallites.

As the most common acidic stones, UA stones ranked only second to CaOx stones, in which UA are accounted for 5% to 40% of urinary stone [4]. In an industrialized city, the incidence of UA stones is approximately 10%. With the increase in aging population, obesity, and patients with type 2 diabetes mellitus in the next decade, the incidence of UA stones possibly increases annually [5].

Sun [4] analyzed 5,248 cases of stone components in Nanjing City (Jiangsu Province of China) from 1999 to 2008 and found that UA stone accounts for 6.12%; the incidence of UA stone accounts for 24.1%. This result may be attributed to geographical location because Guangdong Province is situated in coastal areas, where individuals prefer seafood and animal offal containing high amounts of purines. This substance likely promotes the excretion of UA and causes hyperuricosuria and aciduria in urine. In addition, individuals in Guangdong Province prefer foods with high sugar content, which promotes the intestinal absorption of calcium and oxalic acid; thus, the incidence of urinary stone diseases increases. Concentrated urine, formed because of high temperature and sweat, also poses the risk of stone formation.

Hyperuricosuria, aciduria, and less urine volume are considered as risk factors of CaOx stone formation [6]. Less urine volume can affect the saturation of urine; thus, urine becomes concentrated and promotes stone formation. In this study, the excessive intake of animal protein and offal increased UA content.

Hyperuricosuria can increase substrate concentration to form UA crystallites; furthermore, hyperuricosuria can reduce urine pH and accelerate the crystallization of UA crystals from urine, thereby forming UA stones [7].

Hyperuricosuria is defined as a 24-hour urinary excretion of UA > 700 mg [8]. The 24-hour urinary excretion of UA > 750 mg in women or UA > 800 mg in men indicates the risk of stone formation. Incidence rate reaches 50% at 24-hour urinary excretion of UA > 1100 mg.

Adair [9] investigated the electrophoretic mobility of UA crystals in human urine and revealed the presence of low negative charges in the surface of UA crystals; with these negative charges, UA crystals exhibit a strong adhesion effect on cationic functional groups. Thus, UA crystals can adsorb on CaOx crystals and then induce the formation of COM/UA stone via heterogeneous nucleation because of the positive charges on the surface of CaOx crystals. Therefore, the formation of CaOx stone is often related to hyperuricosuria.

5. Conclusions

XRD and FFT of HRTEM were performed to study the components of urinary crystallites in 109 cases of patients with UA stones. The main components of urinary crystallites in patients with UA stones were UA with a small amount of COM, CaP and ammonium urate. The results showed that the formation of UA stones was closely related to the components of urinary crystallites. The information can be used to provide insight on the appropriate and personalized treatment in clinic.

Acknowledgements

Work supported by grant 81170649 of the National Natural Science Foundation of China.

References

1. S. Verdesca, G.B. Fogazzi, G. Garigali, P. Messa and M. Daudon, Crystalluria: prevalence, different types of crystals and the role of infrared spectroscopy, *Clin. Chem. Lab. Med.* **49**, 515(2011).
2. M. King, W.F. McClure, L.C. Andrews, and M.A. Holomery, Powder diffraction file alphabetical index, inorganic phases/organic phases. (International center for diffraction data, U.S.A , 1992), pp. 19081-2389.
3. J.Y. He, S.P. Deng and J.M. Ouyang, Morphology, size distribution, aggregation, and crystal phase of nanocrystallites in urines of healthy persons and lithogenic patients, *IEEE Trans. Nanobiosci.* **9**, 156 (2010).

4. X.Z. Sun, L.M. Shen, X.M. Cong, H.J. Zhu, L. He and J.L. Lu, Infrared spectroscopic analysis of 5,248 urinary stones from Chinese patients presenting with the first stone episode, *Urol. Res.* **39**, 339 (2011).
5. D. Bertrand, Uric acid nephrolithiasis, *La Revue du praticien* **61**, 389 (2011).
6. M.R. Wiederkehr and O.W. Moe, Uric acid nephrolithiasis: a systemic metabolic disorder, *Clinic. Rev. Bone. Miner. Metab.* **9**, 207 (2011).
7. G.N. Zhang, J.M. Ouyang, J.F. Xue and Y.F. Shang, Property changes of urinary nanocrystallites and urine of uric acid stone formers after taking potassium citrate, *Mater. Sci. Eng. C-Mater, Biol. Appl.* **33**, 4039 (2013).
8. Y.S. Ha, D.U. Tchey, H.W. Kang, Y.J. Kim , S.J. Yun , S.C. Lee and W.J. Kim, Phosphaturia as a promising predictor of recurrent stone formation in patients with urolithiasis, *KJU.* **51**, 54 (2010).
9. J.H. Adair, L.A.G. Aylmore, J.G. Brockis and R.C. Bowyer, An electrophoretic mobility study of uric acid with special reference to kidney stone formation, *J. Coll. Interf. Sci.* **124**, 1 (1988).

Comparative study of nanocrystallites in urine of magnesium ammonium phosphate stone patients and healthy controls by means of XRD, HRTEM and nanoparticle size analyzer

Chong-Yu Zhang, Poonam Bhadja and Jian-Ming Ouyang †

Department of Chemistry, Jinan University,
Guangzhou, 510632, China
E-mail: † toyjm@jnu.edu.cn

To study the formation mechanism of magnesium ammonium phosphate (MAP) stones. X-ray diffraction (XRD), high-resolution transmission electron microscopy (HRTEM), and nanoparticle size analyzer were used to investigate the nanocrystallites in urines of 10 MAP stone patients and 10 healthy persons. Most of the nanocrystallites in healthy urines were uric acid, calcium oxalate dihydrate (COD), ß-$Ca_3(PO_4)_2$ and NH_4MgPO_4. However, there were mainly MAP·$6H_2O$, calcium oxalate monohydrate (COM), hydroxylapatite, ß-$Ca_3(PO_4)_2$ in the urine of MAP stone patients. The sizes of nanocrystallites were ranged from 22-68 nm in healthy urines and 17-92 nm in lithogenic samples. The stability of nanocrystalline in healthy urines was better than that in lithogenic samples. The nanocrystallites in MAP stone patients have a large size variation and a poor stability, easily aggregated to each other and contained more COM crystals which having a strong adhesion ability to cell membrane. All these factors promote the formation of kidney stones.

Keywords: Nanocrystallites; TEM; Nanotechnology; Magnesium Ammonium Phosphate.

1. Introduction

Urinary stone is a worldwide disease, and more common in developed countries. The annual morbidity rate of the urolithiasis is in the range from 0.5 % to 5.3 % and has a tendency to increase. Up to now, there is no effective means for preventing the occurrence of urolithiasis and its mechanism is also unclear yet [1]. Geographical location, environment, living condition, diet habits, metabolic disorder and urinary system disease are the factors causing the urinary stones.

Formation of crystals in urine is the necessary initial step of lithogenesis in all type of urinary stone disease. In the early seventies, some authors proposed crystalluria as an index of stone disease activity [2]. Daudon [3] considered that crystalluria was highly indicative of stone disease. However, because

† Corresponding author.

crystalluria is often found in the urine of healthy and nonlithiasic subjects, hence, a number of authors [3,4] stated that presence of crystalluria is not discriminant between stone formers and non stone formers. Therefore, the clinical relevance of crystalluria in urinary stone formers remains largely debated.

MAP, the most common component of alkaline stones, contains MAP hexahydrate (struvite, $MgNH_4PO_4 \cdot 6H_2O$) and magnesium hydrogen phosphate trihydrate (newberyite, $MgHPO_4 \cdot 3H_2O$). MAP accounts for about 10 % to 15 % of all stone components [5].

2. Experiment

2.1. *Reagents and instruments*

All reagents were analytical regent. TEM was conducted on a HRTEM (JOEL 2100F). To determine the component of urinary nanocrystallites, we performed fast Fourier transformation (FFT) of the HRTEM. XRD results were recorded using a D/max 2400 X-ray diffractometer (Rigaku, Tokyo, Japan). Nanoparticle size analyzer of Zetasizer Nano-ZS (Malvern, England) was also used.

MAP stones were collected from five patients with stones after surgery. XRD results showed that the main components of these five stones were MAP.

Fasting morning urines were collected from subjects. The urine was centrifuged at 4000 r/min for 15 min. The aforementioned treated urine was filtered by a microporous membrane with a pore size of 1.2 μm to remove the proteins in the urines. The filtered urine was stored in clean glassware for examination.

In the preparation of samples for HRTEM detection, the urine sample was first subjected to ultrasound treatment for 5 min, approximately 5 μL of urine was submerged in a copper mesh by a microsyringe. Then the mesh was stored in a desiccator for 2 d prior to HRTEM analyses.

Approximately 100 μL of spare urine sample was placed on clean glass slides by using a microsyringe. After the soluble fractions of sodium chloride and urea were removed, the glass slides were dried for XRD characterization.

3. Results and Discussion

3.1. *Components of urinary nanocrystals in urine of healthy persons*

Figure 1 shows the representative XRD patterns of the urinary nanocrystals in healthy persons. Compared with the standard diagrams, the main peaks were assigned to the (110) crystal plane of $MgNH_4PO_4 \cdot H_2O$; the (121) plane of $MgNH_4PO_4 \cdot 6H_2O$; the (021) and $(\bar{2}11)$ plane of uric acid; and the (310), (411), (501), (323) and (730) planes of calcium oxalate dihydrate (COD) crystals. Thus, urinary crystallites of all healthy persons contained COD crystals.

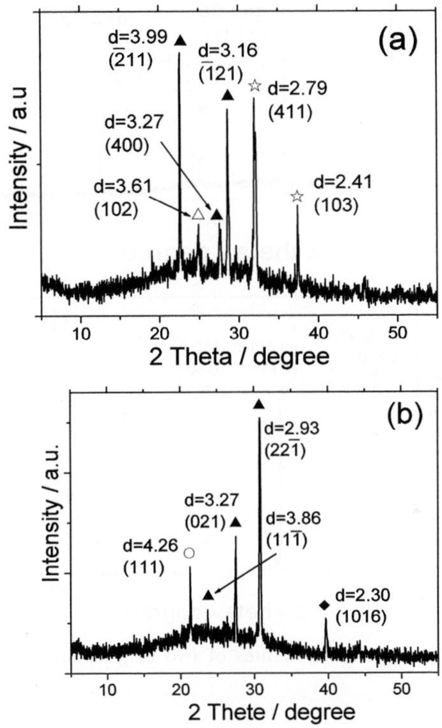

Fig. 1 XRD patterns of urinary nanocrystals in urine of two health persons.
(☆: COD, ▲: uric acid, Δ: ammonium urate, ○: $MgNH_4PO_4 \cdot 6H_2O$, ◆: β-$Ca_3(PO_4)_2$,)

3.2. *Components of urinary nanocrystals in urine of MAP stone formers*

Figure 2 shows the representative XRD patterns of the urinary nanocrystals in MAP stone formers. These peaks were respectively assigned to the (024), (404), and (210) crystal planes of β-$Ca_3(PO_4)_2$; the (040) and (151) planes of $MgNH_4PO_4 \cdot 6H_2O$; the (130) and (040) planes of $MgNH_4PO_4 \cdot 6H_2O$, the (121) plane of COM; the $(\bar{2}11)$ and (021) planes of uric acid (UA); and the (411) and (323) planes of COD crystals.. So that, the main components of urinary nanocrystals were MAP·$6H_2O$, COM, COD, UA, and $Ca_3(PO_4)_2$ etc.

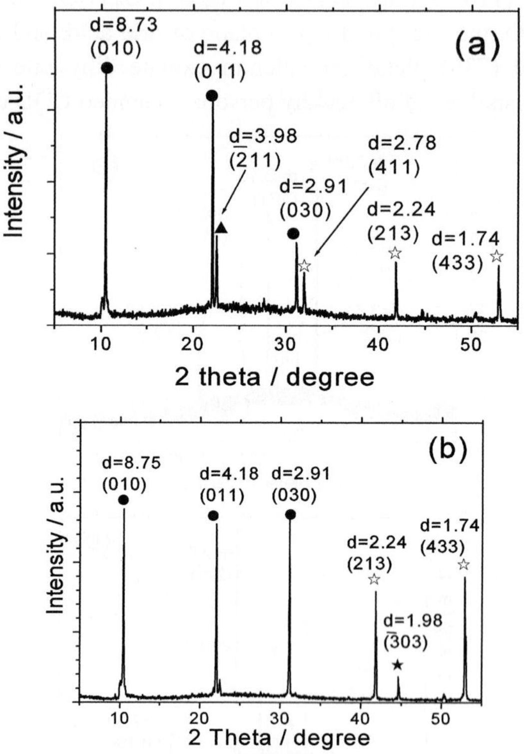

Fig. 2 XRD patterns of urinary nanocrystallites of two magnesium ammonium phosphate stone formers.

3.3. *Size calculation of urinary nanoparticles of MAP stone patients and healthy control based on XRD patterns*

The approximate size of urinary nanoparticles could also be obtained based on the XRD diffraction data by using Scherer formula as shown in Eq. (1) [6]:

$$L = \frac{K\lambda}{\cos\theta \cdot \Delta(2\theta)}$$

, where, L is the size of the nanoparticles, K is constant and dependent on crystallite shape (0.89), λ is the X-ray wavelength (λ = 1.5418 Å in this experiment), $\Delta(2\theta)$ is the radian value of fwhm (full width at half-max), θ is the Bragg angle.

The diameters of urinary nanocrystals in four healthy persons and in four MAP stone formers calculated using Scherer formula was shown in Table 1.

Table 1. Diameters of urinary nanocrystals in four healthy persons (H) and in five MAP stone formers (S) calculated using Scherer formula / nm

Compositions of nanocrystals	H1	H2	H3	H4	S1	S2	S3	S4	S5
COD	31-47	34-52	–	39-50	17-54	40-50	–	–	–
COM	–	–	–	–	–	64	–	19-35	–
hydroxylapatite	–	–	–	–	–	–	55-58	–	–
uric acid	22-68		36-41	–	–	–	–	–	45-51
ammonium urate	54	–	–	–	–	–	–	–	–
β-$Ca_3(PO_4)_2$	–	–	39	–	–	–	34	34-78	–
$NH_4MgPO_4 \cdot H_2O$	–	–	–	35	42-54	71-92	–	–	–
$NH_4MgPO_4 \cdot 6H_2O$	–	–	–	–	–	–	–	90	–

3.4. *HRTEM analysis of urinary nanocrystallites*

Figure 3 shows the images obtained from HRTEM and FFT in different areas of the urinary nanocrystallites. To analyze the clear lattice fringes in random selection, we detected the spacing of lattices plane of d=5.76 (Fig. 3a), which was assigned to (020) plane of $MAP \cdot 6H_2O$; and d=5.38 (Fig. 3b), which was assigned to (020) plane of $MgHPO_4 \cdot 3H_2O$. We also detected lattice fringe at d=4.71 Å, which was assigned to (110) plane of $MAP \cdot H_2O$ in Fig. 3c. In Fig. 3d we also detected the lattice fringes at 4.49 Å, which was assigned to (102) plane of $MgHPO_4 \cdot 3H_2O$. Namely the main components of urinary nanocrystalites in patients with MAP stones were $MAP \cdot H_2O$ and $MgHPO_4 \cdot 3H_2O$, which is consistent with the result obtained from the XRD. Since the crystal water in crystal lattice of COD is more than that of COM, the surface of crystals is easily formed a hydro-layer in aqueous, so that the surface charge of COD is very small in aqueous medium. Therefore, COD has less adsorption to amino acids.

3.5. *Differences of the average diameter of particle amount of urine nanocrystallites with different sizes*

Figure 4 shows the size distribution of urinary nanoparticles detected by nano-particle size analyzer. It can be seen that the particle size distribution in two kinds of urines were significantly different. The particle size distribution of the crystallites in healthy urine was relatively even, and the size of most crystallites was less than 100 nm (Figs. 4a-4c). However, the particle size distribution in lithogenic urine was uneven (Figs. 4d-4g), ranging from about 7 to 500 nm.

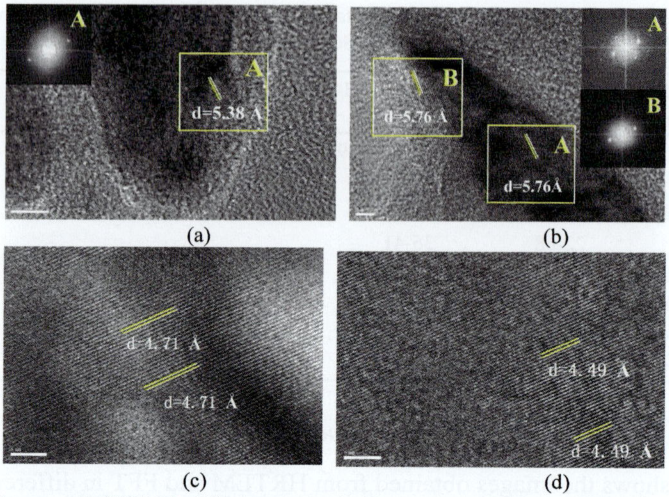

Fig. 3. FFT images of HRTEM in different areas of urinary naocrystallites in MAP stone patients.

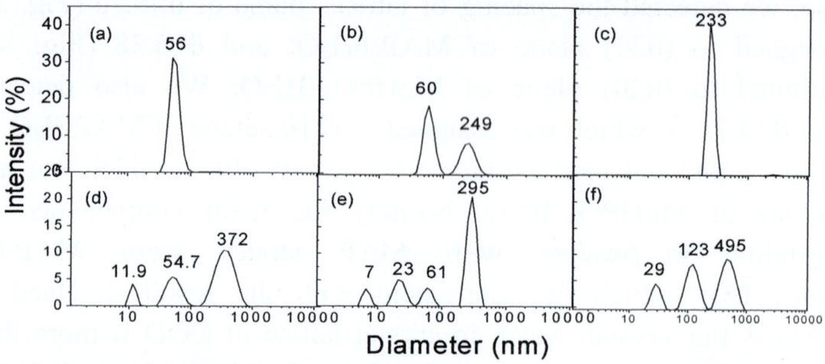

Fig. 4. Size distribution of urinary nanoparticles detected by nano-particle size analyzer. (a-c) three cases of healthy subjects; (d-f) three cases of MAP stone formers.

3.6. *Changes in light intensity autocorrelation curves*

Figure 5 shows the autocorrelation curve of the two types of urinary crystallites. The curves of the urine crystallites in the controls (Figs. 5d-f) were smoother than those in the MAP stone patients (Figs. 5a-c). When the particles in solution are smaller, the Brownian motion is rapid, the changes in the intensity of the scattered light are fast, the correlation of the signal is poor, and the correlation curve rapidly decreases [6]. However, the curve will be steeper and vice versa.

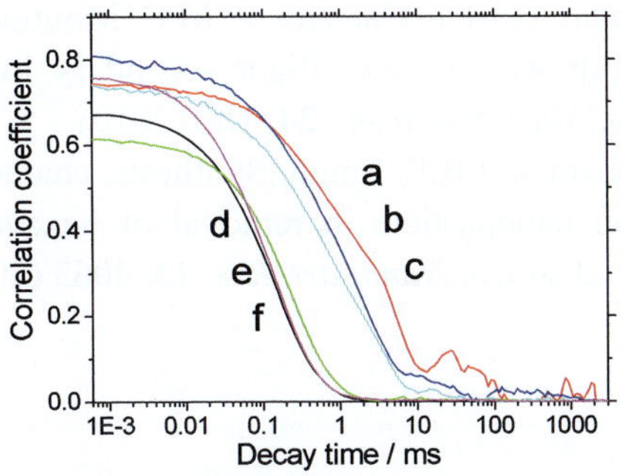

Fig. 5. Autocorrelation curves of the different kinds of urine crystallites of three MAP stone patients (a,b,c) and three healthy controls (d,e,f), respectively.

4. Conclusions

The urinary nanocrystals in healthy persons were mainly COD, uric acid and magnesium ammonium phosphate; but in MAP lithogenic patients it was mainly magnesium ammonium phosphate, β-tricalcium phosphate, COD and COM. The condition of urinary crystallites and the variation of urine components were indicative for the activity of stone disease in stone formers. The formation of MAP stones was closely related to the presence of high amounts of MAP crystallites in urine.

Acknowledgements

Work supported by grant 81170649 of the National Natural Science Foundation of China.

References

1. M. Tsujihata, Mechanism of calcium oxalate renal stone formation and renal tubular cell injury, *Int. J. Urol.* **15**, 115 (2008).
2. W.G. Robertson, M. Peacock and B.E.C. Nordin, Calcium crystalluria in recurrent renal-stone formers, *Lancet* **2**, 21 (1969).
3. M. Robert, A.M. Boularan and O. Delbos, Study of calcium oxalate crystalluria on renal and vesical urines in stone formers and normal subjects, *Urol. Int.* **60**, 41 (1998).
4. M. Robert, A.M. Boularan and O. Delbos, Evaluation of the risk of stone formation: study on crystalluria in patients with recurrent calcium oxalate urolithiasis, *Eur. Urol.* **29**, 456 (1996).

5. C.A.A. Ghumman, O.M.T. Carreira, A.M.C. Moutinho,et al, Identification of human calculi with time-of-flight secondary ion mass spectrometry, *Rapid Commun. Mass Spectrom.* **24**, 185 (2010).
6. R. Singh, V. Misra and R.P. Singh, Synthesis, characterization and role of zero-valent iron nanoparticle in removal of hexavalent chromium from chromium-spiked soil, *J. Nanopart. Res.* **13**, 4063 (2011).

Research on intelligent control strategy of uniform temperature in pulse combustion furnace

Jie-Jia Li* and Hui-Peng Li
Information and Control Engineering college, Shenyang Jianzhu University, Shenyang, Liaoning, China
†E-mail: ljj_0123@sjzu.edu.cn
1174363820@qq.com

In the temperature control, the object has a non-linear, strongly coupled system with large time delay and other characteristics, traditional PID can't achieve good results because the adjustment of the PID controller's parameter cannot be achieved. This paper presents an algorithm for optimization of PID controller's parameters based on fruit flies optimization algorithm, which can achieve optimal control by tuning PID controller's parameters. On this basis, we design a different regional temperature optimal control system, which makes pulse combustion furnace in different areas to achieve uniform temperature. The simulation results show that the organic combination of fruit flies optimization algorithm and the different regional temperature optimal control system improves the accuracy of temperature in furnace.

Keywords: Uniform Temperature Property; The Drosophila Optimization Algorithm; PID Controller; Fuzzy Neural Network.

1. Introduction

In modern industry, pulse combustion furnace plays an important role in the iron and steel technology, whose precision and uniformity of temperature control method has a great influence on the quality of steel. While the metal workpieces are processed in the burner, whose internal organizational structure has changed a lot, finally metal workpiece' s organizational structure has better mechanical properties. However, the pulse combustion furnace will be affected by some things in the industrial scene, and its internal temperature are subject to different air in different areas, thermal radiation and so on. Therefore, this paper uses pulse combustion control based on optimization algorithm of Drosophila.

---

* Corresponding author.

2. Pulse combustion furnace's hardware design of the control system

In this design, this paper uses the 8051 microcontroller as the main chip to process date and calculate algorithms, and design the process of data input and output channel and human-machine interfaces. The main structure contain each temperature detection, process control channel, LED temperature display and alarm, and its principle of operation temperature detection zone is divided into 4 regions firstly, then the system will get 4 temperature signals through thermocouple, the thermometric analogue signals will loop through the amplifier and A/D conversion in turn to insert into CPU after multiplexing switch, CPU compares the detected temperature signal and set values, according to the results of the comparison CPU is calculation of the output control signals, these output control signals are the actuators through the control output channel on. Hardware system structure is showed in Figure 1.

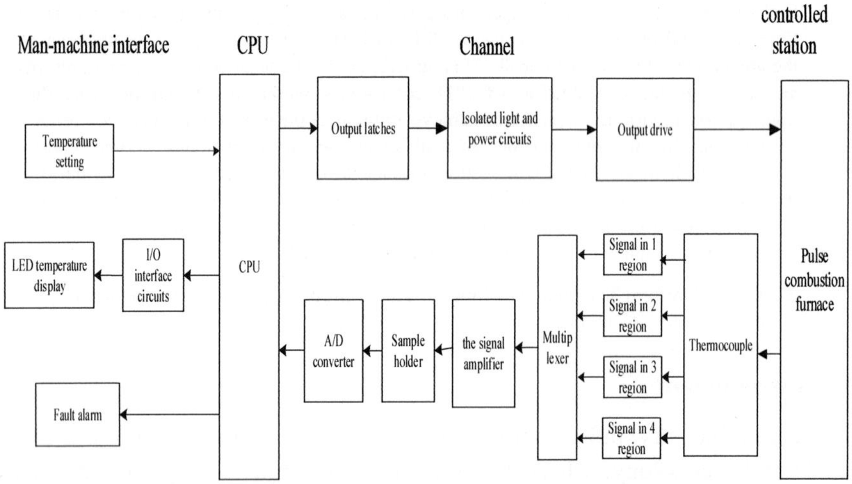

Fig. 1. The architecture diagram of hardware system.

3. Uniform temperature control and optimization design of control system

3.1. *Control system structure*

Within the pulse combustion furnace, according to the difference of the different heating atmospheres, the furnace area proper is divided into 4 parts, each piece features 2 burners and several thermocouples. Each area has the same PID controller based on fruit flies optimization algorithm, optimal algorithm of the control is proposed, four control zones are constituted by the four regional controllers and a regulator to have a uniform temperature, the controller in each region uses fruit flies optimization algorithm to optimize PID controllers, the uniform control regulator calculates additional output date to control the actuators based on fuzzy neural network. The architecture diagram of control strategy is showed in Figure 2.

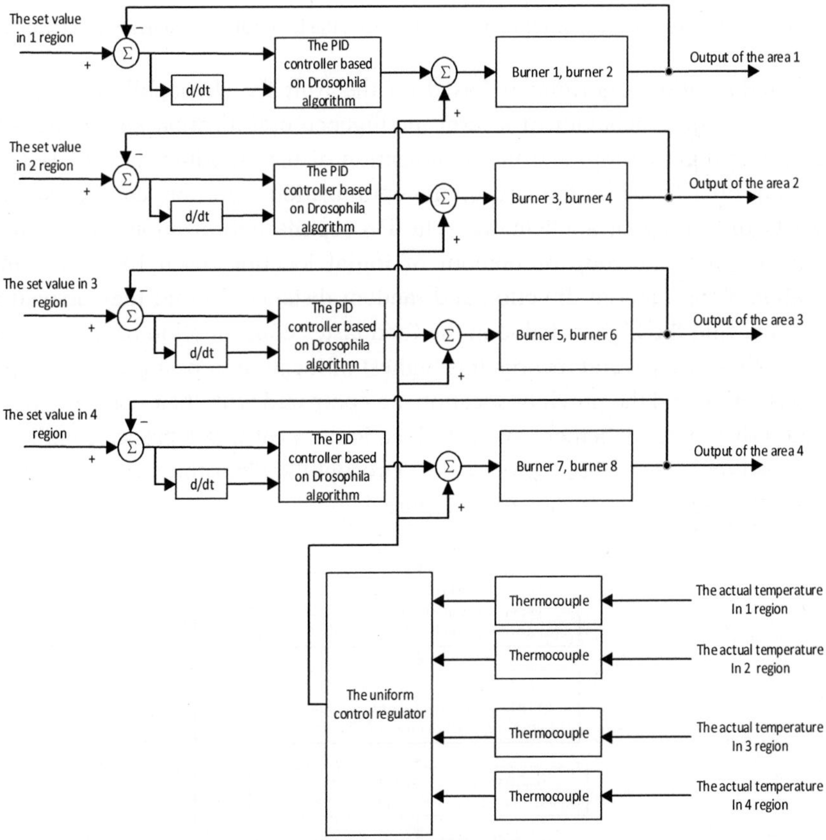

Fig. 2. The architecture diagram of control strategy.

3.2. *The design of PID controller based on the Drosophila optimization algorithm*

The structure diagram of PID controller based on the Drosophila optimization algorithm is showed in Figure 3.

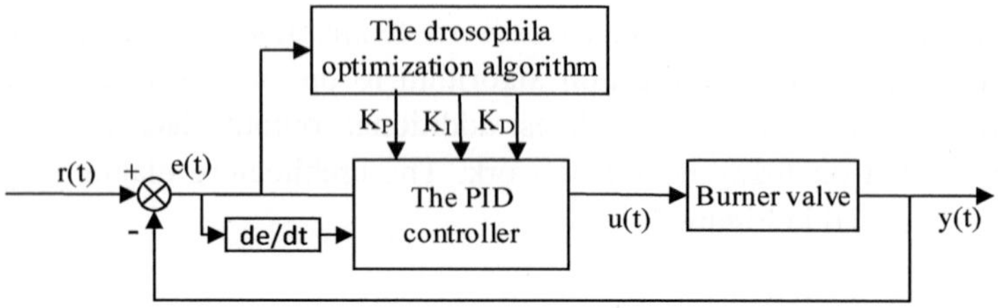

Fig. 3. The structure diagram of PID controller based on the Drosophila optimization algorithm.

Optimization algorithm is based on the fruit fly Drosophila melanogaster species foraging behavior of swarm intelligence optimization algorithm, which has easy programming, and the optimization of the algorithm is very fast, then the controller based on the Drosophila optimization algorithm can be used for a variety of intelligent applications. The Drosophila optimization algorithm first generated a flies group of random of initial location, each flies from initial location along random direction and random distance fly out; they arrived new location, find the highest value in the food tastes concentration of flies locations in this flies groups, and records this highest of location; repeat previous step, the new location and the previous location are compared with their concentrations of food tastes, judge whether save new location; repeat this a process, until found food source. The whole process of Drosophila algorithm to optimize PID controller parameters is showed in Figure 4.

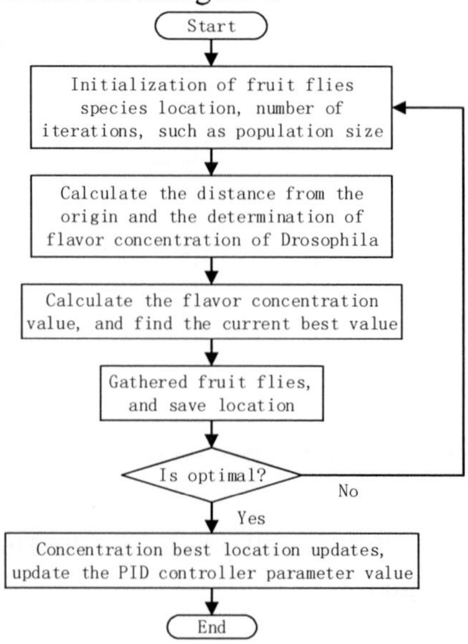

Fig. 4. The whole process of Drosophila algorithm to optimize PID controller parameters.

3.3. *The design of the optimal controller to get uniform temperature*

The optimal controller is based on the fuzzy neural network. The fuzzy neural network as a whole is divided into 5 levels: the 1th floor node is a node of input language variable, achieve the desired signal input and the connection coefficients of neural network is $\omega_{ji}^{(1)} = 1$; the 2nd layer node is a node of input language value, implement the appropriate language value of variable membership function established, this layer is represented by simple single neuron functions (including proportional function, quadratic functions, bell-shaped function); the 3rd layer node is a node which has fuzzy control rules by matching fuzzy logic conditions so as to achieve the distribution of expert knowledge and the operation of logic "and", which can be showed with $f_j^{(3)} = \min(u_1^{(3)}, u_2^{(3)}, \cdots, u_p^{(3)})$, $a_j^{(3)} = f_j^{(3)}$, and the 3rd layer node connects with the 2nd floor through connection coefficients $\omega_{ji}^{(3)} = 1$; the 4th floor is s node of the output language value, output linguistic fuzzy membership function is established and bi-directional transmission of signal patterns, and neural network back propagation learning implemented by this layer; The 5th floor is a node of the output variable too, and which connects to the 4th floor, which achieve the accurate computation of the fuzzy output.

4. The simulation analysis

In this design, the mathematical model of pulse combustion furnace which is studied use the first order delay transfer function to describe approximately according the experience of experts. The inertia parameter in the Model is T=11, the proportionality factor is K = 9, and the delay parameter is R=12, $G(s) = \frac{9}{11s+1}e^{-12s}$. And the desired temperature is set of 800 degrees Celsius. Finally, the whole time of simulation is set for 20 minutes, the initial temperature is room temperature, which is 20 degrees Celsius.

As illustrated in Fig. 5(The blue line is the temperature value for the traditional PID controllers, The red line the temperature value for Drosophila PID controller and the optimal controller), the pulse combustion furnace based on conventional PID controller has its own unique characteristics in each region for their temperature. First area has large oscillations of temperature, rise time of 4 minutes, peak temperature of 860 degrees Celsius, stable for nearly 10 minutes, and the most important thing is that this area is temperature stable at 810 degrees Celsius; Temperature also has large oscillations of the 2nd area, rise time is 4 minutes and 10 seconds, peak temperature of 855 degrees, 12 minutes for temperature stability, end temperature stability of this region at 800 degrees Celsius, so this area does not have steady-state error; 3rd District of oscillation is

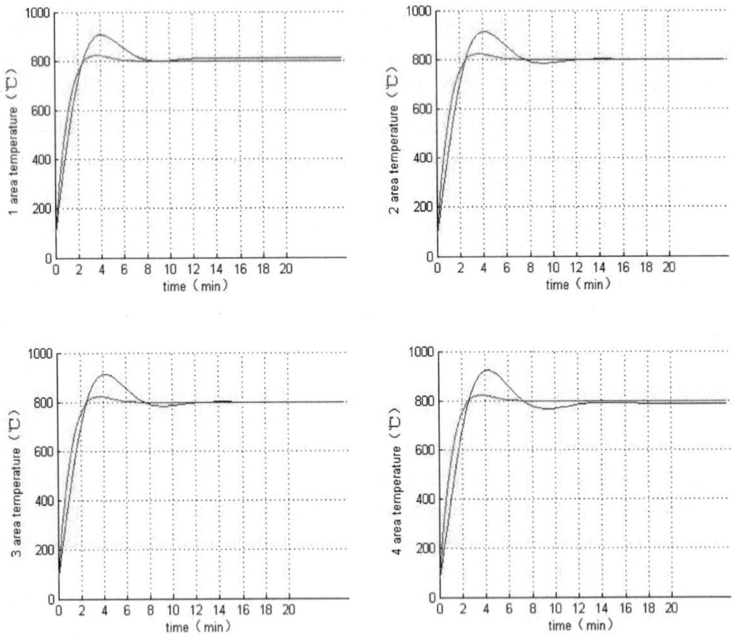

Fig. 5. Simulation of combustion furnace under different temperature controller.

also very obvious, rise time is 4 minutes and 12 seconds, peak temperature of 855 degrees, end temperature stability of this region at 800 degrees Celsius, no steady-state error; Forth oscillation is also evident, rise time is 4 minutes and 22 seconds, peak temperature of 865 degrees Celsius, this regional final temperature stability at 785 degrees Celsius, steady-state error is very small.

The simulation results show that Drosophila algorithm to optimize PID controller combined with optimization of fuzzy neural network control with obvious advantages: rising early, oscillations have little, and there is no steady-state error, pulse combustion furnace atmosphere to achieve the best combustion in each of the actual temperature values reached expectations 800 degrees Celsius, so this design has achieved the aims of temperature.

5. Conclusions

What can be obtained from the above analysis are that the systems improve control where it is using a new type of controller, improve control precision, reducing overshoot of the system, adjust the time decreases, so the controller improve dynamic performance and steady-state performance of the system, while the uniformity of temperature control of the system has been greatly improved.

References

1. Wei Wei, He Yan. Fundamentals of Intelligent Control [M]. Tsinghua University Press, 2008:1-5.
2. Han Junying, LIU Chengzhong. Fruit fly optimization algorithm based on bacterial chemotaxis [J]. Computer Applacations,2013.
3. FU Wenxiu, SU Jie. Optimal PID Governor Tuning of Hydraulic Turbine Generators with Drosophila Optimization [J]. Computer Simulation, 2015.
4. Xiao Jinsong, Yu Linzhan. Application of Sequential Pulse Combustion Control Technique to Tempering Furnace [J]. STEEL PIPE, 2008.
5. YU Jin, LI Xiaofeng. Prediction of the Annealing Furnace Temperature Control Based on T-S Model Fuzzy Neural Network[J]. Journal of Shenyang Jianzhu University(Social Science), 2014(1).
6. P Jin-Xiang, Hybrid Intelligent Control Method for Laminar Cooling Process of Hot Rolled Strip , Journal of Northeastern University, 2009, 30(11):1534-1537.

Improving the production of cyclodextrins from the Paenibacillus polymyxa in *Bacillus subtilis*

Hui-Jie Liu[2], Zhi-Mei Feng, Shan Cao[1] and Yu Li[3]

The College of Biotechnology, Tianjin University of Science and Technology, Tianjin 300457, People's Republic of China

E-mail:[3]liyu@tust.com; [2]liuhuijie@163.com; [1]cs1988@tust.edu.cn

The hydrophobic cavity of cyclodextrin contributes to the wide applications in medicine, pharmaceutical, food, cosmetic industries, agriculture and environmental protection. Cyclodextrin glycosyltransferase (cgt) gene was able to be amplified from *Paenibacillus polymyxa* and inserted into vectors of pWB980. The recombinant vectors were transformed to *Bacillus subtilis* WB600. The results showed that cgt gene was expressed in *Bacillus subtilis* successfully. The optimum conditions for maximum enzyme activity were obtained at a temperature 35◦C, pH 7 and quantity of inoculum 6%. Corn starch and soybean meal were chosen as carbon source and nitrogen source in the process of production. A maximum activity of the CGTase activity reached 9.2 U/mL, which is about 15-fold to that of the original strain *Paenibacillus polymyxa*. β-cyclodextrin is 0.104 mg/mL from 1% (m/m) soluble starch in 10 min (pH 7.0 50 °C). Cyclodextrins can form inclusion complexes with different molecules with the aid of their special chemical (molecular) structures. Physical and chemical properties of molecules can be changed after the formation of complex. This special feature enables the usage of cyclodextrins in different industry areas.

Keywords: Cyclodextrin; Expression; *Paenibacillus Polymyxa*; *Application.*

1. Introduction

Cyclodextrin (CD) is applied widely in medicine, pharmaceutical, food, cosmetic industries, agriculture and environment. CD cyclic-oligosaccharides consisting of six (α-CD), seven (β-CD), eight (γ-CD) or more glucopyranosyl units linked by α-(1,4) bonds[1]. The cyclodextrin molecules have unique structure with hydrophobic cavity and hydrophilic surface. They can form inclusion complex with a wide variety of compounds. On account of these properties, cyclodextrins enhance solubility of insoluble guests, stabilize labile guests against the derivative effects by oxidation[2].

The cyclodextrins are produced by cyclodextrin glucosyltransferase (CGTase EC 2.4.1.19) through cyclization reaction of gelatinized starch. CGTase is an extracellular bacterial enzyme belonging to the α-amylase family [3]. These macro cyclic carbohydrates with a polar internal cavities can solubilize many normally water-insoluble compounds by forming complexes, and their hydrophilic outside surfaces make CDs soluble in water[4].

---

[1] Corresponding author.

The product cyclodextrin catalyzed synthesis by CGTases is usually a mixture of CDs. To date, CGTases of about fifty microorganisms have been classified as α-, β- and γ-CGTases according to the major type of CD produced by the cyclizing reaction. β-CD is mainly produced by *Bacilli alkaliphilic*[5], and CGTase-producing *Paenibacillus polymyxa* is also main source of β-CD.

Recently, we have succeeded in isolation of a CGTase-producing strain, which identified was *Paenibacillus polymyxa.* The enzyme from this strain can mainly convert starch into β-CD. In order to improve CD production, the cgt-gene from *P. polymyxa* was successfully cloned into pWB980 vector and expressed in *B. subtilis* WB600.

2. Materials and Methods

2.1. *Bacterial strains and plasmid*

The cyclodextrin glucanotransferase (cgt) gene was from *P. polymyxa* TCCC 1492. *E. coli* JM109 was used for a cloning host. *B. subtilis* WB600 was used for expressing of CGTase proteins. Plasmid pMD-18 was used for cloning. The vector of pWB980 was used as an expression vector.

2.2. *Clone and expression of CGT gene*

CGT gene was amplified by PCR using genomic DNA from *P. polymyxa* as template. The amplified CGTase gene fragment was purified and cloned into the PMD-18 vector. Optimized terrific broth (TB) was used for the expression. 50 μg/mL kanamycin was added into the medium to allow the growth of the plasmid carrying strain. Then, this culture was inoculated into TB medium. The cells were cultivated at 37∘C on a shaker (200 rpm/min) 40 h.

2.3. *Enzyme assays*

β-CGTase cyclizing activity was determined by the method of phenolphthalein[6]. One unit of β-CGTase activity was defined as the amount of enzyme that formed 1 μmol of β-CD min per min under standard conditions.

2.4. *The optimization of conditions*

In the fermentation process, the enzyme activity is greatly influenced by physical and biological parameters. The optimal design of the culture medium and the environmental conditions are very critical aspects in the metabolic production and development of fermentation processes. We studied the fermentation temperature, initial pH, inoculation quantity, carbon source and nitrogen source, respectively.

3. Results and Discussion

3.1. *Cloning the CGTase gene*

CGTase gene of *P. polymyxa* was cloned as a PCR product containing sequence encoding the mature enzyme, into the blunt-ended linear vector pWB980. The cloning of the pWB980-CGT gene in the host WB600 was confirmed by a restriction digest with the enzyme *Xho*I and *Kpn*I. The stripes of about 3500bp and 2100bp were founded in the agarose gel (0.8% W/V). This suggested that the recombinant plasmid was successful.

3.2. *SDS-PAGE analysis*

The molecular weight of the CGTase was determined by sodium dodecyl sulfate polyacrylamide gel electrophoresis (SDS-PAGE) (Fig. 1). The fermentation liquid of recombinant strains had obvious a stripe at molecular weight of 72 kD, while the negative control had no corresponding bands. It suggested that CGTase secreted successful to the outside of the cell. The secreted CGTase from the recombinant strains showed the characteristic of adaptive enzyme.

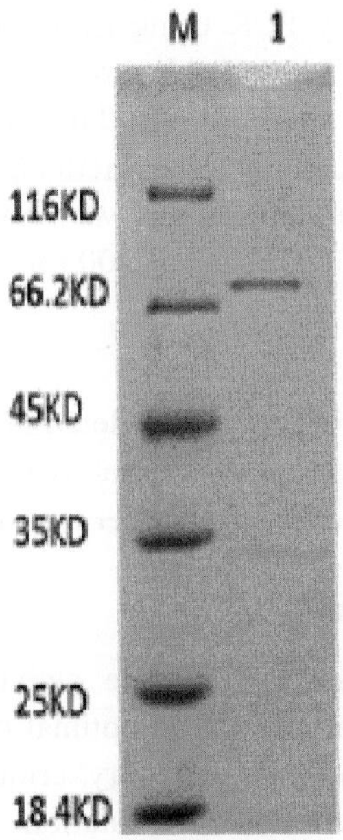

Fig. 1. The molecular weight of CGTase

3.3. *Extracellular secretion of CGTase*

Recombinant *B. subtilis* WB600/pWB980-CGT and WB600/pWB980 were grew into the LB culture medium 12 h as a seed. And then transferred into medium of TB. After 10000 ×*g*, 15 min, the supernatant was taken. Enzyme activity was measured from supernatant fermented liquid of recombinant *B. subtilis*. And the enzyme activity of the negative control had not been detected.

3.4. *Effects of the cultivation conditions on the expression*

In order to enhance the expression level of the extracellular recombinant CGTase, some aspects were applied the optimum culture conditions for the recombinant[7]. The most important physical factors, which affect the fermentation are the temperature, initial pH and quantity of inoculum. In this study, we explored the factors of temperature, initial pH and quantity of inoculum for the effect of enzyme fermentation level.

The growth of bacteria and production of enzyme under the influence of culture temperature, as showed in Fig. 2A, 35°C was the best temperature for expressing. When the temperature was lower than 35°C, enzyme activity increased with temperature. When the temperature was 40°C, the relative activity began to decline to 90%. Therefore the most appropriate temperature was 35°C. The pH of environment is one of the important factors affecting the growth of microorganism and metabolic products. The effects of the pH on the CGTase production showed in Fig. 2B, pH 7.0 was the best pH for expressing. The enzyme activity under pH 6.0 was just 50% when pH 7.0, which is the optimal pH. Relatively higher activity was observed at pH 7.0 as compared to pH 6.0 and 8.0, respectively. The pH condition of neutral was the optimum pH for the shake flask fermentation. As indicated in Fig. 2C, quantity of inoculate also influence the enzyme activity. An increase in the quantity of inoculate with time increased the enzyme activity gradually but at a higher quantity of inoculate is reversed. 6% is the optimum for shake flask fermentation of recombinant enzyme.

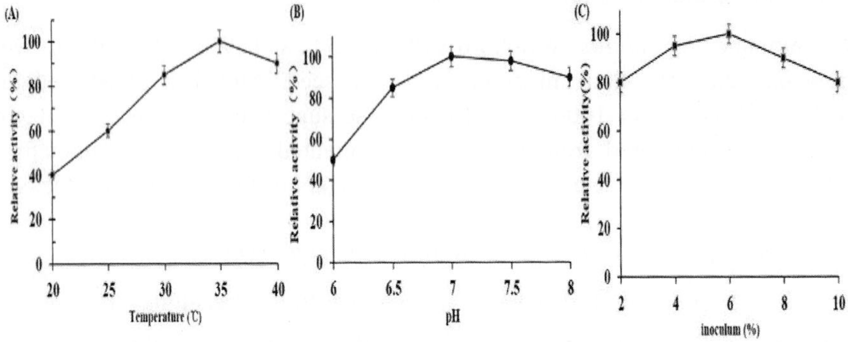

Fig. 2. Effects of the cultivation conditions on the expression

After the optimization of the cultivation conditions, the optimum conditions were obtained at a temperature of 35°C, pH of 7.0 and quantity of inoculum of fermentation of 6%.

3.5. *The optimization of culture medium*

Carbon source can provide energy for the growth of microorganisms, which is also the composition of carbon frame at the same time. As showed Fig. 3, CGTase from *B. subtilis* WB600/pWB980-cgt can be highly-efficient expression in the TB medium. However, glycerin as carbon source of medium, which price is relatively high. Therefore, the carbon source with a cheaper price should be replaced. As indicated in Fig. 3A, the enzyme activity reached 9.2 U/mL using corn starch as a carbon source. Therefore, corn starch was chosen as the optimum carbon source in the process of production.

Nitrogen source has an important impact on the growth of microbial metabolism. In this study, we discussed several kinds of nitrogen source for the influence of shaking flask fermentation. As indicated in Fig. 3B, organic nitrogen source was better than inorganic nitrogen source. Therefore, soybean meal was chosen as optimum nitrogen source in the process of production.

3.6. *Expression of CGTase*

The increase of expression amount of CGTase occurred at the first 20 h after induction was not significant Fig. 4. The low-level of expression for the recombinant CGTase might due to the fact that the first 20 h was the period for proliferation of cells. As a result, with the aids of both secretory *B. subtitles* expression system, the high yield of the extracellular recombinant CGTase was achieved.

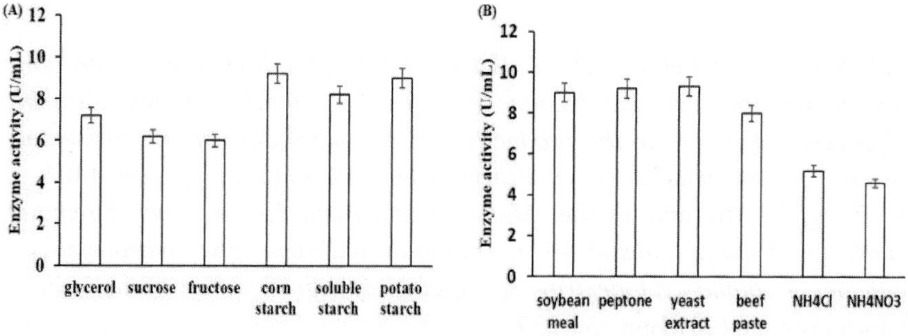

Fig. 3. The optimization of culture medium

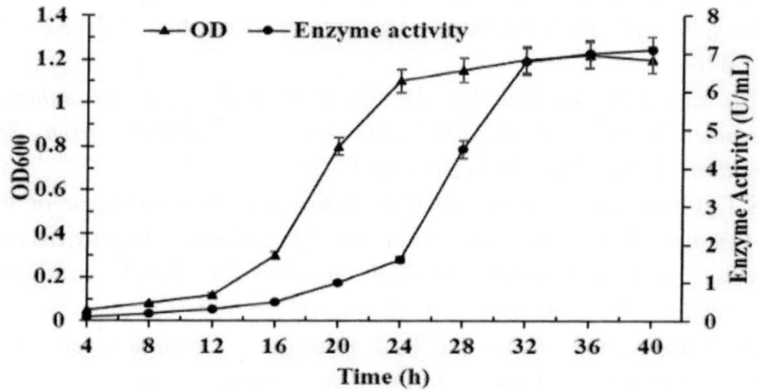

Fig. 4. The growth curve of *B. bacillus* and the enzyme activity of CGTase

4. Conclusion

Cyclodextrins can form inclusion complexes with different molecules with the aid of their special structures. Physical and chemical properties of molecules can be changed after the formation of complex. This special feature expands the usage of cyclodextrins in different industry areas. All the above researches showed that cgt gene of *P. polymyxa* was successfully expressed in *B. subtilis*. The optimum conditions for maximum enzyme activity were obtained at a temperature 35 ◦C, pH 7 and quantity of inoculum 6%. Corn starch, soybean meal were chosen as carbon source and nitrogen source in the process of production. A maximum activity of 9.2 U/mL was obtained. It was seen that β-CD reached 0.104mg/mL, which was converted from 1% (m/m) soluble starch.

Reference

1. S.R. Muria, B. Cheirsilp and S. Kitcha, *Effect of substrate concentration and temperature on the kinetics and thermal stability of cyclodextrin glycosyltransferase for the production of beta-cyclodextrin: Experimental results vs. mathematical model*, Process Biochem. 46 (2011)pp 1399-1404.
2. B. Song, Y. Yue and T. Xie, *Mutation of tyrosine167histidine at remote substrate binding subsite -6 in alpha-cyclodextrin glycosyltransferase enhancing alpha-cyclodextrin specificity by directed evolution*, Mol. Biotechnol. 56 (2014)pp 232-239.
3. R.Z. Han, J.H. Li and H.D. Shin, *Recent advances in discovery, heterologous expression, and molecular engineering of cyclodextrin glycosyltransferase for versatile applications*, Biotechnol Adv, 32 (2014)pp 415-428.
4. C.M. Li, W.W. Li and T.P. Holler, *Polyethylene glycols enhance the thermostability of beta-cyclodextrin glycosyltransferase from Bacillus circulans*, Food Chem, 164 (2014)pp 17-22.
5. F.S.T. Pinto, S.H. Flores and C.E. Schneider, *The Influence of Oxygen Volumetric Mass Transfer Rates on Cyclodextrin Glycosyltransferase Production by Alkaliphilic Bacillus circulans lein Batch and Fed-Batch Cultivations*, Food Bioprocess Tech. 4 (2011)pp 559-565.
6. J.D. Schoffer, M.P. Klein and R.C. Rodrigues, *Continuous production of beta-cyclodextrin from starch by highly stable cyclodextrin glycosyltransferase immobilized on chitosan*, Carbohyd Polym. 98 (2013)pp 1311-1316.
7. D. Zehentgruber, P. Lundemo and D. Svensson, *Substrate complexation and aggregation influence the cyclodextrin glycosyltransferase (CGTase) catalyzed synthesis of alkyl glycosides,* J Biotechnol. 155 (2011)pp 232-235.

Author Index